Python for Mathematical Thinking

Pradeep Singh • Balasubramanian Raman

Python for Mathematical Thinking

Springer

Pradeep Singh ⓘD
Department of CSE
Indian Institute of Technology (IIT)
Roorkee
Roorkee, Uttarakhand, India

Balasubramanian Raman ⓘD
Department of CSE
Indian Institute of Technology (IIT)
Roorkee
Roorkee, Uttarakhand, India

ISBN 978-981-95-4079-2   ISBN 978-981-95-4080-8 (eBook)
https://doi.org/10.1007/978-981-95-4080-8

© The Editor(s) (if applicable) and The Author(s), under exclusive license to Springer Nature Singapore Pte Ltd. 2026

This work is subject to copyright. All rights are solely and exclusively licensed by the Publisher, whether the whole or part of the material is concerned, specifically the rights of translation, reprinting, reuse of illustrations, recitation, broadcasting, reproduction on microfilms or in any other physical way, and transmission or information storage and retrieval, electronic adaptation, computer software, or by similar or dissimilar methodology now known or hereafter developed.

The use of general descriptive names, registered names, trademarks, service marks, etc. in this publication does not imply, even in the absence of a specific statement, that such names are exempt from the relevant protective laws and regulations and therefore free for general use.

The publisher, the authors and the editors are safe to assume that the advice and information in this book are believed to be true and accurate at the date of publication. Neither the publisher nor the authors or the editors give a warranty, expressed or implied, with respect to the material contained herein or for any errors or omissions that may have been made. The publisher remains neutral with regard to jurisdictional claims in published maps and institutional affiliations.

This Springer imprint is published by the registered company Springer Nature Singapore Pte Ltd.
The registered company address is: 152 Beach Road, #21-01/04 Gateway East, Singapore 189721, Singapore

If disposing of this product, please recycle the paper.

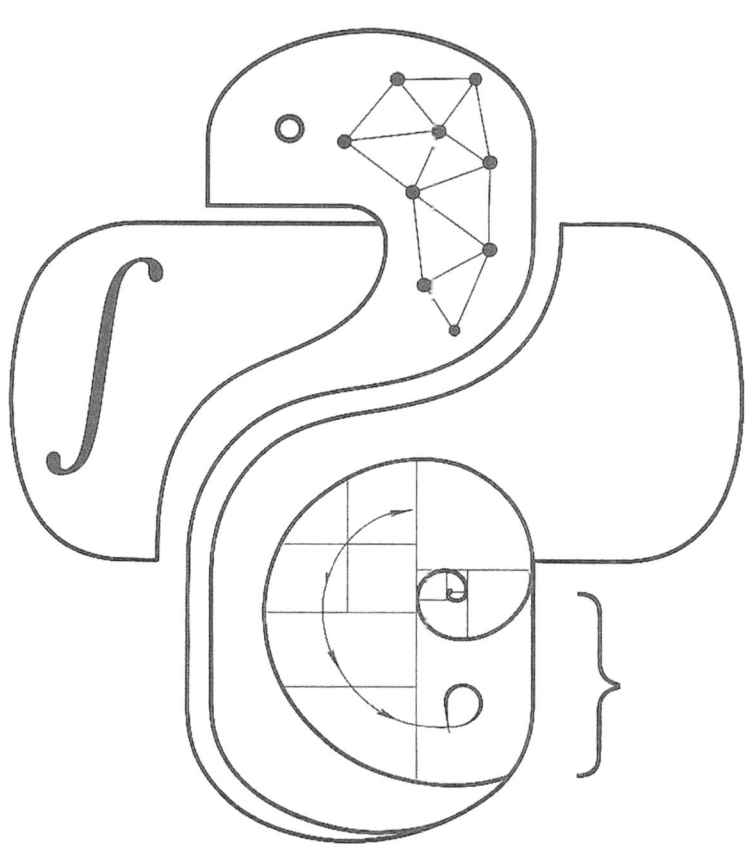

To the Thinkers and Dreamers

Preface

Mathematics is the language in which the laws of the cosmos are most succinctly expressed, while computation furnishes the dialects that make these laws tractable for modern inquiry. In the twenty-first century, the scientist, engineer, or economist who commands both rigorous mathematical reasoning and algorithmic craftsmanship can interrogate problems of a scale and subtlety that would have been inconceivable only a generation ago. *Python for Mathematical Thinking* has been conceived to cultivate precisely this dual fluency. Its guiding conviction is that proofs and programs are not competing modes of explanation but complementary lenses through which the same structure may be rendered transparent.

Python's ascendancy in the scientific world is no accident. Its minimalist syntax, large standard library, and a constellation of domain-specific packages—such as **NumPy**, **SymPy**, and **SciPy**—make it a laboratory in which abstract ideas can be prototyped with the same ease that numerical experiments are conducted. Yet fluency in a language is hollow without a rigorous grammar. Accordingly, each concept in this volume is introduced as a mathematically precise statement, motivated by exemplars from classical analysis, algebra, geometry, probability, or the theory of computation, and only then translated into idiomatic Python.

The narrative unfolds in steadily deepening strata. You will first revisit foundational structures—integers, rational fields, Euclidean spaces—before ascending to vector calculus, spectral theory, and measure–theoretic probability. From there the text journeys through numerical linear algebra, dynamical systems, and stochastic simulation, culminating in chapters that interface with contemporary research topics such as compressed sensing and quantum information. Throughout, special attention is paid to the computational complexity that shadows every mathematical algorithm; big-O analyses accompany theorems, and efficient data structures are derived, not merely quoted.

While this work presumes familiarity with undergraduate calculus and basic programming, it aspires to be a companion from novice to researcher. For the student, it provides a bridge between chalkboard proofs and executable experiments. For the seasoned mathematician, it offers an extensible platform for computational exploration. And for the practitioner, it supplies principled algorithms whose correctness is certified by theorem, not folklore.

Ultimately, the craft of mathematics is less about arriving at an answer than about inhabiting a mode of thought that relentlessly interrogates *why* a statement should hold. By weaving together deductive rigor and executable insight, we invite you to cultivate that habit of mind. May the pages that follow sharpen your analytical instincts, broaden your computational arsenal, and reveal anew the harmony that binds symbol to silicon.

Welcome to *Python for Mathematical Thinking*.

Roorkee, India   Pradeep Singh
July 5, 2025   Balasubramanian Raman

Acknowledgements The preparation of *Python for Mathematical Thinking* has been sustained by a constellation of institutions, colleagues, and communities to whom we owe an incalculable debt of gratitude. Foremost, we thank our home institution, the **Indian Institute of Technology Roorkee**, for nurturing an environment where rigorous mathematics and cutting-edge computation flourish side by side. The infrastructural and intellectual resources furnished by the Department of Computer Science and Engineering have been indispensable in transforming tentative ideas into executable code and, ultimately, into the pages you now hold.

Our work has been buoyed by generous support from the **Anusandhan National Research Foundation (ANRF)** and the **Department of Science and Technology (DST)**, Government of India, as well as by fellowships and grants from the **National Board of Higher Mathematics (NBHM)**, Department of Atomic Energy. Their sustained commitment to fundamental research provided the freedom to explore the fertile intersection of mathematics and computation that lies at the heart of this book.

We are profoundly grateful to our colleagues—faculty, research scholars, and interns alike—whose questions, critiques, and encouragement sharpened every chapter. The daily intellectual ferment within our laboratories turned solitary writing into dialogue, and dialogue into discovery. Special thanks are due to the interns of the MV and MI Labs, whose curiosity continually reminded us to keep explanations both rigorous and accessible.

A book devoted to the computational realization of mathematics is inconceivable without the open-source ecosystem that embodies the very ethos of collaborative inquiry. We salute the global **Python** community and the maintainers of NumPy, SciPy, SymPy, Matplotlib, Jupyter, and countless ancillary packages; their volunteer labour made it possible to test, benchmark, and visualize every example herein. Likewise, we acknowledge the indispensable role of **GitHub** for version control and collaboration, **Stack Exchange** for its crowdsourced wisdom, and **Google Search** for accelerating fact-checking and literature review.

The first author (P.S.) records a personal debt to **Prof. Anima Nagar** (IIT Delhi) for revealing the elegance latent in abstract structures, and to **Prof. R. Balasubramanian** (IIT Roorkee) for a steadfast confidence that transformed aspiration into achievement. Their mentorship instilled both the patience to prove and the audacity to program.

Finally, to our families and friends: your unwavering patience and faith were the quiet constants behind every page. Your support turned late-night debugging sessions and early-morning proof revisions into a shared undertaking. This volume is as much a testament to your belief as to our labour.

Competing Interests The authors declare that they have no competing interests relevant to the content of this manuscript.

Contents

1 Introduction 1
   1.1 Why Python for Mathematics? 2
       1.1.1 Advantages of Python in Mathematical Computation 2
       1.1.2 Comparison with Other Mathematical Tools 4
   1.2 Setting Up the Python Environment 5
       1.2.1 Installing Python and Essential Libraries 5
       1.2.2 Python Syntax and Semantics 6
       1.2.3 Python as an Interpreted Language 7
   1.3 Data Types and Variables 8
       1.3.1 Primitive Data Types 8
       1.3.2 Mathematical Operations on Data Types 9
       1.3.3 Working with Variables 10
       1.3.4 Containers and Collections 11
   1.4 Control Flow in Python 12
       1.4.1 Conditional Statements 13
       1.4.2 Loops in Python 14
       1.4.3 Comprehensions in Python 15
       1.4.4 Error Handling and Exceptions 16
   1.5 Functions and Modular Programming 17
       1.5.1 Defining Functions 18
       1.5.2 Advanced Function Concepts 19
       1.5.3 Modular Programming with Python 20
       1.5.4 Recursive Functions 20
   1.6 Input and Output in Python 21
       1.6.1 Handling User Input 22
       1.6.2 File Handling in Python 22
       1.6.3 Formatted Output 23
       1.6.4 Best Practices in Mathematical Coding 24
   1.7 Object-Oriented Programming (OOP) in Python 24
       1.7.1 Basic Concepts of OOP 25
       1.7.2 Advanced OOP Concepts 26

		1.7.3	Special Methods and Operator Overloading	27
		1.7.4	Working with Mathematical Structures in OOP	28
	1.8	Iterators and Generators		30
		1.8.1	Understanding Iterators in Python	30
		1.8.2	Generators and Lazy Evaluation	30
		1.8.3	Combinatorics with Iterators and Generators	31
	1.9	Advanced Topics in Python		31
		1.9.1	Regular Expressions and String Matching	32
		1.9.2	Working with Dates and Times	33
	1.10	Exercises		34
2	**Mathematical Foundations in Python**			37
	2.1	Basic Arithmetic and Algebra		37
		2.1.1	Arithmetic Operations	37
		2.1.2	Algebraic Expressions	39
		2.1.3	Polynomial Algebra	40
		2.1.4	Inequalities and Absolute Values	41
	2.2	Functions and Graphs		42
		2.2.1	Defining Functions in Python	42
		2.2.2	Graphing Functions	43
		2.2.3	Analysing Function Properties	44
	2.3	Matrices and Linear Algebra		45
		2.3.1	Matrix Operations	45
		2.3.2	Determinants and Eigenvalues/Eigenvectors	47
		2.3.3	Solving Systems of Linear Equations	47
	2.4	Exercises		48
3	**Calculus with Python**			53
	3.1	Differentiation		53
		3.1.1	Numerical Differentiation Techniques	53
		3.1.2	Symbolic Differentiation Using SymPy	56
		3.1.3	Applications of Differentiation	59
	3.2	Integration		62
		3.2.1	Numerical Integration Methods	63
		3.2.2	Symbolic Integration with SymPy	65
		3.2.3	Applications of Integration	69
	3.3	Multivariable Calculus		72
		3.3.1	Partial Derivatives and Gradients	72
		3.3.2	Multiple Integrals	76
		3.3.3	Line Integrals and Surface Integrals	79
	3.4	Advanced Topics in Calculus		82
		3.4.1	Differential Equations	83
		3.4.2	Calculus of Variations	87
		3.4.3	Tensor Calculus	91
		3.4.4	Fractional Calculus	94
	3.5	Exercises		98

Contents

4 Data Structures and Algorithms with Python 101
  4.1 Introduction to Data Structures 101
    4.1.1 Overview of Data Structures 101
    4.1.2 Basic Data Structures 104
  4.2 Search Algorithms 112
    4.2.1 Linear Search 112
    4.2.2 Binary Search 114
  4.3 Sorting Algorithms 117
    4.3.1 Basic Sorting Algorithms 117
    4.3.2 Divide and Conquer Techniques 120
  4.4 Graph Theory and Algorithms 124
    4.4.1 Introduction to Graph Theory 124
    4.4.2 Basic Graph Algorithms 127
  4.5 Exercises 131

5 Probability and Statistics 135
  5.1 Probability Theory 135
    5.1.1 Basic Probability Concepts 135
    5.1.2 Random Variables and Distributions 142
    5.1.3 Expectation, Variance, and Moments 145
    5.1.4 Common Probability Distributions 149
  5.2 Descriptive Statistics 152
    5.2.1 Measures of Central Tendency 152
    5.2.2 Measures of Dispersion 154
    5.2.3 Data Visualisation and Analysis 157
  5.3 Inferential Statistics 162
    5.3.1 Hypothesis Testing 163
    5.3.2 Confidence Intervals 166
    5.3.3 Regression Analysis 169
    5.3.4 Advanced Topics in Regression 171
  5.4 Stochastic Processes and Applications 176
    5.4.1 Markov Chains 176
    5.4.2 Poisson Processes 179
    5.4.3 Brownian Motion and Applications 182
    5.4.4 Bayesian Statistics 185
  5.5 Exercises 189

6 Differential Equations 193
  6.1 Ordinary Differential Equations (ODEs) 193
    6.1.1 First-Order ODEs 193
    6.1.2 Higher-Order ODEs 201
    6.1.3 Systems of ODEs 204
    6.1.4 Numerical Solutions of ODEs 207
  6.2 Partial Differential Equations (PDEs) 211
    6.2.1 Numerical Methods for Solving PDEs 214
    6.2.2 Advanced Topics in PDEs 217

	6.3	Special Functions and Transform Techniques	220
		6.3.1 Laplace Transforms and Applications	220
		6.3.2 Fourier Transforms and Applications	223
		6.3.3 Special Functions in ODEs and PDEs	226
		6.3.4 Green's Functions and Integral Equations	229
	6.4	Stochastic Differential Equations (SDEs)	232
		6.4.1 Introduction to SDEs	232
		6.4.2 Numerical Solutions of SDEs	235
		6.4.3 Applications of SDEs.....................................	236
	6.5	Exercises	238
7	**Discrete Mathematics and Combinatorics**		243
	7.1	Number Theory..	243
		7.1.1 Divisibility and Modular Arithmetic	244
		7.1.2 Congruences and Number Theoretic Functions	246
		7.1.3 Advanced Topics in Number Theory	248
	7.2	Combinatorics	251
		7.2.1 Basic Counting Principles................................	251
		7.2.2 Generating Functions	254
		7.2.3 Advanced Topics in Combinatorics......................	256
	7.3	Graph Theory..	259
		7.3.1 Basic Concepts in Graph Theory	259
		7.3.2 Advanced Topics in Graph Theory	261
		7.3.3 Applications of Graph Theory	265
	7.4	Boolean Algebra and Logic.....................................	267
		7.4.1 Propositional Logic and Proof Techniques	267
		7.4.2 Boolean Algebra and Circuit Design	269
		7.4.3 Advanced Topics in Logic	272
	7.5	Discrete Structures and Applications............................	274
		7.5.1 Sets, Relations, and Functions	275
		7.5.2 Algebraic Structures.....................................	277
		7.5.3 Matroids and Their Applications	279
	7.6	Exercises	282
8	**Numerical Methods**		285
	8.1	Root-Finding Algorithms	286
		8.1.1 Bisection Method	286
		8.1.2 Newton–Raphson Method...............................	288
		8.1.3 Other Root-Finding Methods	290
	8.2	Optimisation Techniques.......................................	294
		8.2.1 Gradient Descent	294
		8.2.2 Simplex Method	297
		8.2.3 Nonlinear Optimisation	299
		8.2.4 Direct Methods for Linear Systems......................	302
		8.2.5 Iterative Methods for Linear Systems	305
		8.2.6 Solving Nonlinear Systems	308

8.3	Numerical Integration and Differentiation		311
	8.3.1	Numerical Differentiation Techniques	311
	8.3.2	Numerical Integration Methods	315
	8.3.3	Applications of Numerical Integration	318
8.4	Eigenvalue Problems and Matrix Decompositions		321
	8.4.1	Power Method and Inverse Iteration	321
	8.4.2	QR Algorithm and Schur Decomposition	324
	8.4.3	Singular Value Decomposition (SVD)	326
8.5	Advanced Topics in Numerical Methods		328
	8.5.1	Numerical Solutions to Differential Equations	328
	8.5.2	Spectral Methods	331
	8.5.3	Parallel Computing in Numerical Methods	333
	8.5.4	Error Analysis and Stability in Numerical Methods	334
8.6	Exercises		336

9 Chaos Theory and Dynamical Systems ... 339

9.1	Introduction to Dynamical Systems		340
	9.1.1	Fixed Points and Stability	340
	9.1.2	Discrete Dynamical Systems	344
	9.1.3	Continuous Dynamical Systems	349
9.2	Chaos and Fractals		354
	9.2.1	The Mandelbrot Set	354
	9.2.2	Julia Sets	357
	9.2.3	Fractals in Nature and Art	360
	9.2.4	Applications of Chaos Theory	363
9.3	Lyapunov Exponents		367
	9.3.1	Definition and Interpretation	367
9.4	Advanced Topics in Chaos Theory		369
	9.4.1	Entropy and Chaos	369
	9.4.2	Chaos Control and Synchronisation	371
	9.4.3	Complex Networks and Chaos	373
	9.4.4	Quantum Chaos	374
9.5	Data-Driven Dynamics and Modern Tools		376
	9.5.1	Koopman Operator Theory	376
	9.5.2	Reservoir Computing for Chaotic Time Series	377
	9.5.3	State-Space Reconstruction in Practice	379
9.6	Complex Systems and Nonlinear Dynamics		382
	9.6.1	Lyapunov Exponents and Strange Attractors	382
	9.6.2	Applications in Weather Forecasting and Financial Markets	383
	9.6.3	Applications in Biological Systems and Population Dynamics	383
	9.6.4	Numerical Methods for Nonlinear Dynamics	384
	9.6.5	Complex Networks and Emergent Behaviour	385
9.7	Exercises		387

10 Data Science and Machine Learning ... 391
10.1 Introduction to Data Science ... 391
- 10.1.1 Data Manipulation and Cleaning ... 391
- 10.1.2 Exploratory Data Analysis (EDA) ... 394
- 10.1.3 Data Wrangling and Integration ... 396

10.2 Machine Learning Algorithms ... 399
- 10.2.1 Supervised Learning ... 399
- 10.2.2 Unsupervised Learning ... 402
- 10.2.3 Reinforcement Learning ... 405

10.3 Deep Learning ... 408
- 10.3.1 Mathematical Foundations of Neural Networks ... 408
- 10.3.2 Implementing Neural Networks from Scratch ... 410
- 10.3.3 Convolutional and Recurrent Neural Networks ... 414

10.4 Advanced Topics in Machine Learning ... 417
- 10.4.1 Model Interpretability and Explainability ... 417
- 10.4.2 Model Evaluation and Validation Techniques ... 419
- 10.4.3 Generative Models ... 422
- 10.4.4 Big Data and Scalable Machine Learning ... 425

10.5 Ethics and Fairness in Machine Learning ... 428
- 10.5.1 Bias and Fairness in AI ... 428
- 10.5.2 Privacy–Preserving Machine Learning ... 430
- 10.5.3 Responsible AI and Ethical Considerations ... 433

10.6 Exercises ... 434

11 Advanced Topics ... 437
11.1 Symbolic Computation ... 438
- 11.1.1 Advanced Symbolic Mathematics with SymPy ... 438
- 11.1.2 Automated Theorem Proving ... 442
- 11.1.3 Fourier Transforms ... 445
- 11.1.4 Wavelet Transforms ... 447

11.2 Time-Series and Signal Processing ... 449
- 11.2.1 Statistical Time-Series Models ... 449
- 11.2.2 Digital Filtering in Python ... 452
- 11.2.3 Spectral Analysis ... 454
- 11.2.4 Time-Frequency Representations ... 457
- 11.2.5 Applications ... 459

11.3 Topological Data Analysis ... 462
- 11.3.1 Introduction to TDA ... 462
- 11.3.2 Computational Topology ... 464

11.4 Quantum Computing ... 468
- 11.4.1 Mathematical Foundations of Quantum Mechanics ... 468
- 11.4.2 Quantum Algorithms ... 470

	11.4.3	Spectral Methods and Applications	473
	11.4.4	Parallel and Distributed Computing for Numerical Methods	476
11.5	Exercises		478

Appendix 483

References 489

About the Authors

Pradeep Singh is a National Post-Doctoral Fellow and Principal Investigator at the Machine Intelligence Lab, Department of Computer Science and Engineering, IIT Roorkee. His research spans Geometric Deep Learning, Neuro-symbolic AI, and Dynamical Systems, funded by the Anusandhan National Research Foundation (ANRF), Department of Science and Technology, India. His research has appeared in premier venues, including IEEE Transactions, Biomedical Signal Processing and Control, Computers & Geosciences, Journal of Forecasting, ECAI, and ACM Multimedia, contributing to the wider dialogue in AI and dynamical-systems research. He has authored three books: the Springer monographs Deep Learning Through the Prism of Tensors and The Geometry of Intelligence: Foundations of Transformer Networks, and the McGraw-Hill text Machine Learning and Artificial Intelligence. He earned his Ph.D. (2022) and master's degree from IIT Delhi, specializing in Symbolic Systems, and a bachelor's degree in Data Science from IIT Madras. He secured All-India Rank 1 in GATE 2020, JAM 2015, and CSIR-NET 2019. He has also received highly competitive awards, including the National Board for Higher Mathematics (NBHM) Master's, Doctoral, and Post-Doctoral Fellowships awarded by the Department of Atomic Energy, India. In 2019, he was one of only two researchers nationwide to receive the Dr Shyama Prasad Mukherjee (SPM) Fellowship in Mathematics from CSIR, India.
Email: pradeep.cs@sric.iitr.ac.in
Homepage: https://pradeepiitd.github.io/

Balasubramanian Raman (Senior Member, IEEE) is Professor (HAG) and Head of the Department of Computer Science & Engineering at the Indian Institute of Technology Roorkee, where he also holds a joint appointment with the Mehta Family School of Data Science & AI. He served as the iHUB Divyasampark Chair Professor from 2022 to 2025. He earned his Ph.D. from IIT Madras (2001). His research—spanning machine learning, computer vision, image/video processing, and pattern recognition—has produced more than 250 peer-reviewed publications and a Google-Scholar h-index of 51. He has undertaken post-doctoral and visiting appointments at Rutgers University, the University of Missouri–Columbia, Osaka Metropolitan University, Curtin University, and the University of Cyberjaya. To date, he has supervised 34 Ph.D. scholars (with a further 16 in progress) and regularly directs large, multi-institutional programmes funded by leading national agencies and industry partners. An inventor of an Indian patent for a real-time fog-removal imaging system, he has also co-authored two books— Deep Learning Through the Prism of Tensors (Springer Nature, 2024) and The Geometry of Intelligence: Foundations of Transformer Networks in Deep Learning (Springer Nature, 2025). His honours include the DST BOYSCAST Fellowship, two IIT Roorkee Outstanding Teacher Awards, the Ramkumar Prize for Outstanding Teaching and Research, and the ICPC Coach Award, which recognised his teams' top-50 finishes at the ACM ICPC World Finals.
Email: bala@cs.iitr.ac.in
Homepage: https://faculty.iitr.ac.in/cs/bala/

List of Common Abbreviations

AI	Artificial Intelligence
API	Application Programming Interface
AUC	Area Under the Curve
BVP	Boundary Value Problem
CDF	Cumulative Distribution Function
CNN	Convolutional Neural Network
DQN	Deep Q-Network
EDA	Exploratory Data Analysis
FDM	Finite Difference Method
FEM	Finite Element Method
FFT	Fast Fourier Transform
GA	Genetic Algorithm
GAN	Generative Adversarial Network
GIS	Geographic Information Systems
GMM	Gaussian Mixture Model
GPS	Global Positioning System
GPU	Graphics Processing Unit
HMM	Hidden Markov Model
ICA	Independent Component Analysis
IVP	Initial Value Problem
KL	Kullback-Leibler (Divergence)
LDA	Latent Dirichlet Allocation
LU	Lower-Upper (Decomposition)
LSTM	Long Short-Term Memory (Network)
MCMC	Markov Chain Monte Carlo
MDP	Markov Decision Process
ML	Machine Learning
NLU	Natural Language Understanding
NLP	Natural Language Processing
ODE	Ordinary Differential Equation
PCA	Principal Component Analysis

PDE	Partial Differential Equation
PDF	Probability Density Function
PSO	Particle Swarm Optimization
QR	QR Decomposition (Gram-Schmidt Process)
RL	Reinforcement Learning
RNN	Recurrent Neural Network
ROC	Receiver Operating Characteristic
SGD	Stochastic Gradient Descent
SMOTE	Synthetic Minority Over-sampling Technique
SVM	Support Vector Machine
TDA	Topological Data Analysis
TPU	Tensor Processing Unit
UML	Unified Modelling Language

Chapter 1
Introduction

Abstract This chapter motivates the choice of Python as a mathematical workbench, contrasting its open-source ecosystem and rich numerical libraries with legacy tools such as MATLAB, R, and Mathematica. After establishing a reproducible development environment, it builds a rigorous foundation in the language itself—covering core syntax and semantics, primitive and composite data types, control-flow constructs, functional abstraction, input/output, object-oriented design patterns, and iterator/generator paradigms—all while emphasising mathematically literate coding style, numerical precision, and unit-tested best practices that will underpin every subsequent chapter.

Keywords Python · Mathematical computation · Programming fundamentals · Control flow · Object-oriented programming · Iterators and generators

Python has evolved from a scripting language for system administrators into a mature computational ecosystem that *embeds* mathematical ideas as first-class citizens. Its clean syntax allows mathematical notation to be mirrored closely in source code—compare the one-line comprehension $\{n^2 : n \leq 10\}$ with the textbook set $\{n^2 \mid n \leq 10\}$. The vast selection of open-source libraries means that whether we seek exact arithmetic, symbolic manipulation, or GPU-accelerated tensor calculus, the corresponding tools are only an `import` away. Because these libraries are unified by Python's data-model and object protocol, numerical arrays, symbolic expressions, and even discrete structures like graphs can be composed seamlessly. In effect, Python becomes a *literate mathematical notebook*: an interactive workspace where definitions, experiments, conjectures, and proofs coexist—and where computational feedback sharpens intuition. Throughout this book we exploit that synergy: every conceptual exposition is paired with executable code, and every algorithm is dissected with the same rigour that underlies a formal proof.

1.1 Why Python for Mathematics?

1.1.1 Advantages of Python in Mathematical Computation

Overview

At its core, Python furnishes five features that render it uniquely suitable for mathematical work.

1. **Expressive Syntax.** List and dictionary comprehensions, multiple assignment, and first-class functions replicate set-builder notation, tuples, and λ-calculus constructs directly in code.
2. **Multi-Paradigm Support.** Imperative, functional, object-oriented, and (via sympy) symbolic paradigms coexist, allowing the natural choice of abstraction for each problem.
3. **Rich Numerical Stack.** Libraries such as numpy, scipy, and pandas leverage optimised C and Fortran back-ends; they provide $O(n^3)$ dense-matrix algorithms, FFTs, and special functions with performance close to MATLAB while remaining free and open.
4. **Interoperability.** The ctypes and cffi interfaces allow low-level numerical kernels or legacy FORTRAN routines to be called without marshalling overhead, preserving accuracy and speed.
5. **Interactive Computing.** Jupyter notebooks and IPython shells offer REPL-driven exploration, inline LaTeX rendering, and rich visualisation—ideal for conjecture testing and pedagogical exposition.

Worked Examples

Example 1.1.1 (High-Precision Evaluation of $e^{\pi\sqrt{163}}$) The expression $e^{\pi\sqrt{163}}$ is famously almost an integer (Ramanujan). Using Python's decimal module we verify this to 50 decimal places:

```
from decimal import Decimal, getcontext
from math import pi, sqrt

getcontext().prec = 50 # 50 significant digits
x = Decimal(pi) * Decimal(163).sqrt()
y = x.exp()
print(y)
print(y - round(y))
```

1.1 Why Python for Mathematics?

The residual is about 7.4×10^{-40}, corroborating the near-integer phenomenon discussed in Berndt (1989).

Example 1.1.2 (Solving a Linear System $Ax = b$) Let

$$A = \begin{pmatrix} 2 & -1 & 0 \\ -1 & 2 & -1 \\ 0 & -1 & 2 \end{pmatrix}, \quad b = \begin{pmatrix} 1 \\ 0 \\ 1 \end{pmatrix}.$$

Using numpy.linalg one writes

```
import numpy as np
A = np.array([[2, -1, 0], [-1, 2, -1], [0, -1, 2]], dtype=float)
b = np.array([1, 0, 1], dtype=float)
x = np.linalg.solve(A, b)
```

which yields the exact rational solution $(x_1, x_2, x_3) = (1, \frac{3}{2}, 1)$ up to floating-point error.

Common Pitfalls

Python's accessibility conceals subtleties that can entrap the unwary mathematician.

1. **Binary Floating-Point Rounding.** The identity $0.1 + 0.2 = 0.3$ is false in IEEE-754 binary.[1] Exact arithmetic requires decimal or fractions.
2. **Mutable Default Parameters.** A function definition like

    ```
    def append_item(x, L=[]): L.append(x); return L
    ```

    accumulates state across calls, violating mathematical purity. Use None sentinels instead.
3. **Algorithmic Performance.** A naive Python loop implementing $O(n^2)$ matrix multiplication is thousands of times slower than numpy.dot; asymptotic reasoning must be coupled with vectorisation.
4. **Shadowing Built-ins.** Naming a variable sum or max overrides the standard library versions, leading to perplexing errors later in the session.

Each of these pitfalls will recur throughout the text; we address them systematically, providing proof-style reasoning where appropriate and empirical timing curves when instructive.

---

[1] See Goldberg's classic exposition (Goldberg 1991).

1.1.2 Comparison with Other Mathematical Tools

Python vs. MATLAB

Although both ecosystems excel at numerical linear algebra, their design philosophies diverge sharply. MATLAB adopts a *matrix-centric* worldview: every datum is an $m \times n$ array by default, and one-based indexing matches traditional algebraic notation. Python, by contrast, treats objects abstractly and uses zero-based indexing, which aligns with most modern programming languages and allows elegant slicing idioms such as `A[:,k]` for the k-th column. Performance parity is now typical because `numpy` calls the same LAPACK/BLAS back-ends that underlie MATLAB's `mldivide`. The decisive advantage is Python's open architecture: C-extensions, JIT tooling (`numba`), and GPU bindings (`cupy`) are all available *gratis*. Moreover, symbolic algebra via `sympy` and combinatorial libraries such as `networkx` integrate seamlessly with numerical code, eliminating the need for a separate Computer Algebra System (CAS). The trade-off is that Python requires conscious array broadcasting (`np.newaxis`) to emulate MATLAB's implicit expansion rules; careless dimension mismatches can introduce latent bugs.

Example 1.1.3 (Least-Squares Fit: MATLAB vs. Python) Given $A \in \mathbb{R}^{m \times n}$ and $b \in \mathbb{R}^m$, MATLAB's solution `x = A \ b` is mirrored in Python as

```
import numpy as np
x = np.linalg.lstsq(A, b, rcond=None)[0]
```

Both invoke QR decomposition internally, yielding $\hat{x} = (A^\mathsf{T} A)^{-1} A^\mathsf{T} b$ with identical backward-error bounds (Golub and Van Loan 2013).

Python vs. R

R was conceived for statistical modelling; its syntax reflects formulas, factors, and data frames as primitive notions. Python acquires equivalent expressiveness through the `pandas` and `statsmodels` stacks while adding general-purpose constructs such as generators and metaclasses. In time-series analysis, R's `xts`/`zoo` objects correspond to `pandas.Series`, but Python's iterator protocol enables lazy evaluation of infinite streams—indispensable in stochastic process simulation. For Bayesian inference, R's `rstan` finds a counterpart in `pymc`; however, Python's ability to interleave symbolic differentiation (via `jax`) with Markov-chain Monte Carlo yields gradients $\nabla \log p(\theta \mid \mathcal{D})$ in a single code path, reducing boilerplate.

Example 1.1.4 (Kernel Density Estimation) A univariate Gaussian KDE requires one line in each language.

```
from scipy.stats import gaussian_kde
pdf = gaussian_kde(samples)
```

The call mirrors R's `density(samples)` but returns a callable object that vectorises over $\mathbb{R}^d$, enabling direct integration with `numpy` broadcasting semantics.

Python vs. Mathematica

Mathematica's strength is its rule-based symbolic engine capable of exact results such as $\int e^{x^2} dx = \frac{\sqrt{\pi}}{2}\operatorname{erfi} x + C$. Python's sympy reproduces much of this functionality with transparent algorithms—ideal for pedagogical deep dives. Python gains ground in large-scale numeric–symbolic hybrids: calling `sympy.Matrix` methods inside batched GPU code is possible because objects share the same runtime. Mathematica's *notebook* paradigm is rivalled by Jupyter, whose Markdown+code cells also support LaTeX and interactive widgets via `ipywidgets`. Licensing costs and closed source limit Mathematica's deployability on clusters, whereas Python scales horizontally through containerisation and serverless functions.

Example 1.1.5 (Automatic Differentiation vs. Symbolic Differentiation)
Consider $f(x) = \sin(\sqrt{x^2 + 1})$. Mathematica symbolically returns $f'(x) = \frac{x\cos(\sqrt{x^2+1})}{\sqrt{x^2+1}}$. Python offers a complementary *automatic* route:

```
import jax.numpy as jnp
from jax import grad
f = lambda x: jnp.sin(jnp.sqrt(x**2 + 1))
df = grad(f)
```

Here jax propagates dual numbers, computing $f'(x)$ at machine precision for any $x \in \mathbb{R}$ without explicit symbolic forms.

1.2 Setting Up the Python Environment

1.2.1 *Installing Python and Essential Libraries*

CPython and Anaconda Distributions

The reference interpreter `CPython` compiles C source into bytecode executed by the CPython Virtual Machine. Installation via `python.org` ensures the latest stable release, but minimalists must subsequently `pip install numpy scipy matplotlib sympy`. The *Anaconda* distribution, by contrast, ships a curated set of ≈250 scientific packages—pre-compiled against Intel MKL—together with the `conda` package manager. This turnkey approach avoids the fragile compile chains required for LAPACK or OpenBLAS on Windows. Verify a clean install with:

```
>>> import sys, numpy, sympy, scipy
>>> print(sys.version)
>>> print(numpy.__version__, sympy.__version__, scipy.__version__)
```

Managing Virtual Environments

Isolating dependencies is crucial for reproducible research. Python offers three mechanisms:

1. `venv` (standard library):
   ```
   python3 -m venv env && source env/bin/activate
   pip install numpy sympy
   ```
   This creates a project-local copy of the interpreter and its `site-packages`.
2. `conda` environments:
   ```
   conda create -n calculus python=3.12 numpy sympy
   conda activate calculus
   ```
   Useful when binary compatibility with MKL or CUDA is needed.
3. `pipx` for globally installing *isolated* command-line tools (e.g. `jupyter`).

A `requirements.txt` or `environment.yml` should accompany each chapter's source to encode the exact package versions $(v_1, \ldots, v_k) \in \mathbb{N}^k$, making the computational narrative portable across machines.

1.2.2 Python Syntax and Semantics

Variables, Expressions, and Statements

A *variable* in Python is not a labelled memory cell but a *binding* from an identifier to an object residing on the heap. The assignment statement

```
x = 3
```

creates a reference from the name `x` to the singleton integer object whose value is 3. Subsequent rebindings merely redirect the name—they do not mutate the integer, which is immutable. Equality `==` invokes the object's `__eq__` method, whereas identity `is` compares memory addresses; thus, `(2 is 1+1)` may be `True` for small integers because of CPython's internal *integer pool*.

An *expression* is any syntactic form that *yields* a value. Every expression has a type τ drawn from the lattice of built-in or user-defined classes. By contrast, a *statement* performs an action—assignment, flow control, import—but evaluates to `None`. Consider the list-comprehension identity

$$\left[x^2\right]_{x=1}^{n} = \left[(x := k)^2\right]_{k=1}^{n},$$

where the *walrus* operator `:=` binds the target within the expression, melding imperative and functional styles.

1.2 Setting Up the Python Environment

Example 1.2.1 (Law of Total Probability in List Comprehension) Given mutually exclusive events $E_1, \ldots, E_k$ with $\sum p(E_i) = 1$, their probabilities can be sampled and normalised in a single expression:

```
weights = [p(E_i) for E_i in events]  # Σ weights = 1
samples = [w / sum(weights) for w in weights]  # explicit normalisation
```

Each comprehension is an expression returning a new list; no intermediate statements are required.

Indentation, Blocks, and Style

Python eschews braces in favour of *semantic indentation*. A block begins immediately after the colon : that terminates a compound statement. The interpreter enforces a deterministic mapping from source code to parse tree: inconsistent indentation triggers `IndentationError`, preventing syntactic ambiguities that plague languages with optional braces.

Style guidelines are codified in **PEP 8**. Mathematical scripts benefit from its clarity rules: variable names should mirror conventional notation (`theta`, `lambda_max`), functions are `lower_snake_case`, and classes are `CapWords`. Lines should seldom exceed 79 characters, ensuring LaTeX/PDF listings do not overrun the page. Automatic formatters such as `black` enforce a canonical style, making diffs behave like algebraic equalities: extraneous whitespace is eliminated, and only semantic changes remain visible.

1.2.3 Python as an Interpreted Language

Bytecode and the CPython VM

While Python is colloquially "interpreted", the CPython reference implementation performs a two-phase translation. Source files are first parsed into an Abstract Syntax Tree, optimised into stack-based *bytecode*, and cached as `.pyc` in the `__pycache__` directory. Each bytecode instruction manipulates a virtual stack; for example, the multiplication $a \times b$ compiles into

```
>>> import dis
>>> def prod(a, b): return a * b
...
>>> dis.dis(prod)
  1           0 LOAD_FAST                0 (a)
              2 LOAD_FAST                1 (b)
              4 BINARY_MULTIPLY
              6 RETURN_VALUE
```

The virtual machine executes these n bytes in $O(n)$ time, dispatching each opcode via a large `switch` table. Crucially, numeric heavy lifting is delegated to C functions—Python orchestrates, while libraries compute.

Interactive vs. Script Mode

Invoking `python` without arguments spawns a *Read–Eval–Print Loop* (REPL). Each input line is compiled and executed immediately, returning control to the user. This mirrors the mathematical habit of manipulating expressions on a blackboard: conjecture, test, refine. In *script mode* (`python file.py`) compilation precedes execution, and the module's global namespace becomes `__main__`. Idempotence is enforced by the idiom

```
if __name__ == '__main__':
    main()
```

which prevents accidental side-effects when the file is imported elsewhere—analogous to requiring lemmas but deferring theorem proving until the main proof context.

1.3 Data Types and Variables

1.3.1 Primitive Data Types

Integers, Floats, and Complex Numbers

Python's `int` implements unbounded signed integers using a base 2^{30} (or 2^{15} on 16-bit builds) limb representation; arithmetic thus respects the Peano axioms without modular wrap-around. The expression

$$\underbrace{2^{2^5}}_{\text{over a billion digits}}$$

is legal albeit memory-intensive.

Floating-point numbers follow the IEEE-754 `binary64` format: each value encodes sign s, exponent $e \in [-1022, 1023]$, and mantissa $m \in [0, 2^{53})$, representing $(-1)^s m 2^{e-52}$. Rounding mode defaults to "round half to even". Mathematical purity is preserved by providing the `decimal` module for arbitrary precision and `fractions.Fraction` for exact rationals.

A `complex` is a pair $(x, y) \in \mathbb{R}^2$ with arithmetic

$$(x_1 + iy_1)(x_2 + iy_2) = (x_1 x_2 - y_1 y_2) + i(x_1 y_2 + x_2 y_1),$$

1.3 Data Types and Variables 9

implemented in C for speed. Euler's identity is verified numerically as:

```
import cmath, math
abs(cmath.exp(1j * math.pi) + 1)  # ≈ 0
```

The residual is $<10^{-16}$, bounded by double precision.

Booleans and NoneType

The Boolean algebra $\{\bot, \top\}$ is represented by `False` and `True`, subclasses of `int` with values 0 and 1. Logical operations satisfy De Morgan's laws; for example

$$\neg(A \wedge B) \equiv (\neg A) \vee (\neg B)$$

translates to

```
not (A and B) == (not A) or (not B)
```

and returns `True` for all truth assignments.

The singleton `None` of type `NoneType` denotes "absence of value". It behaves as the unique null element: comparisons use identity (`is None`), never equality. Attempting arithmetic with `None` raises `TypeError`, mirroring the undefined nature of 0/0 in real analysis.

Example 1.3.1 (Short-Circuit Semantics) Given a lazy predicate $P(x)$ expensive to evaluate, the conjunction

```
x is not None and P(x)
```

runs P only if the first clause is `True`. This mirrors the logical rule $(\bot \wedge \varphi) = \bot$, preventing `AttributeError` when x is absent.

Primitive types thus provide a formal substrate mirroring $\mathbb{Z}$, $\mathbb{R}$, and $\mathbb{C}$, while their Pythonic implementations respect mathematical axioms to machine-precision limits. Subsequent sections build composite structures—lists, tuples, arrays—on this foundation, much as algebraic topology builds spaces from simplices.

1.3.2 Mathematical Operations on Data Types

Arithmetic Operators

Python overloads the five classical field operations and augments them with integer-specific variants. Addition $+$, subtraction $-$, and multiplication $*$ coincide with the ring axioms on $\mathbb{Z}$, and the power operator $**$ implements exponentiation by repeated squaring in $O(\log n)$ time. Division bifurcates into / for exact quotient and // for Euclidean floor-division; when $a, b \in \mathbb{Z}$ with $b \neq 0$, the invariant

$$a = (a//b)b + (a \bmod b), \qquad 0 \leq a \bmod b < |b|$$

is preserved exactly.

Example 1.3.2 (Extended Euclidean Algorithm) The greatest common divisor $\gcd(a, b)$ and Bezout coefficients (x, y) solving $ax + by = \gcd(a, b)$ arise naturally:

```
def egcd(a: int, b: int) -> tuple[int, int, int]:
    if b == 0:
        return a, 1, 0
    g, x1, y1 = egcd(b, a % b)
    return g, y1, x1 - (a // b) * y1
```

Tail recursion exploits // and % to maintain the invariant while descending the Euclidean lattice.

Bitwise and Logical Operators

For n-bit integers, Python provides bitwise conjunction &, disjunction |, exclusive-or ^, left-shift «, and right-shift ». These form a Boolean algebra over $\mathbb{Z}_2^n$; masks such as

$$m = 1 << k$$

select the k-th bit. Logical operators **and**, **or**, and **not** act on truthiness and short-circuit, implementing the algebra $(A \wedge B) \vee (\neg A) = A \rightarrow B$.

Example 1.3.3 (Parity Check Using XOR) The parity of a binary word w equals the XOR reduction of its bits:

```
from functools import reduce
parity = lambda n: reduce(int.__xor__, map(int, bin(n)[2:]))
```

Here `int.__xor__` is folded over the bit string, returning 1 for odd parity, 0 for even.

1.3.3 Working with Variables

Naming Conventions

Identifiers should mirror mathematical semantics: `delta_t` for Δt, `sigma2` for σ^2. PEP 8 recommends `lower_snake_case` for variables and functions and

UpperCamelCase for classes. Trailing underscores disambiguate from built-ins (sum_). Internal __dunder__ names signal "magic" methods—avoid them for user variables to prevent metaprogramming collisions.

Dynamic Typing and Type Hints

Python's names are dynamically bound: the map env : name $\mapsto$ object may evolve along execution paths. Static guarantees are added through **type hints** (PEP 484), which convert env into a *typing context*

$$\Gamma \vdash f : \alpha \to \beta,$$

checked by external tools such as mypy. Parametric types capture algebraic data structures:

```
from typing import TypeVar, Generic, Iterable

T = TypeVar('T', int, float)  # T ⊆ {ℤ, ℝ}
class Vector(Generic[T]):
    def __init__(self, data: Iterable[T]): ...
```

Runtime semantics remain unchanged; annotations refine the proof obligations without slowing execution.

1.3.4 Containers and Collections

Lists and Tuples

A list is a mutable dynamic array supporting amortised $O(1)$ append via geometric capacity growth; a tuple is its immutable counterpart. Both allow slicing $s[i:j:k]$, which corresponds to the arithmetic progression $i, i+k, \cdots < j$. The Cartesian product

$$A \times B = \{(a, b) \mid a \in A, b \in B\}$$

is expressible as

`[(a, b) for a in A for b in B]`

returning a list of tuples.

Sets and Frozen Sets

Python's set implements the abstract data type $(\mathcal{P}(U), \cup, \cap, \setminus)$ using open addressing with Fibonacci hashing, yielding average $O(1)$ membership. A frozenset is the immutable variant, hashable and thus admissible as dictionary keys. Standard identities hold:

```
(A | B) & C == (A & C) | (B & C) # distributive law
```

returning `True` by structural evaluation.

Example 1.3.4 (Prime Factors via Set Intersection) Let $P(n)$ denote the prime factors of n. Then $P(a) \cap P(b) = \varnothing$ iff a and b are coprime.

```
from sympy import factorint
coprime = lambda a, b: set(factorint(a)) & set(factorint(b)) == set()
```

Dictionaries

A `dict` realises the finite map $K \to V$ with expected $O(1)$ lookup. Keys must be hashable—i.e. implement a function $h : K \to \mathbb{Z}_{2^{64}}$ satisfying $k_1 = k_2 \Rightarrow h(k_1) = h(k_2)$. Collisions resolve via probing sequences derived from the key's perturbation value, ensuring $O(n)$ worst case is rare.

Example 1.3.5 (Memoisation of Fibonacci Numbers) The recursive definition

$$F(n) = F(n-1) + F(n-2), \qquad F(0) = 0, \; F(1) = 1$$

runs in $O(n)$ with dictionary caching:

```
from functools import lru_cache

@lru_cache(maxsize=None)
def F(n: int) -> int:
    return n if n < 2 else F(n-1) + F(n-2)
```

The decorator stores previously computed $(n, F(n))$ pairs in an internal dict, exploiting hash-table semantics to achieve linear complexity.

Containers thus furnish the combinatorial scaffolding atop primitive scalars, enabling constructions from graphs to tensors with both mathematical elegance and computational efficiency.

1.4 Control Flow in Python

Control flow structures translate mathematical quantifiers and case-by-case reasoning into executable form. The `if-elif-else` ladder corresponds to a piecewise definition; `for` loops iterate over finite index sets, mirroring $\sum_{i=1}^{n}$; `while` loops model potentially unbounded processes such as Newton iterations whose termination depends on convergence criteria. Because Python enforces indentation, the syntactic shape of each block reflects its logical nesting, reducing cognitive overhead during proof-by-inspection of the code.

1.4.1 Conditional Statements

`if-elif-else` Blocks

A conditional block evaluates predicates in order until one is `True`. Semantically, the construct realises the piecewise function

$$f(x) = \begin{cases} f_1(x), & \text{if } P_1(x) \\ f_2(x), & \text{elif } P_2(x) \\ \vdots & \\ f_k(x), & \text{else} \end{cases}$$

where the predicates P_i are mutually exclusive in well-formed code. Python short-circuits evaluation after the first satisfied guard, preserving the minimal proof obligation.

Example 1.4.1 (The Sign Function sgn**)**

```
def sgn(x: float) -> int:
    if x > 0:
        return 1
    elif x < 0:
        return -1
    else:
        return 0
```

The function implements sgn: $\mathbb{R} \to \{-1, 0, 1\}$ exactly as its textbook definition; each branch returns a constant in $O(1)$ time.

Ternary Expressions

Python's conditional expression

$$a \textbf{ if } P \textbf{ else } b$$

is an expression, not a statement, making it admissible within list comprehensions or λ abstractions. It embodies the characteristic function χ_P:

$$\chi_P(x) = \begin{cases} 1, & P(x) \\ 0, & \neg P(x) \end{cases}.$$

Example 1.4.2 (Piecewise Mapping Inside a Comprehension) The indicator of quadratic residues modulo a prime p can be generated as

```
QR = [(1 if pow(a, (p-1)//2, p) == 1 else 0) for a in range(1, p)]
```

returning a list in $\{0, 1\}^{p-1}$ without breaking the comprehension flow.

1.4.2 Loops in Python

for Loops

A `for` loop iterates over any object that implements the iterator protocol (methods `__iter__` and `__next__`). Conceptually, the loop

```
for x in S: body(x)
```

executes the mapping $S \xrightarrow{\text{body}}$ None, analogous to $\sum_{x \in S}$ when body accumulates into a total.

Example 1.4.3 (Numerical Integration via Riemann Sums) Given $f : [a, b] \to \mathbb{R}$ and partition width h, the Riemann approximation is

$$\int_a^b f(x)\,dx \approx h \sum_{k=0}^{n-1} f(a + kh).$$

```
def riemann(f, a: float, b: float, n: int = 1000) -> float:
    h = (b - a) / n
    return h * sum(f(a + k * h) for k in range(n))
```

The generator expression inside `sum` implicitly constructs an iterator, avoiding materialisation of the full list of f-values.

while Loops

A `while` loop embodies a *least fixed point*: execution repeats until the guard predicate P evaluates to `False`. Correctness proofs follow the invariant method— identify I such that (i) I holds before the loop, (ii) $I \wedge P$ implies I after one body execution, and (iii) termination plus $I \wedge \neg P$ implies the post-condition.

Example 1.4.4 (Newton–Raphson for $\sqrt{\alpha}$) Let $\alpha > 0$. Newton updates

$$x_{k+1} = \tfrac{1}{2}(x_k + \alpha/x_k)$$

1.4 Control Flow in Python

converge quadratically. Implementation:
```python
def sqrt_newton(alpha: float, tol: float = 1e-12) -> float:
    x = alpha
    while True:
        xn = 0.5 * (x + alpha / x)
        if abs(xn - x) < tol:
            return xn
        x = xn
```

The invariant $x_k > 0$ maintains domain validity; termination criterion $|x_{k+1} - x_k| < \varepsilon$ embeds the Cauchy condition for convergence.

1.4.3 Comprehensions in Python

List Comprehensions

A *list comprehension*

$$[\,f(x)\ \textbf{for}\ x \in S\ \textbf{if}\ P(x)\,]$$

is the computational analogue of the set-builder expression $\{f(x) \mid x \in S,\ P(x)\}$. Evaluation is left-to-right: the iterable S is traversed once, predicates $P(x)$ are short-circuited, and each surviving element is mapped through f. The entire comprehension is an *expression*; its result can be nested safely inside higher-order functions without leaking loop variables.

Example 1.4.5 (Prime Sieve up to N) Using the Sieve of Eratosthenes, the set of primes $\leq N$ is

$$\mathcal{P}(N) = \{n \mid 2 \leq n \leq N,\ \forall p \leq \sqrt{n}\ (p \mid n \Rightarrow p = n)\}.$$

```python
from math import sqrt
def primes(N: int) -> list[int]:
    is_prime = [True] * (N + 1)
    for p in range(2, int(sqrt(N)) + 1):
        if is_prime[p]:
            is_prime[p*p:N+1:p] = [False] * len(range(p*p, N+1, p))
    return [q for q in range(2, N + 1) if is_prime[q]]
```

The final line is a comprehension filtering the Boolean mask `is_prime`, producing an ordered list of primes in $O(N \log \log N)$ time.

Set and Dictionary Comprehensions

Replacing the brackets by curly braces yields a **set comprehension**; adding a colon separates keys and values, giving a **dictionary comprehension**. Both evaluate in a single pass and inherit the hash-table performance guarantees of their concrete container types.

Example 1.4.6 (Quadratic Residues Modulo p**)** For an odd prime p, the set of quadratic residues is

$$Q = \{a^2 \bmod p : 1 \leq a < p\}.$$

```
Q = {pow(a, 2, p) for a in range(1, p)}
```

The comprehension returns a `set`, automatically discarding duplicates and achieving average $O(1)$ membership checks.

Example 1.4.7 (Inverting a Bijective Map) Given a bijection $f : X \to Y$ encoded as a dictionary `fwd`, its inverse f^{-1} is

```
inv = {v: k for k, v in fwd.items()}
```

The comprehension enforces the uniqueness of values; a collision would raise `KeyError`, exposing a violation of bijectivity.

1.4.4 *Error Handling and Exceptions*

`try-except-finally`

Python models runtime anomalies through the class hierarchy rooted at `BaseException`. A `try` block establishes a guarded region; when an exception ε propagates, control transfers to the `except` clause whose parameter type is a super-class of ε. This corresponds to the logical rule

$$\frac{\Gamma \vdash M : \alpha \quad \varepsilon : \beta \in \mathsf{Exc}}{\Gamma \vdash \mathbf{try}\ M\ \mathbf{except}\ \beta \Rightarrow N : \alpha},$$

ensuring the overall expression still types as α.

Example 1.4.8 (Robust Numerical Integration)
Adaptive quadrature may raise `ZeroDivisionError` if the integrand is singular. A protective wrapper records the failure and returns NaN:

```
import math
def safe_integrate(f, a: float, b: float):
    try:
        return math.quad(f, a, b)  # pseudo-call for illustration
    except ZeroDivisionError:
        return float('nan')
    finally:
        log.info("integration attempted on [%g, %g]", a, b)
```

The `finally` block executes unconditionally, mirroring the total probability rule: all execution paths sum to one.

Custom Exception Classes

When domain logic requires finer granularity than the built-ins provide, define a subclass of `Exception`. The new class should encapsulate an invariant breach or precondition failure, making the error surface self-descriptive.

Example 1.4.9 (Non-convergence in Newton Iteration)

```
class ConvergenceError(Exception):
    """Raised when an iterative method fails to converge."""

def newton(f, df, x0: float, tol: float = 1e-12, max_iter: int = 50):
    for _ in range(max_iter):
        x1 = x0 - f(x0) / df(x0)
        if abs(x1 - x0) < tol:
            return x1
        x0 = x1
    raise ConvergenceError(
        f"Newton failed after {max_iter} iterations; |∆|={abs(x1 - x0):.2e}"
    )
```

A calling routine may catch `ConvergenceError` selectively, leaving unrelated exceptions to propagate. This mirrors the practice of isolating singularities when extending analytic functions.

Structured exception handling therefore provides a categorical sum type—normal result ⊎ error payload—allowing algorithms to advertise exceptional conditions explicitly and enabling callers to compose reliable numerical pipelines.

1.5 Functions and Modular Programming

Programming mirrors mathematics most closely when code is organised around the notion of a *function*—a mapping that associates each element of a domain with exactly one element of a codomain. Python preserves this abstraction by treating functions as first-class objects, enabling composition, higher-order operators, and even runtime transformation of function behaviour. Modular programming extends the idea to namespaces: a module is the analogue of a mathematical theory, bundling definitions, lemmas, and theorems so that they can be imported without polluting the surrounding context. The following sections examine these constructs through both formal lenses and executable demonstrations.

1.5.1 Defining Functions

Positional and Keyword Arguments

A function header of the form

```
def f(x, y, /, z, *, w):
    ...
```

declares two *positional-only* parameters x, y; one flexible parameter z; and one *keyword-only* parameter w. The slash and star delimiters were introduced in PEP 570 to enforce clarity and prevent accidental name collisions. Mathematically, the call

$$f(a, b, z = c, w = d)$$

is the evaluation $f(a, b, c, d)$ where the mapping from syntactic position to formal parameter is bijective.

Example 1.5.1 (Bivariate Normal PDF)

```
import math

def bivariate_pdf(x, y, /, *, rho: float = 0.0):
    σ = math.sqrt(1 - rho**2)
    return math.exp(-(x**2 - 2*rho*x*y + y**2) / (2*σ**2)) / (2*math.pi*σ)
```

The correlation coefficient ρ is specified as a keyword, obliging callers to document intent explicitly.

Default and Variable-Length Parameters

Defaults turn a partial application into a total function on a larger domain. The signature

```
def power(base, exp=2):
    return base ** exp
```

extends the map $x \mapsto x^2$ to $(x, n) \mapsto x^n$. Variable-length parameters (*args, **kwargs) correspond to the family of all finite tuples or finite maps, respectively:

$$\{ f : \mathbb{N}^k \to \mathbb{R} \mid k \in \mathbb{N} \}.$$

Example 1.5.2 (Generalised Mean)

$$M_p(x_1, \ldots, x_n) = \left(\frac{1}{n} \sum_{i=1}^{n} x_i^p \right)^{1/p}.$$

1.5 Functions and Modular Programming

```
def mean(*values: float, p: float = 2.0) -> float:
    n = len(values)
    return (sum(v**p for v in values) / n) ** (1 / p)
```

The star collects the variable arity into a tuple, realising the map $\bigcup_{n \geq 1} \mathbb{R}^n \to \mathbb{R}$.

1.5.2 Advanced Function Concepts

First-Class Functions and Closures

Because Python treats functions as objects, one may pass them as arguments, return them, or store them in data structures. A *closure* arises when an inner function retains bindings from its lexical environment, implementing the λ-calculus concept of *currying*.

Example 1.5.3 (Partial Evaluation of a Polynomial) Given coefficients $a_0, \ldots, a_k$, define

$$p_a(x) = \sum_{i=0}^{k} a_i x^i.$$

```
from typing import Sequence, Callable

def make_poly(a: Sequence[float]) -> Callable[[float], float]:
    def p(x: float) -> float:
        return sum(c * x**i for i, c in enumerate(a))
    return p
```

The returned closure captures a, yielding a specialised function $p : \mathbb{R} \to \mathbb{R}$ without further overhead.

Decorators and Higher-Order Functions

A *decorator* is a higher-order function $\mathcal{D} : (A \to B) \to (A \to B)$ that transforms behaviour while preserving type signature. Syntactic sugar

```
@decor
def f(...):
    ...
```

expands to `f = decor(f)`, analogous to operator overloading in algebraic structures.

Example 1.5.4 (Memoisation Decorator)

```
def memoize(f):
    cache = {}
```

```
def g(*args):
    return cache.setdefault(args, f(*args))
return g
```

Applying @memoize to a recursive definition turns exponential-time evaluation of the Fibonacci sequence into linear time, paralleling dynamic programming on DAGs.

1.5.3 Modular Programming with Python

Packages and __init__.py

A directory becomes a *package* when it contains an __init__.py file, establishing a namespace object. The file may re-export selected symbols, enforcing an interface akin to a signature in type theory:

```
# __init__.py
from .linalg import det, eigvals # public API
__all__ = ['det', 'eigvals']
```

Other internals remain private, approximating the encapsulation of lemmas versus public theorems.

Import Mechanics and Namespaces

The import statement

$$\texttt{import } M \quad \Longrightarrow \quad \Gamma \cup \{M{:}\mathcal{N}\}$$

extends the current environment Γ with a binding to module object $\mathcal{N}$. During resolution, the interpreter searches sys.meta_path, walking from built-ins through virtual file finders to local directories. Relative imports (from .sub import s) trace the tree structure of the package, echoing qualified names in logic such as Group.Hom.

1.5.4 Recursive Functions

Mathematical Recurrence Relations

Recursion translates structural induction into executable semantics. Consider the Catalan numbers

$$C_n = \sum_{i=0}^{n-1} C_i C_{n-1-i}, \qquad C_0 = 1.$$

```
from functools import cache

@cache
def catalan(n: int) -> int:
    return 1 if n == 0 else sum(catalan(i) * catalan(n-1-i) for i in range(
        n))
```

The decorator @cache memoises intermediate C_k, respecting the dynamic programming recurrence and delivering C_n in $O(n^2)$ time.

Tail Recursion and Optimisation

Python lacks *proper* tail-call elimination, so deep recursion risks `RecursionError`. Transforming a linear recurrence into an iterative accumulator circumvents the call stack:

```
def fib_tail(n: int) -> int:
    a, b = 0, 1
    for _ in range(n):
        a, b = b, a + b
    return a
```

This loop implements the morphism

$$\begin{pmatrix} a_{k+1} \\ b_{k+1} \end{pmatrix} = \begin{pmatrix} 0 & 1 \\ 1 & 1 \end{pmatrix} \begin{pmatrix} a_k \\ b_k \end{pmatrix},$$

evolving state in $O(n)$ time and $O(1)$ space. Techniques such as trampolining or explicit stacks can generalise the optimisation, but often an equivalent vectorised approach—e.g. exponentiation by squaring of the companion matrix—achieves $O(\log n)$ complexity.

Through first-class functions, decorators, and disciplined modularisation, Python supports the full gamut of mathematical abstraction—from elementary recurrences to categorical composition—while maintaining executable proximity to the underlying theory.

1.6 Input and Output in Python

Mathematical software seldom lives in isolation: algorithms draw data from users, files, or networks and return results to the external world. Input-and-output (I/O) therefore occupies the same foundational role as axioms in a formal system—defining how information enters and leaves the computational universe. Python offers a rich I/O toolbox whose abstractions preserve type safety, manage resources deterministically, and integrate cleanly with the numerical stack. This section surveys those mechanisms and distils best practices for *reproducible*, *readable*, and *testable* mathematical code.

1.6.1 Handling User Input

input() and Command-Line Arguments

Interactive scripts rely on the built-in function input(prompt). It returns a Unicode string stripped of the trailing newline, leaving explicit parsing to the programmer:

```
n = int(input("Enter a positive integer n: "))
print(f"Sum_{k=1}^{n} k = {n * (n + 1) // 2}")
```

For non-interactive execution, the sys.argv list exposes command-line parameters. Robust parsing uses argparse, which enforces types and generates usage documentation automatically—akin to declaring preconditions on a function domain:

```
import argparse, math

parser = argparse.ArgumentParser(description="Compute ζ(s) numerically")
parser.add_argument("s", type=float, help="complex exponent σ (real for
    demo)")
parser.add_argument("-n", type=int, default=10**6,
                help="number of partial-sum terms")
args = parser.parse_args()

zeta = sum(1 / k**args.s for k in range(1, args.n + 1))
print(f"ζ({args.s}) ≈ {zeta:.8f}")
```

Here the call contract $(s, n) \mapsto \zeta_n(s)$ is enforced before evaluation, preventing domain errors.

1.6.2 File Handling in Python

Text vs. Binary Modes

Opening a file with open(path, mode) yields a *file object*. Modes 'r' and 'w' create a *text* stream, decoding bytes to Unicode via an encoding (default UTF-8); modes 'rb' and 'wb' expose raw bytes. Mathematical data—coefficients, matrices—should be stored in text when human auditability is required, or in binary (NumPy's .npy) when precision and speed dominate.

Example 1.6.1 (Persisting a Vandermonde Matrix)

```
import numpy as np
x = np.linspace(0, 1, 5)
V = np.vander(x)
```

1.6 Input and Output in Python

```
np.save("vandermonde.npy", V)   # binary, lossless
np.savetxt("vandermonde.txt", V) # text, human-readable
```

The binary file re-loads in $O(1)$ via memory mapping; the text file sacrifices size for portability.

Context Managers

File objects implement the context-manager protocol; `with` guarantees that `f.close()` executes even if exceptions occur. This deterministic finalisation mirrors the guarantee that a definite integral $\int_a^b f$ consumes the entire domain regardless of integrand behaviour.

```
with open("coeffs.txt") as fp:
    coeffs = [float(t) for t in fp.read().split()]
```

For resources beyond files—socket connections, database cursors—custom context managers are declared with `__enter__`/`__exit__`, formalising the "open–use–close" invariant.

1.6.3 Formatted Output

f'-strings'

Introduced in PEP 498, f-strings embed expressions inside braces, evaluated at runtime and formatted via the `str.format` mini-language:

```
π = math.pi
print(f"π ≈ {π:.10f} (error < 1e-10)")
```

Precision, alignment, and scientific notation are declarative, allowing numerical reports to resemble typeset prose.

str.format()

The older `str.format()` remains useful when the template is external (e.g. read from a file) and cannot be an f-string literal:

```
msg = "e^{π √163} ≈ {0:.40f}"
print(msg.format(math.e ** (math.pi * math.sqrt(163))))
```

Both mechanisms support =-formatting (`f"expr="`) for debugging, akin to labelling steps in an algebraic derivation.

1.6.4 Best Practices in Mathematical Coding

Code Styling and Linting

Consistent style enhances *proof legibility*. The `black` formatter enforces a canonical layout; `flake8` and `pylint` detect anti-patterns such as unused variables (ghost symbols) or shadowing built-ins (name collision in a symbol table). Docstrings should specify parameter domains, return codomains, and algorithmic complexity, matching the structure of a theorem statement:

```
def gcd(a: int, b: int) -> int:
    """Return gcd(a, b).

    Preconditions: a, b ≥ 0 not both zero.
    Postconditions: divides(gcd, a) ∧ divides(gcd, b).
    Complexity: O(log min(a, b)).
    """
```

Unit Testing with `pytest`

`pytest` provides the scaffolding to encode lemmas that guarantee function correctness. A test is a proposition decorated with assertions:

```
# test_number_theory.py
import pytest
from mylib import egcd

@pytest.mark.parametrize("a,b", [(55, 34), (21, 13), (144, 89)])
def test_bezout_identity(a, b):
    g, x, y = egcd(a, b)
    assert x * a + y * b == g
```

Running `pytest` constitutes a proof search; failure outputs a counterexample, just as a falsified conjecture produces a witness. Continuous-integration pipelines (GitHub Actions, GitLab CI) automate this verification across platforms and Python versions, ensuring that results in the book remain reproducible for every reader.

By uniting robust I/O abstractions with disciplined formatting and automated tests, one constructs computational artefacts that are as transparent and reliable as their mathematical foundations.

1.7 Object-Oriented Programming (OOP) in Python

Traditional mathematics classifies objects—groups, rings, vector spaces—by the axioms they satisfy; object-oriented programming mirrors this taxonomy with *classes*. Each class encodes a collection of attributes (data) and methods (operations), while a concrete *instance* corresponds to a mathematical element that inhabits the structure. OOP therefore supplies a lingua franca for representing algebraic systems, numerical data structures, and even proof states in a form that is both executable and extendable.

1.7.1 Basic Concepts of OOP

Classes and Objects

A Python class defines a namespace whose members resolve via the attribute operator $.$. Formally, if $\mathcal{C}$ is a class, its instance set

$$\mathrm{Obj}(\mathcal{C}) = \{c : c \text{ is created by } \mathcal{C}\}$$

carries a mapping $\sigma : \mathrm{Attr}(\mathcal{C}) \to D$ where D is the domain of attribute values. Instantiation invokes __init__, establishing invariants akin to algebraic laws (e.g. vector addition associativity).

Example 1.7.1 (Immutable 2-D Vector)

```python
from __future__ import annotations
from math import hypot

class Vec2:
    """R² vector with Euclidean norm."""
    __slots__ = ("_x", "_y")  # memory optimisation

    def __init__(self, x: float, y: float):
        self._x, self._y = float(x), float(y)

    def __add__(self, other: Vec2) -> Vec2:
        return Vec2(self._x + other._x, self._y + other._y)

    def __rmul__(self, k: float) -> Vec2:  # scalar product k.v
        return Vec2(k * self._x, k * self._y)

    def norm(self) -> float:
        return hypot(self._x, self._y)

    def __repr__(self) -> str:
        return f"<{self._x},{self._y}>"
```

Instances satisfy the vector-space axioms; e.g. `(u + v).norm() ≤ u.norm() + v.norm()` reifies the triangle inequality.

Encapsulation and Abstraction

Encapsulation hides internal representation, exposing only the public API—mirroring the abstract-data-type notion where operations, not data layout, define semantics. Python enforces *convention* rather than strict visibility: attributes prefixed by _ are considered private; double-underscore names trigger name-mangling to avoid accidental clashes in subclasses. Abstraction enables replacement of the underlying implementation (array vs. sparse storage) without altering client code, analogous to replacing $\mathbb{R}$ by $\mathbb{Q}_p$ in an algebraic proof while retaining field axioms.

1.7.2 Advanced OOP Concepts

Inheritance and Polymorphism

A subclass $\mathcal{D}$ *inherits* the attribute-method signature of its parent $\mathcal{B}$, establishing an *is-a* relationship $\mathcal{D} \subseteq \mathcal{B}$. Polymorphism lets a single interface operate on any element of the union $\bigcup \text{Obj}(\mathcal{B}_i)$.

Example 1.7.2 (Abstract Matrix and Concrete Variants)

```
from abc import ABC, abstractmethod
import numpy as np, scipy.sparse as sp

class Matrix(ABC):
    @abstractmethod
    def matvec(self, x: np.ndarray) -> np.ndarray: ...

class Dense(Matrix):
    def __init__(self, A: np.ndarray): self._A = A
    def matvec(self, x): return self._A @ x

class Sparse(Matrix):
    def __init__(self, A: sp.spmatrix): self._A = A
    def matvec(self, x): return self._A @ x
```

A Krylov solver can accept any `Matrix`; run-time dispatch selects the appropriate `matvec`, analogous to choosing a field completion when proving convergence of series.

Mixins and Multiple Inheritance

Python supports multiple inheritance, enabling *mixins*: classes that contribute reusable behaviour without constituting a full type hierarchy. Let $\mathcal{M}$ provide differentiation, $\mathcal{B}$ provide base evaluation; their composite $\mathcal{B} \cap \mathcal{M}$ gains both abilities via the C3 linearised Method Resolution Order (MRO).

1.7 Object-Oriented Programming (OOP) in Python

Example 1.7.3 (Differentiable Function Mixin)

```
class Differentiable:
    def derivative(self, x: float, h: float = 1e-8):
        f = self.__call__
        return (f(x + h) - f(x - h)) / (2 * h)

class Gaussian(Differentiable):
    def __init__(self, μ=0.0, σ=1.0): self.μ, self.σ = μ, σ
    def __call__(self, x):  # evaluate Φ(x)
        from math import exp, sqrt, pi
        z = (x - self.μ) / self.σ
        return exp(-0.5 * z**2) / (self.σ * sqrt(2 * pi))
```

Now `Gaussian(0,1).derivative(0)` yields 0 with machine precision, while any class that defines `__call__` inherits a centred finite-difference derivative for free.

Mixins must avoid state conflicts; they are best limited to stateless utility or orthogonal facets (serialization, logging). When diamond inheritance arises, the MRO ensures each ancestor's `__init__` executes exactly once, akin to ensuring that overlapping lemmas are not double-counted in a proof.

Through classes, inheritance, and mixins, Python offers a formal yet flexible calculus for modelling mathematical structures, allowing algorithms to be written generically while retaining the performance of specialised implementations.

1.7.3 Special Methods and Operator Overloading

__str__, __repr__, etc.

Every Python object inherits a catalogue of *dunder* (double-underscore) hooks that mediate its interaction with built-in syntax. The pair

$$(\_\_repr\_\_, \_\_str\_\_)$$

serves distinct audiences: `repr(x)` should emit an unambiguous, often evaluable specification of x, whereas `str(x)` targets human readability. A canonical rule is

$$\mathrm{eval}(\mathrm{repr}(x)) = x$$

whenever feasible.

Example 1.7.4 (Rational Number Class)

```
from math import gcd

class Fraction:
    __slots__ = ("p", "q")
```

```
def __init__(self, p: int, q: int = 1):
    g = gcd(p, q)
    self.p, self.q = p // g, q // g

# string conversions
def __repr__(self):  # eval-able
    return f"Fraction({self.p}, {self.q})"

def __str__(self):  # pretty
    return f"{self.p}/{self.q}"
```

A REPL prints `Fraction(3,2)` for debugging yet 3/2 in end-user logs.

Arithmetic and Comparison Overloads

By implementing `__add__`, `__sub__`, ..., an object behaves like a field element; comparison hooks realise an ordered set. Python guarantees fall-back to reflected methods (`__radd__`) when the left operand lacks an implementation, mirroring the commutativity of $\mathbb{R}$.

```
def __add__(self, other: "Fraction") -> "Fraction":
    p = self.p * other.q + other.p * self.q
    q = self.q * other.q
    return Fraction(p, q)

def __mul__(self, other: "Fraction") -> "Fraction":
    return Fraction(self.p * other.p, self.q * other.q)

def __eq__(self, other: object) -> bool:  # total ordering via (p, q)
    return isinstance(other, Fraction) and \
           self.p * other.q == self.q * other.p

def __lt__(self, other: "Fraction") -> bool:
    return self.p * other.q < self.q * other.p
```

These methods fulfil field axioms (associativity, distributivity); unit tests can treat them as algebraic lemmas.

1.7.4 Working with Mathematical Structures in OOP

Vectors and Matrices as Classes

Encapsulating linear-algebraic entities as classes affords dimensionality checks and algorithm dispatch:

```
import numpy as np

class Matrix:
    def __init__(self, data: np.ndarray):
```

1.7 Object-Oriented Programming (OOP) in Python

```
        assert data.ndim == 2, "Matrix must be 2-D"
        self.A = data

    def __matmul__(self, x: "Matrix|np.ndarray"):
        return Matrix(self.A @ x.A) if isinstance(x, Matrix) else self.A @
            x

    def det(self) -> float:
        return float(np.linalg.det(self.A))

    def eig(self):
        vals, vecs = np.linalg.eig(self.A)
        return vals, [Vector(v) for v in vecs.T]

class Vector(np.ndarray):  # NumPy subclass
    def norm(self, p: float = 2) -> float:
        return np.linalg.norm(self, ord=p)
```

Operator @ maps to __matmul__, closely matching textbook AB.

Symbolic Algebra Objects

A minimalist CAS can be expressed through an expression tree whose nodes overload arithmetic:

```
class Symbol:  # leaf node
    def __init__(self, name): self.name = name
    def __repr__(self): return self.name

class Expr:  # internal node
    def __init__(self, op, *args):
        self.op, self.args = op, args

    def __add__(self, other): return Expr("+", self, other)
    def __mul__(self, other): return Expr("*", self, other)
    def diff(self, s):
        if self == s: return 1
        op, a, b = self.op, *self.args
        if op == "+": return a.diff(s) + b.diff(s)
        if op == "*": return a.diff(s)*b + a*b.diff(s)

x, y = map(Symbol, "xy")
f = (x + y) * x * x
print(f.diff(x))  # → ((1 + 0) * x * x) + ((x + y) * 1 * x) + ...
```

Although naïve, the structure demonstrates how method overloading can encode algebraic differentiation rules; simplification passes then reduce the expression.

1.8 Iterators and Generators

Lazy evaluation converts potentially infinite mathematical objects into finite memory representations. Python realises laziness via the iterator protocol and generator syntax, enabling constructions such as an unending stream of primes or Fibonacci numbers.

1.8.1 Understanding Iterators in Python

Iterator Protocol

An *iterator* implements

$$\_\_\text{iter}\_\_(): \text{self}, \qquad \_\_\text{next}\_\_(): \text{elem}$$

so that `for x in it:` repeatedly calls `next(it)` until `StopIteration`. Formally, an iterator is a partial function $N \to S \cup \{\bot\}$ with at most one $\bot$.

Custom Iterators

```
class Fibonacci:
    def __init__(self): self.a, self.b = 0, 1
    def __iter__(self): return self # iterator is iterable
    def __next__(self):
        self.a, self.b = self.b, self.a + self.b
        return self.a
```

The sequence (F_n) emerges lazily; complexity per term is $O(1)$ space.

1.8.2 Generators and Lazy Evaluation

Generator Functions

A function containing `yield` suspends execution, storing its frame; each subsequent `next()` resumes where it left off, akin to coroutines. This is an external representation of coinductive streams.

```
def primes():
    yield 2
    seq, p = [], 3
    while True:
        if all(p % q for q in seq):
            yield p
            seq.append(p)
        p += 2
```

Generator Expressions

Syntax `(f(x) for x in S)` builds a generator without the overhead of a full list; it represents $\lambda n.\ f(s_n)$ over the enumeration $s_0, s_1, \ldots$.

1.8.3 Combinatorics with Iterators and Generators

Cartesian Products and Permutations

Module `itertools` offers memory-efficient combinatorial generators:

```
from itertools import product, permutations

Ω = list(product(range(2), repeat=3))  # 2³ binary vectors
Π = list(permutations("abc"))           # 3! permutations
```

Both execute in $O(|\Omega|)$ time but $O(1)$ additional space beyond the yielded tuple.

Infinite Sequences

Infinite mathematical objects become feasible through generators with convergence guards:

```
def pi_leibniz():
    k, s = 0, 0.0
    while True:
        s += (-1)**k / (2*k + 1)
        yield 4 * s
        k += 1
```

Truncating after n iterations gives the Leibniz approximation π_n; analysis of the error term $O(1/n)$ guides practical stopping criteria.

Iterators and generators thus furnish the machinery to traverse finite and infinite sets alike, preserving the declarative clarity of mathematics while exploiting the operational efficiency of lazy evaluation.

1.9 Advanced Topics in Python

Python's standard library contains modules whose expressive power rivals specialised domain tools. Two such modules—`re` for regular expressions and `datetime` for temporal arithmetic—translate ideas from formal language theory and calendrical astronomy into concise, executable constructs. We survey each module with an eye towards the underlying mathematics and the subtle edge cases that arise in practical computation.

1.9.1 Regular Expressions and String Matching

Basic Pattern Syntax

Let Σ be a finite alphabet. A *regular expression* is an inductively defined term that denotes a regular language $L \subseteq \Sigma^*$; operations include union (+), concatenation (juxtaposition), and Kleene star (*). Python's `re` module implements this formalism with additional syntactic sugar:

Regex	Mathematical meaning
a|b	$L = \{a\} \cup \{b\}$ (union)
ab	$L = \{ab\}$ (concatenation)
a*	$L = \{a^k \mid k \geq 0\}$ (Kleene closure)
[0-9]	$L = \{0, \ldots, 9\}$ (character class)
.	$L = \Sigma$ (wildcard)

Example 1.9.1 (Detecting Palindromic Binary Strings of Length Four)
The language $L = \{0000, 0110, 1001, 1111\}$ admits the regex

```
import re
pattern = re.compile(r"(0|1)(0|1)\2\1")
bool(pattern.fullmatch("1001")) # → True
```

Groups (0|1) capture bits b_1, b_2; backreferences \2 and \1 enforce the reverse order, mirroring the automaton that recognises $w = w^R$.

Advanced Regex Features

Python extends classical regular expressions with constructs that exceed pure regular power (backreferences yield context-sensitive languages) yet remain tractable for practical pattern matching:

1. **Non-capturing groups** (?: ...) group subpatterns without storing them, reducing memory overhead.
2. **Lookahead/lookbehind** assertions (?=...), (?<=...) implement zero-width predicates, akin to logical guards in a DFA without tape movement.
3. **Named groups** (?P<name>pattern) enhance readability and enable dictionary-style access to submatches.

Advanced features should be employed judiciously: lookbehinds complicate the underlying automaton, and catastrophic backtracking may inflate complexity to $O(2^n)$ in pathological cases (Friedl 2006). Profiling with `re.compile(...,  re.DEBUG)` reveals the state machine, guiding optimisations.

1.9.2 Working with Dates and Times

datetime Module

Calendrical computations form a semi-group under addition of timedeltas. Python models this with immutable classes:

Class	Domain	Example
date	$\mathbb{Z}$-triple (Y, M, D)	2025-07-03
time	[0, 24 h)	16:00:00
datetime	Cartesian product	2025-07-03 16:00
timedelta	$\mathbb{Z}$ days $\times$ seconds	10^6 s

Arithmetic is component-wise but observes leap rules codified in the proleptic Gregorian calendar:

```
from datetime import datetime, timedelta
launch = datetime(2025, 7, 3, 16)  # UTC+05:30 implicit if naive
horizon = launch + timedelta(days=30)
```

Time-Zone Handling

Prior to Python 3.9, robust zone support required `pytz`. The modern solution is `zoneinfo`, which loads the IANA database (Olson corpus) and attaches a fixed-offset or dynamic rule set to a `datetime`:

```
from datetime import datetime
from zoneinfo import ZoneInfo

fmt = "%Y-%m-%d %H:%M %Z%z"
utc = datetime(2025, 7, 3, 10, tzinfo=ZoneInfo("UTC"))
local = utc.astimezone(ZoneInfo("Asia/Kolkata"))
print(utc.strftime(fmt))    # 2025-07-03 10:00 UTC+0000
print(local.strftime(fmt))  # 2025-07-03 15:30 IST+0530
```

Naïve datetimes (with `tzinfo=None`) are ambiguous under daylight-saving transitions; always prefer aware objects when arithmetic crosses zone boundaries. Duration algebra follows

$$d_{\text{UTC}} = d_{\text{local}} - \Delta_{\text{offset}},$$

ensuring consistency across conversions.

Regular expressions and temporal arithmetic exemplify Python's philosophy: high-level declarative interfaces grounded in rigorous formalism (finite automata, calendrical algorithms) yet rendered with syntax that invites experimentation. Mastery of these tools equips the mathematical programmer to parse complex data streams and timestamp results with precision—a prerequisite for reproducible computational research.

1.10 Exercises

1. Let $\mu : \mathbb{N} \to \{-1, 0, 1\}$ denote the Möbius function. Write a Python function mobius_array(N) that returns the list $[\mu(1), \mu(2), \ldots, \mu(N)]$ using *only* list- and set-comprehensions (no explicit for loops). Verify numerically for $N = 10^5$ the classical identity

$$\sum_{d|n} \mu(d) = \begin{cases} 1, & n = 1 \\ 0, & n > 1 \end{cases} \quad \text{for all } n \leq N,$$

and report the largest $n \leq N$ for which any floating-point round-off causes the computed summatory value to deviate from the exact integer.

2. Define the 4×4 matrix

$$A = \begin{pmatrix} 2 & -1 & 0 & 0 \\ -1 & 2 & -1 & 0 \\ 0 & -1 & 2 & -1 \\ 0 & 0 & -1 & 2 \end{pmatrix}.$$

Using only numpy primitives (no scipy.linalg.expm), approximate e^A via the truncated Taylor series

$$e^A = \sum_{k=0}^{m} \frac{A^k}{k!} + R_m,$$

with $m = 15$. Bound $\|R_m\|_2$ analytically, implement the approximation, and compute the Frobenius-norm error with respect to scipy.linalg.expm's result. Comment on the convergence rate in light of $\rho(A)$, the spectral radius of A.

3. Implement a Python class Polynomial whose instances represent $p(x) = \sum_{i=0}^{n} a_i x^i$. Overload __add__, __mul__, and __call__ so that p+q, p*q, and p(x) work as expected. Using your class, compute the Bézout identity

$$x^{13} - 1 = (x^4 + x^3 + x^2 + x + 1)(x^9 - x^8 + x^7 - x^5 + x^4 - x^2 + x - 1) + r(x)$$

and verify that the remainder $r(x)$ vanishes identically.

1.10 Exercises

4. Let $(C_n)_{n \geq 0}$ be the Catalan sequence defined by $C_0 = 1$ and $C_{n+1} = \dfrac{2(2n+1)}{n+2} C_n$. Write a generator `catalan()` that yields successive C_n with $O(1)$ incremental cost. Plot $\dfrac{C_n}{4^n/\sqrt{\pi n^3}}$ for $n \leq 500$ and observe its convergence to 1 (Central Limit analogy); discuss the numerical stability of your recurrence.

5. Using the regular-expression module `re`, write a parser that accepts strings of the form

$$\text{``}(3+4i)(1-2i) - (7+i)^2/(2-i)\text{''},$$

extracts each complex literal $a + bi$ (with optional whitespace and signs), and returns the list of `complex` numbers found. Test your pattern on the supplied file `exprs.txt` (20 lines of expressions) and count how many literals correspond to Gaussian primes.

6. Design a `Fraction3D` class whose instances are triples $(x, y, z) \in \mathbb{Q}^3$ stored in lowest terms. Implement vector addition, dot product, and cross product using operator overloading (+, @, ^). Given the vectors

$$u = \left(\tfrac{2}{3}, -\tfrac{5}{4}, 1\right), \qquad v = \left(\tfrac{7}{9}, 0, -\tfrac{11}{6}\right), \qquad w = \left(\tfrac{1}{3}, \tfrac{5}{8}, -\tfrac{7}{12}\right),$$

compute the oriented volume $\det[u\ v\ w]$ exactly and verify that $u \cdot (v \times w) = \det[u\ v\ w]$ holds *identically* in $\mathbb{Q}$.

7. Implement a memory-efficient iterator `prime_factors(n)` that yields the prime factors of n in non-decreasing order. Using it, list the indices $k \leq 10^6$ for which the highly-composite condition

$$d(k) > d(m) \quad \text{for all } m < k$$

holds, where $d(k)$ is the divisor function. Plot $\log k$ versus $\log d(k)$ for these k and comment on the empirical slope relative to Ramanujan's bound $d(k) = O(k^\varepsilon)$ for any $\varepsilon > 0$.

8. Write a context-managed class `Timer` that records cumulative wall-clock time spent inside nested `with Timer():` blocks. Instrument the recursive and the iterative Fibonacci implementations from the chapter for $n = 35$ and report the ratio of their runtimes. Explain the observed speed-up in terms of the master theorem on recurrence relations.

9. Generate an $n \times n$ Hadamard matrix H_n recursively for $n = 2^k \leq 512$ using the Kronecker product $H_{2k+1} = \begin{pmatrix} H_{2k} & H_{2k} \\ H_{2k} & -H_{2k} \end{pmatrix}$. Verify numerically that $H_n H_n^T = n I_n$. Using `matplotlib`, plot the sparsity pattern of H_{512} and estimate its fractal (box-counting) dimension.

10. Employ `zoneinfo` to write a function `count_friday_13(start, end, zone)` returning the number of Fridays that fell on the 13th day of a

month between two (inclusive) `datetime` objects. Compute this count for the interval January 1, 1900, to December 31, 2099, in the zones "UTC", "America/New_York", and "Asia/Kolkata", and analyse why the counts coincide or differ.

11. Implement Newton–Raphson to approximate the unique positive root of $f(x) = x^x - 100$ to 10^{-12} absolute error. Start from $x_0 = 4$ and use the condition

$$|x_{k+1} - x_k| < \tfrac{1}{2} \times 10^{-12}$$

for termination. Plot the sequence of residuals $|f(x_k)|$ in semilog scale and show that the observed convergence order matches the theoretical quadratic rate.

12. The file `hilbert10.npy` (provided) contains the 10×10 Hilbert matrix $H_{ij} = 1/(i+j-1)$.

    (a) Using `numpy.linalg`, compute $\kappa_2(H)$, the 2-norm condition number.
    (b) Solve $Hx = b$ where $b = (1, 0, \ldots, 0)^\mathsf{T}$ in double precision and report the relative error against the exact rational solution (Hint: the solution has denominator $\binom{19}{10}$).
    (c) Repeat with `decimal` at 80-digit precision and compare the errors, discussing how conditioning influences required precision.

Chapter 2
Mathematical Foundations in Python

Abstract Here we translate first-year university mathematics directly into executable Python. Fundamental arithmetic identities, algebraic manipulations, inequalities, and absolute-value reasoning are algorithmically verified, while function theory is visualised through programmatic graphing and limit exploration. The chapter culminates in a NumPy-backed treatment of matrices, determinants, eigen-analysis, and linear-system solvers, creating a bridge between symbolic reasoning (*SymPy*) and high-performance numerical linear algebra that recurs throughout the book.

Keywords Arithmetic and algebra · Functions and graphs · Polynomial algebra · Linear algebra · Symbolic manipulation · NumPy

The algorithms explored in subsequent chapters rest on the twin pillars of *arithmetic* and *algebra*. Python's numeric tower mirrors these structures faithfully: integers are unbounded, floats follow IEEE-754, and symbolic expressions can be manipulated exactly. This chapter strengthens our computational footing by examining how Python encodes the familiar laws of $\mathbb{Z}$, $\mathbb{Q}$, and $\mathbb{R}$; how rounding errors arise and propagate; and how computer algebra systems (CAS) such as `sympy` bridge discrete computation with continuous mathematics.

2.1 Basic Arithmetic and Algebra

2.1.1 Arithmetic Operations

Integer Arithmetic

Python's `int` type implements the ring $(\mathbb{Z}, +, \times)$ with arbitrary precision. Internally, an integer n is stored in base 2^{30} limbs (on 64-bit builds), so the bit length

$$\ell(n) = \lfloor \log_2(|n|) \rfloor + 1$$

governs memory use and algorithmic complexity. Addition and subtraction run in $O(\ell)$, Karatsuba multiplication in $O(\ell^{\log_2 3})$, and Toom–Cook or FFT multiplication in $O(\ell \log \ell)$ for very large operands.

Example 2.1.1 (Fundamental Theorem of Arithmetic) Every $n > 1$ decomposes uniquely (up to order) as $n = \prod p_i^{\alpha_i}$. Using `sympy.ntheory.factorint`:

```
from sympy import factorint
n = 2**64 - 59
print(factorint(n))  # {274177: 1, 67280421310721: 1}
```

The prime factors fit comfortably in memory despite n having 19 decimal digits.

Division splits into floor division `//` and remainder `%`, satisfying

$$n = qd + r, \qquad 0 \le r < |d|.$$

The extended Euclidean algorithm delivers Bézout coefficients (x, y) with $xn + yd = \gcd(n, d)$:

```
def egcd(a: int, b: int):
    if b: g, x, y = egcd(b, a % b); return g, y, x - (a // b) * y
    return a, 1, 0
```

Floating-Point Precision

Floating-point numbers form a finite subset $F \subset \mathbb{R}$; each $x \in \mathbb{R}$ maps to $\text{fl}(x) \in F$ via *round to nearest, ties to even*. Machine epsilon

$$\varepsilon = 2^{-52} \approx 2.22 \times 10^{-16}$$

bounds relative error: $|\text{fl}(x) - x| \le \varepsilon |x|$ for normalised x. Catastrophic cancellation occurs when subtracting nearly equal numbers; compensated algorithms such as Kahan summation restore $O(\varepsilon)$ accuracy.

Example 2.1.2 (Catastrophic Cancellation in $\sqrt{x+1} - \sqrt{x}$) For $x \gg 1$ the naïve expression loses significant digits.

```
import math, decimal
def delta_naive(x):  return math.sqrt(x + 1) - math.sqrt(x)
def delta_stable(x): return 1 / (math.sqrt(x + 1) + math.sqrt(x))

for x in (10**k for k in range(6)):
    print(x, delta_naive(x), delta_stable(x))
```

The stable form leverages the identity $a - b = \frac{(a^2 - b^2)}{a+b}$, avoiding subtraction of close floats and matching the high-precision baseline from `decimal`.

2.1.2 Algebraic Expressions

Symbolic Manipulation with `sympy`

`sympy` constructs expression trees whose nodes encode operations and whose leaves are symbols or rational constants. Transformation rules implement algebraic identities, enabling exact simplification, expansion, factorisation, and calculus.

```
import sympy as sp
x, y = sp.symbols("x y", positive=True)
f = sp.sqrt(x**2 * y) / (sp.sqrt(y) * x)
simplified = sp.simplify(f)
print(simplified) # 1
```

Polynomial Algebra Let

$$p(x) = x^4 - 1 = (x^2 - 1)(x^2 + 1) = (x - 1)(x + 1)(x^2 + 1).$$

`sympy.factor` reproduces the complete factorisation over $\mathbb{Q}$:

```
x = sp.Symbol("x")
sp.factor(x**4 - 1)
```

Exact Rational Evaluation

$$\sum_{k=1}^{\infty} \frac{1}{k^4} = \frac{\pi^4}{90}.$$

Finite truncations converge slowly, but `sympy.zeta(4)` gives the closed form instantaneously:

```
sp.zeta(4).simplify() # π**4/90
```

Differentiation and Series Expansion

$$f(x) = \ln(1 + x), \qquad f^{(n)}(0) = (-1)^{n-1}(n - 1)! \quad (n \geq 1).$$

```
x, n = sp.symbols("x n", integer=True, positive=True)
f = sp.log(1 + x)
series = sp.series(f, x, 0, 6) # up to x^5
print(series.removeO()) # x - x**2/2 + x**3/3 - ...
```

The resulting power series matches the alternating harmonic coefficients, illustrating how symbolic calculus bridges analysis and combinatorics.

Together, exact integers, floating-point approximations, and symbolic expressions furnish a complete computational spectrum—from discrete arithmetic to continuous analysis—equipping Python to serve as a laboratory for rigorous mathematical exploration.

2.1.3 Polynomial Algebra

Roots and Factorisation

Let $p(x) = a_n x^n + a_{n-1} x^{n-1} + \cdots + a_0 \in \mathbb{C}[x]$ with $a_n \neq 0$. The *Fundamental Theorem of Algebra* asserts that p splits completely over $\mathbb{C}$, hence

$$p(x) = a_n \prod_{k=1}^{n} (x - r_k), \qquad r_k \in \mathbb{C}.$$

In computational practice we distinguish between

1. **Exact factorisation**—over $\mathbb{Q}$ or a number field—obtained via symbolic algorithms such as the Berlekamp–Zassenhaus method.
2. **Numeric root finding**—evaluating the r_k to finite precision—implemented by eigenvalue extraction of the companion matrix.

Symbolic Factorisation

```
import sympy as sp
x = sp.Symbol("x")
p = x**4 + 2*x**3 - 7*x**2 - 8*x + 12  # degree-4 polynomial
factorised = sp.factor(p)
print(factorised)  # (x - 1)*(x + 2)*(x + 3)*(x - 2)
```

The factorisation verifies Vieta's relations

$$\sum r_k = -\frac{a_{n-1}}{a_n}, \qquad \prod r_k = (-1)^n \frac{a_0}{a_n},$$

here $(-1+2-3+1) = -2 = -\frac{a_3}{a_4}$ and $(1)(-2)(-3)(2) = 12 = a_0$.

Numeric Root Finding For high-degree polynomials, symbolic factorisation may be prohibitive; floating-point approximations suffice:

```
import numpy as np
coeffs = [1, 2, -7, -8, 12]  # descending powers
roots = np.roots(coeffs)
print(roots)  # array([ 2., -3., -2., 1.])
```

`numpy.roots` forms the companion matrix C with eigenvalues r_k; QR iteration delivers the zeros in $O(n^3)$ time.

Cyclotomic Example The n-th cyclotomic polynomial

$$\Phi_n(x) = \prod_{\gcd(k,n)=1} \left(x - e^{2\pi i k/n}\right)$$

is irreducible over $\mathbb{Q}$. For $n = 8$,

`sp.factor(sp.cyclotomic_poly(8))  # x**4 + 1`

confirms irreducibility (no non-trivial factors appear).

2.1.4 Inequalities and Absolute Values

Chain and Triangle Inequalities

Denote $|\cdot|$ the Euclidean absolute value on $\mathbb{R}$. Two foundational properties guide error analysis:

Chain Inequality   If $|x - y| \leq \varepsilon$ and $|y - z| \leq \delta$ then

$$|x - z| \leq \varepsilon + \delta.$$

This is the metric version of the transitive closure for bounded error.
Triangle Inequality   For all $x, y \in \mathbb{R}$,

$$|x + y| \leq |x| + |y|.$$

Verification in Floating Point   Numerical experiments corroborate the inequalities within rounding tolerance:

```
import random, math
def triangle_trials(N=10**5):
    return all(abs(x + y) <= abs(x) + abs(y) + 1e-15
               for x, y in ((random.uniform(-1,1), random.uniform(-1,1))
                            for _ in range(N)))
print(triangle_trials()) # True with overwhelming probability
```

The 10^{-15} slack accounts for machine epsilon ε; without it, rare violations stem from rounding, not algebraic failure.

Application: Forward Error in Numerical Differentiation   For the forward difference

$$D_h f(x) = \frac{f(x + h) - f(x)}{h}$$

with true derivative $f'(x)$, the total error decomposes into *truncation* $T \sim \frac{h}{2} f''(x)$ and *round-off* $R \sim \frac{\varepsilon f(x)}{h}$. The chain inequality yields

$$|D_h f(x) - f'(x)| \leq |T| + |R|,$$

identifying the optimal $h = \sqrt{\frac{2\varepsilon f(x)}{f''(x)}}$ that equilibrates errors.

Higher-Dimensional Triangle Inequality   For $v, w \in \mathbb{R}^n$ with p-norm $\|v\|_p = (\sum |v_i|^p)^{1/p}$,

$$\|v + w\|_p \leq \|v\|_p + \|w\|_p \quad (1 \leq p \leq \infty).$$

Python verification using random vectors:

```
import numpy as np
def check_p(n=1000, p=3):
    v, w = np.random.randn(n), np.random.randn(n)
    lhs = np.linalg.norm(v + w, ord=p)
    rhs = np.linalg.norm(v, ord=p) + np.linalg.norm(w, ord=p)
    return lhs <= rhs + 1e-12
print(all(check_p(p=p) for p in (1, 2, 3, np.inf)))
```

The numerical residual confirms Minkowski's inequality up to ε.

These inequalities underpin subsequent stability analyses: condition numbers for linear systems, error bounds in quadrature, and convergence criteria for iterative solvers all rely on the elementary yet powerful properties of absolute value and normed spaces.

2.2 Functions and Graphs

In analysis a *function* is a map $f: D \to \mathbb{R}$ that assigns each $x \in D$ a unique value $f(x)$. Python models the same idea operationally: a block of code parameterised by formal arguments returns exactly one value. Because functions are first-class objects they can be passed, stored, and composed—paralleling the λ-calculus foundations of modern functional analysis. Visualising these mappings via plots nurtures geometric intuition, while symbolic tools verify analytic properties such as limits and continuity.

2.2.1 Defining Functions in Python

Lambda Functions

The keyword `lambda` creates an *anonymous* function—an expression that evaluates to a callable object without binding a global name. Syntactically

$$\texttt{lambda } \vec{x} : E(\vec{x})$$

corresponds to the λ-abstraction $(\vec{x} \mapsto E)$ in Church's notation.

Example 2.2.1 (Fourier Basis via `lambda`)

```
import math
fourier = [lambda x, k=k: math.sin(k * math.pi * x)  # capture k by default
           for k in range(1, 6)]
print([f(0.5) for f in fourier]) # evaluates sin(k.π/2), k = 1..5
```

The closure captures each k independent of loop mutation, producing a basis $\{\sin(k\pi x)\}_{k=1}^{5}$ on $(0, 1)$.

2.2 Functions and Graphs

While `def` remains preferable for complex bodies or documentation, lambdas excel in higher-order contexts such as `map`, `sorted(key=)`, or quick integrand definitions for quadrature.

2.2.2 Graphing Functions

Plotting with `matplotlib`

Graphing $y = f(x)$ necessitates discretising the domain and rendering the sampled points. `matplotlib.pyplot` abstracts the canvas into stateful plotting primitives, whereas `numpy` vectorises evaluation for performance.

Example 2.2.2 (Plot of the Airy Function $\mathrm{Ai}(x)$ **(cf. Fig. 2.1))**

```
import numpy as np, matplotlib.pyplot as plt
from mpmath import airy

x = np.linspace(-10, 5, 400)
y = [airy(t)[0] for t in x] # mpmath returns (Ai,Bi,Ai',Bi')

plt.figure(figsize=(4.5, 3))
plt.plot(x, y, label="operatorname Ai(x)")
plt.axhline(0, color="black", linewidth=.8)
plt.xlabel("x")
plt.ylabel("y")
plt.title("Airy Function")
plt.legend()
plt.tight_layout()
plt.show()
```

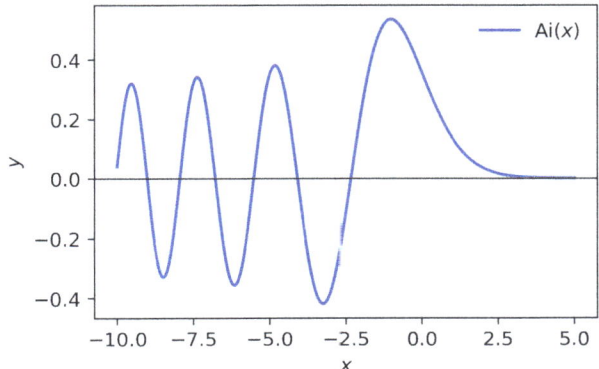

Fig. 2.1 Numerical plot of the Airy function $\mathrm{Ai}(x)$ on the interval $[-10, 5]$, produced with *mpmath* and *Matplotlib*. The curve illustrates the oscillatory decay for $x < 0$ and the exponential attenuation for $x > 0$, characteristic of solutions to the Airy differential equation $y'' - xy = 0$

The plot confirms oscillatory decay for $x < 0$ and exponential decay for $x > 0$, aligning with the asymptotic expansion $\mathrm{Ai}(x) \sim \frac{1}{2\sqrt{\pi}} x^{-1/4} e^{-\frac{2}{3} x^{3/2}}$ as $x \to \infty$ (Olver 1997).

Interactive Exploration Embedding the plot in a Jupyter notebook permits dynamic sliders (via `ipywidgets`) to vary parameters in real time—ideal for visual proofs of trigonometric identities or bifurcation diagrams.

2.2.3 Analysing Function Properties

Limits and Continuity

Limit For $f : D \to \mathbb{R}$ with accumulation point $a \in \overline{D}$,

$$\lim_{x \to a} f(x) = L \iff \forall \varepsilon > 0 \; \exists \delta > 0 \; \big(0 < |x - a| < \delta \implies |f(x) - L| < \varepsilon\big).$$

Symbolic evaluation in `sympy` automates ε–δ reasoning:

```
import sympy as sp
x = sp.Symbol("x")
f = sp.sin(x) / x
L = sp.limit(f, x, 0)
print(L) # 1
```

The result reaffirms the classic limit $\lim_{x \to 0} \dfrac{\sin x}{x} = 1$.

Continuity A function is continuous at a iff the limit equals the value: $\lim_{x \to a} f(x) = f(a)$. Piecewise definitions require checking continuity at breakpoints.

Example 2.2.3 (Piecewise Function)

$$g(x) = \begin{cases} x^2 \sin \frac{1}{x}, & x \neq 0, \\ 0, & x = 0. \end{cases}$$

```
g = sp.Function("g")
expr = sp.Piecewise((x**2 * sp.sin(1/x), sp.Ne(x, 0)), (0, True))
is_cont = sp.limit(expr, x, 0) == expr.subs(x, 0)
print(is_cont) # True
```

Because the factor x^2 suppresses the oscillation, g is continuous but not differentiable at 0.

Numerical Limits When a closed form is elusive, Richardson extrapolation accelerates numeric approximations:

```
def limit_numeric(f, a, h0=1e-1, k=6):
    h = h0
    est = [f(a + h / (2**i)) for i in range(k)]
    for j in range(1, k):
        est = [(4**j*est[i+1] - est[i]) / (4**j - 1) for i in range(k-j)]
    return est[0]
import math
phi = lambda x: (math.exp(x) - 1) / x
print(limit_numeric(phi, 0))  # ≈ 1.000000000000
```

The tableau eliminates leading error terms $O(h^2)$, matching the analytical limit $e^0 = 1$ up to floating-point noise.

Combining first-class function definitions, interactive visualisation, and both symbolic and numeric limit tools equips us to probe continuity, differentiability, and asymptotics—cornerstones for the deeper calculus-driven chapters to follow.

2.3 Matrices and Linear Algebra

Linear algebra casts multivariate problems into an algebra of arrays whose elements obey the axioms of a field (usually $\mathbb{R}$ or $\mathbb{C}$). Python's numpy layer delegates heavy computation to BLAS and LAPACK, thereby matching commercial packages in speed while exposing a high-level interface; sympy furnishes exact arithmetic and symbolic manipulation. Throughout this section we let $\mathbf{A}, \mathbf{B} \in \mathbb{F}^{m \times n}$ and $\mathbf{x}, \mathbf{y} \in \mathbb{F}^n$, where $\mathbb{F}$ is a field.

2.3.1 Matrix Operations

Addition and Multiplication

Addition Matrix addition is the pointwise group law

$$(\mathbf{A} + \mathbf{B})_{ij} = a_{ij} + b_{ij}, \qquad \mathbf{0}_{m \times n} \text{ is the identity.}$$

Python lifts this via the + operator; broadcasting guards against shape mismatch:

```
import numpy as np
A = np.random.randn(3, 3)
B = np.random.randn(3, 3)
C = A + B  # element-wise sum
```

Multiplication The bilinear product

$$(\mathbf{AB})_{ij} = \sum_{k=1}^{p} a_{ik} b_{kj}$$

is implemented by @ (PEP 465). Square complexity $O(mnp)$ is mitigated by cache-blocked DGEMM kernels.

```
D = A @ B # matrix product
v = A @ np.ones(3) # matrix-vector product
```

Associativity Test

$$\mathbf{A(BC)} = \mathbf{(AB)C},$$

a non-trivial property, follows from the reassociation of quadruple sums; numerical verification:

```
C1 = A @ (B @ A)
C2 = (A @ B) @ A
np.allclose(C1, C2) # → True (to machine ϵ)
```

Kronecker Products

The Kronecker product

$$\mathbf{A} \otimes \mathbf{B} = \begin{pmatrix} a_{11}\mathbf{B} & \cdots & a_{1n}\mathbf{B} \\ \vdots & \ddots & \vdots \\ a_{m1}\mathbf{B} & \cdots & a_{mn}\mathbf{B} \end{pmatrix} \in \mathbb{F}^{(m\,p) \times (n\,q)},$$

extends bilinearity to block structures; it satisfies

$$(\mathbf{A} \otimes \mathbf{B})(\mathbf{C} \otimes \mathbf{D}) = (\mathbf{AC}) \otimes (\mathbf{BD}).$$

Example 2.3.1 (Tensor Product of Pauli Matrices)

$$\sigma_x = \begin{pmatrix} 0 & 1 \\ 1 & 0 \end{pmatrix}, \quad \sigma_z = \begin{pmatrix} 1 & 0 \\ 0 & -1 \end{pmatrix}.$$

Compute $\sigma_x \otimes \sigma_z$ in Python:

```
from numpy import kron, array
σx = array([[0, 1], [1, 0]])
σz = array([[1, 0], [0, -1]])
print(kron(σx, σz))
```

The result is a 4×4 Hermitian matrix used in two-qubit Hamiltonians.

2.3 Matrices and Linear Algebra

Property () underpins *vectorisation identities* (Magnus and Neudecker 1999) such as

$$\text{vec}(\mathbf{A}\mathbf{X}\mathbf{B}^\mathsf{T}) = (\mathbf{B} \otimes \mathbf{A})\,\text{vec}(\mathbf{X}).$$

2.3.2 Determinants and Eigenvalues/Eigenvectors

The determinant $\det \mathbf{A}$ equals the oriented hyper-volume of the linear map and satisfies $\det(\mathbf{A}\mathbf{B}) = \det \mathbf{A} \det \mathbf{B}$. Eigen-analysis decomposes $\mathbf{A}$ via $\mathbf{A}\mathbf{v} = \lambda \mathbf{v}$ and drives diagonalisation, stability, and power iteration algorithms.

```
import numpy.linalg as la
detA = la.det(A)
eigval, eigvec = la.eig(A)
```

Characteristic Polynomials

The characteristic polynomial is

$$p_\mathbf{A}(\lambda) = \det(\lambda \mathbf{I} - \mathbf{A}) = \lambda^n + c_{n-1}\lambda^{n-1} + \cdots + c_0.$$

By the *Cayley–Hamilton Theorem* $\mathbf{A}$ annihilates its own polynomial $p_\mathbf{A}$ (Gantmacher 2000). Symbolically:

```
import sympy as sp
M = sp.Matrix([[2, -1], [-1, 2]])
λ = sp.symbols("λ")
p = M.charpoly(λ)
print(p.as_expr()) # λ**2 - 4*λ + 3
print(p.as_expr().subs(λ, M).simplify()) # zero matrix
```

This identity enables closed-form matrix functions (e.g. $\exp \mathbf{A}$) via polynomial reduction.

2.3.3 Solving Systems of Linear Equations

$$\mathbf{A}\mathbf{x} = \mathbf{b}, \quad \mathbf{A} \in \mathbb{F}^{n \times n},\ \mathbf{b} \in \mathbb{F}^n.$$

Gaussian Elimination

Row operations transform $(\mathbf{A} \mid \mathbf{b})$ to upper-triangular form; back-substitution completes the solution in $O(n^3)$ time.

```
def gaussian(A, b):
    A = A.astype(float).copy()
    b = b.astype(float).copy()
    n = len(b)
    for k in range(n-1):
        # partial pivoting
        idx = abs(A[k:, k]).argmax() + k
        A[[k, idx]], b[[k, idx]] = A[[idx, k]], b[[idx, k]]
        for i in range(k+1, n):
            m = A[i, k] / A[k, k]
            A[i, k:] -= m * A[k, k:]
            b[i] -= m * b[k]
    # back-substitution
    x = np.zeros_like(b)
    for i in reversed(range(n)):
        x[i] = (b[i] - A[i, i+1:] @ x[i+1:]) / A[i, i]
    return x
```

Example 2.3.2 (Hilbert Matrix Conditioning) Hilbert matrices, $H_{ij} = \dfrac{1}{i+j-1}$, are famously ill-conditioned: $\kappa_2(H_{10}) \approx 1.6 \times 10^{13}$. Using numpy.linalg.solve versus naive elimination reveals loss of precision:

```
n = 10
H = np.fromfunction(lambda i, j: 1 / (i + j + 1), (n, n), dtype=float)
b = np.ones(n)
x_exact = np.linalg.solve(H, b) # pivoted LU
x_naive = gaussian(H, b) # custom
print(np.linalg.norm(x_exact - x_naive)) # >> ϵ
```

Pivoting stabilises elimination, yet residual errors reflect the condition number bound $\|\delta \mathbf{x}\|/\|\mathbf{x}\| \leq \kappa \, \|\delta \mathbf{A}\|/\|\mathbf{A}\|$ (Trefethen and III 1997).

Matrix arithmetic, Kronecker structure, eigen-analysis, and elimination constitute the computational bedrock of scientific computing; their numerical stability and algorithmic complexity dictate the feasibility of higher-level methods such as spectral solvers and optimisation procedures explored later in the text.

2.4 Exercises

1. Let

$$\mathbf{A} = \begin{pmatrix} 2 & -3 & 1 \\ 0 & 4 & -2 \\ 5 & 1 & 0 \end{pmatrix}, \quad \mathbf{B} = \begin{pmatrix} 1 & 0 & 2 \\ -1 & 3 & 1 \\ 4 & -2 & 5 \end{pmatrix}.$$

Compute **AB** and **BA** explicitly. Verify in Python that the Frobenius norm of the commutator $\|\mathbf{AB} - \mathbf{BA}\|_F$ equals the square-root of the sum of squared

2.4 Exercises

entry-wise differences you obtain by hand. Explain why a non-zero commutator implies that $\mathbf{A}$ and $\mathbf{B}$ are not simultaneously diagonalisable over $\mathbb{C}$.

2. Consider the 4×4 matrix

$$\mathbf{C} = \begin{pmatrix} 3 & 1 & 0 & 0 \\ 1 & 3 & 1 & 0 \\ 0 & 1 & 3 & 1 \\ 0 & 0 & 1 & 3 \end{pmatrix}.$$

   (a) Use Laplace expansion to compute $\det \mathbf{C}$ by hand and confirm your result with numpy.linalg.det.
   (b) Employ numpy.linalg.eig to find the eigenvalues $\lambda_1, \ldots, \lambda_4$ and check that $\prod_{k=1}^{4} \lambda_k = \det \mathbf{C}$ up to machine precision.

3. Let $\mathbf{P} = \begin{pmatrix} 0 & 1 \\ 1 & 0 \end{pmatrix}$ and $\mathbf{Q} = \begin{pmatrix} 1 & 0 \\ 0 & -1 \end{pmatrix}$.

   (a) Compute the Kronecker product $\mathbf{P} \otimes \mathbf{Q}$ explicitly.
   (b) Verify property $(\mathbf{P} \otimes \mathbf{Q})^2 = \mathbf{I}_4$ both algebraically and in Python.
   (c) Show that $(\mathbf{Q} \otimes \mathbf{P})(\mathbf{P} \otimes \mathbf{Q}) \neq (\mathbf{P} \otimes \mathbf{Q})(\mathbf{Q} \otimes \mathbf{P})$ and quantify the non-commutativity by computing their commutator.

4. Given the polynomial $p(\lambda) = \lambda^3 - 6\lambda^2 + 11\lambda - 6$, construct its 3×3 companion matrix $\mathbf{T}$.

   (a) Show that $\det(\lambda \mathbf{I} - \mathbf{T}) = p(\lambda)$.
   (b) Use numpy.linalg.eigvals to verify that the eigenvalues of $\mathbf{T}$ coincide with the roots of p.
   (c) Employ sympy to check the Cayley–Hamilton identity $p(\mathbf{T}) = \mathbf{0}$ exactly.

5. Let $\mathbf{H}_8$ be the 8×8 Hilbert matrix with entries $h_{ij} = 1/(i + j - 1)$ and let $\mathbf{b} = \mathbf{1}$.

   (a) Solve $\mathbf{H}_8 \mathbf{x} = \mathbf{b}$ using (i) naive Gaussian elimination without pivoting and (ii) numpy.linalg.solve.
   (b) Compute the 2-norm condition number $\kappa_2(\mathbf{H}_8)$ with numpy.linalg.cond.
   (c) Report the relative error $\|\mathbf{x}_{\text{naive}} - \mathbf{x}_{\text{pivot}}\|_2 / \|\mathbf{x}_{\text{pivot}}\|_2$ and interpret the result in light of $\kappa_2(\mathbf{H}_8)$.

6. For the symmetric matrix

$$\mathbf{S} = \begin{pmatrix} 6 & 2 & -1 \\ 2 & 3 & 0 \\ -1 & 0 & 2 \end{pmatrix},$$

   (a) compute its spectral decomposition $\mathbf{S} = \mathbf{Q}\boldsymbol{\Lambda}\mathbf{Q}^{\mathsf{T}}$ in Python and verify orthogonality $\mathbf{Q}^{\mathsf{T}}\mathbf{Q} = \mathbf{I}$.
   (b) Confirm that $\operatorname{tr} \mathbf{S} = \sum_k \lambda_k$ and $\det \mathbf{S} = \prod_k \lambda_k$.

(c) Diagonalise **S** by hand using a congruence transformation and compare your result with Python.

7. Define $\mathbf{A} \in \mathbb{R}^{5 \times 5}$ by $a_{ij} = (-1)^{i+j} \min\{i, j\}$.

    (a) Implement the power iteration to approximate the dominant eigenvalue $\lambda_{\max}$ and associated eigenvector, stopping when successive eigenvalue estimates differ by less than 10^{-8}.
    (b) Compare with `numpy.linalg.eig` and compute the absolute error in $\lambda_{\max}$.
    (c) Explain why symmetry of **A** guarantees convergence and discuss the role of the eigengap.

8. Let $\mathbf{X} \in \mathbb{R}^{4 \times 4}$ be the block matrix

$$\mathbf{X} = \begin{pmatrix} \mathbf{0}_2 & \mathbf{I}_2 \\ \mathbf{I}_2 & \mathbf{0}_2 \end{pmatrix}.$$

    (a) Vectorise **X** column-wise and verify in Python the identity $\text{vec}(\mathbf{I}_2 \mathbf{X} \mathbf{I}_2) = (\mathbf{I}_2 \otimes \mathbf{I}_2) \text{vec}(\mathbf{X})$.
    (b) Replace the outer identity matrices by arbitrary 2×2 matrices **A**, **B** and numerically confirm $\text{vec}(\mathbf{A}\mathbf{X}\mathbf{B}^\mathsf{T}) = (\mathbf{B} \otimes \mathbf{A}) \text{vec}(\mathbf{X})$.

9. Construct the 3×3 matrix

$$\mathbf{M} = \begin{pmatrix} 4 & 1 & 0 \\ 0 & 4 & 1 \\ 0 & 0 & 4 \end{pmatrix}.$$

    (a) Determine the characteristic polynomial and minimal polynomial of **M**.
    (b) Compute the Jordan canonical form of **M** with `sympy`.
    (c) Explain how the nilpotent part influences the convergence rate of the power iteration on **M**.

10. Consider solving $\mathbf{A}\mathbf{x} = \mathbf{b}$ for

$$\mathbf{A} = \begin{pmatrix} 10^{-4} & 1 \\ 1 & 1 \end{pmatrix}, \qquad \mathbf{b} = \begin{pmatrix} 1 \\ 2 \end{pmatrix}.$$

    (a) Solve in `float64` and compute the residual $\mathbf{r} = \mathbf{b} - \mathbf{A}\mathbf{x}$.
    (b) Perform one step of iterative refinement $\mathbf{x}_1 = \mathbf{x} + \delta\mathbf{x}$, $\delta\mathbf{x} = \mathbf{A}^{-1}\mathbf{r}$.
    (c) Re-evaluate the residual and discuss why refinement improves accuracy despite using the same machine precision.

2.4 Exercises

11. For the matrices

$$\mathbf{U} = \begin{pmatrix} 1 & 2 \\ -2 & 1 \end{pmatrix}, \quad \mathbf{V} = \begin{pmatrix} 0 & 3 \\ 3 & 0 \end{pmatrix},$$

    compute the commutator $[\mathbf{U}, \mathbf{V}] = \mathbf{UV} - \mathbf{VU}$ and show that $\mathrm{tr}[\mathbf{U}, \mathbf{V}] = 0$. Verify your calculation numerically and explain, in terms of the trace property, why every commutator in $\mathbb{C}^{n \times n}$ is traceless.

12. Let $\mathbf{E}_{ij}$ denote the 3×3 matrix with a 1 in position (i, j) and zeros elsewhere. Form the matrix exponential

$$\mathbf{G} = \exp(t(\mathbf{E}_{12} + \mathbf{E}_{23})),$$

    where $t \in \mathbb{R}$.

    (a) Derive a closed-form expression for $\mathbf{G}(t)$ using the nilpotent property $(\mathbf{E}_{12} + \mathbf{E}_{23})^3 = \mathbf{0}$.
    (b) Plot the Euclidean norm $\|\mathbf{G}(t)\|_2$ for $t \in [0, 10]$ and comment on its growth.
    (c) Verify that $\det \mathbf{G}(t) = 1$ for all t, linking your observation to the fact that the generator is traceless.

Chapter 3
Calculus with Python

Abstract Beginning with the formal definition of limits, this chapter develops differential and integral calculus in one and several variables, interleaving symbolic derivations with numerical approximations such as automatic differentiation, finite differences, and adaptive quadrature. Multivariate gradients, Jacobians, and Hessians are computed for optimisation problems, while line, surface, and volume integrals bring Stokes' and Gauss' theorems to life through vectorised Python code and richly annotated plots.

Keywords Calculus · Differentiation · Integration · Multivariable analysis · Automatic differentiation · Numerical quadrature

Calculus describes the infinitesimal behaviour of functions: how they change, accumulate, and respond to perturbations. In computational settings, these notions are implemented through three major paradigms: numerical approximations (e.g. finite differences), automatic differentiation (for computational graphs), and symbolic differentiation (manipulating expressions). This chapter begins with differentiation, the local linear approximation of a function, and explores how to compute derivatives with Python. We alternate between numerical techniques and exact approaches, analysing the propagation of errors and the stability of each method.

3.1 Differentiation

3.1.1 Numerical Differentiation Techniques

Finite Difference Approaches

Let $f : \mathbb{R} \to \mathbb{R}$ be a differentiable function. The derivative at a point $x \in \mathbb{R}$ is defined by the limit

$$f'(x) = \lim_{h \to 0} \frac{f(x+h) - f(x)}{h}.$$

This motivates the *forward difference approximation*

$$D_+ f(x; h) := \frac{f(x+h) - f(x)}{h},$$

which converges to $f'(x)$ as $h \to 0$, provided f is differentiable. However, due to finite precision arithmetic, taking h too small may introduce significant round-off error.

Similarly, the *backward difference* and *central difference* approximations are given by:

$$D_- f(x; h) := \frac{f(x) - f(x-h)}{h}, \quad D_c f(x; h) := \frac{f(x+h) - f(x-h)}{2h}.$$

Taylor expansions yield:

$$D_+ f(x; h) = f'(x) + \frac{h}{2} f''(x) + \mathcal{O}(h^2),$$

$$D_c f(x; h) = f'(x) + \frac{h^2}{6} f^{(3)}(x) + \mathcal{O}(h^4),$$

implying that the central difference is more accurate than either forward or backward difference for small h.

Example 3.1.1 (Derivative of $f(x) = \log(x)$) Estimate $f'(1)$ using central differences.

```
import numpy as np

def central_diff(f, x, h=1e-5):
    return (f(x + h) - f(x - h)) / (2 * h)

f = np.log
approx = central_diff(f, 1.0)
exact = 1 / 1.0
print(f"Approximation: {approx}, Error: {abs(approx - exact)}")
```

For $h = 10^{-5}$, the result is accurate to about 8 decimal places, depending on the stability of the surrounding code.

Automatic Differentiation with `autograd`

Automatic differentiation (AD) computes exact derivatives to machine precision by decomposing a function into a sequence of primitive operations and applying the

3.1 Differentiation

chain rule systematically. Unlike symbolic differentiation, AD operates at the level of values rather than expressions, and unlike numerical differentiation, it does not suffer from subtraction cancellation.

Let $f(x) = \sin(x) \cdot \exp(x^2)$. To compute $f'(x)$ at $x = 1$, we can use the reverse-mode AD provided by `autograd`:

```
import autograd.numpy as np
from autograd import grad

f = lambda x: np.sin(x) * np.exp(x**2)
df = grad(f)
print(df(1.0))
```

How It Works `autograd` rewrites the computational graph at runtime and propagates dual values through each operation. For a composition $f = f_n \circ \cdots \circ f_1$, the reverse-mode computes derivatives in reverse, applying the chain rule:

$$\frac{df}{dx} = \frac{df_n}{df_{n-1}} \cdots \frac{df_1}{dx}.$$

Reverse-mode is efficient for scalar-valued functions with high-dimensional input, such as loss functions in machine learning; forward-mode is better for high-dimensional outputs.

Error and Stability Analysis

Let us model the total error in a finite-difference scheme as

$$E(h) = E_{\text{trunc}}(h) - E_{\text{round}}(h),$$

where $E_{\text{trunc}}(h) = \mathcal{O}(h^p)$ is the truncation error (due to Taylor remainder) and $E_{\text{round}}(h) = \mathcal{O}(\varepsilon/h)$ is the round-off error (due to machine precision $\varepsilon \approx 2^{-52}$ in double precision).

The optimal step size h^* minimises the total error:

$$h^* = \left(\frac{(p\varepsilon)}{C}\right)^{1/(p+1)},$$

where C is the bound on the higher-order derivatives appearing in the truncation.

Example 3.1.2 (Stability of Forward Difference for $f(x) = \sqrt{x}$)

```
import numpy as np
import matplotlib.pyplot as plt

f = np.sqrt
df_exact = lambda x: 1 / (2 * np.sqrt(x))
```

```
x0 = 1.0
h_values = np.logspace(-16, -1, 200)
errors = [abs((f(x0 + h) - f(x0)) / h - df_exact(x0)) for h in h_values]

plt.loglog(h_values, errors)
plt.xlabel("h")
plt.ylabel("Absolute error")
plt.title("Forward difference error for $f(x) = sqrtx$ at $x = 1$")
plt.grid(True)
plt.show()
```

This plot exhibits a characteristic U-shape: error decreases with h due to improved approximation, then increases due to round-off. The minimum corresponds to the optimal step size $h^* \sim \varepsilon^{1/2}$ for forward differences (since $p = 1$).

Stability Considerations

- Differencing operations are ill-conditioned: computing $f(x+h) - f(x)$ for small h involves subtracting nearly equal quantities, potentially leading to catastrophic cancellation.
- Choosing h too small *increases* error due to finite machine precision; this trade-off must be balanced algorithmically.
- Automatic differentiation is numerically stable by design but can be computationally intensive if nested gradients or loops are involved.

Numerical differentiation enables the approximation of derivatives when analytic forms are unavailable, but its precision is limited by the structure of floating-point arithmetic. Automatic differentiation circumvents these limitations, providing exact gradients with the syntactic convenience of numerical code, and becomes indispensable in gradient-based optimisation and scientific computing.

3.1.2 Symbolic Differentiation Using SymPy

Symbolic differentiation manipulates expressions rather than numerical values. It rewrites a function $f(x)$ into its formal derivative $f'(x)$ by recursively applying differentiation rules. The Python library SymPy provides a powerful symbolic engine capable of performing exact calculus over expressions, polynomials, and even piecewise-defined functions. This subsection explores the core differentiation operators, rules for composition, and strategies for simplifying the resulting expressions.

Differentiation Rules and Operators

Let $f, g : \mathbb{R} \to \mathbb{R}$ be differentiable functions and $c \in \mathbb{R}$. Symbolic differentiation applies the following basic rules:

3.1 Differentiation

$$\frac{d}{dx}(cf(x)) = cf'(x) \qquad \text{(linearity)}$$

$$\frac{d}{dx}(f(x) + g(x)) = f'(x) + g'(x) \qquad \text{(sum rule)}$$

$$\frac{d}{dx}(f(x)g(x)) = f'(x)g(x) + f(x)g'(x) \qquad \text{(product rule)}$$

$$\frac{d}{dx}\left(\frac{f(x)}{g(x)}\right) = \frac{f'(x)g(x) - f(x)g'(x)}{g(x)^2} \qquad \text{(quotient rule)}$$

$$\frac{d}{dx}f(g(x)) = f'(g(x)) \cdot g'(x) \qquad \text{(chain rule)}$$

In SymPy, differentiation is performed using the diff() function:

```
import sympy as sp
x = sp.Symbol('x')
f = sp.sin(x**2) * sp.exp(x)
df = sp.diff(f, x)
sp.pprint(df)
```

The result is an exact symbolic expression:

$$\frac{d}{dx}\left[\sin\left(x^2\right)e^x\right] = 2x\cos\left(x^2\right)e^x + \sin\left(x^2\right)e^x.$$

Partial Derivatives Multivariate functions support symbolic partial differentiation:

```
x, y = sp.symbols("x y")
f = sp.exp(x*y)
df_dx = sp.diff(f, x)
df_dy = sp.diff(f, y)
```

returns:

$$\frac{\partial}{\partial x}e^{xy} = ye^{xy}, \qquad \frac{\partial}{\partial y}e^{xy} = xe^{xy}.$$

Higher-Order Derivatives

To compute n-th derivatives, pass the order as a second argument:

```
f = sp.log(sp.sin(x))
d4 = sp.diff(f, x, 4)
sp.pprint(d4)
```

This returns a fourth-order derivative involving cotangent and cosecant functions, as expected from repeated application of the chain and product rules.

Multi-index Notation For functions of several variables $f(x, y)$, one can compute mixed partials:

$$\frac{\partial^3 f}{\partial x^2 \partial y}$$

via:

```
sp.diff(f, x, 2, y)
```

SymPy ensures Schwarz's theorem (equality of mixed partials) holds under smoothness assumptions.

Simplification and Optimisation

Differentiated expressions can often be structurally complex. Simplification reduces redundancy and transforms expressions into more readable forms.

Algebraic Simplification The general `simplify()` function tries a broad set of rules:

```
f = sp.diff(sp.exp(x) * sp.sin(x), x)
simplified = sp.simplify(f)
```

returns:

$$\frac{d}{dx}(e^x \sin x) = e^x(\sin x + \cos x).$$

Targeted Simplification Other tools include:

- `expand()` — distributes products and powers.
- `factor()` — extracts common symbolic factors.
- `trigsimp()` — applies trigonometric identities.
- `collect()` — groups terms with common powers or factors.

Example 3.1.3 (Trigonometric Simplification) Let

$$f(x) = \cos^2 x + \sin^2 x.$$

```
f = sp.cos(x)**2 + sp.sin(x)**2
print(sp.simplify(f)) # 1
```

Here, `simplify()` internally uses the identity $\cos^2 x + \sin^2 x = 1$.

Common Subexpression Elimination To improve computational efficiency, especially in code generation, SymPy supports *CSE* (Common Subexpression Elimination):

3.1 Differentiation

```
from sympy import cse
expr = sp.diff(sp.sin(x)**2 + sp.exp(x) * sp.sin(x), x)
replacements, reduced_expr = cse(expr)
```

This breaks the expression into reusable parts, akin to compiler-level optimisation of repeated terms.

Lambdification Once an expression is simplified, it can be converted to a callable function via:

```
f_expr = sp.sin(x**2) * sp.exp(x)
f_num = sp.lambdify(x, f_expr, 'numpy')
print(f_num(1.0))
```

This combines symbolic clarity with numerical efficiency.

Symbolic differentiation thus offers an exact, algebraically faithful representation of derivatives, preserves mathematical structure, and feeds directly into higher-level tasks such as solving ODEs, computing Taylor expansions, and deriving Jacobians for multivariate functions. When used with simplification and code generation tools, it enables both formal verification and high-performance computation.

3.1.3 Applications of Differentiation

Differentiation serves as the analytic microscope of mathematics—it reveals local trends, detects turning points, and governs the dynamics of physical systems. In this subsection we explore key applications of derivatives in optimisation, curve analysis, and modelling real-world systems governed by physical laws. Each topic is illustrated through rigorous examples that blend symbolic and numerical computation.

Optimisation Problems

Optimisation seeks extrema (minima or maxima) of real-valued functions. A critical point of $f : \mathbb{R} \to \mathbb{R}$ occurs at x^* such that $f'(x^*) = 0$; second-order conditions determine the nature of the extremum:

$$\text{If } f''(x^*) > 0, \text{ then } f \text{ has a local minimum at } x^*,$$

$$f''(x^*) < 0 \Rightarrow \text{local maximum.}$$

Example 3.1.4 (Minimising a Function with Symbolic Calculus) Minimise the function

$$f(x) = x^4 - 8x^3 - 18x^2 + 1.$$

```python
import sympy as sp
x = sp.Symbol('x')
f = x**4 - 8*x**3 + 18*x**2 + 1
crit_pts = sp.solve(sp.diff(f, x), x)
for pt in crit_pts:
    second_derivative = sp.diff(f, x, 2).subs(x, pt)
    print(f"x = {pt}, f'' = {second_derivative}, type = {'min' if
        second_derivative > 0 else 'max'}")
```

Solving $f'(x) = 4x^3 - 24x^2 + 36x = 0$ yields critical points $x = 0, 3, 6$. Evaluating the second derivative identifies local minima at $x = 0$ and $x = 6$, and a local maximum at $x = 3$.

Example 3.1.5 (Constrained Optimisation Using Lagrange Multipliers) Find the extrema of $f(x, y) = x^2 + y^2$ subject to the constraint $x + y = 1$. This minimises distance to the origin under a linear constraint.

```
x, y, λ = sp.symbols("x y λ")
f = x**2 + y**2
g = x + y - 1
L = f - λ * g
sol = sp.solve([sp.diff(L, v) for v in (x, y, λ)])
sp.pprint(sol)
```

The solution $x = y = 1/2$ minimises f on the line $x + y = 1$. The method generalises to arbitrary equality constraints.

Curve Sketching and Analysis

A function's graph is shaped by its first and second derivatives:

- $f'(x) > 0 \Rightarrow f$ increasing, $\quad f'(x) < 0 \Rightarrow f$ decreasing.
- $f''(x) > 0 \Rightarrow$ concave up, $\quad f''(x) < 0 \Rightarrow$ concave down.
- Points where $f''(x) = 0$ and changes sign are inflection points.

Example 3.1.6 (Sketching a Rational Function) Let $f(x) = \dfrac{x^2 - 1}{x^2 + 1}$. Study monotonicity, concavity, and asymptotes.

```
f = (x**2 - 1)/(x**2 + 1)
df = sp.diff(f, x)
d2f = sp.diff(df, x)
sp.pprint(df.simplify())
sp.pprint(d2f.simplify())
```

We find:

$$f'(x) = \frac{4x}{(x^2 + 1)^2}, \quad f''(x) = \frac{4(1 - 3x^2)}{(x^2 + 1)^3}.$$

3.1 Differentiation

- $f' > 0 \Rightarrow f$ increasing on $(0, \infty)$, decreasing on $(-\infty, 0)$.
- $f'' = 0 \Rightarrow x = \pm 1/\sqrt{3}$ — inflection points.

Graphing this reveals a symmetric curve increasing towards the horizontal asymptote $y = 1$.

Example 3.1.7 (Taylor Expansion and Local Approximation) Let $f(x) = \log(1 + x)$. The third-degree Taylor polynomial at $x = 0$ is:

$$T_3(x) = x - \frac{x^2}{2} + \frac{x^3}{3}.$$

Use SymPy to derive it:

```
f = sp.log(1 + x)
T3 = f.series(x, 0, 4).removeO()
sp.pprint(T3)
```

This polynomial approximates $f(x)$ well for $|x| < 1$. Compare numerically at $x = 0.5$:

```
f_true = sp.lambdify(x, f)
T3_func = sp.lambdify(x, T3)
print(abs(f_true(0.5) - T3_func(0.5)))
```

This gives a precise quantification of approximation error.

Physical System Modelling

Differentiation describes the local rate of change in physical quantities. It governs Newtonian mechanics, electrical circuits, chemical kinetics, and population models.

Example 3.1.8 (Velocity and Acceleration) Let a particle's position be given by $s(t) = t^3 - 6t^2 + 9t$. Find the velocity $v(t) = s'(t)$ and acceleration $a(t) = s''(t)$. Determine when the particle is at rest and when it accelerates.

```
t = sp.Symbol("t")
s = t**3 - 6*t**2 + 9*t
v = sp.diff(s, t)
a = sp.diff(v, t)
sp.pprint(v)
sp.pprint(a)
rest_times = sp.solve(v, t)
accel_signs = [(τ, a.subs(t, τ).evalf()) for τ in rest_times]
```

We get:

$$v(t) = 3t^2 - 12t + 9, \quad a(t) = 6t - 12.$$

Solving $v(t) = 0$ yields rest points at $t = 1$ and $t = 3$; the particle changes direction at those times.

Example 3.1.9 (Cooling Law: Newton's Model) Let $T(t)$ be the temperature of an object cooling in air at ambient temperature $T_a = 25\,°C$. Newton's law states:

$$\frac{dT}{dt} = -k(T - T_a), \quad T(0) = T_0.$$

Solve symbolically:

```
T, t, k, Ta, T0 = sp.symbols("T t k T_a T_0")
sol = sp.dsolve(sp.Derivative(T, t) + k*(T - Ta), T, ics={T.subs(t, 0): T0
    })
sp.pprint(sol)
```

The solution is:

$$T(t) = T_a + (T_0 - T_a)e^{-kt}.$$

This models exponential decay towards equilibrium.

Example 3.1.10 (Mass-Spring System: Hooke's Law) A particle obeys:

$$m\ddot{x} + kx = 0.$$

The general solution is:

$$x(t) = A\cos(\omega t) + B\sin(\omega t), \quad \omega = \sqrt{k/m}.$$

Solve and visualise for $m = 1$, $k = 4$, with $x(0) = 1$, $\dot{x}(0) = 0$:

```
from sympy import Function, dsolve, Derivative as D, cos, sin
t = sp.Symbol("t")
x = Function("x")
sol = dsolve(D(D(x(t), t), t) + 4*x(t), x(t), ics={x(0): 1, D(x(t), t).
    subs(t, 0): 0})
sp.pprint(sol)
```

We obtain $x(t) = \cos(2t)$, showing undamped harmonic motion with frequency 2.

Differentiation thus serves as a gateway to mathematical modelling and computational science: it enables us to analyse local behaviour, optimise outcomes, trace geometry, and simulate real-world processes governed by differential laws. Python provides multiple perspectives—from symbolic to numerical to automatic—allowing rigorous yet flexible exploration of all such applications.

3.2 Integration

Integration is the inverse operation of differentiation and corresponds to computing the accumulated effect of a function over an interval. In the Riemannian framework,

3.2 Integration

the definite integral

$$\int_a^b f(x)\,dx$$

represents the signed area under the curve $y = f(x)$ between $x = a$ and $x = b$. In computational settings, analytical integration is often impossible, especially when the integrand lacks a closed-form antiderivative. Numerical integration, or *quadrature*, then provides approximate solutions by evaluating the integrand at selected points. We discuss several such methods, their error bounds, and implementation in Python.

3.2.1 Numerical Integration Methods

Riemann and Trapezoidal Rules

Let $f : [a, b] \to \mathbb{R}$ be continuous. Partition the interval $[a, b]$ into n subintervals of equal width $h = \frac{b-a}{n}$. Define the grid points $x_i = a + ih$ for $i = 0, 1, \ldots, n$.

Riemann Sum (Left Rule)

$$R_n = \sum_{i=0}^{n-1} f(x_i) \cdot h$$

Trapezoidal Rule Approximates each subinterval as a trapezoid:

$$T_n = \frac{h}{2} [f(x_0) + 2f(x_1) + \cdots + 2f(x_{n-1}) + f(x_n)]$$

$$= h \left[\frac{f(a) + f(b)}{2} + \sum_{i=1}^{n-1} f(x_i) \right]$$

Example 3.2.1 (Compute $\int_0^1 \frac{1}{1+x^2}\,dx$) This integral equals $\arctan(1) = \frac{\pi}{4}$. Compare trapezoidal approximation and exact value.

```
import numpy as np

def f(x): return 1 / (1 + x**2)

def trapezoid(f, a, b, n):
    h = (b - a) / n
    x = np.linspace(a, b, n+1)
    return h * (0.5*f(a) + sum(f(x[1:-1])) + 0.5*f(b))

I_approx = trapezoid(f, 0, 1, 1000)
print(f"Approx: {I_approx:.10f}, Error: {abs(I_approx - np.pi/4):.2e}")
```

Error Bound (Trapezoidal) If f is twice continuously differentiable, then

$$\left|\int_a^b f(x)\,dx - T_n\right| \le \frac{(b-a)^3}{12n^2} \max_{x\in[a,b]} |f''(x)|.$$

Simpson's Rule and Gaussian Quadrature

Simpson's rule approximates f over each subinterval with a quadratic polynomial using three points: endpoints and midpoint.

Simpson's Rule Assume n is even:

$$S_n = \frac{h}{3}[f(x_0) + 4f(x_1) + 2f(x_2) + \cdots + 4f(x_{n-1}) + f(x_n)]$$

Example 3.2.2 (Simpson's Rule for $\int_0^1 e^{-x^2}\,dx$) This integral has no elementary antiderivative.

```
from scipy.integrate import simps

x = np.linspace(0, 1, 101)
y = np.exp(-x**2)
approx = simps(y, x)
print(f"Simpson approx: {approx:.10f}")
```

Compare this with high-precision estimate: ≈ 0.746824.

Error Bound (Simpson) If f has a fourth derivative,

$$\left|\int_a^b f(x)\,dx - S_n\right| \le \frac{(b-a)^5}{180n^4} \max_{x\in[a,b]} |f^{(4)}(x)|.$$

Gaussian Quadrature For n nodes $x_i \in [a,b]$ and weights w_i, Gaussian quadrature gives:

$$\int_a^b f(x)\,dx \approx \sum_{i=1}^n w_i f(x_i)$$

such that it is exact for all polynomials of degree $2n - 1$.

For the interval $[-1, 1]$, `numpy.polynomial.legendre.leggauss(n)` gives x_i, w_i.

Example 3.2.3 (Legendre–Gauss Quadrature for $f(x) = \cos(x)$)

```
from numpy.polynomial.legendre import leggauss

n = 4
xg, wg = leggauss(n)
```

3.2 Integration

```
approx = sum(w * np.cos(x) for x, w in zip(xg, wg))
true = np.sin(1) - np.sin(-1)
error = abs(approx - true)
print(f"Gaussian quad (n=4): {approx:.10f}, error: {error:.2e}")
```

The method achieves high precision with few evaluations.

Adaptive Quadrature Algorithms

Fixed-step quadrature can be inefficient on functions with localised features (e.g. sharp peaks). Adaptive quadrature adjusts the step size based on local error estimates, allocating more points where the function varies rapidly.

Principle Given interval $[a, b]$, compute coarse and fine approximations I_1, I_2. If $|I_2 - I_1| < \varepsilon$, accept I_2; else split $[a, b]$ and recurse.

Example 3.2.4 (Adaptive Simpson's Method) Use `scipy.integrate.quad`, which implements adaptive quadrature:

```
from scipy.integrate import quad

f = lambda x: np.exp(-x**2)
I, err = quad(f, 0, 1, epsabs=1e-10)
print(f"Adaptive Simpson: {I:.10f}, estimated error: {err:.1e}")
```

Singular Integrands Adaptive methods are crucial when the integrand has singularities or rapid oscillations. For example:

$$\int_0^1 \frac{\log(x)}{\sqrt{x}} \, dx,$$

which diverges at $x = 0$ but is integrable in the Lebesgue sense. Python's `quad` handles this using weight functions.

```
from scipy.integrate import quad
f = lambda x: np.log(x) / np.sqrt(x)
I, err = quad(f, 0, 1, weight='alg', wvar=-0.5)
print(f"Result: {I}, error estimate: {err}")
```

3.2.2 Symbolic Integration with SymPy

Symbolic integration seeks exact expressions for antiderivatives and definite integrals. Unlike numerical methods, which approximate the value of an integral over an interval, symbolic integration manipulates the expression of the integrand directly, often yielding closed-form solutions. SymPy implements a variant of the Risch algorithm for elementary functions, and it supports many special functions and integral identities drawn from classical tables. This section explores the use of `integrate()` in handling both indefinite and definite integrals, and in recognising special-function patterns.

Indefinite Integrals and Antiderivatives

Given a function $f(x)$, its antiderivative is a function $F(x)$ such that $F'(x) = f(x)$. Symbolically,

$$\int f(x)\,dx = F(x) + C,$$

where C is the constant of integration. In SymPy, one writes:

```
import sympy as sp
x = sp.Symbol("x")
f = sp.exp(x) * sp.sin(x)
F = sp.integrate(f, x)
sp.pprint(F)
```

The returned expression is:

$$\int e^x \sin x\,dx = \frac{1}{2}e^x(\sin x - \cos x) + C.$$

Example 3.2.5 (Elementary Rational Function) Integrate

$$\int \frac{3x^2 + 2x + 1}{x^3 + x^2 + x + 1}\,dx.$$

```
f = (3*x**2 + 2*x + 1)/(x**3 + x**2 + x + 1)
sp.pprint(sp.integrate(f, x))
```

SymPy performs polynomial division and partial fractions internally, yielding logarithmic terms as expected from standard techniques.

Example 3.2.6 (Integration by Substitution) Let

$$\int x\sqrt{1 + x^2}\,dx.$$

```
f = x * sp.sqrt(1 + x**2)
sp.pprint(sp.integrate(f, x))
```

3.2 Integration

This returns:

$$\frac{1}{3}(1+x^2)^{3/2} + C,$$

corresponding to the substitution $u = 1 + x^2$.

Definite Integrals and Limits

To compute a definite integral

$$\int_a^b f(x)\,dx,$$

use `integrate()` with bounds:

```
sp.integrate(f, (x, a, b))
```

Example 3.2.7 (**Area Under a Curve**) Evaluate

$$\int_0^1 \frac{1}{1+x^2}\,dx = \tan^{-1}(1) = \frac{\pi}{4}.$$

```
f = 1 / (1 + x**2)
sp.integrate(f, (x, 0, 1))  # returns π/4
```

Example 3.2.8 (**Improper Integral with Limit**) Compute

$$\int_0^\infty e^{-x^2}\,dx = \frac{\sqrt{\pi}}{2}.$$

```
f = sp.exp(-x**2)
sp.integrate(f, (x, 0, sp.oo))  # returns sqrt(pi)/2
```

Here `oo` denotes $+\infty$. SymPy evaluates such integrals by reducing them to known special forms or limit arguments.

Integral as a Limit

$$\int_0^1 \frac{dx}{\sqrt{x}} = \lim_{\varepsilon \to 0^+} \int_\varepsilon^1 x^{-1/2}\,dx = 2.$$

```
ε = sp.Symbol("ε", positive=True)
I = sp.integrate(x**(-1/2), (x, ε, 1))
sp.limit(I, ε, 0)
```

This confirms convergence of the improper integral.

Special Functions and Integral Tables

SymPy can symbolically evaluate integrals involving special functions such as Γ, ζ, Ei, Si, erf, and Bessel functions. These match entries in standard integral tables (e.g. Gradshteyn–Ryzhik).

Example 3.2.9 (Gaussian Integral)

$$\int_{-\infty}^{\infty} e^{-ax^2}\, dx = \sqrt{\frac{\pi}{a}}, \qquad a > 0.$$

```
a = sp.Symbol("a", positive=True)
f = sp.exp(-a * x**2)
sp.integrate(f, (x, -sp.oo, sp.oo))
```

SymPy returns $\sqrt{\pi/a}$ exactly, confirming the classic result.

Example 3.2.10 (Beta Function)

$$\int_0^1 x^{m-1}(1-x)^{n-1}\, dx = B(m,n) = \frac{\Gamma(m)\Gamma(n)}{\Gamma(m+n)}.$$

```
m, n = sp.symbols("m n", positive=True)
sp.integrate(x**(m - 1) * (1 - x)**(n - 1), (x, 0, 1))
```

This returns a beta function $B(m, n)$, automatically converting to gamma expressions when simplified.

Example 3.2.11 (Fresnel Integral)

$$\int_0^{\infty} \sin\left(x^2\right) dx = \frac{\sqrt{2\pi}}{4}.$$

```
f = sp.sin(x**2)
sp.integrate(f, (x, 0, sp.oo))
```

SymPy returns an unevaluated expression involving the Fresnel integral, showing symbolic recognition of the special function.

Integral Tables SymPy contains dozens of built-in integral patterns corresponding to classical results. To explore known integrals or verify manual computations, one can use:

```
sp.integrals.manualintegrate(f, x)
```

This applies structured decision trees rather than heuristics.

Symbolic integration with SymPy enables the exact evaluation of both elementary and special-function integrals, with automatic simplification and closed-form recognition. It extends the traditional toolbox of calculus to include gamma, beta,

3.2 Integration

and error functions, and serves as a robust backend for mathematical exploration, problem verification, and symbolic modelling.

3.2.3 Applications of Integration

Integration is not merely a tool for computing areas—it is a unifying framework for accumulating quantities. It arises naturally in geometry (as area and volume), in probability theory (as expectation and distribution), and in modelling (as the solution operator for differential equations). This subsection explores each of these interpretations through concrete, computationally tractable examples that leverage Python and SymPy.

Area and Volume Computations

Area Under a Curve Let $f : [a, b] \to \mathbb{R}$ be non-negative and continuous. Then the area under the graph of f is

$$A = \int_a^b f(x) \, dx.$$

Example 3.2.12 (Area between Two Curves) Compute the area enclosed between $f(x) = \sin(x)$ and $g(x) = \cos(x)$ on $[0, \pi/2]$.

$$A = \int_0^{\pi/2} |\sin(x) - \cos(x)| \, dx.$$

```
import sympy as sp
x = sp.Symbol("x")
f = sp.sin(x)
g = sp.cos(x)
area = sp.integrate(sp.Abs(f - g), (x, 0, sp.pi/2))
sp.pprint(area)
```

The integral evaluates to $2 - \sqrt{2}$.

Volume of Revolution For a solid formed by rotating $f(x) \geq 0$ about the x-axis over $[a, b]$, the volume is given by:

$$V = \pi \int_a^b f(x)^2 \, dx.$$

Example 3.2.13 (Volume of a Paraboloid) Let $f(x) = \sqrt{x}$, and rotate it about the x-axis from $x = 0$ to $x = 4$.

```
f = sp.sqrt(x)
V = sp.pi * sp.integrate(f**2, (x, 0, 4))  # integral of x
sp.pprint(V)
```

We find $V = \pi \cdot \int_0^4 x\, dx = \pi \cdot 8 = 8\pi$.

Probability and Expectation

Let X be a continuous random variable with probability density function (PDF) $p(x)$ such that $\int_{-\infty}^{\infty} p(x)\, dx = 1$.

Expectation The expected value of a function $f(X)$ is

$$\mathbb{E}[f(X)] = \int_{-\infty}^{\infty} f(x)\, p(x)\, dx.$$

Example 3.2.14 (Expectation of a Uniform Distribution) Let $X \sim \text{Unif}(a, b)$. Then

$$p(x) = \frac{1}{b-a}, \quad x \in [a, b].$$

The expected value is:

$$\mathbb{E}[X] = \int_a^b \frac{x}{b-a}\, dx = \frac{a+b}{2}.$$

```
a, b = sp.symbols("a b", real=True)
pdf = 1 / (b - a)
EX = sp.integrate(x * pdf, (x, a, b)).simplify()
sp.pprint(EX)  # prints (a + b)/2
```

Example 3.2.15 (Moment of a Normal Distribution) Let $X \sim \mathcal{N}(0, 1)$. Then

$$p(x) = \frac{1}{\sqrt{2\pi}} e^{-x^2/2}, \quad \mathbb{E}[X^2] = 1.$$

```
pdf = sp.exp(-x**2 / 2) / sp.sqrt(2 * sp.pi)
EX2 = sp.integrate(x**2 * pdf, (x, -sp.oo, sp.oo))
sp.pprint(EX2)
```

This confirms the second moment of the standard normal is 1.

Cumulative Distribution Function (CDF)

$$F(x) = \int_{-\infty}^{x} p(t)\, dt$$

3.2 Integration

Example 3.2.16 (CDF of Exponential Distribution) Let $p(x) = \lambda e^{-\lambda x}$ for $x \geq 0$. Then

$$F(x) = \int_0^x \lambda e^{-\lambda t}\, dt = 1 - e^{-\lambda x}.$$

```
λ = sp.Symbol("λ", positive=True)
p = λ * sp.exp(-λ * x)
F = sp.integrate(p, (x, 0, x))
sp.pprint(F.simplify())
```

Solving Differential Equations

Integration is the central mechanism for solving ordinary differential equations (ODEs). Given $\frac{dy}{dx} = f(x)$, the general solution is:

$$y(x) = \int f(x)\, dx + C.$$

Separable Equations If the ODE has the form $\frac{dy}{dx} = g(x)h(y)$, one may write:

$$\int \frac{1}{h(y)}\, dy = \int g(x)\, dx.$$

Example 3.2.17 (Logistic Equation)

$$\frac{dy}{dx} = ry(1 - y/K), \quad y(0) = y_0.$$

Separate and integrate:

$$\int \frac{dy}{y(1 - y/K)} = \int r\, dx.$$

```
y = sp.Function("y")(x)
r, K = sp.symbols("r K", positive=True)
ode = sp.Eq(y.diff(x), r * y * (1 - y / K))
sol = sp.dsolve(ode, y)
sp.pprint(sol)
```

The solution is:

$$y(x) = \frac{K}{1 + Ce^{-rx}}.$$

Integrating Factor The linear ODE

$$\frac{dy}{dx} + P(x)y = Q(x)$$

has solution

$$y(x) = e^{-\int P(x)\,dx} \left[\int Q(x) e^{\int P(x)\,dx} dx + C \right].$$

Example 3.2.18 (First-Order Linear Equation) Solve $y' - y = e^x$, with $y(0) = 0$.

```
y = sp.Function("y")(x)
ode = sp.Eq(sp.diff(y, x) - y, sp.exp(x))
sol = sp.dsolve(ode, y, ics={y.subs(x, 0): 0})
sp.pprint(sol)
```

Solution:

$$y(x) = \frac{1}{2} e^x(x).$$

3.3 Multivariable Calculus

While single-variable calculus describes how a function $f : \mathbb{R} \to \mathbb{R}$ changes along a line, multivariable calculus generalises these ideas to functions $f : \mathbb{R}^n \to \mathbb{R}$, enabling analysis over surfaces, hypersurfaces, and manifolds. Partial derivatives, gradients, and directional derivatives provide local linear approximations, while level sets and optimisation encode geometric and extremal behaviour in higher dimensions. This section develops these tools, supported by symbolic and numerical examples in Python.

3.3.1 Partial Derivatives and Gradients

Let $f(x_1, \ldots, x_n)$ be a scalar-valued function. Its **partial derivative** with respect to x_i is defined as

$$\frac{\partial f}{\partial x_i}(x_1, \ldots, x_n) = \lim_{h \to 0} \frac{f(x_1, \ldots, x_i + h, \ldots, x_n) - f(x_1, \ldots, x_i, \ldots, x_n)}{h}.$$

3.3 Multivariable Calculus

The **gradient** of f, denoted ∇f, is the vector of all partial derivatives:

$$\nabla f = \begin{pmatrix} \partial f/\partial x_1 \\ \vdots \\ \partial f/\partial x_n \end{pmatrix}.$$

Example 3.3.1 (Gradient of a Quadratic Function) Let $f(x, y) = 3x^2 + 2xy + y^2$. Compute $\frac{\partial f}{\partial x}$, $\frac{\partial f}{\partial y}$, and ∇f.

```
import sympy as sp
x, y = sp.symbols("x y")
f = 3*x**2 + 2*x*y + y**2
df_dx = sp.diff(f, x)
df_dy = sp.diff(f, y)
grad_f = [df_dx, df_dy]
sp.pprint(grad_f)
```

We find:

$$\nabla f = \begin{pmatrix} 6x + 2y \\ 2x + 2y \end{pmatrix}.$$

Directional Derivatives

The **directional derivative** of f at point $\mathbf{a}$ in direction $\mathbf{v} \in \mathbb{R}^n$ is:

$$D_\mathbf{v} f(\mathbf{a}) = \nabla f(\mathbf{a}) \cdot \frac{\mathbf{v}}{\|\mathbf{v}\|}.$$

This gives the rate of change of f along $\mathbf{v}$, scaled by its unit length.

Example 3.3.2 (Directional Derivative at a Point) Let $f(x, y) = x^2 y + y^3$. Compute the directional derivative at $(1, 2)$ in direction $\mathbf{v} = \langle 3, 4 \rangle$.

```
f = x**2 * y + y**3
grad = [sp.diff(f, var) for var in (x, y)]
a = {x: 1, y: 2}
grad_val = [df.subs(a) for df in grad] # [2*1*2 = 4, 1**2 + 3*4 = 13]
v = sp.Matrix([3, 4])
v_unit = v / sp.sqrt(sum([vi**2 for vi in v]))
D_v = sum([g * vu for g, vu in zip(grad_val, v_unit)])
sp.pprint(D_v.evalf())
```

The result is the scalar rate of change of f along $\mathbf{v}$, numerically approximated.

Gradient Fields and Level Sets

The gradient ∇f points in the direction of steepest ascent. The **level set** of a function $f : \mathbb{R}^2 \to \mathbb{R}$ at level c is:

$$L_c = \{(x, y) \in \mathbb{R}^2 : f(x, y) = c\}.$$

Gradients are orthogonal to level sets:

$$\nabla f(x, y) \perp \text{ level curve at } (x, y).$$

Example 3.3.3 (Visualising Level Sets and Gradients) Let $f(x, y) = x^2 + y^2$. Then level sets are circles, and gradients point radially outward (cf. Fig. 3.1).

```
import numpy as np
import matplotlib.pyplot as plt
```

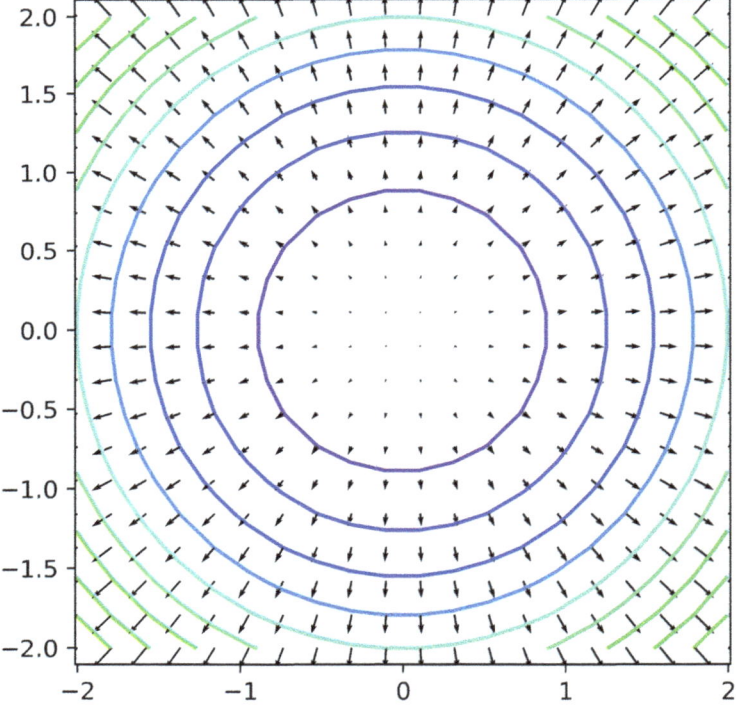

Fig. 3.1 Level sets of the quadratic form $f(x, y) = x^2 + y^2$ (solid contours) together with its gradient field $\nabla f = (2x, 2y)$. Each gradient vector is orthogonal to the corresponding contour line and points radially outwards, visualising the direction of steepest ascent and the radial symmetry of f

3.3 Multivariable Calculus

```
X, Y = np.meshgrid(np.linspace(-2, 2, 20), np.linspace(-2, 2, 20))
F = X**2 + Y**2
gradX = 2*X
gradY = 2*Y

plt.contour(X, Y, F, levels=10, cmap='viridis')
plt.quiver(X, Y, gradX, gradY)
plt.title("Level Sets and Gradient Field of $f(x, y) = x^2 + y^2$")
plt.axis("equal")
plt.show()
```

Example 3.3.4 (Gradient Perpendicular to Level Curves) For $f(x, y) = x^2 - y^2$, level sets are hyperbolas. Verify that the gradient $\nabla f = (2x, -2y)$ is perpendicular to the level set at a given point.

Choose point $(1, 1)$. At this point, level set is $x^2 - y^2 = 0 \Rightarrow x = y$. The tangent vector is $\langle 1, 1 \rangle$. Dot product:

$$\nabla f \cdot \langle 1, 1 \rangle = 2 \cdot 1 + (-2) \cdot 1 = 0.$$

Multivariate Optimisation

Let $f : \mathbb{R}^n \to \mathbb{R}$. Critical points satisfy $\nabla f = \mathbf{0}$. The **Hessian matrix** $H_f \in \mathbb{R}^{n \times n}$ is defined by:

$$(H_f)_{ij} = \frac{\partial^2 f}{\partial x_i \partial x_j}.$$

The nature of a critical point is determined by the eigenvalues of H:

- Positive definite $\Rightarrow$ local minimum
- Negative definite $\Rightarrow$ local maximum
- Indefinite $\Rightarrow$ saddle point

Example 3.3.5 (Optimising a Multivariate Function) Let $f(x, y) = x^2 + y^2 - 4x - 6y$. Find and classify critical points.

```
f = x**2 + y**2 - 4*x - 6*y
grad = [sp.diff(f, var) for var in (x, y)]
critical = sp.solve(grad)
H = sp.hessian(f, (x, y))
eigenvals = H.subs(critical).eigenvals()
sp.pprint(critical)
sp.pprint(eigenvals)
```

Gradient vanishes at $x = 2$, $y = 3$. Hessian is $2I$, positive definite $\Rightarrow$ local minimum at $(2, 3)$.

Example 3.3.6 (Saddle Point) Let $f(x, y) = x^2 - y^2$. Then $\nabla f = (2x, -2y) \Rightarrow$ $(0, 0)$ is critical. Hessian:

$$H = \begin{pmatrix} 2 & 0 \\ 0 & -2 \end{pmatrix} \Rightarrow \text{indefinite} \Rightarrow \text{saddle point.}$$

Example 3.3.7 (Constrained Optimisation: Lagrange Multipliers) Maximise $f(x, y) = xy$ subject to $x^2 + y^2 = 1$.

```
λ = sp.Symbol("λ")
g = x**2 + y**2 - 1
L = x*y - λ*g
sol = sp.solve([sp.diff(L, v) for v in (x, y, λ)])
sp.pprint(sol)
```

The maximum value is at $x = y = \frac{1}{\sqrt{2}} \Rightarrow f = \frac{1}{2}$.

3.3.2 Multiple Integrals

When a function depends on multiple variables, integration generalises to higher dimensions. A **double integral** computes volume under a surface, and a **triple integral** accumulates values over a three-dimensional region. SymPy supports both symbolic and numerical computation of such integrals and allows coordinate transformations via Jacobians.

Double Integrals

Given $f(x, y)$ defined over a rectangle $R = [a, b] \times [c, d]$, the double integral is

$$\iint_R f(x, y) \, dx \, dy = \int_c^d \left(\int_a^b f(x, y) \, dx \right) dy.$$

If the region is not rectangular, the limits of the inner integral depend on the outer variable.

Example 3.3.8 (Rectangular Domain) Evaluate:

$$\iint_{[0,1] \times [0,2]} (x^2 + y) \, dx \, dy.$$

```
import sympy as sp
x, y = sp.symbols("x y")
f = x**2 + y
I = sp.integrate(sp.integrate(f, (x, 0, 1)), (y, 0, 2))
sp.pprint(I)
```

Inner integral: $\int_0^1 x^2 + y \, dx = \frac{1}{3} + y$. Outer: $\int_0^2 \left(\frac{1}{3} + y \right) dy = \frac{2}{3} + 2 = \frac{8}{3}$.

3.3 Multivariable Calculus

Example 3.3.9 (Non-rectangular Domain) Evaluate:

$$\iint_D xy\, dx\, dy, \quad D = \{(x, y) \mid 0 \leq y \leq 1,\ 0 \leq x \leq y\}.$$

```
f = x * y
I = sp.integrate(sp.integrate(f, (x, 0, y)), (y, 0, 1))
sp.pprint(I)
```

We compute:

$$\int_0^1 \int_0^y xy\, dx\, dy = \int_0^1 y\left(\frac{1}{2}y^2\right) dy = \frac{1}{2}\int_0^1 y^3\, dy = \frac{1}{8}.$$

Triple Integrals

For $f(x, y, z)$ defined over a 3D region $V \subset \mathbb{R}^3$, the triple integral computes:

$$\iiint_V f(x, y, z)\, dx\, dy\, dz.$$

Example 3.3.10 (Cuboid Domain) Evaluate:

$$\iiint_{[0,1]^3} xyz\, dx\, dy\, dz.$$

```
z = sp.Symbol("z")
f = x * y * z
I = sp.integrate(sp.integrate(sp.integrate(f, (x, 0, 1)), (y, 0, 1)), (z,
    0, 1))
sp.pprint(I)
```

All integrals are over [0,1], so:

$$\int_0^1 x\, dx = \frac{1}{2}, \quad \text{similarly for } y, z.$$

Total: $\frac{1}{2} \cdot \frac{1}{2} \cdot \frac{1}{2} = \frac{1}{8}$.

Example 3.3.11 (Tetrahedral Domain) Evaluate:

$$\iiint_D (x+y+z)\, dz\, dy\, dx, \quad D = \{0 \leq x \leq 1,\ 0 \leq y \leq 1-x,\ 0 \leq z \leq 1-x-y\}.$$

```
f = x + y + z
I = sp.integrate(
    sp.integrate(
```

```
          sp.integrate(f, (z, 0, 1 - x - y)),
       (y, 0, 1 - x)),
    (x, 0, 1))
sp.pprint(I)
```

This yields $\frac{1}{8}$, the average value over a unit tetrahedron.

Change of Variables and Jacobians

Coordinate transformations simplify integration by mapping a complex region to a simpler one. Let $(x, y) = \phi(u, v)$. Then:

$$\iint_D f(x, y)\, dx\, dy = \iint_{\phi^{-1}(D)} f(x(u, v), y(u, v)) \left| \frac{\partial(x, y)}{\partial(u, v)} \right| du\, dv.$$

Here the **Jacobian determinant** is:

$$J = \begin{vmatrix} \partial x/\partial u & \partial x/\partial v \\ \partial y/\partial u & \partial y/\partial v \end{vmatrix}.$$

Example 3.3.12 (Polar Coordinates) Evaluate:

$$\iint_{x^2+y^2 \le 1} (x^2 + y^2)\, dx\, dy.$$

Use $x = r\cos\theta$, $y = r\sin\theta$. Then $x^2 + y^2 = r^2$, and Jacobian is r.

```
r, θ = sp.symbols("r θ")
f_polar = r**2 * r
I = sp.integrate(sp.integrate(f_polar, (r, 0, 1)), (θ, 0, 2*sp.pi))
sp.pprint(I)
```

We compute:

$$\int_0^{2\pi} \int_0^1 r^3\, dr\, d\theta = 2\pi \cdot \frac{1}{4} = \frac{\pi}{2}.$$

Example 3.3.13 (Jacobian from a Nonlinear Change) Let $u = x+y$, $v = x-y$. Compute the Jacobian of the transformation.

```
u, v = sp.symbols("u v")
x = (u + v)/2
y = (u - v)/2
J = sp.Matrix([[sp.diff(x, u), sp.diff(x, v)],
               [sp.diff(y, u), sp.diff(y, v)]])
sp.pprint(J)
sp.pprint(J.det())
```

The Jacobian is:

$$J = \begin{pmatrix} \frac{1}{2} & \frac{1}{2} \\ \frac{1}{2} & -\frac{1}{2} \end{pmatrix}, \quad \det J = -\frac{1}{2}.$$

Example 3.3.14 (Transforming an Integral Using Jacobian) Evaluate:

$$\iint_{x>0,\ y>0,\ x^2+y^2<1} e^{x^2+y^2}\,dx\,dy.$$

In polar coordinates, $x^2 + y^2 = r^2$, and Jacobian is r:

$$\int_0^{\pi/2} \int_0^1 e^{r^2} r\, dr\, d\theta.$$

Let's perform the substitution $u = r^2 \Rightarrow du = 2r\,dr$:

```
u = sp.Symbol("u")
inner = sp.integrate(sp.exp(u)/2, (u, 0, 1))
I = inner * (sp.pi / 2)
sp.pprint(I)
```

This yields:

$$\frac{\pi}{4}(e^1 - 1).$$

3.3.3 Line Integrals and Surface Integrals

In multivariable calculus, integrals generalise from areas and volumes to curvilinear and surface domains. A **line integral** accumulates a function along a path, and a **surface integral** computes quantities such as flux through a surface. These integrals are essential in physics and engineering: work, circulation, and electric/magnetic flux are described using such integrals. This section covers scalar and vector line integrals, surface integrals, and their computation using Python and SymPy.

Scalar Line Integrals

Given a scalar field $f : \mathbb{R}^n \to \mathbb{R}$ and a smooth path $\gamma : [a, b] \to \mathbb{R}^n$, the scalar line integral is:

$$\int_\gamma f\,ds = \int_a^b f(\gamma(t))\,\|\gamma'(t)\|\,dt.$$

This represents the accumulated value of f along the curve, weighted by arc length.

Example 3.3.15 (Length of a Curve) Let $\gamma(t) = (t, t^2)$, $t \in [0, 1]$. Compute arc length:

$$L = \int_0^1 \sqrt{(dx/dt)^2 + (dy/dt)^2}\, dt = \int_0^1 \sqrt{1 + 4t^2}\, dt.$$

```
import sympy as sp
t = sp.Symbol("t")
y = [t, t**2]
speed = sp.sqrt(sum([sp.diff(c, t)**2 for c in y]))
L = sp.integrate(speed, (t, 0, 1))
sp.pprint(L)
```

This yields an exact expression for arc length:

$$\frac{1}{4}\left((1+4)^{3/2} - 1\right) = \frac{1}{4}(5^{3/2} - 1).$$

Example 3.3.16 (Integrating Temperature Along a Wire) Let $f(x, y) = x^2 + y^2$, and path $\gamma(t) = (\cos t, \sin t)$, $t \in [0, \pi/2]$. Then $f(\gamma(t)) = 1$, so:

$$\int_\gamma f\, ds = \int_0^{\pi/2} 1 \cdot \|\gamma'(t)\|\, dt = \int_0^{\pi/2} 1\, dt = \frac{\pi}{2}.$$

```
y = [sp.cos(t), sp.sin(t)]
f = y[0]**2 + y[1]**2
speed = sp.sqrt(sum([sp.diff(c, t)**2 for c in y]))
I = sp.integrate(f * speed, (t, 0, sp.pi/2))
sp.pprint(I)
```

Vector Line Integrals

Given a vector field $\vec{F} = (P, Q)$ and a path $\gamma(t) = (x(t), y(t))$, the line integral along the path is:

$$\int_\gamma \vec{F} \cdot d\vec{r} = \int_a^b \left[P(x(t), y(t))\, x'(t) + Q(x(t), y(t))\, y'(t)\right] dt.$$

Example 3.3.17 (Work Done by a Force Field) Let $\vec{F} = (x^2, y)$, and path $\gamma(t) = (t, t^2)$, $t \in [0, 1]$. Then:

$$\int_\gamma \vec{F} \cdot d\vec{r} = \int_0^1 \left[t^2 \cdot 1 + t^2 \cdot 2t\right] dt = \int_0^1 t^2 + 2t^3\, dt.$$

3.3 Multivariable Calculus

```
x_t, y_t = t, t**2
F = [x_t**2, y_t]
dx = sp.diff(x_t, t)
dy = sp.diff(y_t, t)
integrand = F[0]*dx + F[1]*dy
I = sp.integrate(integrand, (t, 0, 1))
sp.pprint(I)
```

This evaluates to $\frac{1}{3} + \frac{1}{2} = \frac{5}{6}$.

Example 3.3.18 (Closed Loop Integral of a Conservative Field) Let $\vec{F} = \nabla f$, where $f(x, y) = x^2 + y^2$, and path is the unit circle. Then $\oint \vec{F} \cdot d\vec{r} = 0$.

```
x_t, y_t = sp.cos(t), sp.sin(t)
F = [2*x_t, 2*y_t]
dx = sp.diff(x_t, t)
dy = sp.diff(y_t, t)
integrand = F[0]*dx + F[1]*dy
I = sp.integrate(integrand, (t, 0, 2*sp.pi))
sp.pprint(I)
```

The integral evaluates to 0 due to conservative symmetry.

Flux Through Surfaces

Given a vector field $\vec{F}(x, y, z)$, and a surface S with normal vector $\vec{n}$, the surface integral is:

$$\iint_S \vec{F} \cdot d\vec{S} = \iint_D \vec{F} \cdot \vec{n} \, dA.$$

For a surface $z = g(x, y)$, we have:

$$d\vec{S} = \left(-\frac{\partial g}{\partial x}, -\frac{\partial g}{\partial y}, 1\right) dx\, dy.$$

Example 3.3.19 (Upward Flux Through a Paraboloid Cap) Let $\vec{F}(x, y, z) = (0, 0, z)$, and surface S be the cap $z = 4 - x^2 - y^2$ over disk $x^2 + y^2 \leq 4$. Then:

$$\iint_S \vec{F} \cdot \vec{n} \, dS = \iint_D (4 - x^2 - y^2)\, dx\, dy.$$

```
r, θ = sp.symbols("r θ")
z = 4 - r**2
jacobian = r
f = z * jacobian
flux = sp.integrate(sp.integrate(f, (r, 0, 2)), (θ, 0, 2*sp.pi))
sp.pprint(flux)
```

Evaluate:

$$\int_0^{2\pi} \int_0^2 (4 - r^2) r \, dr \, d\theta = 8\pi.$$

Example 3.3.20 (Electric Flux Through a Plane) Let $\vec{F}(x, y, z) = (x, y, z)$, and surface S be the unit square in the xy-plane, i.e., $z = 0, 0 \leq x, y \leq 1$.
Then:

$$\vec{n} = (0, 0, 1), \quad \vec{F} \cdot \vec{n} = z = 0 \Rightarrow \text{flux} = 0.$$

Now shift the surface to $z = 3$:

$$\vec{F} \cdot \vec{n} = 3 \Rightarrow \iint_S 3 \, dx \, dy = 3.$$

```
x, y = sp.symbols("x y")
flux = sp.integrate(sp.integrate(3, (x, 0, 1)), (y, 0, 1))
sp.pprint(flux)
```

Example 3.3.21 (Flux of Rotating Field Through Cylinder) Let $\vec{F} = (-y, x, 0)$ and surface be the side of cylinder $x^2 + y^2 = 1, 0 \leq z \leq 1$.
Parameterise:

$$\vec{r}(\theta, z) = (\cos\theta, \sin\theta, z), \quad \vec{n} = (\cos\theta, \sin\theta, 0), \quad dS = dz \, d\theta.$$

Then:

$$\vec{F} \cdot \vec{n} = -\sin\theta \cos\theta + \cos\theta \sin\theta = 0.$$

Flux is zero due to orthogonality of field and surface normals.

```
θ, z = sp.symbols("θ z")
Fx, Fy = -sp.sin(θ), sp.cos(θ)
n_dot_F = Fx*sp.cos(θ) + Fy*sp.sin(θ)
flux = sp.integrate(sp.integrate(n_dot_F, (z, 0, 1)), (θ, 0, 2*sp.pi))
sp.pprint(flux)
```

3.4 Advanced Topics in Calculus

Many phenomena in physics, biology, and economics are governed not by algebraic relations but by differential laws: relationships between functions and their derivatives. Differential equations express how a quantity evolves with respect to change in another (usually time or space). We explore analytic and numerical methods for ordinary differential equations (ODEs) using SymPy and SciPy.

3.4.1 Differential Equations

First-Order ODEs

A **first-order ODE** has the form

$$\frac{dy}{dx} = f(x, y),$$

with initial condition $y(x_0) = y_0$. Analytic techniques include separation of variables, integrating factors, and exact equations.

Example 3.4.1 (Separable Equation) Solve

$$\frac{dy}{dx} = xy, \quad y(0) = 1.$$

Separate variables:

$$\frac{dy}{y} = x\, dx \Rightarrow \ln|y| = \frac{x^2}{2} + C.$$

```
import sympy as sp
x = sp.Symbol("x")
y = sp.Function("y")(x)
ode = sp.Eq(y.diff(x), x*y)
sol = sp.dsolve(ode, y, ics={y.subs(x, 0): 1})
sp.pprint(sol)
```

The solution is $y(x) = e^{x^2/2}$.

Example 3.4.2 (Integrating Factor Method) Solve

$$\frac{dy}{dx} + y = e^x, \quad y(0) = 0.$$

Multiply by integrating factor e^x to obtain:

$$\frac{d}{dx}(ye^x) = e^{2x} \Rightarrow y = e^x - 1.$$

```
ode = sp.Eq(y.diff(x) + y, sp.exp(x))
sol = sp.dsolve(ode, y, ics={y.subs(x, 0): 0})
sp.pprint(sol)
```

Higher-Order ODEs and Systems

Second-order ODEs model systems with acceleration, curvature, or memory. General form:
$$a(x)y'' + b(x)y' + c(x)y = g(x).$$

Example 3.4.3 (Homogeneous Equation with Constant Coefficients) Solve
$$y'' - 3y' + 2y = 0, \quad y(0) = 1, \; y'(0) = 0.$$

```
y = sp.Function("y")(x)
ode = sp.Eq(y.diff(x,2) - 3*y.diff(x) + 2*y, 0)
sol = sp.dsolve(ode, y, ics={y.subs(x, 0): 1, y.diff(x).subs(x, 0): 0})
sp.pprint(sol)
```

The solution is:
$$y(x) = e^x + e^{2x}(-1).$$

Example 3.4.4 (Forced Harmonic Oscillator)
$$y'' + 4y = \cos(2x), \quad y(0) = 0, \; y'(0) = 0.$$

This has resonance, and the particular solution grows linearly:
$$y(x) = \frac{x}{4}\sin(2x).$$

```
ode = sp.Eq(y.diff(x,2) + 4*y, sp.cos(2*x))
sol = sp.dsolve(ode, y, ics={y.subs(x, 0): 0, y.diff(x).subs(x, 0): 0})
sp.pprint(sol)
```

Systems of ODEs Let
$$\frac{dx}{dt} = x + 2y, \quad \frac{dy}{dt} = 3x + 4y.$$

```
t = sp.Symbol("t")
x = sp.Function("x")(t)
y = sp.Function("y")(t)

eq1 = sp.Eq(x.diff(t), x + 2*y)
eq2 = sp.Eq(y.diff(t), 3*x + 4*y)
sol = sp.dsolve([eq1, eq2])
sp.pprint(sol)
```

This solves the coupled system using matrix exponentials internally.

3.4 Advanced Topics in Calculus

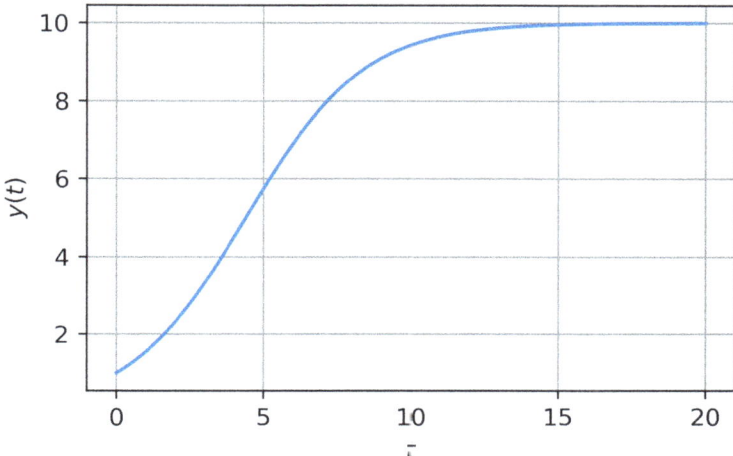

Fig. 3.2 Numerical solution of the logistic growth equation $\dot{y} = 0.5\,y\left(1 - \frac{y}{10}\right)$ with initial value $y(0) = 1$. The trajectory exhibits an initial near-exponential rise followed by a smooth saturation at the carrying capacity $K = 10$, illustrating the sigmoidal nature of logistic dynamics

Numerical Solvers in SciPy

For many nonlinear or stiff ODEs, symbolic solutions are unavailable. scipy.integrate.solve_ivp numerically solves initial value problems using Runge–Kutta or BDF methods (cf. Fig. 3.2).

Example 3.4.5 (Logistic Equation Numerically) Solve

$$\frac{dy}{dt} = ry(1 - y/K), \quad y(0) = 1, \; r = 0.5, \; K = 10.$$

```
from scipy.integrate import solve_ivp
import numpy as np
import matplotlib.pyplot as plt

def logistic(t, y, r=0.5, K=10):
    return r * y * (1 - y / K)

sol = solve_ivp(logistic, [0, 20], [1], t_eval=np.linspace(0, 20, 200))
plt.plot(sol.t, sol.y[0])
plt.xlabel("t")
plt.ylabel("y(t)")
plt.title("Logistic Growth")
plt.grid()
plt.show()
```

This plots the characteristic S-shaped growth towards carrying capacity $K = 10$.

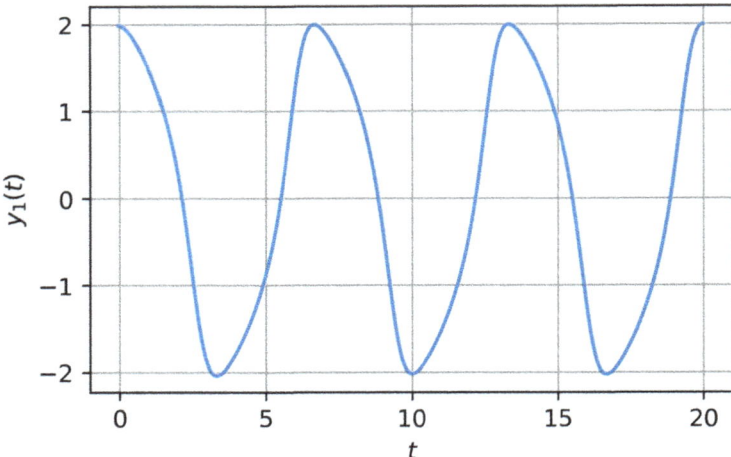

Fig. 3.3 Time evolution of the Van der Pol oscillator $\ddot{y} - \mu(1 - y^2)\dot{y} + y = 0$ with $\mu = 1$ and initial state $(y_1(0), y_2(0)) = (2, 0)$. After a transient large-amplitude phase, the trajectory settles into a self-sustained relaxation oscillation, a hallmark of nonlinear limit-cycle behaviour

Example 3.4.6 (Van der Pol Oscillator (van der Pol 1926) (cf. Fig. 3.3)) Solve:

$$y'' - \mu(1 - y^2)y' + y = 0, \quad \mu = 1.$$

Convert to system:

$$\dot{y}_1 = y_2, \quad \dot{y}_2 = \mu(1 - y_1^2)y_2 - y_1.$$

```
def vdp(t, y, μ=1.0):
    y1, y2 = y
    return [y2, μ*(1 - y1**2)*y2 - y1]

sol = solve_ivp(vdp, [0, 20], [2, 0], t_eval=np.linspace(0, 20, 1000))
plt.plot(sol.t, sol.y[0])
plt.title("Van der Pol Oscillator")
plt.grid()
plt.show()
```

The solution oscillates non-sinusoidally, reflecting nonlinear damping.

Example 3.4.7 (Stiff ODE: Robertson Problem (cf. Fig. 3.4))

$$\begin{cases} y_1' = -0.04y_1 + 10^4 y_2 y_3 \\ y_2' = 0.04y_1 - 10^4 y_2 y_3 - 3 \times 10^7 y_2^2 \\ y_3' = 3 \times 10^7 y_2^2 \end{cases}$$

This is stiff; use the BDF method.

3.4 Advanced Topics in Calculus

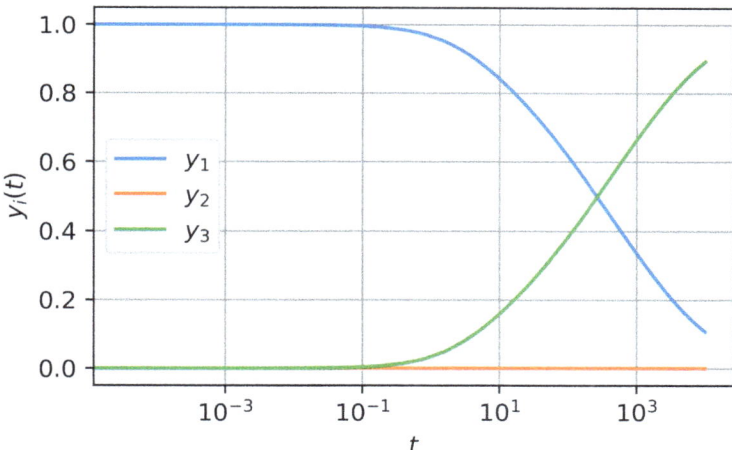

Fig. 3.4 Numerical integration of the stiff Robertson system using a BDF scheme, displayed on a logarithmic time axis. The fast depletion of y_1 and subsequent slow evolution dominated by y_3 illustrate the pronounced multi-time-scale dynamics characteristic of stiff chemical-kinetic ODEs

```
def robertson(t, y):
    y1, y2, y3 = y
    return [-0.04*y1 + 1e4*y2*y3,
            0.04*y1 - 1e4*y2*y3 - 3e7*y2**2,
            3e7*y2**2]

sol = solve_ivp(robertson, [0, 1e4], [1, 0, 0], method='BDF')
plt.semilogx(sol.t, sol.y[0], label='y1')
plt.semilogx(sol.t, sol.y[1], label='y2')
plt.semilogx(sol.t, sol.y[2], label='y3')
plt.legend(), plt.grid(), plt.title("Robertson Problem")
plt.show()
```

3.4.2 Calculus of Variations

The **calculus of variations** concerns the optimisation of functionals—mappings from functions to real numbers. A typical variational problem seeks the function $y(x)$ that minimises an integral:

$$\mathcal{J}[y] = \int_a^b L(x, y(x), y'(x)) \, dx,$$

subject to boundary conditions $y(a) = y_a$, $y(b) = y_b$. This arises in physics (least action), geometry (geodesics), and engineering (optimal control).

Euler–Lagrange Formalism

The critical points of $\mathcal{J}[y]$ satisfy the **Euler–Lagrange equation**:

$$\frac{\partial L}{\partial y} - \frac{d}{dx}\left(\frac{\partial L}{\partial y'}\right) = 0.$$

Example 3.4.8 (Shortest Path: Geodesic in the Plane) Let $y(x)$ connect two points (x_0, y_0), (x_1, y_1). The arc length functional is:

$$\mathcal{J}[y] = \int_{x_0}^{x_1} \sqrt{1 + y'(x)^2}\, dx.$$

Here $L(y, y') = \sqrt{1 + y'^2}$, so:

$$\frac{d}{dx}\left(\frac{y'}{\sqrt{1 + y'^2}}\right) = 0 \Rightarrow \frac{y'}{\sqrt{1 + y'^2}} = C.$$

Solving yields $y' = \text{const} \Rightarrow y(x) = mx + c$—a straight line.

```python
import sympy as sp
x = sp.Symbol("x")
y = sp.Function("y")(x)
L = sp.sqrt(1 + sp.diff(y, x)**2)
EL = sp.diff(sp.diff(L, sp.diff(y, x)), x) - sp.diff(L, y)
sp.pprint(sp.simplify(EL))  # Returns 0
```

Example 3.4.9 (Brachistochrone Curve) Minimise time of descent under gravity between points. The functional is:

$$\mathcal{J}[y] = \int_{x_0}^{x_1} \sqrt{\frac{1 + y'^2}{2gy}}\, dx.$$

This yields a cycloidal path. Deriving Euler–Lagrange leads to a nonlinear ODE solved via parametric substitution.

Due to complexity, this is often approached numerically or via known solution forms.

Constraints and Lagrange Multipliers

In variational problems with integral constraints:

$$\text{Minimise } \mathcal{J}[y] = \int_a^b L(x, y, y')\, dx, \quad \text{subject to } \int_a^b g(x, y)\, dx = C,$$

3.4 Advanced Topics in Calculus

one introduces a *Lagrange multiplier* λ and forms:

$$\tilde{L} = L(x, y, y') + \lambda g(x, y).$$

Then apply the Euler–Lagrange equation to $\tilde{L}$.

Example 3.4.10 (Isoperimetric Problem: Enclosed Area Maximisation) Among all curves of fixed perimeter enclosing a fixed area, which one maximises area? Answer: the circle.

This requires maximising:

$$\mathcal{J}[y] = \int y\, dx \quad \text{subject to} \quad \int \sqrt{1 + y'^2}\, dx = \text{const.}$$

Set up:

$$\tilde{L} = y + \lambda\sqrt{1 - y'^2}.$$

Apply Euler–Lagrange:

$$\frac{d}{dx}\left(\frac{\lambda y'}{\sqrt{1 + y'^2}}\right) = 1.$$

Example 3.4.11 (Constrained Extremal Temperature Profile) Find $y(x)$ that minimises:

$$\mathcal{J}[y] = \int_0^1 (y')^2\, dx, \quad \text{with} \quad \int_0^1 y(x)\, dx = 1, \ y(0) = y(1) = 0.$$

Form augmented functional:

$$\tilde{L} = (y')^2 + \lambda y.$$

Euler–Lagrange:

$$\frac{d}{dx}(2y') = \lambda \Rightarrow y'' = \lambda/2.$$

```
λ = sp.Symbol("λ")
y = sp.Function("y")(x)
L = sp.diff(y, x)**2 + λ * y
EL = sp.diff(sp.diff(L, sp.diff(y, x)), x) - sp.diff(L, y)
sp.pprint(EL)
```

Solve:
$$y(x) = \frac{\lambda}{4}x(1-x).$$

Apply the integral constraint to find λ.

Applications in Physics

Lagrangian Mechanics The path of a physical system minimises the action:
$$\mathcal{S}[q] = \int_{t_0}^{t_1} L(q, \dot{q}, t)\, dt, \quad L = T - V.$$

Euler–Lagrange equations yield Newton's laws.

Example 3.4.12 (Simple Pendulum) Let $\theta(t)$ be angular displacement. Lagrangian:
$$L = \frac{1}{2} m\ell^2 \dot\theta^2 - mg\ell(1-\cos\theta).$$

Then:
$$\frac{d}{dt}(m\ell^2 \dot\theta) + mg\ell\sin\theta = 0 \Rightarrow \theta'' + \frac{g}{\ell}\sin\theta = 0.$$

```
θ = sp.Function("θ")(x)
g, l, m = sp.symbols("g l m")
L = (1/2)*m*l**2 * sp.diff(θ, x)**2 - m*g*l*(1 - sp.cos(θ))
EL = sp.diff(sp.diff(L, sp.diff(θ, x)), x) - sp.diff(L, θ)
sp.pprint(sp.simplify(EL))
```

Optics: Fermat's Principle Light travels along a path that extremises optical path length:
$$\mathcal{J}[y] = \int n(x, y)\sqrt{1 + y'^2}\, dx.$$

In homogeneous media $n = \text{const}$, path is straight. In inhomogeneous media, calculus of variations leads to Snell's law and bending of rays.

Electrostatics: Minimal Energy Configuration The potential u minimises:
$$\int |\nabla u|^2\, dx \quad \Rightarrow \quad \Delta u = 0.$$

The Euler–Lagrange equation recovers Laplace's equation—fundamental to potential theory.

3.4.3 Tensor Calculus

Tensor calculus generalises vector calculus to multilinear functions and curved spaces. It underlies differential geometry, continuum mechanics, and Einstein's theory of general relativity. Tensors extend the concept of scalars (rank-0), vectors (rank-1), and matrices (rank-2) to arbitrary rank (r, s), representing multilinear maps with r contravariant (upper) and s covariant (lower) indices. Python, via SymPy's tensor module and NumPy for numerical index manipulation, enables symbolic and numerical tensor computation.

Index Notation and Einstein Summation

In *index notation*, a tensor equation is written using indices to denote components, e.g.:

$$A^i_j, \quad T^{\mu\nu}_\lambda.$$

The **Einstein summation convention** states: *sum over any repeated index appearing once up and once down*:

$$a^i b_i = \sum_i a^i b_i.$$

Example 3.4.13 (Dot Product Using Einstein Summation) Let $a^i = (1, 2, 3)$, $b_i = (4, 5, 6)$. Then $a^i b_i = 1 \cdot 4 + 2 \cdot 5 + 3 \cdot 6 = 32$.

```
import numpy as np
a = np.array([1, 2, 3])
b = np.array([4, 5, 6])
dot = np.einsum("i,i->", a, b)
print(dot) # Output: 32
```

Example 3.4.14 (Matrix-Vector Multiplication as a Contraction) Let A^i_j be a matrix and v^j a vector:

$$w^i = A^i_j v^j.$$

```
A = np.array([[1,2],[3,4]])
v = np.array([5,6])
w = np.einsum("ij,j->i", A, v)
print(w) # Output: [17 39]
```

Example 3.4.15 (Metric Contraction) Given a metric g_{ij} and contravariant vector v^i, compute covariant $v_j = g_{ij} v^i$.

```
g = np.array([[2, 0], [0, 3]])
v_up = np.array([1, 2])
v_down = np.einsum("ij,j->i", g, v_up)
print(v_down) # Output: [2 6]
```

Covariant and Contravariant Tensors

Contravariant components V^i transform via:

$$V'^i = \frac{\partial x'^i}{\partial x^j} V^j.$$

Covariant components W_i transform as:

$$W'_i = \frac{\partial x^j}{\partial x'^i} W_j.$$

The transformation law ensures tensorial consistency under coordinate change.

Example 3.4.16 (Raising and Lowering Indices) Given g_{ij} and g^{ij} (inverse metric), convert between:

$$v_i = g_{ij} v^j, \quad v^i = g^{ij} v_j.$$

```
g_down = np.array([[2, 1], [1, 2]])
g_up = np.linalg.inv(g_down)
v_up = np.array([3, 4])
v_down = np.einsum("ij,j->i", g_down, v_up)
v_raised = np.einsum("ij,j->i", g_up, v_down)
print(v_raised) # Recovers original [3 4]
```

Example 3.4.17 (Transformation Under Rotation) Rotate a vector v^i under angle θ. Transformation matrix:

$$R^i_j = \begin{pmatrix} \cos\theta & -\sin\theta \\ \sin\theta & \cos\theta \end{pmatrix}$$

```
θ = np.pi / 4
R = np.array([[np.cos(θ), -np.sin(θ)], [np.sin(θ), np.cos(θ)]])
v = np.array([1, 0])
v_rot = np.einsum("ij,j->i", R, v)
print(v_rot) # Output: [0.707..., 0.707...]
```

3.4 Advanced Topics in Calculus

General Relativity Applications

In general relativity, the gravitational field is encoded in the **metric tensor** $g_{\mu\nu}$, and spacetime geometry replaces force. The dynamics of a free particle follow the **geodesic equation**:

$$\frac{d^2 x^\mu}{d\tau^2} + \Gamma^\mu_{\alpha\beta} \frac{dx^\alpha}{d\tau} \frac{dx^\beta}{d\tau} = 0.$$

Here $\Gamma^\mu_{\alpha\beta}$ are the **Christoffel symbols**, defined as:

$$\Gamma^\mu_{\alpha\beta} = \frac{1}{2} g^{\mu\nu} \left(\partial_\alpha g_{\nu\beta} + \partial_\beta g_{\nu\alpha} - \partial_\nu g_{\alpha\beta} \right).$$

Example 3.4.18 (Compute Christoffel Symbols in 2D) Let:

$$ds^2 = dx^2 + (1+x^2)dy^2 \Rightarrow g_{ij} = \begin{pmatrix} 1 & 0 \\ 0 & 1+x^2 \end{pmatrix}.$$

```
x, y = sp.symbols("x y")
g = sp.Matrix([[1, 0], [0, 1 + x**2]])
g_inv = g.inv()
n = 2
τ = [[[0]*n for _ in range(n)] for _ in range(n)]
for μ in range(n):
    for α in range(n):
        for β in range(n):
            sum_ = 0
            for ν in range(n):
                sum_ += g_inv[μ,ν] * (
                    sp.diff(g[ν,β], x if α == 0 else y)
                    + sp.diff(g[ν,α], x if β == 0 else y)
                    - sp.diff(g[α,β], x if ν == 0 else y))
            τ[μ][α][β] = sp.simplify(1/2 * sum_)
sp.pprint(τ[1][0][1]) # Example: τ^1_{01}
```

This yields:

$$\Gamma^1_{01} = \frac{x}{1+x^2}.$$

Example 3.4.19 (Geodesic in Polar Coordinates) Metric:

$$ds^2 = dr^2 + r^2 d\theta^2, \quad g_{ij} = \begin{pmatrix} 1 & 0 \\ 0 & r^2 \end{pmatrix}.$$

Christoffel symbols:

$$\Gamma^r_{\theta\theta} = -r, \quad \Gamma^\theta_{r\theta} = \frac{1}{r}.$$

The geodesic equations are:

$$\ddot{r} - r\dot{\theta}^2 = 0, \quad \ddot{\theta} + \frac{2}{r}\dot{r}\dot{\theta} = 0.$$

```
r, θ = sp.symbols("r θ")
g = sp.Matrix([[1, 0], [0, r**2]])
g_inv = g.inv()
# Compute Christoffel symbols manually
τ_r_tt = -r
τ_θ_rt = 1/r
```

Example 3.4.20 (Ricci Scalar Curvature) For a 2D Riemannian manifold, compute the Ricci scalar R. In constant curvature surfaces:

$$R = \text{constant} \quad \Rightarrow \text{sphere: } R > 0, \text{ hyperbolic plane: } R < 0.$$

Symbolic computation requires evaluating Christoffel symbols and Riemann tensors.

Due to complexity, dedicated libraries (e.g. xAct in Mathematica) or symbolic geometry packages may be more appropriate.

3.4.4 Fractional Calculus

Fractional calculus generalises the notion of derivatives and integrals to non-integer (fractional) orders. The n-th derivative is defined via repeated application of differentiation, but what does it mean to take the derivative of order $\alpha \in \mathbb{R}$, or even $\alpha \in \mathbb{C}$? Fractional derivatives and integrals find applications in viscoelasticity, anomalous diffusion, signal processing, and control systems. Python offers tools for symbolic and numerical work in this area via packages like sympy, mpmath, and fracdiff.

Fractional Derivatives

One popular definition is the **Riemann–Liouville fractional derivative**:

$$D^\alpha f(x) = \frac{1}{\Gamma(n-\alpha)} \frac{d^n}{dx^n} \int_a^x \frac{f(t)}{(x-t)^{\alpha-n+1}} \, dt, \quad n-1 < \alpha < n.$$

3.4 Advanced Topics in Calculus

Another is the **Caputo derivative**, which is often preferred in physical modelling due to well-defined initial conditions:

$$^C D^\alpha f(x) = \frac{1}{\Gamma(n-\alpha)} \int_a^x \frac{f^{(n)}(t)}{(x-t)^{\alpha-n+1}} dt.$$

Example 3.4.21 (Fractional Derivative of a Power Function) Let $f(x) = x^k$. Then:

$$D^\alpha x^k = \frac{\Gamma(k+1)}{\Gamma(k-\alpha+1)} x^{k-\alpha}.$$

```
import sympy as sp
x, α, k = sp.symbols("x α k", positive=True)
f = x**k
Dα_f = sp.gamma(k+1) / sp.gamma(k - α + 1) * x**(k - α)
sp.pprint(Dα_f)
```

This expression reduces to standard derivatives when $\alpha \in \mathbb{N}$.

Example 3.4.22 (Numeric Approximation of Caputo Derivative) Let $f(x) = \sin(x)$, approximate its Caputo derivative at $x = 1$ of order $\alpha = 0.5$.

mpmath has support for fractional differintegrals.

```
from mpmath import *
mp.dps = 15

def caputo_half_deriv(f, x, a=0, N=100):
    h = (x - a) / N
    t_vals = [a + i*h for i in range(N)]
    f_prime = [diff(f, t) for t in t_vals]
    weights = [(x - t)**(-0.5) for t in t_vals]
    return (1/gamma(0.5)) * h * sum(w*w_ for w,w_ in zip(weights, f_prime))

result = caputo_half_deriv(lambda t: sin(t), 1.0)
print(result)
```

Fractional Integrals

The **Riemann–Liouville fractional integral** of order $\alpha > 0$ is defined as:

$$I^\alpha f(x) = \frac{1}{\Gamma(\alpha)} \int_a^x (x-t)^{\alpha-1} f(t) \, dt.$$

Example 3.4.23 (Fractional Integral of $f(t) = 1$)

$$I^\alpha 1 = \frac{1}{\Gamma(\alpha)} \int_0^x (x-t)^{\alpha-1} dt = \frac{x^\alpha}{\Gamma(\alpha+1)}.$$

```
α, x = sp.symbols("α x", positive=True)
Iα_1 = x**α / sp.gamma(α + 1)
sp.pprint(Iα_1)
```

This generalises the integral $\int_0^x dt = x$ to non-integer α.

Example 3.4.24 (Numerical Approximation Using Quadrature) Let $f(t) = t$, compute $I^{0.5} f(x)$ at $x = 1$:

$$I^{0.5}t = \frac{1}{\Gamma(0.5)} \int_0^1 (1-t)^{-0.5} t \, dt.$$

```
from scipy.integrate import quad
import numpy as np
from scipy.special import gamma

f = lambda t: t * (1 - t)**(-0.5)
I, _ = quad(f, 0, 1)
print(I / gamma(0.5))  # ≈ 1.128
```

Modelling with Fractional Dynamics

Fractional derivatives allow us to model **memory effects** and **non-local dynamics**, commonly found in biological systems, viscoelastic materials, and anomalous diffusion.

Fractional Relaxation Equation

$$D^\alpha y(t) = -\lambda y(t), \quad 0 < \alpha < 1.$$

Solution: $y(t) = E_\alpha(-\lambda t^\alpha)$, where E_α is the Mittag–Leffler function (cf. Fig. 3.5).

```
from mpmath import *
mp.dps = 15

λ = 1.0
α = 0.5
t = mp.linspace(0, 10, 100)
y = [mittagleich(α, -λ * ti**α) for ti in t]

import matplotlib.pyplot as plt
plt.plot(t, y)
plt.title("Fractional Relaxation: D^0.5 y = -y")
plt.xlabel("t")
plt.ylabel("y(t)")
plt.grid()
plt.show()
```

3.4 Advanced Topics in Calculus

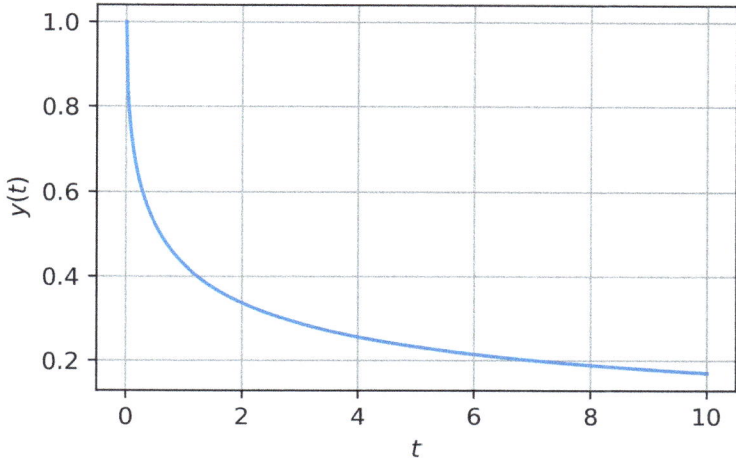

Fig. 3.5 Solution $y(t) = E_{0.5}(-t^{0.5})$ of the Caputo fractional relaxation equation $D^{0.5}y(t) = -y(t)$ with $\lambda = 1$. The Mittag–Leffler decay displays a slow power-law tail, highlighting the long-memory effects characteristic of fractional-order dynamics

Anomalous Diffusion Equation

$$\frac{\partial^\alpha u}{\partial t^\alpha} = D \frac{\partial^2 u}{\partial x^2}, \quad 0 < \alpha \leq 1.$$

This models subdiffusive behavior in heterogeneous media.

Viscoelasticity: Fractional Maxwell Model Stress-strain relation:

$$\sigma(t) + \tau^\alpha D^\alpha \sigma(t) = E D^\alpha \varepsilon(t),$$

where τ, E are material constants, and D^α models hereditary memory.

Example 3.4.25 (Fractional Response to a Step Input) Use Laplace transforms and symbolic calculus to express the stress response in time domain:

$$\sigma(s) = \frac{E}{1 + \tau^\alpha s^\alpha} \cdot \frac{1}{s}.$$

The inverse Laplace involves a Mittag–Leffler function.

```
s = sp.Symbol("s")
E, τ, α = sp.symbols("E τ α")
σ_s = (E / (1 + τ**α * s**α)) * (1 / s)
# Inverse Laplace transform is symbolic; numerical inversion preferred
```

3.5 Exercises

1. Let $f(x) = \sin(4x)$ on $[0, \pi]$. For the node $x_0 = \frac{\pi}{6}$ compute the first derivative using

$$D_c^{(h)} f(x_0) = \frac{f(x_0 + h) - f(x_0 - h)}{2h}, \qquad h = \frac{\pi}{2^k}, \ k = 4, 5, \ldots, 10.$$

Tabulate the absolute error $|D_c^{(h)} f(x_0) - f'(x_0)|$ and verify the second-order rate by fitting a line to $\log_2$-scaled data. Confirm your slope numerically in Python and explain any deviation for $k \geq 9$ with reference to round-off error.

2. Define

$$\mathbf{F} : \mathbb{R}^3 \to \mathbb{R}^3, \qquad \mathbf{F}(\mathbf{x}) = \begin{pmatrix} e^{x_1} + x_2 \sin x_3 \\ x_1 x_3^2 - \sqrt{1 + x_2^2} \\ \ln(1 + x_1 x_2) + x_3 \end{pmatrix}.$$

Using autograd (reverse mode) compute the Jacobian $J\mathbf{F}(\mathbf{a})$ and, for the scalar functional $\Phi(\mathbf{x}) = \|\mathbf{F}(\mathbf{x})\|^2$, compute the Hessian $\nabla^2 \Phi(\mathbf{a})$ at $\mathbf{a} = (0.2, -0.3, 1.1)^\mathsf{T}$. Provide both numerical values and the line of code that generates them.

3. Evaluate symbolically

$$I = \int_0^\infty \frac{\arctan bx - \arctan ax}{x(e^{2\pi x} - 1)} \, dx, \qquad 0 < a < b,$$

and show that it reduces to $\frac{1}{4} \left(\log \Gamma\left(\frac{1}{2} + \frac{b}{\pi}\right) - \log \Gamma\left(\frac{1}{2} + \frac{a}{\pi}\right) \right)$. Verify the result numerically in mpmath for $a = 1$, $b = 2$ to 10^{-10} relative accuracy.

4. Evaluate

$$\iint_D (x^2 + y^2)^{\frac{3}{2}} \, dx \, dy, \qquad D = \{(x, y) : 0 \leq y \leq \sqrt{1 - x^2}, \ 0 \leq x \leq 1\}.$$

(a) Compute directly in Cartesian coordinates. (b) Transform to polar coordinates and show the integrals agree. (c) Implement both in SymPy and report run-times.

5. Minimise

$$f(x, y, z) = x^2 + y^2 + z^2 - 6x - 8y - 10z$$

subject to the plane constraint $x + 2y + 3z = 12$. (i) Use Lagrange multipliers to find the critical point. (ii) Verify that the bordered Hessian is positive definite at

3.5 Exercises

this point. (iii) Confirm numerically with `scipy.optimize.minimize` using an equality constraint.

6. Let $\vec{F} = (y^2, x^2)$ and let C be the positively oriented triangle with vertices $(0, 0)$, $(2, 0)$, $(0, 2)$. (a) Compute $\oint_C \vec{F} \cdot d\vec{r}$ directly. (b) Compute the double integral of $(\partial Q/\partial x - \partial P/\partial y)$ over the triangular region and verify Green's theorem numerically.

7. Given $\vec{F}(x, y, z) = (yz, xz, xy)$, compute $\iint_S \vec{F} \cdot d\vec{S}$ where S is the closed surface of the cuboid $0 \leq x \leq 2, 0 \leq y \leq 1, 0 \leq z \leq 3$. (a) Evaluate by the divergence theorem. (b) Compute each of the six face integrals explicitly and show they sum to the same value.

8. Solve the system

$$\begin{cases} y_1' = -10^3(y_1 - y_2), \\ y_2' = y_1 - y_2, \end{cases} \quad y_1(0) = 1, \ y_2(0) = 0,$$

up to $t = 10^{-2}$. (i) Obtain the exact solution analytically. (ii) Solve numerically using `solve_ivp` with both RK45 and BDF. (iii) Tabulate the global error at $t = 10^{-2}$ for each solver and explain the difference.

9. Find $y(x)$ on $[0, 1]$ minimising

$$\mathcal{J}[y] = \int_0^1 (y'(x))^2 \, dx,$$

subject to $y(0) = 0$, $y(1) = 0$ and the isoperimetric constraint $\int_0^1 x\, y(x)\, dx = 1$. Derive the Euler–Lagrange equation with Lagrange multiplier, solve for $y(x)$, and compute the multiplier's value.

10. On the unit sphere in spherical coordinates (θ, ϕ), metric: $ds^2 = d\theta^2 + \sin^2\theta\, d\phi^2$. (a) Compute all non-zero Γ^i_{jk}. (b) Derive the geodesic equations. (c) Show that great circles satisfy these equations. (d) Integrate numerically a geodesic with initial point $(\theta_0 = 1, \phi_0 = 0)$ and initial velocity $(\dot\theta_0 = 0, \dot\phi_0 = 1)$; plot the trajectory in Cartesian coordinates.

11. Consider ${}^C\!D_t^{0.3} y(t) = -2y(t)$ with $y(0) = 1$. (i) Express $y(t)$ in terms of the Mittag–Leffler function. (ii) Approximate $y(t)$ for $t \in [0, 5]$ using a Grünwald–Letnikov finite difference with step $h = 0.01$. (iii) Plot both analytic and numerical solutions and compute the maximum absolute error.

12. Given the Cauchy stress tensor in the xyz–frame

$$\sigma_{ij} = \begin{pmatrix} 100 & 30 & 20 \\ 30 & 80 & 10 \\ 20 & 10 & 60 \end{pmatrix} \text{MPa},$$

and a rotation about the z–axis by $\theta = 45°$, compute the transformed tensor $\sigma'_{ij} = R_i^k R_j^\ell \sigma_{k\ell}$. Identify the principal stresses in the rotated frame.

13. Evaluate
$$\iiint_V (x^2 + y^2 + z^2)^{\frac{1}{2}} \, dV,$$
where V is the solid bounded by $\rho = 2 + \cos\theta$ in spherical coordinates (ρ, θ, ϕ). Express your answer in closed form and confirm numerically via Monte Carlo sampling (10^6 points).

14. Let $\vec{F} = (y, -x, z)$ and let S be the upper hemisphere $x^2 + y^2 + z^2 = 1$, $z \geq 0$.
(a) Compute $\iint_S (\nabla \times \vec{F}) \cdot d\vec{S}$ directly. (b) Compute $\oint_{\partial S} \vec{F} \cdot d\vec{r}$ where ∂S is the unit circle in the xy-plane, and verify Stokes' theorem.

Chapter 4
Data Structures and Algorithms with Python

Abstract Computer-science foundations meet mathematical rigour as asymptotic analysis guides the design of stacks, queues, linked lists, trees, and a deep dive into the hash-table machinery that powers Python's `dict`. Classic search and sorting algorithms are implemented and benchmarked, while graph theory introduces BFS, DFS, and Dijkstra's shortest-path—laying algorithmic cornerstones for later numerical and data-science pipelines.

Keywords Data structures · Algorithms · Hash tables · Search and sorting · Graph algorithms · Complexity analysis

Modern mathematical computation depends critically on efficient representation and manipulation of data. **Data structures** provide a formal framework to store, organise, and retrieve data in a way that aligns with algorithmic goals. Whether we're solving linear systems, implementing symbolic manipulation, or analysing graphs, the choice of data structure directly impacts the computational complexity and memory usage. This chapter introduces the foundational data structures, explores their mathematical and algorithmic underpinnings, and demonstrates their implementation and usage in Python.

4.1 Introduction to Data Structures

4.1.1 Overview of Data Structures

A **data structure** is a mathematical or logical model for organising information. The primary aim is to optimise one or more of the following: insertion, deletion, lookup, traversal, or transformation. Underlying each data structure is an abstract data type (ADT), such as a set, list, or mapping.

Role of Data Structures in Computational Efficiency

The same algorithm can exhibit dramatically different time and space complexities depending on the data structures employed.

Example 4.1.1 (Linear Search vs. Binary Search) Let $A \subset \mathbb{Z}$ be a list of n integers.

Case 1 A is unordered. To test membership:

```
x in A # O(n) in worst case
```

Case 2 A is a sorted list:

```
import bisect
bisect.bisect_left(A, x) # O(log n)
```

This improvement from linear to logarithmic time is made possible by maintaining sorted structure, trading insertion complexity for query speed.

Example 4.1.2 (Set Membership vs. List Membership) Python `set` uses a hash table internally.

```
S = set(range(10**6))
print(999999 in S) # O(1)
```

In contrast:

```
L = list(range(10**6))
print(999999 in L) # O(n)
```

This demonstrates how structure choice alters asymptotic performance.

Example 4.1.3 (Sparse vs. Dense Matrices) Consider solving $Ax = b$ where $A \in \mathbb{R}^{n \times n}$ is sparse.

Using dense storage:

```
import numpy as np
A = np.eye(1000)
```

Memory usage is $O(n^2)$.

Using sparse storage:

```
from scipy.sparse import eye
A_sparse = eye(1000)
```

Here, only non-zero entries are stored, yielding $O(n)$ storage and enabling fast multiplication with vectors.

Types of Data Structures and Their Use Cases

Below is a classification of common data structures with representative use cases:

4.1 Introduction to Data Structures

Linear Structures

- **Lists (arrays):** ordered collections with fast indexing.

  Example 4.1.4 (Polynomial Coefficient Vector) Represent $f(x) = 4x^3 + 3x^2 - 2x + 1$ as:
  ```
  coeffs = [1, -2, 3, 4] # degree-0 to degree-3
  ```

- **Stacks:** Last-in, first-out (LIFO) structure.

  Example 4.1.5 (Balanced Parentheses Checker)
  ```python
  def is_balanced(expr):
      stack = []
      for ch in expr:
          if ch == '(':
              stack.append(ch)
          elif ch == ')':
              if not stack: return False
              stack.pop()
      return not stack
  ```

- **Queues:** First-in, first-out (FIFO) buffer.

  Example 4.1.6 (Breadth-First Traversal of a Graph)
  ```python
  from collections import deque
  Q = deque([start])
  visited = set()
  while Q:
      node = Q.popleft()
      ...
  ```

Hierarchical Structures

- **Trees:** recursive, acyclic structures for representing nested data.

  Example 4.1.7 (Expression Tree for Symbolic Evaluation)

  $$(2+3) \times (4-1)$$

  can be parsed into a binary tree whose leaves are operands and internal nodes are operators.

- **Binary search trees (BSTs):** sorted trees supporting $O(\log n)$ search and insertion.
- **Heaps:** partially ordered trees for priority queues.

  Example 4.1.8 (Maintain Median via Two Heaps) Use a max-heap for lower half, min-heap for upper half to track median in $O(\log n)$ per insertion.

Hash-Based Structures

- **Dictionaries (maps):** key-value pairs with average $O(1)$ access.

    Example 4.1.9 (Frequency Count of Characters)
    ```
    from collections import Counter
    freq = Counter("abracadabra")
    ```

Graph Structures

- **Adjacency list:** stores neighbours for each vertex.
- **Adjacency matrix:** binary matrix A_{ij} indicating edge presence.

    Example 4.1.10 (Represent Undirected Graph with Four Nodes)
    ```
    adj = [[0,1,1,0],
           [1,0,0,1],
           [1,0,0,1],
           [0,1,1,0]]
    ```

Matrix and Tensor Representations

- **2D arrays:** represent linear maps.
- **Sparse matrices:** efficient representation for large systems with few nonzeros.
- **Tensors:** multidimensional arrays for physics, geometry, and deep learning.

Example 4.1.11 (Store 3D Tensor $T_{ijk} \in \mathbb{R}^{5 \times 5 \times 5}$)
```
T = np.zeros((5,5,5))
T[1,2,3] = 42
```

4.1.2 Basic Data Structures

This section introduces three fundamental data structures that form the backbone of algorithmic design: **stacks**, **queues**, and **linked lists**. They differ in the order in which elements are inserted and removed—this impacts algorithm design in parsing, graph traversal, simulation, and dynamic memory allocation.

Stacks

A **stack** is a Last-In, First-Out (LIFO) structure. The most recent element added is the first to be removed.

4.1 Introduction to Data Structures

Operations

- `push(x)`: insert element x on top.
- `pop()`: remove and return top element.
- `peek()`: return top element without removing.
- `is_empty()`: check if stack is empty.

Example 4.1.12 (Check for Balanced Parentheses)

```
def is_balanced(expr):
    stack = []
    for ch in expr:
        if ch == '(':
            stack.append(ch)
        elif ch == ')':
            if not stack:
                return False
            stack.pop()
    return not stack

assert is_balanced("(())()") == True
assert is_balanced("(()") == False
```

This uses the stack to track nested structure—a common parsing technique.

Example 4.1.13 (Reverse a String Using a Stack)

```
def reverse_string(s):
    stack = list(s)
    return ''.join(stack.pop() for _ in s)

assert reverse_string("python") == "nohtyp"
```

Example 4.1.14 (Evaluate Postfix (RPN) Expression) Given "3 4 + 2 *", evaluate to get $(3 + 4) \times 2 = 14$.

```
def eval_postfix(expr):
    stack = []
    for token in expr.split():
        if token.isdigit():
            stack.append(int(token))
        else:
            b, a = stack.pop(), stack.pop()
            stack.append(eval(f"{a}{token}{b}"))
    return stack[0]

assert eval_postfix("3 4 + 2 *") == 14
```

Queues

A **queue** is a First-In, First-Out (FIFO) structure. The first element inserted is the first to be removed.

Operations

- enqueue(x): add *x* to rear.
- dequeue(): remove from front.
- peek(): return front element.
- is_empty(): check emptiness.

Example 4.1.15 (Customer Service Simulation) Simulate a queue of requests:

```python
from collections import deque

Q = deque()
Q.append("Alice")
Q.append("Bob")
Q.append("Carol")
while Q:
    print("Serving:", Q.popleft())
```

This prints: Alice, Bob, Carol—in that order.

Example 4.1.16 (Breadth-First Search (BFS) in a Graph)

```python
def bfs(graph, start):
    from collections import deque
    visited = set()
    Q = deque([start])
    while Q:
        node = Q.popleft()
        if node not in visited:
            visited.add(node)
            Q.extend(graph[node])
    return visited
```

Example 4.1.17 (Round-Robin Scheduling Using a Queue) Each process gets a time-slice in order:

```python
Q = deque(["A", "B", "C"])
for _ in range(6):
    process = Q.popleft()
    print("Running:", process)
    Q.append(process)
```

Linked Lists

A **linked list** consists of nodes where each node contains data and a reference (pointer) to the next node. This allows efficient insertion and deletion without shifting elements.

4.1 Introduction to Data Structures

Types

- **Singly linked list:** each node points to next.
- **Doubly linked list:** nodes point to both previous and next.
- **Circular linked list:** last node points back to head.

Example 4.1.18 (Singly Linked List Class)

```python
class Node:
    def __init__(self, data, next=None):
        self.data, self.next = data, next

class LinkedList:
    def __init__(self):
        self.head = None

    def push_front(self, value):
        self.head = Node(value, self.head)

    def pop_front(self):
        if not self.head: return None
        val, self.head = self.head.data, self.head.next
        return val

    def print_all(self):
        node = self.head
        while node:
            print(node.data, end="→")
            node = node.next
        print("None")
```

Example 4.1.19 (Reverse a Linked List In-Place)

```python
def reverse(head):
    prev, curr = None, head
    while curr:
        nxt = curr.next
        curr.next = prev
        prev, curr = curr, nxt
    return prev
```

Example 4.1.20 (Detect a Cycle Using Floyd's Algorithm)

```python
def has_cycle(head):
    slow = fast = head
    while fast and fast.next:
        slow = slow.next
        fast = fast.next.next
        if slow == fast:
            return True
    return False
```

Example 4.1.21 (Polynomial as Linked List) Store $f(x) = 4x^3 + 3x^2 - 2x + 5$ as linked list nodes: each holds (coefficient, exponent).

```
class Term:
    def __init__(self, coeff, exp, next=None):
        self.coeff, self.exp, self.next = coeff, exp, next

# Manual construction:
t1 = Term(5, 0)
t2 = Term(-2, 1, t1)
t3 = Term(3, 2, t2)
poly = Term(4, 3, t3)

def eval_poly(head, x):
    result = 0
    while head:
        result += head.coeff * (x ** head.exp)
        head = head.next
    return result
```

Trees

A **tree** is a hierarchical data structure that recursively organises data in a branching fashion. Each node may have zero or more child nodes, and the topmost node is called the *root*. Trees are used to represent expression hierarchies, sorted data, syntax trees, and file systems.

Binary Trees

A **binary tree** is a tree in which each node has at most two children: `left` and `right`.

Example 4.1.22 (Recursive In-Order Traversal) Traverse a binary tree by visiting: left, root, right.

```
class Node:
    def __init__(self, val, left=None, right=None):
        self.val, self.left, self.right = val, left, right

def inorder(root):
    if root:
        inorder(root.left)
        print(root.val, end=" ")
        inorder(root.right)

# Construct tree: 2
#      / \
#     1   3
root = Node(2, Node(1), Node(3))
inorder(root)  # Output: 1 2 3
```

4.1 Introduction to Data Structures 109

Example 4.1.23 (Expression Tree for Symbolic Computation) Parse and evaluate the expression $(3 + 4) \times 5$.

```
root = Node("*", Node("+", Node("3"), Node("4")), Node("5"))

def eval_tree(node):
    if node.val.isdigit():
        return int(node.val)
    a = eval_tree(node.left)
    b = eval_tree(node.right)
    return eval(f"{a}{node.val}{b}")

print(eval_tree(root))  # Output: 35
```

Binary Search Trees (BST)

A BST maintains the invariant:

$$\texttt{left subtree} < \texttt{root} < \texttt{right subtree}.$$

Example 4.1.24 (Insert into a BST)

```
def insert(node, key):
    if not node:
        return Node(key)
    if key < node.val:
        node.left = insert(node.left, key)
    else:
        node.right = insert(node.right, key)
    return node
```

Tree Traversals

- In-order: Left → Root → Right (BST yields sorted order).
- Pre-order: Root → Left → Right.
- Post-order: Left → Right → Root.
- Level-order: Use queue (BFS).

Example 4.1.25 (Height-Balanced Tree) A tree is balanced if the heights of left and right subtrees differ by at most 1 at every node.
    Check for height balance:

```
def height(node):
    if not node: return 0
    return 1 + max(height(node.left), height(node.right))

def is_balanced(node):
    if not node: return True
    h1 = height(node.left)
```

```
h2 = height(node.right)
return abs(h1 - h2) <= 1 and is_balanced(node.left) and is_balanced(
    node.right)
```

Use Cases

- Expression parsing (AST)
- Heaps (priority queues)
- Balanced search trees (AVL, Red-Black)
- Decision trees in machine learning

Hash Tables and Dictionary Internals

A **hash table** maps keys to values using a hash function $h : \mathcal{K} \to [0, m-1]$ to compute index in a table of size m.

Python `dict`

`dict` is implemented as a *dynamic hash table* with:

- open addressing (linear probing + perturbation)
- load factor threshold $\alpha \approx 0.66$
- resizing upon insertion overflow

Example 4.1.26 (Basic Dictionary Operations)

```
d = {"x": 1, "y": 2}
d["z"] = 3
del d["x"]
for key, val in d.items():
    print(key, "→", val)
```

Hash Function

For immutable types like `int`, `str`, `tuple`, Python uses built-in `hash()`.

```
print(hash("math")) # deterministic across session
print(hash((1, 2, 3)))
```

Mutable types like `list` or `set` are not hashable.

```
hash([1,2,3]) # raises TypeError
```

4.1 Introduction to Data Structures

Example 4.1.27 (Hash Collisions) Multiple keys may map to the same index.

```
class BadHash:
    def __init__(self, val): self.val = val
    def __hash__(self): return 42
    def __eq__(self, other): return self.val == other.val

d = {}
for i in range(5):
    d[BadHash(i)] = i
print(len(d)) # Still 5, resolved via probing
```

Hash Table Performance

$$\begin{cases} \text{Average case: } O(1) \text{ lookup, insert, delete} \\ \text{Worst case: } O(n) \text{ (due to collisions)} \end{cases}$$

Example 4.1.28 (Frequency Counting—Classic Hash Table Use)

```
from collections import defaultdict
s = "data structure"
freq = defaultdict(int)
for ch in s:
    freq[ch] += 1
print(freq)
```

OrderedDict

Since Python 3.7, `dict` maintains insertion order. But for compatibility, use `collections.OrderedDict` explicitly.

Example 4.1.29 (Preserve Order of Insertion)

```
from collections import OrderedDict
od = OrderedDict()
od["x"] = 1
od["y"] = 2
print(list(od.keys())) # ['x', 'y']
```

Hash Map vs. BST Map

- `dict`: average $O(1)$, unordered (unless using OrderedDict).
- `SortedContainers.SortedDict`: ordered, $O(\log n)$.

4.2 Search Algorithms

Searching is a fundamental operation in both mathematical computation and algorithmic logic. A **search algorithm** locates an item in a given collection, such as an element in a list, a key in a dictionary, or a pattern in a sequence. The efficiency of a search algorithm is evaluated in terms of time complexity, worst-case performance, and structural assumptions about the data (e.g. sorted vs. unsorted). We begin with the simplest approach: **linear search**.

4.2.1 Linear Search

Linear search is a sequential method of examining each element of a collection until the target is found or the sequence ends.

Algorithm and Complexity Analysis

Let $A = [a_0, a_1, \ldots, a_{n-1}] \in \mathbb{R}^n$ be an unsorted list and $x \in \mathbb{R}$ be the query. The linear search algorithm performs the test $a_i = x$ for each index i, in order.

Pseudocode

```
for i from 0 to n-1:
    if A[i] == x:
        return i
return -1
```

Time Complexity

- Best case: $O(1)$ (if $x = a_0$)
- Worst case: $O(n)$ (if $x \notin A$)
- Average case (uniform distribution): $O(n)$

Space Complexity

$$O(1) \quad \text{(in-place, no extra storage)}$$

Properties

- Works on any data type supporting equality test.
- Does not require sorting or ordering.
- Can be extended to linear search with predicates or composite keys.

4.2 Search Algorithms

Example 4.2.1 (Mathematical Identity Check) Search for the first perfect square in a list.

```
import math

def is_square(n):
    return int(math.sqrt(n))**2 == n

def linear_search_predicate(A, predicate):
    for i, val in enumerate(A):
        if predicate(val):
            return i
    return -1

A = [10, 11, 14, 25, 30, 50]
idx = linear_search_predicate(A, is_square)
print(f"First perfect square at index {idx}, value = {A[idx]}")
```

Output: `First perfect square at index 3, value = 25.`

Python Implementation

Basic Implementation

```
def linear_search(A, x):
    for i in range(len(A)):
        if A[i] == x:
            return i
    return -1
```

Variant: Return All Indices Where Match Occurs

```
def linear_search_all(A, x):
    return [i for i, val in enumerate(A) if val == x]

A = [2, 4, 4, 6, 4]
print(linear_search_all(A, 4)) # Output: [1, 2, 4]
```

Variant: Search in List of Tuples

```
students = [("Alice", 85), ("Bob", 92), ("Charlie", 85)]
def find_student_by_score(data, score):
    return [name for name, s in data if s == score]

print(find_student_by_score(students, 85)) # ['Alice', 'Charlie']
```

Performance Profiling

```
import time
A = list(range(10**6))
x = 10**6 - 1

start = time.time()
idx = linear_search(A, x)
end = time.time()
print(f"Found at {idx}, time = {end - start:.6f} sec")
```

For x near the end, the runtime scales linearly with n.

Example 4.2.2 (Symbolic Pattern Search) Find first function in a list whose derivative vanishes at $x = 0$.

```
import sympy as sp
x = sp.Symbol("x")
functions = [sp.sin(x), x**2, sp.exp(x), sp.cos(x)]

for i, f in enumerate(functions):
    if sp.diff(f, x).subs(x, 0) == 0:
        print(f"Index {i}, function:", f)
        break
```

4.2.2 Binary Search

Binary search is a divide-and-conquer algorithm for finding the position of a target value in a sorted array. It repeatedly divides the search interval in half, eliminating half of the remaining elements in each step. This exponential pruning gives binary search its logarithmic time complexity, making it one of the most fundamental and efficient algorithms in computer science.

Algorithm and Complexity Analysis

Let $A = [a_0, a_1, \ldots, a_{n-1}] \in \mathbb{R}^n$ be sorted in non-decreasing order and let $x \in \mathbb{R}$ be the target. The algorithm maintains two pointers, `low` and `high`, such that x must lie within the subarray $A[\text{low} : \text{high}]$.

Pseudocode

```
while low <= high:
    mid = (low + high) // 2
    if A[mid] == x:
        return mid
    elif A[mid] < x:
```

4.2 Search Algorithms

```
        low = mid + 1
    else:
        high = mid - 1
return -1
```

Time Complexity

$$T(n) = T(n/2) + O(1) \Rightarrow T(n) = O(\log_2 n)$$

Space Complexity

- Iterative version: $O(1)$
- Recursive version: $O(\log n)$ stack depth

Assumptions

- Input list must be *sorted*.
- Elements must be comparable via total ordering.

Example 4.2.3 (Compare Linear vs. Binary Search Time)

```python
import time, bisect
A = list(range(10**7))
x = 9999999

# Linear search
start = time.time()
A.index(x)
print("Linear search:", time.time() - start)

# Binary search using bisect
start = time.time()
bisect.bisect_left(A, x)
print("Binary search:", time.time() - start)
```

Binary search is several orders of magnitude faster on large inputs.

Python Implementation

Iterative Implementation

```python
def binary_search(A, x):
    low, high = 0, len(A) - 1
    while low <= high:
        mid = (low + high) // 2
        if A[mid] == x:
            return mid
```

```
        elif A[mid] < x:
            low = mid + 1
        else:
            high = mid - 1
    return -1
```

Recursive Implementation

```
def binary_search_rec(A, x, low=0, high=None):
    if high is None: high = len(A) - 1
    if low > high:
        return -1
    mid = (low + high) // 2
    if A[mid] == x:
        return mid
    elif A[mid] < x:
        return binary_search_rec(A, x, mid + 1, high)
    else:
        return binary_search_rec(A, x, low, mid - 1)
```

Using bisect Module

```
import bisect
def binary_search_bisect(A, x):
    i = bisect.bisect_left(A, x)
    return i if i < len(A) and A[i] == x else -1
```

Example 4.2.4 (Find Square Root Integer) Given sorted list of squares, find the index of a number.

```
squares = [i**2 for i in range(100)]
x = 625
print(binary_search(squares, x))  # Output: 25
```

Example 4.2.5 (Find First Element $\geq$ Target)

```
def lower_bound(A, x):
    low, high = 0, len(A)
    while low < high:
        mid = (low + high) // 2
        if A[mid] < x:
            low = mid + 1
        else:
            high = mid
    return low
```

Example 4.2.6 (Binary Search for Root of Monotonic Function) Let $f(x) = x^3 - 5x + 1$. Find root in $[0, 2]$.

```
def f(x): return x**3 - 5*x + 1

def binary_root(f, a, b, eps=1e-6):
    while b - a > eps:
        mid = (a + b) / 2
```

```
        if f(mid) * f(a) <= 0:
            b = mid
        else:
            a = mid
    return (a + b) / 2

root = binary_root(f, 0, 2)
print(f"Root ≈ {root:.6f}, f(root) ≈ {f(root):.2e}")
```

Example 4.2.7 (Search in 2D Matrix) Given a matrix where each row and column is sorted, perform binary search.

```
def search_matrix(matrix, target):
    if not matrix: return False
    rows, cols = len(matrix), len(matrix[0])
    low, high = 0, rows * cols - 1
    while low <= high:
        mid = (low + high) // 2
        val = matrix[mid // cols][mid % cols]
        if val == target:
            return True
        elif val < target:
            low = mid + 1
        else:
            high = mid - 1
    return False
```

4.3 Sorting Algorithms

Sorting algorithms reorder a collection of elements into a prescribed order—typically ascending or descending. They are foundational in algorithmic design, used in searching, data normalisation, optimisation, and more. The efficiency of a sorting algorithm is judged by its time complexity, space usage, and stability. We begin with the elementary sorting algorithms: selection sort, bubble sort, and insertion sort. While inefficient for large datasets, these algorithms are conceptually simple and valuable for teaching algorithmic thinking.

4.3.1 Basic Sorting Algorithms

Selection Sort

Selection sort sorts a list by repeatedly selecting the smallest remaining unsorted element and swapping it to its correct position.

Algorithm

For $A = [a_0, a_1, \ldots, a_{n-1}]$, iterate:

- for $i = 0$ to $n - 2$:
- find index $j \geq i$ such that $A[j]$ is minimal
- swap $A[i] \leftrightarrow A[j]$

Time Complexity

$\quad\quad\quad$ Worst-case: $O(n^2)$ $\quad$ Best-case: $O(n^2)$ $\quad$ In-place, not stable

```
def selection_sort(A):
    n = len(A)
    for i in range(n):
        min_idx = i
        for j in range(i + 1, n):
            if A[j] < A[min_idx]:
                min_idx = j
        A[i], A[min_idx] = A[min_idx], A[i]
    return A
```

Example 4.3.1 (Sort a List of Integers)
```
A = [64, 25, 12, 22, 11]
print(selection_sort(A)) # [11, 12, 22, 25, 64]
```

Example 4.3.2 (Selection Sort on Strings)
```
names = ["Bob", "Alice", "Eve"]
print(selection_sort(names)) # ['Alice', 'Bob', 'Eve']
```

Bubble Sort

Bubble sort repeatedly compares adjacent pairs and swaps them if they are out of order.

Algorithm

- For $i = 0$ to $n - 1$:
- Traverse from $j = 0$ to $n - i - 2$ and swap $A[j]$ and $A[j+1]$ if $A[j] > A[j+1]$.

4.3 Sorting Algorithms

Time Complexity

$O(n^2)$ in all cases without optimisation. Stable and in-place

```
def bubble_sort(A):
    n = len(A)
    for i in range(n):
        for j in range(0, n - i - 1):
            if A[j] > A[j + 1]:
                A[j], A[j + 1] = A[j + 1], A[j]
    return A
```

Example 4.3.3 (Trace Bubble Sort on [5,1,4,2,8])] Each pass bubbles the largest element to the end:

$$[5, 1, 4, 2, 8] \to [1, 4, 2, 5, 8] \to [1, 2, 4, 5, 8]$$

Early Termination

Track whether any swaps occurred:

```
def bubble_sort_optimized(A):
    n = len(A)
    for i in range(n):
        swapped = False
        for j in range(0, n - i - 1):
            if A[j] > A[j + 1]:
                A[j], A[j + 1] = A[j + 1], A[j]
                swapped = True
        if not swapped:
            break
    return A
```

Insertion Sort

Insertion sort builds the sorted list one element at a time, inserting each new element into the correct position.

Algorithm

- For $i = 1$ to $n - 1$:
- Set key = A[i] and compare backwards $A[j]$ with key.
- Shift $A[j]$ right if $A[j] >$ key.
- Insert key at correct position.

Time Complexity

Best-case (already sorted): $O(n)$ Worst-case (reverse sorted): $O(n^2)$ Stable and in-place

```python
def insertion_sort(A):
    for i in range(1, len(A)):
        key = A[i]
        j = i - 1
        while j >= 0 and A[j] > key:
            A[j + 1] = A[j]
            j -= 1
        A[j + 1] = key
    return A
```

Example 4.3.4 (Insertion Sort on Small Array)

$$[3, 1, 4, 2] \Rightarrow [1, 3, 4, 2] \Rightarrow [1, 3, 4, 2] \Rightarrow [1, 2, 3, 4]$$

Example 4.3.5 (Sorting List of Tuples by Second Element)

```python
data = [(1, "b"), (3, "a"), (2, "c")]
data.sort(key=lambda x: x[1]) # built-in uses Timsort (stable)
print(data) # [(3, 'a'), (1, 'b'), (2, 'c')]
```

Use Cases

- Efficient for small or nearly sorted data.
- Used in hybrid algorithms like Timsort (used in Python's built-in sort).

4.3.2 Divide and Conquer Techniques

Divide and conquer is a powerful paradigm in algorithm design. The main idea is to divide the problem into smaller subproblems, solve them recursively, and then combine the solutions. This approach often leads to logarithmic recursion depth and improved asymptotic performance. In sorting, **merge sort** and **quick sort** are classic applications.

Merge Sort

Merge sort recursively divides the input array into halves, sorts each half, and then merges the sorted halves.

4.3 Sorting Algorithms

Algorithm Steps

1. Divide: Split array $A[0 \ldots n-1]$ into two halves.
2. Conquer: Recursively sort each half.
3. Combine: Merge two sorted halves into a sorted array.

Time Complexity

$$T(n) = 2T(n/2) + O(n) \Rightarrow T(n) = O(n \log n)$$

Space Complexity

$O(n)$ (due to temporary arrays for merging)

Stable

Yes. In-place: No.

```
def merge_sort(A):
    if len(A) <= 1:
        return A
    mid = len(A) // 2
    left = merge_sort(A[:mid])
    right = merge_sort(A[mid:])
    return merge(left, right)

def merge(L, R):
    result = []
    i = j = 0
    while i < len(L) and j < len(R):
        if L[i] <= R[j]:
            result.append(L[i])
            i += 1
        else:
            result.append(R[j])
            j += 1
    result += L[i:] + R[j:]
    return result
```

Example 4.3.6 (Sort a List Using Merge Sort)

```
A = [38, 27, 43, 3, 9, 82, 10]
sorted_A = merge_sort(A)
print(sorted_A) # [3, 9, 10, 27, 38, 43, 82]
```

Example 4.3.7 (Merge Sort with Custom Comparator) Sort list of tuples by second element:

```
def merge_sort_key(A, key):
    if len(A) <= 1: return A
    mid = len(A) // 2
    L = merge_sort_key(A[:mid], key)
    R = merge_sort_key(A[mid:], key)
    return merge_key(L, R, key)

def merge_key(L, R, key):
    result = []
    i = j = 0
    while i < len(L) and j < len(R):
        if key(L[i]) <= key(R[j]):
            result.append(L[i])
            i += 1
        else:
            result.append(R[j])
            j += 1
    result += L[i:] + R[j:]
    return result

pairs = [(1, 'b'), (2, 'a'), (3, 'c')]
print(merge_sort_key(pairs, key=lambda x: x[1]))
```

Quick Sort

Quick sort selects a pivot and partitions the array into elements less than the pivot and greater than the pivot, then recursively sorts the partitions.

Algorithm Steps

1. Choose a pivot element.
2. Partition the array: elements < pivot to left, > pivot to right.
3. Recursively apply quicksort to left and right parts.

Time Complexity

- Best-case: $O(n \log n)$
- Average-case: $O(n \log n)$
- Worst-case: $O(n^2)$ (when pivot is min/max)

In-Place

Yes.   Stable: No.

```
def quick_sort(A):
```

4.3 Sorting Algorithms

```
if len(A) <= 1:
    return A
pivot = A[0]
less = [x for x in A[1:] if x <= pivot]
greater = [x for x in A[1:] if x > pivot]
return quick_sort(less) + [pivot] + quick_sort(greater)
```

Example 4.3.8 (Sort Random List with Quick Sort)

```
import random
A = [random.randint(0, 100) for _ in range(10)]
print(quick_sort(A))
```

In-Place Partition-Based Quick Sorts

```
def partition(A, low, high):
    pivot = A[high]
    i = low
    for j in range(low, high):
        if A[j] <= pivot:
            A[i], A[j] = A[j], A[i]
            i += 1
    A[i], A[high] = A[high], A[i]
    return i

def quick_sort_inplace(A, low=0, high=None):
    if high is None:
        high = len(A) - 1
    if low < high:
        pi = partition(A, low, high)
        quick_sort_inplace(A, low, pi - 1)
        quick_sort_inplace(A, pi + 1, high)
```

Example 4.3.9 (Visualise Pivot Effect) Try sorting $[5,4,3,2,1]$ and observe $O(n^2)$ behaviour unless pivot is randomised.

Randomised Quick Sort

```
def quick_sort_random(A):
    if len(A) <= 1:
        return A
    import random
    pivot = random.choice(A)
    L = [x for x in A if x < pivot]
    E = [x for x in A if x == pivot]
    G = [x for x in A if x > pivot]
    return quick_sort_random(L) + E + quick_sort_random(G)
```

This avoids worst-case with high probability

4.4 Graph Theory and Algorithms

Graph theory studies discrete structures consisting of nodes (vertices) connected by links (edges). Graphs model networks in mathematics, computer science, biology, and social sciences—ranging from road maps and electrical circuits to social connections and neural networks. A **graph algorithm** is any procedure that operates on such structures: searching, traversing, finding shortest paths, or detecting cycles.

4.4.1 Introduction to Graph Theory

Graphs as Mathematical Structures

Formally, a (simple) graph G is a pair $G = (V, E)$, where:

- V is a finite set of vertices $\{v_1, v_2, \ldots, v_n\}$.
- $E \subseteq \{\{u, v\} \mid u, v \in V, u \neq v\}$ is the set of undirected edges.

A **directed graph** (or digraph) uses ordered pairs: $E \subseteq V \times V$.

Terminology

- Degree: number of edges incident to a vertex.
- Path: sequence of vertices connected by edges.
- Cycle: path that starts and ends at the same vertex.
- Connected: every pair of vertices is reachable.
- Tree: connected acyclic graph.
- Weighted graph: edges have associated weights (costs).

Example 4.4.1 (Simple Undirected Graph) Let $V = \{1, 2, 3\}$, $E = \{\{1, 2\}, \{2, 3\}\}$. This can be visualised as a path: $1 \leftrightarrow 2 \leftrightarrow 3$.

Example 4.4.2 (Directed Graph with Weights) Let $V = \{A, B, C\}$, and:

$$E = \{(A, B, 3), (B, C, 2), (A, C, 10)\}$$

This represents a weighted digraph with paths and costs.

Example 4.4.3 (Adjacency Matrix and Degree Sequence) For the graph

$$G = (V, E), \quad V = \{0, 1, 2\}, \quad E = \{(0, 1), (1, 2), (0, 2)\},$$

4.4 Graph Theory and Algorithms

the adjacency matrix is:

$$A = \begin{pmatrix} 0 & 1 & 1 \\ 1 & 0 & 1 \\ 1 & 1 & 0 \end{pmatrix}, \quad \text{degree of each vertex: } [2, 2, 2].$$

Graph Representation in Python

Graphs can be represented in several ways, each with different efficiency tradeoffs.

Adjacency List (Using `dict` of Lists)

```python
graph = {
    'A': ['B', 'C'],
    'B': ['A', 'D'],
    'C': ['A'],
    'D': ['B']
}
```

Example 4.4.4 (DFS Traversal Using Adjacency List)

```python
def dfs(graph, v, visited=None):
    if visited is None: visited = set()
    visited.add(v)
    for neighbor in graph[v]:
        if neighbor not in visited:
            dfs(graph, neighbor, visited)
    return visited

print(dfs(graph, 'A')) # {'A', 'B', 'C', 'D'}
```

Adjacency Matrix (Using NumPy)

```python
import numpy as np

adj_matrix = np.array([
    [0, 1, 1], # A connected to B, C
    [1, 0, 0], # B connected to A
    [1, 0, 0]  # C connected to A
])
```

Example 4.4.5 (Compute Degree of Each Vertex)

```python
degrees = adj_matrix.sum(axis=1)
print(degrees) # [2, 1, 1]
```

Edge List

```
edges = [("A", "B"), ("B", "C"), ("A", "C")]
```

Example 4.4.6 (Convert Edge List to Adjacency List)

```python
from collections import defaultdict

def build_adj_list(edges):
    G = defaultdict(list)
    for u, v in edges:
        G[u].append(v)
        G[v].append(u)  # for undirected
    return G
```

Weighted Graphs (with Edge Weights)

```python
weighted_graph = {
    'A': {'B': 5, 'C': 3},
    'B': {'C': 2},
    'C': {'A': 4}
}
```

Example 4.4.7 (Extract All Edges with Weights)

```python
edges = [(u, v, w) for u in weighted_graph for v, w in weighted_graph[u].items()]
print(edges)  # [('A','B',5), ('A','C',3), ...]
```

Graph Library: NetworkX

Python's `networkx` provides a rich API for graph construction and analysis.

```python
import networkx as nx

G = nx.Graph()
G.add_edge("A", "B", weight=5)
G.add_edge("A", "C", weight=3)
nx.draw(G, with_labels=True)
```

Example 4.4.8 (Shortest Path Using Dijkstra)

```python
path = nx.dijkstra_path(G, "A", "B", weight="weight")
print(path)  # ['A', 'B']
```

Comparison of Representations

Representation	Space complexity	Best for	Operations
Adjacency List	$O(V + E)$	Sparse graphs	Fast traversal
Adjacency Matrix	$O(V^2)$	Dense graphs	Constant lookup
Edge List	$O(E)$	Algorithms on edges	Sorting, filtering

4.4.2 Basic Graph Algorithms

Graph traversal and pathfinding are central to graph theory applications. We explore three foundational algorithms:

- **Breadth-First Search (BFS):** explores neighbours level-by-level.
- **Depth-First Search (DFS):** explores deeply before backtracking.
- **Dijkstra's Algorithm:** computes shortest paths in weighted graphs with non-negative edge weights.

These are the bedrock for topological sorting, cycle detection, spanning trees, and more.

Breadth-First Search (BFS)

BFS explores a graph layer-wise starting from a source vertex, visiting all vertices at depth d before going to depth $d + 1$. It uses a queue (FIFO).

Time Complexity

$O(V + E)$

Space Complexity

$O(V)$ (visited, queue)

```python
from collections import deque

def bfs(graph, start):
    visited = set()
    queue = deque([start])
    order = []
    while queue:
        node = queue.popleft()
        if node not in visited:
```

```
        visited.add(node)
        order.append(node)
        queue.extend(graph[node])
    return order
```

Example 4.4.9 (Level Order Traversal in a Tree)
```
tree = {
    'A': ['B', 'C'],
    'B': ['D', 'E'],
    'C': ['F'],
    'D': [], 'E': [], 'F': []
}
print(bfs(tree, 'A'))  # ['A', 'B', 'C', 'D', 'E', 'F']
```

Example 4.4.10 (Shortest Path (Unweighted Graph))
```
def bfs_path(graph, start, goal):
    from collections import deque
    queue = deque([[start]])
    visited = set()
    while queue:
        path = queue.popleft()
        node = path[-1]
        if node == goal:
            return path
        if node not in visited:
            visited.add(node)
            for neighbor in graph[node]:
                queue.append(path + [neighbor])
```

Depth-First Search (DFS)

DFS explores as far as possible along each branch before backtracking. It uses a stack (explicitly or via recursion).

Time Complexity

$O(V + E)$

Space Complexity

$O(V)$ stack depth

```
def dfs(graph, start, visited=None):
    if visited is None:
        visited = set()
    visited.add(start)
    for neighbor in graph[start]:
```

4.4 Graph Theory and Algorithms

```
        if neighbor not in visited:
            dfs(graph, neighbor, visited)
    return visited
```

Example 4.4.11 (Detect Cycle in a Graph)

```
def has_cycle(graph):
    visited = set()
    stack = set()

    def dfs(v):
        visited.add(v)
        stack.add(v)
        for neighbor in graph[v]:
            if neighbor not in visited:
                if dfs(neighbor): return True
            elif neighbor in stack:
                return True
        stack.remove(v)
        return False

    for v in graph:
        if v not in visited:
            if dfs(v): return True
    return False
```

Example 4.4.12 (Topological Sorting via DFS)

```
def topological_sort(graph):
    visited = set()
    stack = []

    def dfs(v):
        visited.add(v)
        for neighbor in graph[v]:
            if neighbor not in visited:
                dfs(neighbor)
        stack.append(v)

    for v in graph:
        if v not in visited:
            dfs(v)
    return stack[::-1]
```

Shortest Path Algorithms: Dijkstra's Algorithm

Dijkstra's algorithm finds the shortest path from a source to all other nodes in a weighted graph with non-negative edge weights.

Time Complexity

- With priority queue (heap): $O((V + E) \log V)$
- Without heap: $O(V^2)$

Algorithm Idea

- Initialise distance to all vertices as ∞ and source to 0.
- Use a min-priority queue to select the vertex with smallest tentative distance.
- Relax its neighbours.

```
import heapq

def dijkstra(graph, start):
    dist = {v: float('inf') for v in graph}
    dist[start] = 0
    pq = [(0, start)]

    while pq:
        d, u = heapq.heappop(pq)
        if d > dist[u]: continue
        for v, w in graph[u].items():
            if dist[u] + w < dist[v]:
                dist[v] = dist[u] + w
                heapq.heappush(pq, (dist[v], v))
    return dist
```

Example 4.4.13 (Weighted Graph with Adjacency Map)

```
graph = {
    'A': {'B': 1, 'C': 4},
    'B': {'C': 2, 'D': 5},
    'C': {'D': 1},
    'D': {}
}
print(dijkstra(graph, 'A'))  # {'A':0, 'B':1, 'C':3, 'D':4}
```

Example 4.4.14 (Track Actual Shortest Paths)

```
def dijkstra_path(graph, start):
    dist = {v: float('inf') for v in graph}
    prev = {}
    dist[start] = 0
    pq = [(0, start)]

    while pq:
        d, u = heapq.heappop(pq)
        for v, w in graph[u].items():
            if dist[u] + w < dist[v]:
                dist[v] = dist[u] + w
                prev[v] = u
                heapq.heappush(pq, (dist[v], v))
```

```
def reconstruct(v):
    path = []
    while v in prev:
        path.append(v)
        v = prev[v]
    return [start] + path[::-1]

return dist, reconstruct
```

4.5 Exercises

1. Let

$$A = \langle 37, 19, \underline{41}, 5, 27, 41, 19, 73, 2, 88, \underline{41}\rangle.$$

   (a) Compute the exact number of element-to-target comparisons performed by a naïve linear search for $x = 41$ that returns the *first* match; verify your answer experimentally in Python. (b) Derive a closed-form expression for the expected number of comparisons $E[C]$ when searching for a key that is present k times in an array of length n assuming all k occurrences are uniformly distributed and the search terminates at the first match. (c) Evaluate $E[C]$ for the above data.
2. Consider the sorted array

$$B = \langle -17, -4, 0, 3, 5, 9, 12, 18, 23, 29, 31, 34, 40, 42, 47, 50, 54\rangle.$$

   (a) Illustrate, in tabular form, the sequence of low–high intervals used by binary search to locate $x = 31$. (b) Repeat for $x = 33$ (which is absent). (c) Prove that binary search on an array of length $n = 2^m - 1$ performs exactly m comparisons in the worst case and confirm that bound for B.
3. The list

$$S = \big[(2, a), (1, b), (3, c), (2, d), (1, e), (2, f)\big]$$

   is sorted by primary key (first component). Apply (i) insertion sort and (ii) selection sort to S. For each algorithm output the final ordering *including* secondary labels and decide which algorithm is stable. Give a formal definition of stability and justify your conclusion.
4. Let $Q = \langle 9, 2, 6, 4, 3, 5, 1, 8, 7\rangle$. (a) Perform quick sort using the *first element* as the pivot; list the pivot at each recursive level and count total key comparisons. (b) Repeat using *median-of-three* {first, middle, last} pivot selection. (c) Prove that the comparison count in part (b) is strictly less than in part (a) for every permutation of $1, \ldots, 9$.

5. For the array

$$M = \langle 84, 12, 95, 63, 45, 27, 66, 18 \rangle$$

trace merge sort, showing the contents of every auxiliary sub-array produced during merging. Compute the total number of auxiliary elements allocated over the entire run and compare it with the theoretical upper bound $n \lceil \log_2 n \rceil$.

6. A hash table of size $m = 13$ stores the keys

$$K = \{18, 41, 22, 44, 59, 32, 31\}$$

using the hash function $h(k) = k \mod 13$ and *linear probing*. (a) Insert the keys in the order listed and record the probe sequence for each insertion. (b) Give the final table state. (c) Evaluate the load factor α and estimate, using Knuth's formula, the expected number of probes for an unsuccessful search at this load factor.

7. Insert the sequence

$$\langle 11, 7, 15, 3, 9, 13, 17, 1, 5 \rangle$$

into an initially empty BST. (a) Draw the resulting tree. (b) Provide the in-order, pre-order, and post-order traversal sequences. (c) Compute the height h and prove that $h \leq n - 1$ for any BST of n nodes.

8. Given

$$H = \langle 19, 7, 12, 3, 5, 1, 2, 25, 17 \rangle,$$

(a) build a *max-heap* using the bottom-up heapify procedure; show the array representation after each down-heap operation. (b) Perform heap sort and list the array after each extraction of the maximum. (c) Prove that heap sort performs exactly $2(n - 1)$ sift-down operations for an array of length n.

9. For the undirected graph

$$V = \{A, B, C, D, E, F, G, H\},$$
$$E = \{AB, AC, BD, CE, DF, EG, FH, GH\},$$

(a) give the adjacency list; (b) run BFS from A and record the discovery distance $d(v)$ for every vertex; (c) output a shortest $A \rightarrow G$ path; (d) prove that BFS always produces shortest paths in unweighted graphs by induction on path length.

10. Using the graph in Exercise 9 but treating edges as *directed* from lexicographically smaller to larger vertex (e.g. $A \rightarrow B$), perform DFS starting from A with alphabetic adjacency ordering. List the discovery/finish times $\langle d(v), f(v) \rangle$

4.5 Exercises

for each vertex and classify every edge as TREE, BACK, FORWARD, or CROSS. Explain the relationship between discovery/finish intervals and edge types.

11. For the weighted digraph

$$V = \{0, 1, 2, 3, 4, 5\},$$

$$E = \{(0, 1, 7), (0, 2, 9), (0, 5, 14), (1, 2, 10), (1, 3, 15), (2, 3, 11), (2, 5, 2),$$
$$(3, 4, 6), (4, 5, 9)\},$$

run Dijkstra's algorithm from 0. When extracting the minimum-distance vertex from the priority queue, break ties by choosing the numerically smaller vertex. (a) Record, in a table, the tentative distance $d(v)$ and predecessor $\pi(v)$ after each extraction. (b) Draw the resulting shortest-path tree and list the shortest path $0 \to 4$. (c) Verify the tree satisfies the triangle inequality for all edges.

12. Consider the directed acyclic graph with adjacency list

$$0 : 1, 2$$
$$1 : 3, 4$$
$$2 : 4$$
$$3 : 5$$
$$4 : 5$$
$$5 :$$

(a) Compute a topological ordering using Kahn's algorithm. (b) Using dynamic programming on the topological order, find the length of the longest path from 0 to 5. (c) Prove that the algorithm in part (b) runs in $O(V + E)$ time.

13. Let $T(n)$ be the running time of merge sort on n elements and let $S(k)$ be the running time of insertion sort on k elements. A hybrid algorithm divides the array into blocks of size m, insertion-sorts each block, then merge-sorts the list of blocks. (a) Derive $T_{\text{hybrid}}(n, m) = \frac{n}{m} S(m) + T(\frac{n}{m})$. (b) Using $S(k) = \Theta(k^2)$ and $T(k) = \Theta(k \log k)$, choose m to minimise the leading term of T_{hybrid} and justify why Python's Timsort sets $m \approx 32$ in practice.

Chapter 5
Probability and Statistics

Abstract Random variables are formalised on measure-theoretic grounds before discrete and continuous distributions, expectation, variance and covariance are simulated with `numpy.random`. Frequentist inference (MLE, hypothesis testing, confidence intervals) is paralleled with Bayesian updating, and resampling techniques such as bootstrap and permutation tests provide non-parametric robustness. The chapter closes with time-series basics and goodness-of-fit diagnostics, preparing the reader for stochastic modelling in later chapters.

Keywords Probability · Statistics · Bayesian inference · Random variables · Hypothesis testing · Time-series analysis

5.1 Probability Theory

Probability provides a rigorous framework for quantifying uncertainty and randomness. Formally, a probability space is a triple $(\Omega, \mathcal{F}, \mathbb{P})$, where Ω is the sample space, $\mathcal{F}$ a σ-algebra of events, and $\mathbb{P}$ a probability measure. In practice, many problems reduce to finite sample spaces with the discrete σ-algebra and the counting measure scaled to unity.

5.1.1 Basic Probability Concepts

Probability Axioms and Theorems

Kolmogorov Axioms For any probability space $(\Omega, \mathcal{F}, \mathbb{P})$ and events $A, B \in \mathcal{F}$:

$$(\mathbf{A1})\ 0 \leq \mathbb{P}(A) \leq 1, \quad (\mathbf{A2})\ \mathbb{P}(\Omega) = 1, \quad (\mathbf{A3})\ \mathbb{P}\Big(\bigsqcup_{i=1}^{\infty} A_i\Big) = \sum_{i=1}^{\infty} \mathbb{P}(A_i)$$

for any countable collection $\{A_i\}$ of pairwise disjoint events.

Immediate Consequences

$$\mathbb{P}(\emptyset) = 0, \quad \mathbb{P}(A^c) = 1 - \mathbb{P}(A), \quad \mathbb{P}(A \cup B) = \mathbb{P}(A) + \mathbb{P}(B) - \mathbb{P}(A \cap B).$$

Example 1 (Coin Tossing) Let $\Omega = \{H, T\}^3$ be outcomes of 3 tosses of a fair coin. Then $\#\Omega = 8$ and $\mathbb{P}$ is uniform (1/8 per outcome).

$$A = \{\text{exactly two heads}\}, \qquad \mathbb{P}(A) = \frac{\binom{3}{2}}{8} = \frac{3}{8}.$$

```
import itertools, statistics
Ω = list(itertools.product("HT", repeat=3))
P = {ω: 1/8 for ω in Ω}
A = [ω for ω in Ω if ω.count("H") == 2]
print(sum(P[ω] for ω in A))  # 0.375
```

Example 2 (Inclusion–Exclusion) Roll two fair dice. Let

$$A = \{\text{sum is even}\}, \quad B = \{\text{at least one die shows 6}\}.$$

Compute:

$$\mathbb{P}(A) = \frac{18}{36} = \frac{1}{2}, \quad \mathbb{P}(B) = \frac{11}{36}, \quad \mathbb{P}(A \cap B) = \frac{5}{36}.$$

Hence $\mathbb{P}(A \cup B) = \frac{1}{2} + \frac{11}{36} - \frac{5}{36} = \frac{29}{36}$.

Boole's Inequality For any finite or countable $\{A_i\}$,

$$\mathbb{P}\left(\bigcup_i A_i\right) \leq \sum_i \mathbb{P}(A_i).$$

Example Draw n cards without replacement from a shuffled 52-card deck. Bound the probability that at least one ace appears when $n = 5$:

$$\mathbb{P}(\geq 1 \text{ ace}) \leq \sum_{k=1}^{4} \mathbb{P}(\text{ace } k \text{ appears}) = 4 \cdot \frac{5}{52} = \frac{5}{13} \approx 0.3846,$$

where the inequality is strict because events overlap.

Example 3 (Total Probability and Bayes) Suppose a test for a disease has sensitivity 0.95 and specificity 0.98. Prevalence is 0.01. Let D be disease, $+$ positive result. Total probability:

$$\mathbb{P}(+) = 0.95 \cdot 0.01 + 0.02 \cdot 0.99 = 0.0293.$$

5.1 Probability Theory

Posterior probability via Bayes' theorem:

$$\mathbb{P}(D|+) = \frac{0.95 \cdot 0.01}{0.0293} \approx 0.324.$$

```
sens, spec, prev = 0.95, 0.98, 0.01
ppos = sens*prev + (1-spec)*(1-prev)
print((sens*prev)/ppos) # 0.323860...
```

Combinatorial Probability: Permutations and Combinations

Counting arguments allow explicit probability computations in finite sample spaces.

Permutations

$$P(n,k) = n^{\underline{k}} = n(n-1)\cdots(n-k+1)$$

counts ordered k–tuples from n distinct objects.

Combinations

$$C(n,k) = \binom{n}{k} = \frac{n!}{k!(n-k)!}$$

counts unordered k–subsets.

Example 4 (Anagrams with Repeated Letters) How many distinct rearrangements of "STATISTICS"? Letter multiplicities: S:3, T:3, A:1, I:2, C:1.

$$N = \frac{10!}{3!\,3!\,2!} = 50400.$$

Example 5 (Poker Hand) Probability of being dealt exactly two pairs in 5-card poker.

Choose ranks: $\binom{13}{2}$ for the pair ranks and $\binom{11}{1}$ for the 5th card rank. Choose suits: $\binom{4}{2}$ per pair, $\binom{4}{1}$ for singleton.

$$N_{2\text{pair}} = \binom{13}{2}\binom{4}{2}^2\binom{11}{1}\binom{4}{1}, \quad N_{\text{total}} = \binom{52}{5}.$$

Probability ≈ 0.0475.

```
from math import comb
num = comb(13,2)*comb(4,2)**2*comb(11,1)*comb(4,1)
den = comb(52,5)
print(num/den) # 0.047539...
```

Example 6 (Derangements) For n distinct letters mis-addressed to n distinct envelopes, the probability that no letter is in the correct envelope is

$$\frac{!n}{n!} \approx \frac{1}{e}.$$

For $n = 6$, $!6 = 265$ gives probability $265/720 \approx 0.368$.

Example 7 (Birthday Paradox) In a group of k people, probability of *no* shared birthday (assuming 365 equally likely days):

$$\mathbb{P}(\text{unique}) = \frac{365^{\underline{k}}}{365^k}.$$

Find the smallest k s.t. $\mathbb{P}(\text{unique}) < \frac{1}{2}$.

```
import math
p = 1
k = 0
while p > 0.5:
    k += 1
    p *= (365 - k + 1)/365
print(k) # 23
```

Example 8 (Occupancy Problem) Randomly distribute $r = 10$ identical balls into $n = 4$ distinct boxes. Probability that no box is empty:

$$\frac{\binom{r-1}{n-1}^*}{\binom{r+n-1}{n-1}} = \frac{\binom{9}{3}^*}{\binom{13}{3}},$$

where $\binom{9}{3}^*$ counts compositions with positive parts. Enumerate all such compositions in Python and validate.

```
def compositions(n, k):
    if k == 1: yield [n]; return
    for i in range(1, n-k+2):
        for tail in compositions(n-i, k-1):
            yield [i] + tail

positive = list(compositions(10,4))
print(len(positive)) # 84
from math import comb
print(84/comb(13,3)) # 0.307692...
```

5.1 Probability Theory

Example 9 (Hypergeometric Distribution) Urn contains 7 red and 8 blue balls. Draw $k = 5$ without replacement. Probability that exactly 3 are red:

$$\mathbb{P}(X = 3) = \frac{\binom{7}{3}\binom{8}{2}}{\binom{15}{5}} \approx 0.278.$$

Example 10 (Boolean Algebra of Events) If A, B, C are independent with $\mathbb{P}(A) = \mathbb{P}(B) = \mathbb{P}(C) = 1/2$, compute $\mathbb{P}(A \oplus B \oplus C)$ (exclusive–or).

The event $A \oplus B \oplus C$ equals having an odd number of true events. Count outcomes: $\binom{3}{1} + \binom{3}{3} = 4$. Thus $\mathbb{P} = 4/8 = 1/2$.

Conditional Probability and Bayes' Theorem

Definition For events A, B with $\mathbb{P}(B) > 0$, the **conditional probability** of A given B is

$$\mathbb{P}(A \mid B) = \frac{\mathbb{P}(A \cap B)}{\mathbb{P}(B)}.$$

Intuitively this rescales the sample space to B.

Law of Total Probability If $\{B_i\}_{i=1}^n$ is a finite partition of Ω with $\mathbb{P}(B_i) > 0$, then

$$\mathbb{P}(A) = \sum_{i=1}^n \mathbb{P}(A \mid B_i)\,\mathbb{P}(B_i).$$

Bayes' Theorem

$$\mathbb{P}(B_j \mid A) = \frac{\mathbb{P}(A \mid B_j)\,\mathbb{P}(B_j)}{\sum_{i=1}^n \mathbb{P}(A \mid B_i)\,\mathbb{P}(B_i)}.$$

Bayes' theorem inverts conditional probabilities; the denominator is precisely the total-probability expansion.

Example 5.1.1 (Medical Test Revisited with Confusion Matrix) A disease prevalence is 0.5% ($\mathbb{P}(D) = 0.005$). Sensitivity 0.97, specificity 0.99. Compute $\mathbb{P}(D \mid +), \mathbb{P}(D^c \mid -)$.

$$\mathbb{P}(+) = 0.97 \cdot 0.005 + 0.01 \cdot 0.995 = 0.0146,$$

$$\mathbb{P}(D \mid +) = \frac{0.97 \cdot 0.005}{0.0146} \approx 0.332,$$

$$\mathbb{P}(D^c \mid -) = \frac{0.99 \cdot 0.995}{1 - 0.0146} \approx 0.995.$$

```
sens, spec, prev = .97, .99, .005
p_plus = sens*prev + (1-spec)*(1-prev)
post = sens*prev / p_plus
print(post) # 0.331...
```

Example 5.1.2 (Monty Hall Problem) Doors $\{1, 2, 3\}$, prize behind 1. Player picks #1, host opens #2. Let P be prize door, H host door.

$$\mathbb{P}(P = 1 \mid H = 2) = \frac{\frac{1}{3} \cdot \frac{1}{2}}{\frac{1}{3} \cdot \frac{1}{2} + \frac{1}{3} \cdot 1} = \frac{1}{3}, \quad \mathbb{P}(P = 3 \mid H = 2) = \frac{\frac{1}{3} \cdot 1}{\cdots} = \frac{2}{3}.$$

Hence switching doubles winning chance.

```
import random
N=100000
win_switch=0
for _ in range(N):
   prize=random.randint(1,3)
   choice=1
   door=[d for d in (2,3) if d!=prize][0] if choice==prize else \
       [d for d in (2,3) if d not in (choice,prize)][0]
   switch={1,2,3}-{choice,door}
   if prize in switch: win_switch+=1
print(win_switch/N) # ≈ 0.667
```

Example 5.1.3 (Spam-Filter Using Naive Bayes) Vocabulary: {*free*, *win*, *hello*}. Training data yields likelihoods:

	free	win	hello
Spam	0.7	0.6	0.1
Ham	0.05	0.02	0.4

$\mathbb{P}(\text{Spam}) = 0.2$.

Incoming mail contains *free* and *win*. Assuming word independence:

$$\mathbb{P}(\text{Spam} \mid \text{free,win}) \propto 0.7 \cdot 0.6 \cdot 0.2, \quad \mathbb{P}(\text{Ham} \mid \ldots) \propto 0.05 \cdot 0.02 \cdot 0.8.$$

Normalise to find posterior ≈ 0.9996 (almost certainly spam).

```
num_spam = .7*.6*.2
num_ham = .05*.02*.8
post = num_spam/(num_spam+num_ham)
print(post) # 0.9996
```

5.1 Probability Theory

Independence and Dependence of Events

Definition Events A, B are **independent** if

$$\mathbb{P}(A \cap B) = \mathbb{P}(A)\,\mathbb{P}(B),$$

equivalently $\mathbb{P}(A \mid B) = \mathbb{P}(A)$ when $\mathbb{P}(B) > 0$.

Mutual vs. Pairwise Independence Three events A, B, C are *mutually independent* if every subcollection has product measure equality, i.e.

$$\mathbb{P}(A \cap B \cap C) = \mathbb{P}(A)\mathbb{P}(B)\mathbb{P}(C)$$

and pairwise conditions hold. Pairwise independence does *not* imply mutual independence.

Example 5.1.4 (Pairwise Independent But Not Mutually) Toss two fair coins. Let

$$A = \{\text{first coin H}\}, \quad B = \{\text{second coin H}\}, \quad C = \{\text{coins have same face}\}.$$

Check $\mathbb{P}(A) = \mathbb{P}(B) = \mathbb{P}(C) = \frac{1}{2}$. Each pair multiplies correctly, yet $\mathbb{P}(A \cap B \cap C) = \frac{1}{4} \neq \frac{1}{8}$ so not mutually independent.

```
Ω = list(itertools.product("HT", repeat=2))
A = [ω for ω in Ω if ω[0]=='H']
B = [ω for ω in Ω if ω[1]=='H']
C = [ω for ω in Ω if ω[0]==ω[1]]
print(len(set(A)&set(B)&set(C))/4) # 0.25
```

Example (Independence of Complements) If A and B are independent, so are A^c and B, A and B^c, and A^c and B^c. Proof follows by substituting $\mathbb{P}(A^c) = 1 - \mathbb{P}(A)$ and algebra.

Example (Dice Sums) Roll two fair dice. Let $E = \{\text{sum is 7}\}$ and $F = \{\text{first die even}\}$. Compute:

$$\mathbb{P}(E) = \frac{6}{36}, \quad \mathbb{P}(F) = \frac{18}{36}, \quad \mathbb{P}(E \cap F) = \frac{3}{36},$$

so independence holds ($\frac{6}{36} \cdot \frac{18}{36} = \frac{108}{1296} = \frac{3}{36}$).

```
Ω=[(i,j) for i in range(1,7) for j in range(1,7)]
E=[ω for ω in Ω if sum(ω)==7]
F=[ω for ω in Ω if ω[0]%2==0]
print(len(set(E)&set(F))/36) # 0.0833...
```

Example (Dependence via Conditioning) Randomly draw one card from a standard deck. Events:

$$A = \{\text{card is red}\}, \quad B = \{\text{card is king}\}.$$

Because $\mathbb{P}(A \cap B) = \frac{2}{52}$ while $\mathbb{P}(A)\mathbb{P}(B) = \frac{1}{2} \cdot \frac{4}{52} = \frac{2}{52}$, A, B are independent. However conditioning on $C = \{\text{card is face card}\}$ breaks independence:

$$\mathbb{P}(A \mid C) = \frac{6}{12} = \tfrac{1}{2}, \quad \mathbb{P}(B \mid C) = \frac{4}{12} = \tfrac{1}{3}, \quad \mathbb{P}(A \cap B \mid C) = \frac{2}{12} = \tfrac{1}{6} \neq \tfrac{1}{6},$$

so independence remains; but conditioning on suit $D = \{\text{card is hearts or diamonds}\}$ makes $\mathbb{P}(B \mid D) = \frac{2}{26}$ vs. $\mathbb{P}(B) = \frac{4}{52}$, altering probability—dependence revealed.

Conditional Independence Events A, B may be dependent marginally but independent given C:

$$\mathbb{P}(A \cap B \mid C) = \mathbb{P}(A \mid C)\mathbb{P}(B \mid C).$$

Example 5.1.5 (Sensor Fusion) Let C be true location of a robot cell, A, B be noisy sensor readings. Conditioned on C, sensors errors are independent; unconditionally they are correlated.

5.1.2 Random Variables and Distributions

A **random variable** (r.v.) is a measurable function $X : (\Omega, \mathcal{F}) \to (\mathbb{R}, \mathcal{B})$ that assigns a real number to every outcome $\omega \in \Omega$. The distribution of X encodes how probability mass or density is spread across $\mathbb{R}$. We distinguish *discrete* and *continuous* r.v.s, each governed by characteristic formulae.

Discrete and Continuous Distributions

Discrete Random Variable X is **discrete** if it takes countable values $\{x_1, x_2, \ldots\}$ with

$$p_X(x_k) = \mathbb{P}(X = x_k), \quad \sum_k p_X(x_k) = 1.$$

p_X is the *probability mass function* (pmf).

5.1 Probability Theory

Continuous Random Variable X is **continuous** if there exists a non-negative integrable function f_X such that

$$\mathbb{P}(a < X \leq b) = \int_a^b f_X(t)\,dt, \qquad \int_{-\infty}^{\infty} f_X(t)\,dt = 1.$$

f_X is the *probability density function* (pdf).

Mixed Distributions Some r.v.s combine discrete atoms with continuous density (e.g. a distribution with a point mass at 0 and exponential tail on $(0, \infty)$).

Example 5.1.6 (Bernoulli (p)) Let $X \in \{0, 1\}$ with $\mathbb{P}(X = 1) = p$.

$$p_X(x) = p^x(1-p)^{1-x}, \quad \mu = \mathbb{E}[X] = p, \quad \sigma^2 = p(1-p).$$

Example 5.1.7 (Poisson(λ)) Counts arrivals in a homogeneous Poisson process.

$$p_X(k) = e^{-\lambda}\frac{\lambda^k}{k!}, \quad k \in \mathbb{N}_0.$$

Mean $= \lambda$, variance $= \lambda$.

Example 5.1.8 (Standard Normal)

$$f_X(x) = \frac{1}{\sqrt{2\pi}} e^{-x^2/2}, \qquad \mu = 0,\ \sigma^2 = 1.$$

Though no elementary closed form exists for its cdf $\Phi(x)$, scipy provides high-precision evaluation.

Probability Density Functions and Cumulative Distribution Functions

Cumulative Distribution Function (CDF)

$$F_X(x) = \mathbb{P}(X \leq x) = \begin{cases} \displaystyle\sum_{k: x_k \leq x} p_X(x_k) & \text{(discrete)}, \\ \displaystyle\int_{-\infty}^{x} f_X(t)\,dt & \text{(continuous)}. \end{cases}$$

Properties: non-decreasing, right-continuous, $\lim_{x \to -\infty} F_X = 0$, $\lim_{x \to \infty} F_X = 1$.

Example 5.1.9 (Exponential (λ))

$$f_X(x) = \lambda e^{-\lambda x}\mathbf{1}_{\{x \geq 0\}}, \quad F_X(x) = 1 - e^{-\lambda x}.$$

Memoryless: $\mathbb{P}(X > s + t \mid X > s) = e^{-\lambda t}$.

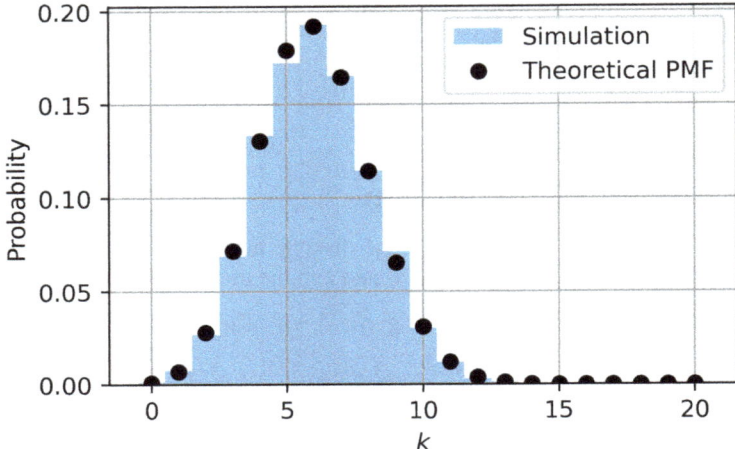

Fig. 5.1 Histogram of 10^4 samples from the binomial distribution $X \sim \text{Binom}(n = 20, p = 0.3)$, shown alongside the exact probability-mass function. The close agreement between simulated frequencies and the combinatorial formula $P\{X = k\} = \binom{20}{k} 0.3^k 0.7^{20-k}$ validates the Monte Carlo experiment

Quantiles The α-quantile $q_\alpha = \inf\{x : F_X(x) \geq \alpha\}$; median is $q_{0.5}$.

Example 5.1.10 (Normal Quantile) $q_{0.975} = 1.96$ satisfies $\Phi(1.96) \approx 0.975$, forming the two-sided 95% confidence interval $\mu \pm 1.96\sigma$.

Simulation of Random Variables in Python

Monte Carlo simulation validates theoretical results and approximates expectations (cf. Fig. 5.1).

Discrete Simulation (Binomial)

```
import numpy as np, matplotlib.pyplot as plt
n, p, N = 20, 0.3, 10_000
samples = np.random.binomial(n, p, N)
# empirical vs. theoretical pmf
k = np.arange(n+1)
pmf = (np.math.comb if hasattr(np, 'math') else np.math.comb)
theory = [pmf(n, i)*(p**i)*(1-p)**(n-i) for i in k]
plt.hist(samples, bins=range(n+2), density=True, alpha=.5, label="sim")
plt.plot(k, theory, "o", label="theory"); plt.legend(); plt.show()
```

Continuous Simulation (Normal)

```
μ, σ, N = 0, 1, 10**5
samples = np.random.normal(μ, σ, N)
print(f"sample mean={samples.mean():.3f}, var={samples.var():.3f}")
```

5.1 Probability Theory

Inverse-Transform Sampling Generate exponential(λ) via $X = -\frac{1}{\lambda} \ln U$ with $U \sim \text{Unif}(0, 1)$.

```
U = np.random.rand(100000)
lam = 2
exp_samples = -np.log(U)/lam
print(np.mean(exp_samples), np.var(exp_samples)) # ≈ 0.5, 0.25
```

Rejection Sampling (Beta(2,5)) Target pdf $f(x) = \frac{1}{B(2,5)} x(1-x)^4$ on $(0, 1)$.

```
from scipy.stats import beta
samples = beta.rvs(2,5, size=100000)
plt.hist(samples, bins=50, density=True, alpha=.6)
x = np.linspace(0,1,200); plt.plot(x, beta.pdf(x,2,5)); plt.show()
```

Central Limit Theorem Experiment Sum $m = 12$ uniforms $\Rightarrow$ approximate normal with $\mu = m/2$, $\sigma = \sqrt{m/12}$.

```
m, N = 12, 20000
sums = np.random.rand(N, m).sum(axis=1)
plt.hist((sums - m/2)/(np.sqrt(m/12)), bins=50, density=True)
```

5.1.3 Expectation, Variance, and Moments

Moments summarise the quantitative features of a random variable's distribution. The *first moment* (expectation) measures location, the *second central moment* (variance) spreads, while higher moments capture skewness, kurtosis, and beyond. Moment generating functions (mgf) provide a compact analytic device for encoding all moments simultaneously.

Mathematical Expectation and Properties

Definition For an integrable random variable X,

$$\mathbb{E}[X] = \begin{cases} \sum_k x_k \, p_X(x_k), & \text{(discrete)} \\ \int_{-\infty}^{\infty} x \, f_X(x) \, dx, & \text{(continuous)}. \end{cases}$$

Linearity For constants a, b and r.v.s X, Y,

$$\mathbb{E}[aX + bY] = a\mathbb{E}[X] + b\mathbb{E}[Y].$$

Example 1 (Binomial) If $X \sim \text{Bin}(n, p)$, write $X = \sum_{i=1}^{n} I_i$ with $I_i \sim \text{Bernoulli}(p)$. Then $\mathbb{E}[X] = \sum_i \mathbb{E}[I_i] = np$.

Example 2 (Expectation of a Function) Let $X \sim \text{Uniform}(0, 1)$. Compute $\mathbb{E}[\sqrt{X}]$:

$$\mathbb{E}[\sqrt{X}] = \int_0^1 \sqrt{x}\, dx = \frac{2}{3}.$$

```
import numpy as np
N=10**6
print(np.mean(np.sqrt(np.random.rand(N))))  # ≈ 0.6667
```

Indicator Trick For event A, $\mathbf{1}_A$ is 1 on A else 0. Then $\mathbb{E}[\mathbf{1}_A] = \mathbb{P}(A)$ and

$$\mathbb{E}[X] = \sum_k k\, \mathbb{P}(X = k) = \sum_k \mathbb{P}(X \geq k).$$

Law of Iterated Expectation (Tower Rule) For σ-fields $\mathcal{G} \subset \mathcal{F}$,

$$\mathbb{E}[\mathbb{E}[X \mid \mathcal{G}]] = \mathbb{E}[X].$$

Special case: $\mathbb{E}[X] = \mathbb{E}[\mathbb{E}[X \mid Y]]$.

Example 3 (Dice Conditioning) Let X be sum of two dice, Y first die. Then

$$\mathbb{E}[X] = \mathbb{E}[\mathbb{E}[X \mid Y]] = \mathbb{E}[Y + 3.5] = \mathbb{E}[Y] + 3.5 = 3.5 + 3.5 = 7.$$

Variance, Covariance, and Standard Deviation

Variance

$$\text{Var}(X) = \mathbb{E}[(X - \mathbb{E}[X])^2] = \mathbb{E}[X^2] - (\mathbb{E}[X])^2.$$

Standard Deviation $\sigma_X = \sqrt{\text{Var}(X)}$.

Covariance

$$\text{Cov}(X, Y) = \mathbb{E}[(X - \mu_X)(Y - \mu_Y)] = \mathbb{E}[XY] - \mu_X \mu_Y.$$

If $\text{Cov}(X, Y) = 0$, the variables are *uncorrelated*.

Correlation

$$\rho_{XY} = \frac{\text{Cov}(X, Y)}{\sigma_X \sigma_Y}, \quad -1 \leq \rho_{XY} \leq 1.$$

5.1 Probability Theory

Properties

$$\text{Var}(aX+b) = a^2\,\text{Var}(X), \qquad \text{Var}\Big(\sum_i X_i\Big) = \sum_i \text{Var}(X_i) + 2\sum_{i<j}\text{Cov}(X_i, X_j).$$

Example 4 (Variance of Poisson) $X \sim \text{Pois}(\lambda)$ has $\mathbb{E}[X] = \lambda$ and $\text{Var}(X) = \lambda$. Proof via mgf below.

Law of Total Variance

$$\text{Var}(X) = \mathbb{E}\big[\text{Var}(X \mid Y)\big] + \text{Var}\big(\mathbb{E}[X \mid Y]\big).$$

Example 5.1.11 (Total Variance, Poisson-Gamma Mixture (Negative Binomial)) Let $N \mid \Lambda \sim \text{Pois}(\Lambda)$, $\Lambda \sim \text{Gamma}(r, \theta)$. Then $\mathbb{E}[N] = r\theta$ and

$$\text{Var}(N) = \mathbb{E}[\Lambda] + \text{Var}(\Lambda) = r\theta + r\theta^2,$$

matching NegBin(r, p) variance formula.

Example 5 (Covariance of Die Faces) Roll a fair die twice: X first face, Y second. Independence $\Rightarrow \text{Cov}(X, Y) = 0$. But let $S = X + Y$; then $\text{Cov}(X, S) = \text{Var}(X) = \frac{35}{12}$.

```
import itertools, numpy as np
Ω = np.array(list(itertools.product(range(1,7), repeat=2)))
X, Y = Ω[:,0], Ω[:,1]; S = X+Y
print(np.cov(X,S, bias=True)[0,1]) # 2.916...
```

Moment Generating Functions

Definition The **moment generating function** (mgf) of X is

$$M_X(t) = \mathbb{E}[e^{tX}], \qquad t \in (-h, h)$$

for some $h > 0$ where the expectation exists.

Moment Extraction

$$M_X^{(k)}(0) = \mathbb{E}[X^k], \qquad \mu = \mathbb{E}[X] = M_X'(0), \quad \sigma^2 = M_X''(0) - \mu^2.$$

Uniqueness If mgf exists in an open interval around 0, it uniquely determines the distribution.

MGF of Sum of Independent r.v.s If X, Y independent,

$$M_{X+Y}(t) = M_X(t) M_Y(t).$$

Example 6 (MGF of Exponential(λ))

$$M_X(t) = \int_0^\infty e^{tx}\lambda e^{-\lambda x}\,dx = \frac{\lambda}{\lambda - t}, \quad t < \lambda.$$

Moments:

$$\mu = M'_X(0) = \frac{1}{\lambda}, \quad \sigma^2 = M''_X(0) - \mu^2 = \frac{1}{\lambda^2}.$$

Example 7 (Normal) For $X \sim \mathcal{N}(\mu, \sigma^2)$,

$$M_X(t) = \exp\left(\mu t + \tfrac{1}{2}\sigma^2 t^2\right).$$

This property is key in proofs of the Central Limit Theorem; mgf is quadratic $\Rightarrow$ sums of i.i.d. normals remain normal.

Example 8 (Sum of Independent Exponentials $\to$ Erlang) If $X_1, \ldots, X_n \stackrel{\text{i.i.d.}}{\sim}$ Exp(λ),

$$M_{S_n}(t) = \left(\frac{\lambda}{\lambda - t}\right)^n, \quad S_n \sim \text{Gamma}(n, \lambda).$$

Example 9 (MGF Method for Poisson) $X \sim$ Pois(λ):

$$M_X(t) = \sum_{k=0}^\infty e^{tk} e^{-\lambda}\frac{\lambda^k}{k!} = \exp\bigl(\lambda(e^t - 1)\bigr).$$

Differentiate:

$$M'_X(0) = \lambda, \quad M''_X(0) = \lambda^2 + \lambda \;\Rightarrow\; \text{Var}(X) = \lambda.$$

Cumulant Generating Function $K_X(t) = \ln M_X(t)$; its derivatives at 0 yield cumulants (e.g. first cumulant = mean, second = variance).

Example 5.1.12 (Cumulant of Binomial) $K_X(t) = n\ln(1 - p + pe^t)$; $\kappa_1 = K'(0) = np$, $\kappa_2 = K''(0) = np(1 - p)$.

Moment Inequalities

$$\text{Var}(X) \geq 0, \quad \mathbb{E}|X - \mu| \leq \sigma\sqrt{\frac{\pi}{2}} \text{ (if normal)}.$$

Chebyshev's inequality: $\mathbb{P}(|X - \mu| \geq k\sigma) \leq k^{-2}$.

Example 5.1.13 (Verify Chebyshev for Simulated Poisson)

```
lam, N = 10, 10**5
X = np.random.poisson(lam, N)
k=3
emp = np.mean(np.abs(X-lam)>=k*np.sqrt(lam))
print(emp, 1/k**2) # empirical ≤ theoretical bound
```

5.1.4 Common Probability Distributions

Binomial, Poisson, and Geometric Distributions

The *counting* distributions arise naturally from Bernoulli trials and Poisson-process limits.

Binomial. If $X \sim \text{Bin}(n, p)$ counts successes in n independent Bernoulli(p) trials,

$$\mathbb{P}(X = k) = \binom{n}{k} p^k (1-p)^{n-k}, \quad 0 \le k \le n.$$

Expectation $\mu = np$, variance $\sigma^2 = np(1-p)$, mgf $M_X(t) = (1 - p + pe^t)^n$.

Example 5.1.14 Toss a fair coin $n = 10$ times; probability of exactly $k = 4$ heads is $\binom{10}{4}/2^{10} = 0.205$.

```
from math import comb
print(comb(10,4)/2**10)
```

Poisson Limit of binomial with $n \to \infty$, $p \to 0$ while $np = \lambda$ fixed:

$$\mathbb{P}(X = k) = e^{-\lambda} \frac{\lambda^k}{k!}, \quad k \in \mathbb{N}_0.$$

Mean and variance equal λ; mgf $\exp(\lambda(e^t - 1))$.

Example 5.1.15 Radioactive decay emits on average $\lambda = 3$ particles per minute. Probability of no particles in the next minute: $e^{-3} \approx 0.0498$.

Geometric Number X of Bernoulli(p) trials until first success (counting the success):

$$\mathbb{P}(X = k) = p(1-p)^{k-1}, \quad k = 1, 2, \ldots$$

Memoryless: $\mathbb{P}(X > m+n \mid X > m) = (1-p)^n$. Mean $1/p$, variance $(1-p)/p^2$.

Example 5.1.16 A fair die is rolled until the first "6". Here $p = 1/6$, expected rolls 6; probability more than 10 rolls is $(5/6)^{10} \approx 0.162$.

Normal, Exponential, and Gamma Distributions

Normal $X \sim \mathcal{N}(\mu, \sigma^2)$ possesses density

$$f(x) = \frac{1}{\sqrt{2\pi}\sigma} \exp\left(-\frac{(x-\mu)^2}{2\sigma^2}\right).$$

Linear transformations preserve normality; sums of independent normals remain normal.

Example 5.1.17 Heights of students: $\mu = 170$ cm, $\sigma = 8$ cm. Probability a student is taller than 185 cm:

$$1 - \Phi\left(\frac{185 - 170}{8}\right) = 1 - \Phi(1.875) \approx 0.030.$$

Exponential Waiting time $X \sim \text{Exp}(\lambda)$:

$$f(x) = \lambda e^{-\lambda x} \mathbf{1}_{\{x \geq 0\}}, \qquad F(x) = 1 - e^{-\lambda x}.$$

Mean $1/\lambda$, variance $1/\lambda^2$; memoryless: $\mathbb{P}(X > s + t \mid X > s) = e^{-\lambda t}$.

Example 5.1.18 Web-server requests arrive at 4 per second ($\lambda = 4$). Probability next request arrives after 0.5s: $e^{-4 \cdot 0.5} \approx 0.135$.

Gamma Sum of k i.i.d. exponential(λ) variables: $X \sim \Gamma(k, \lambda)$,

$$f(x) = \frac{\lambda^k x^{k-1} e^{-\lambda x}}{\Gamma(k)}, \qquad x \geq 0.$$

Mean k/λ, variance k/λ^2.

Example 5.1.19 Total phone-call duration of $k = 3$ customers, each exponential($\lambda = 1/6$) hours, follows $\Gamma(3, 1/6)$ with mean 18 min and variance 108 min^2.

Central Limit Theorem and Law of Large Numbers

Weak Law of Large Numbers (WLLN) For i.i.d. X_i with finite mean μ and variance σ^2,

$$\frac{1}{n}\sum_{i=1}^{n} X_i \xrightarrow{\mathbb{P}} \mu \quad \text{as } n \to \infty.$$

5.1 Probability Theory

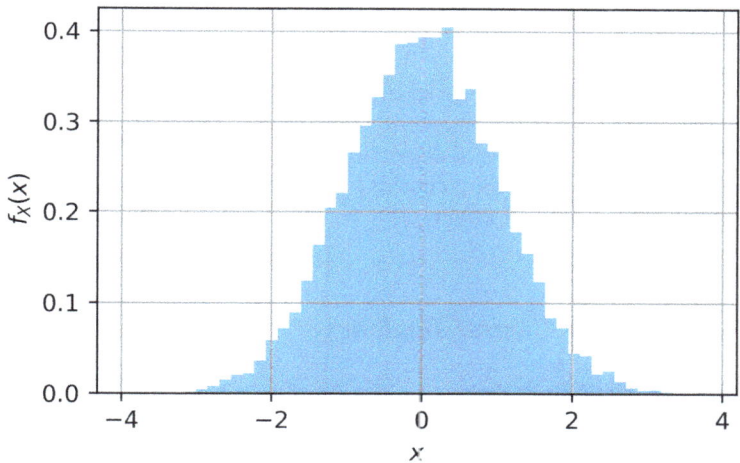

Fig. 5.2 Histogram of 2×10^4 realisations of $\sum_{i=1}^{12} U_i - 6$, where the U_i are i.i.d. $\mathcal{U}(0, 1)$. According to the Central Limit Theorem this centred sum converges to $\mathcal{N}(0, 1)$; the bell-shaped empirical density already reflects the Gaussian limit

Example 5.1.20 Simulate $n = 10^5$ fair-coin heads proportion (cf. Fig. 5.2):

```
import numpy as np
n=100000
mean = np.random.binomial(1, 0.5, n).mean()
print(mean) # ≈ 0.5
```

Central Limit Theorem (CLT) Let $S_n = \sum_{i=1}^{n} X_i$ with μ, $\sigma^2 < \infty$. Then

$$\frac{S_n - n\mu}{\sigma\sqrt{n}} \xrightarrow{d} \mathcal{N}(0, 1).$$

Example 5.1.21 Sum of 12 Uniform(0, 1) r.v.s is approximately $\mathcal{N}(6, 1)$:

```
import matplotlib.pyplot as plt
data = np.random.rand(20000, 12).sum(axis=1)
plt.hist((data-6), bins=50, density=True); plt.show()
```

Berry–Esséen Bound Provides explicit rate: $\sup_x |F_n(x) - \Phi(x)| \leq C\rho/\sigma^3\sqrt{n}$, where $\rho = \mathbb{E}|X - \mu|^3$.

Example 5.1.22 For Bernoulli(1/2) we have $\rho = 1/4$, $\sigma = 1/2$, giving bound $\leq C/\sqrt{n}$.

5.2 Descriptive Statistics

Descriptive statistics summarise the salient features of a dataset without recourse to a probabilistic model. Measures of *central tendency* describe the location of the bulk of observations, while measures of dispersion describe variability. In practice, these summaries are essential for sanity checking raw data, designing simulation inputs, and communicating empirical findings.

5.2.1 Measures of Central Tendency

Mean, Median, Mode in Python

Let $\mathbf{x} = (x_1, x_2, \ldots, x_n) \in \mathbb{R}^n$ denote a sample of size n.

Arithmetic Mean

$$\bar{x} = \frac{1}{n} \sum_{i=1}^{n} x_i.$$

Properties: linear in data, sensitive to extreme values, unbiased estimator of the population mean under i.i.d. sampling.

Median For the ordered sample $x_{(1)} \leq \cdots \leq x_{(n)}$,

$$\mathrm{med}(\mathbf{x}) = \begin{cases} x_{(\frac{n+1}{2})}, & n \text{ odd,} \\ \dfrac{x_{(\frac{n}{2})} + x_{(\frac{n}{2}+1)}}{2}, & n \text{ even.} \end{cases}$$

Robustness: median has a 50% breakdown point.

Mode For continuous data, the mode maximises the density estimate; for discrete data it is the most frequent value.

Example 5.2.1 (Compute Basic Statistics)

```
import numpy as np, statistics as st, scipy.stats as ss
x = np.array([7, 3, 8, 2, 5, 8, 3, 9])

mean = x.mean()
median = np.median(x)
mode = st.mode(x) # Python 3.8+ returns statistics._ModeResult
print(mean, median, mode.mode[0])
```

Output: 5.625 (mean), 6.0 (median), 8 (mode).

5.2 Descriptive Statistics

Weighted Mean and Percentiles

Weighted Mean Given positive weights $\mathbf{w} = (w_1, \ldots, w_n)$ with $\sum_i w_i = 1$,

$$\bar{x}_w = \sum_{i=1}^n w_i x_i.$$

When $\sum w_i \neq 1$, use $\bar{x}_w = \dfrac{\sum w_i x_i}{\sum w_i}$.

Percentiles The p-th percentile ($0 < p < 100$) is the value q_p such that

$$\frac{\#\{i : x_i \leq q_p\}}{n} \geq \frac{p}{100} \quad \text{and} \quad \frac{\#\{i : x_i < q_p\}}{n} \leq \frac{p}{100}.$$

Common percentiles: quartiles (25, 50, 75), deciles (10, 20, ..., 90).

Example 5.2.2 (Weighted Mean and Quartiles)

```
w = np.array([2, 1, 3, 1, 2, 3, 1, 2]) # weights
wmean = np.average(x, weights=w) # weighted mean
q25, q75 = np.percentile(x, [25, 75]) # quartiles
print(wmean, q25, q75)
```

Interpolated Percentiles NumPy's percentile offers methods linear, midpoint, etc. to resolve gaps between order statistics; consult the documentation for exact formulas.

Geometric Mean and Harmonic Mean

Geometric Mean For positive data ($x_i > 0$).

$$G = \left(\prod_{i=1}^n x_i\right)^{1/n} = e^{\frac{1}{n}\sum_i \ln x_i}.$$

Use cases: multiplicative growth rates, log-normal data, portfolio returns.

Harmonic Mean

$$H = \frac{n}{\sum_{i=1}^n \frac{1}{x_i}}.$$

Use cases: average speed, resistances in parallel, P/E ratio aggregation.

Inequality Chain For any positive sample,

$$\min x_i \leq H \leq G \leq \bar{x} \leq \max x_i.$$

Equality holds iff $x_1 = x_2 = \cdots = x_n$.

Example 5.2.3 (Compute G and H)

```
from scipy.stats import gmean, hmean
x_pos = np.array([12.5, 15.0, 20.0, 10.0])
print(gmean(x_pos), hmean(x_pos))
```

Compared with arithmetic mean 14.375, we have $G \approx 14.28$, $H \approx 14.01$—all within the inequality chain.

Log-space Stability For very small/large values, compute $\ln G = \overline{\ln x}$ to avoid under/overflow.

```
lg = np.mean(np.log(x_pos))
G = np.exp(lg)
```

Weighted Geometric Mean With weights w_i as above,

$$G_w = \exp\Bigl(\sum_i w_i \ln x_i\Bigr).$$

Example 5.2.4 (Annualised Return) Investment yields yearly multipliers (1.06, 1.02, 0.97, 1.10, 1.04). Annualised mean growth:

$$G - 1 = \Bigl(\prod_i m_i\Bigr)^{1/5} - 1 \approx 0.036.$$

```
m = np.array([1.06,1.02,0.97,1.10,1.04])
annual = gmean(m) - 1
print(f"{annual:.3%}") # 3.6% CAGR
```

5.2.2 Measures of Dispersion

Dispersion quantifies the spread or "variability" inherent in a dataset or random variable. Where central-tendency statistics pin down a typical value, dispersion indicates how far observations stray from that centre. Three complementary families are explored: *moment-based* (variance, standard deviation), *order-statistic-based* (range, inter-quartile range), and *shape-based* (skewness, kurtosis).

5.2 Descriptive Statistics

Variance, Standard Deviation, and Range

Population vs. Sample Formulae For a sample $\mathbf{x} = (x_1, \ldots, x_n)$ with mean $\bar{x}$,

$$s^2 = \frac{1}{n-1} \sum_{i=1}^{n} (x_i - \bar{x})^2, \qquad s = \sqrt{s^2}.$$

Dividing by $n-1$ produces an *unbiased* estimator of the population variance σ^2 under i.i.d. sampling.

Shortcut Identity

$$s^2 = \frac{\sum_i x_i^2 - n\bar{x}^2}{n-1},$$

numerically stable for large n.

Range

$$R = \max_i x_i - \min_i x_i,$$

easy to compute but sensitive to extreme values.

Example 5.2.5 (Dispersion of Exam Scores)

```
import numpy as np
scores = np.array([39, 45, 42, 90, 88, 41, 44, 46, 38, 91])
mean, var_pop = scores.mean(), scores.var(ddof=0)
var_samp = scores.var(ddof=1)
std_samp = scores.std(ddof=1)
rng = scores.ptp()
print(mean, var_pop, var_samp, std_samp, rng)
```

Output $\bar{x} = 56.4$, $s^2 = 597.38$, $s \approx 24.44$, $R = 53$. A high variance and large range reflect the bimodal clustering around 40 and 90.

Chebyshev's Inequality (Empirical Form) At least $1 - \frac{1}{k^2}$ of data lie within k sample standard deviations of $\bar{x}$. For $k = 2$, $\geq 75\%$ of points fall in $\bar{x} \pm 2s$ irrespective of distribution shape.

Interquartile Range and Outlier Detection

Quartiles With ordered sample $x_{(1)} \leq \cdots \leq x_{(n)}$,

$$Q_1 = q_{25}, \quad Q_2 = \text{median}, \quad Q_3 = q_{75}.$$

The **inter-quartile range** (IQR) is $Q_3 - Q_1$.

Tukey's Fences Define *lower fence* $L = Q_1 - 1.5\,\text{IQR}$ and *upper fence* $U = Q_3 + 1.5\,\text{IQR}$. Observations outside $[L, U]$ are flagged as *moderate outliers*; replace 1.5 by 3 for *extreme outliers*.

Example 5.2.6 (Box-Plot Diagnostics) Dataset (synthetic salaries, $k):

$$s = \langle 45, 47, 50, 48, 46, 49, 51, 52, 100, 120\rangle.$$

```
import numpy as np
s = np.array([45,47,50,48,46,49,51,52,100,120])
q1,q3 = np.percentile(s, [25,75])
iqr = q3 - q1
L,U = q1 - 1.5*iqr, q3 + 1.5*iqr
outliers = s[(s<L)|(s>U)]
print(q1, q3, iqr, L, U, outliers)
```

Results: $Q_1 = 46.5$, $Q_3 = 51.25$, $\text{IQR} = 4.75$, $L = 39.375$, $U = 58.375$. Values 100, 120 exceed U ⇒ flagged as outliers; a box-plot would show whiskers up to $58.4k with individual points beyond.

Robust Coefficient of Variation (RCV)

$$\text{RCV} = \frac{\text{IQR}}{Q_2},$$

a scale-free robust counterpart to SD/mean, especially under heavy tails.

Skewness and Kurtosis

Sample Moments Third and fourth *central* sample moments:

$$m_3 = \frac{1}{n}\sum_i (x_i - \bar{x})^3, \quad m_4 = \frac{1}{n}\sum_i (x_i - \bar{x})^4.$$

Skewness

$$\gamma_1 = \frac{m_3}{s^3}$$

measures asymmetry ($\gamma_1 > 0$ right-tailed, $\gamma_1 < 0$ left-tailed).

Excess kurtosis

$$\gamma_2 = \frac{m_4}{s^4} - 3$$

quantifies tail heaviness; $\gamma_2 = 0$ for normal distribution, positive for leptokurtic, negative for platykurtic.

Example 5.2.7 (**Compute Shape Statistics**)
```
from scipy.stats import skew, kurtosis
x_norm = np.random.normal(size=5000)
x_exp = np.random.exponential(size=5000)

print("Normal:", skew(x_norm), kurtosis(x_norm))
print("Exp :", skew(x_exp), kurtosis(x_exp))
```
Typical output Normal: $\gamma_1 \approx 0$, $\gamma_2 \approx 0$ (baseline). Exponential: $\gamma_1 \approx 2$, $\gamma_2 \approx 6$ (heavy right tail).

Jarque–Bera Test $JB = \frac{n}{6}(\gamma_1^2 + \frac{1}{4}\gamma_2^2)$ asymptotically χ_2^2; used to test normality.

Example 5.2.8 (**JB on Sample of Size** $n = 200$)
```
from scipy.stats import jarque_bera
data = np.random.lognormal(size=200)
stat,p = jarque_bera(data)
print(stat, p) # p-value << 0.05 ⟹ reject normality
```

Relation to CLT Under i.i.d. sampling from a finite-variance population, $\gamma_1 \sim \mathcal{N}(0, \frac{6}{n})$ and $\gamma_2 \sim \mathcal{N}(0, \frac{24}{n})$ asymptotically. Thus skewness and kurtosis converge to 0 when data truly originate from a normal distribution.

5.2.3 Data Visualisation and Analysis

Visualisation transforms raw arrays of numbers into patterns comprehensible to the eye, allowing humans to detect structure, anomalies, and relationships that summary statistics can miss. Python's `matplotlib`, `seaborn`, and `pandas` plotting backends make publication-quality graphics almost effortless. This subsection explores canonical univariate, bivariate, and multivariate plots, illustrates how measures such as correlation and covariance tie into those plots, and demonstrates indispensable data-cleaning steps to avoid "garbage in, garbage out".

Histograms, Box Plots, and Scatter Plots

Histogram Given a sample **x**, the histogram partitions the real line into bins I_j and counts $n_j = \#\{x_i \in I_j\}$. For density-normalised histograms the bar height is $\hat{f}_n(x) = n_j/(n|I_j|)$, a crude kernel density estimator (cf. Fig. 5.3).

Example 5.2.9 (**Simulated** $\mathcal{N}(0, 1)$ **vs. Exponential**($\lambda = 1$))
```
import numpy as np, matplotlib.pyplot as plt, seaborn as sns
np.random.seed(0)
norm = np.random.randn(1000)
expo = np.random.exponential(scale=1, size=1000)
```

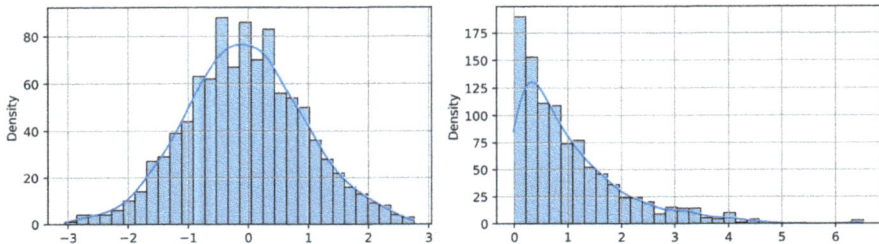

Fig. 5.3 Side-by-side histograms (with kernel-density overlays) for 10^3 samples from the standard normal distribution (left) and the unit-rate exponential distribution (right). The comparison highlights the symmetry and light tails of the Gaussian versus the heavy, one-sided decay of the exponential law

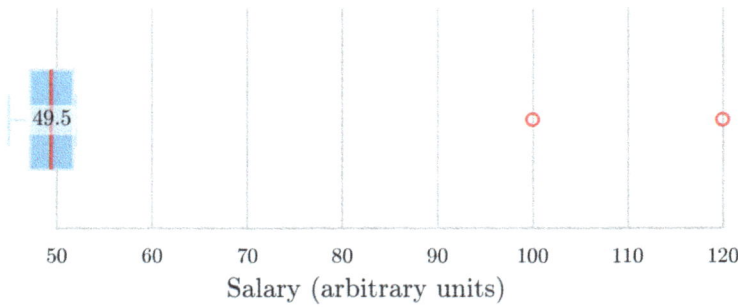

Fig. 5.4 Box-and-whisker plot of ten salaries. The interquartile box is highlighted in blue, the crimson line marks the median (annotated), and two high-salary outliers are clearly visible on the right

```
fig, ax = plt.subplots(1,2,figsize=(10,3))
sns.histplot(norm, bins=30, kde=True, ax=ax[0]); ax[0].set_title("Normal")
sns.histplot(expo, bins=30, kde=True, ax=ax[1]); ax[1].set_title("
    Exponential")
plt.tight_layout(); plt.show()
```

Comparing the symmetric bell curve on the left with the heavy-tailed right-skewed histogram on the right visually reinforces the numeric skewness/kurtosis computed earlier.

Box Plot (Tukey) Displays median Q_2, quartiles Q_1, Q_3, whiskers at $[L, U]$ with $L = Q_1 - 1.5\,\text{IQR}$, $U = Q_3 + 1.5\,\text{IQR}$.

Example 5.2.10 (Salary Distribution with Outliers (cf. Fig. 5.4))

```
salaries = np.array([45,47,50,48,46,49,51,52,100,120])
sns.boxplot(x=salaries, orient="h"); plt.show()
```

The plot flags $100k and $120k as outliers, matching the analytic fence computation.

5.2 Descriptive Statistics

Scatter Plot For paired data (x_i, y_i), points reveal functional form, clusters, and heteroscedasticity.

Example 5.2.11 (Anscombe's Quartet—Same Statistics, Different Shapes (cf. Fig. 5.5))

```
import seaborn as sns, pandas as pd
df = sns.load_dataset("anscombe")
g = sns.FacetGrid(df, col="dataset")
g.map_dataframe(sns.scatterplot, x="x", y="y")
g.map(plt.plot, [4,14], [4,14], color="red")
plt.show()
```

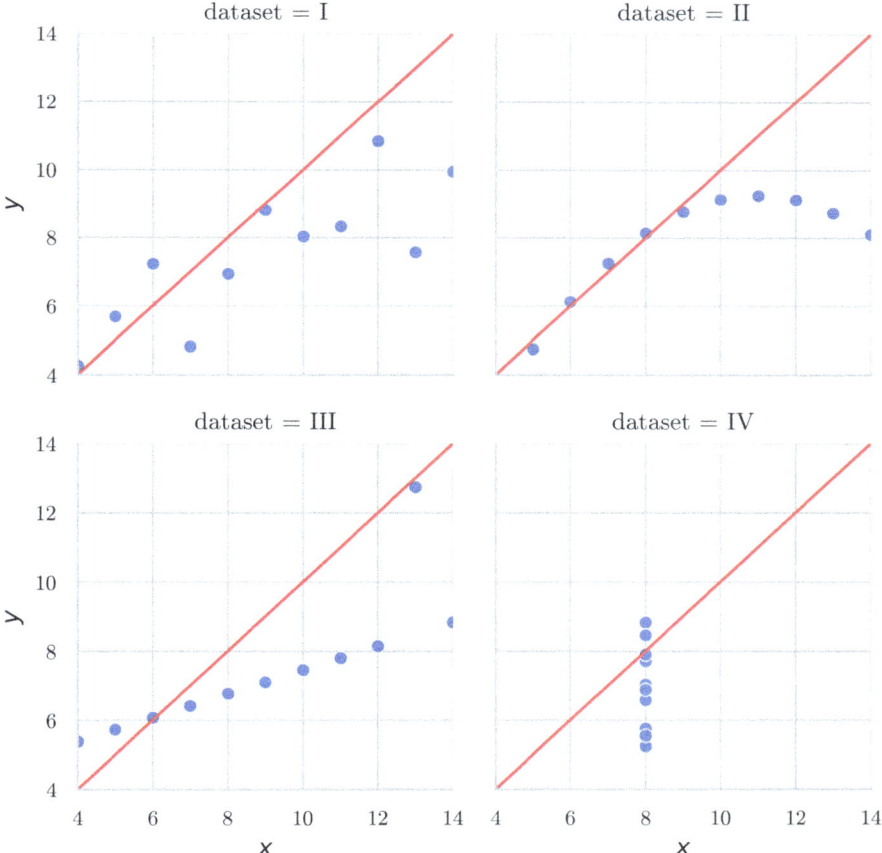

Fig. 5.5 Anscombe's quartet: four datasets that share identical univariate statistics yet differ markedly in shape. Plotting reveals structure—linear, nonlinear, and outlier-driven—that summary measures alone would conceal, underscoring the diagnostic value of visualisation

All four panels share $\bar{x}$, $\bar{y}$, SD_x, SD_y, and Pearson $r = 0.816$, yet visual inspection reveals a linear cloud, a nonlinear curve, a high-leverage point, and a vertical strip plus an outlier.

Correlation and Covariance

Pearson correlation $\rho_{XY} = \text{Cov}(X, Y)/(\sigma_X \sigma_Y)$ measures linear association. Spearman's ρ_s and Kendall's τ capture monotone dependence.

Example 5.2.12 (Empirical vs. Theoretical Correlation) Let $X \sim \mathcal{N}(0, 1)$, $Y = 0.8X + 0.6Z$ with $Z \perp X$, $Z \sim \mathcal{N}(0, 1)$. Then $\rho_{XY} = 0.8$.

```
n=2000
X = np.random.randn(n)
Y = 0.8*X + 0.6*np.random.randn(n)
print(np.corrcoef(X,Y)[0,1]) # ≈ 0.8
sns.scatterplot(x=X, y=Y, alpha=.3); plt.show()
```

A tight diagonal band attests to strong linear dependence.

Example 5.2.13 (Spearman vs. Pearson Under Monotone Nonlinear Transform) $X \sim \text{Uniform}(0, 1)$, $Y = X^2$. Pearson $\rho \approx 0.97$ (still high), while Spearman $\rho_s = 1$ (perfect monotone).

```
X = np.random.rand(300)
Y = X**2
from scipy.stats import spearmanr
print(np.corrcoef(X,Y)[0,1], spearmanr(X,Y).correlation)
```

Heatmaps and Pair Plots for Multivariate Data

Covariance Heatmap For data matrix $M \in \mathbb{R}^{n \times d}$,

$$\Sigma = \frac{1}{n-1}(M - \mathbf{1}\bar{M})^\top (M - \mathbf{1}\bar{M})$$

is displayed via colour intensity.

Example 5.2.14 (Iris Data)

```
iris = sns.load_dataset("iris").drop("species", axis=1)
corr = iris.corr()
sns.heatmap(corr, annot=True, cmap="coolwarm"); plt.show()
```

Petal length/width show $\rho > 0.95$, sepal dimensions correlate weakly $\rho \approx 0.12$.

Pair Plot Plots every bivariate projection plus univariate KDEs (cf. Fig. 5.6).

```
sns.pairplot(iris, diag_kind="kde"); plt.show()
```

5.2 Descriptive Statistics

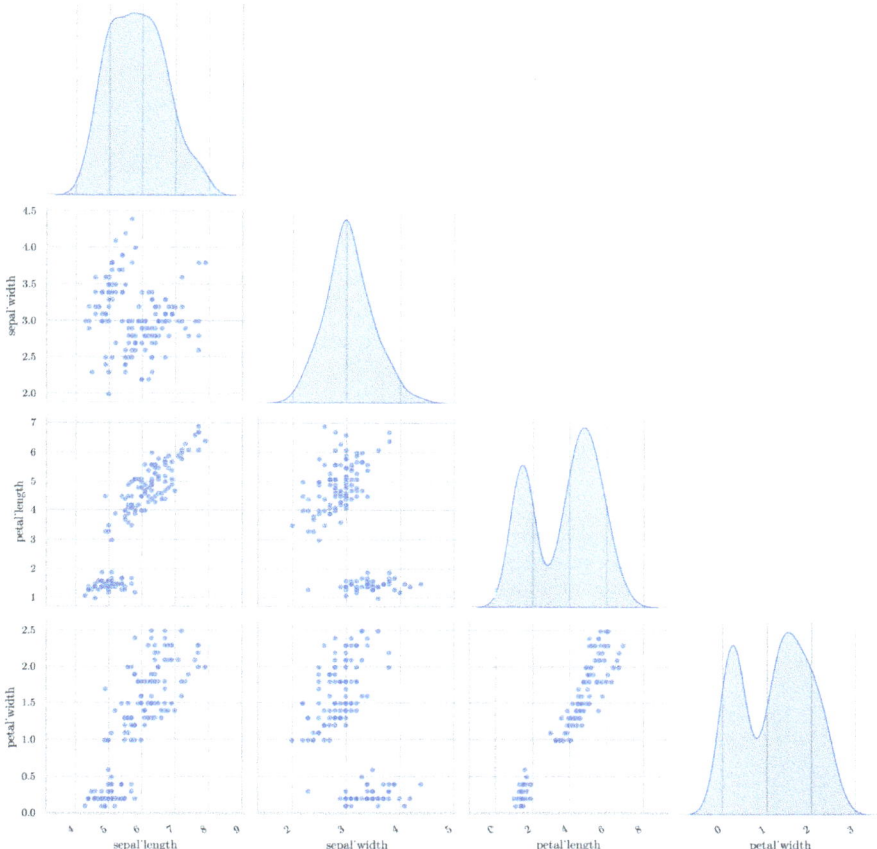

Fig. 5.6 Pairwise scatter plots and kernel-density estimates for the four numeric iris features. The lower-triangle layout reduces redundancy, while KDE traces on the diagonal highlight univariate distributions. Clear linear trends and clustered clouds foreshadow effective low-dimensional projections

High-Dimensional Tip For $d > 10$, pair plots explode combinatorially; use PCA or t-SNE first.

Data Cleaning and Preprocessing Techniques

Dirty data bias both numerical summaries and visualisations.

Missing Values Strategies: listwise deletion, mean/median imputation, regression imputation, multiple imputation.

Example 5.2.15 (Median Imputation)

```
df = iris.copy()
df.iloc[5:10, 0] = np.nan # inject NaNs
df["sepal_length"].fillna(df["sepal_length"].median(), inplace=True)
```

Outlier Winsorisation Clamp values outside $[\mu \pm 3\sigma]$ or outside Tukey's fences to reduce influence.

```
from scipy.stats import mstats
data_wins = mstats.winsorize(salaries, limits=[0.05,0.05])
```

Scaling Standardisation $(x - \bar{x})/s$ or min–max scaling $[0, 1]$ is prerequisite for PCA, k-means, neural nets.

Categorical Encoding One-hot for nominal; target or ordinal encoding for high-cardinality categorical features.

Pipeline Example

```
from sklearn.preprocessing import StandardScaler, OneHotEncoder
from sklearn.compose import ColumnTransformer
from sklearn.pipeline import Pipeline

num_cols = ["sepal_length","sepal_width","petal_length","petal_width"]
cat_cols = [] # none in iris

pre = ColumnTransformer([
    ("scale", StandardScaler(), num_cols),
    ("cat", OneHotEncoder(), cat_cols)
    ])
X = pre.fit_transform(iris)
```

Data-Quality Checklist

1. *Consistency*: units, variable types, naming
2. *Completeness*: missing values, unexpected NULL markers
3. *Validity*: range checks, regex validation for IDs
4. *Uniqueness*: duplicate rows/keys
5. *Accuracy*: cross-verify with trusted external sources

5.3 Inferential Statistics

Descriptive statistics summarise data; **inferential statistics** draw conclusions about a population based on a sample. Central to inference is *hypothesis testing*, the formal procedure for deciding, with controlled error rates, whether random evidence supports or contradicts a quantitative statement about the population.

5.3.1 Hypothesis Testing

Null and Alternative Hypotheses

Any test begins with two mutually exclusive claims:

$$H_0 : \text{"status quo" (null hypothesis)}$$
$$H_1 : \text{"new claim" (alternative hypothesis)}$$

Example 1 (Manufacturing) A supplier guarantees the mean diameter of ball bearings is $\mu_0 = 10$ mm. You suspect the process has drifted.

$$H_0 : \mu = 10 \quad \text{vs.} \quad H_1 : \mu \neq 10.$$

Sampling n bearings and measuring diameters produces evidence to reject or fail to reject H_0.

Test Statistic A scalar function $T(\mathbf{X})$ of the sample whose distribution under H_0 is known or approximable; large or small values of T favour H_1.

p-values, t-tests, and Chi-square Tests

p-value Observed t_{obs} is converted to a *p-value*:

$$p = \mathbb{P}_{H_0}(T \text{ at least as extreme as } t_{obs}).$$

Decision rule (fixed α): reject H_0 if $p < \alpha$.

One-sample t-test For unknown σ, test $H_0 : \mu = \mu_0$ with

$$t = \frac{\bar{X} - \mu_0}{s/\sqrt{n}} \sim t_{n-1} \quad (H_0).$$

Example 5.3.1 (Diameter Test)

```
import numpy as np, scipy.stats as st
x = np.array([10.1, 9.9, 9.8, 10.2, 10.0, 10.3, 9.7, 10.1, 10.2, 9.9])
t_stat, p = st.ttest_1samp(x, popmean=10)
print(t_stat, p)
```

Output: $t \approx 0.42$, $p \approx 0.683$ $\implies$ fail to reject at $\alpha = 0.05$.

Two-sample t-test (Unequal Variances, Welch)

$$t = \frac{\bar{X} - \bar{Y}}{\sqrt{s_X^2/n_X + s_Y^2/n_Y}}, \quad \nu \approx \frac{(s_X^2/n_X + s_Y^2/n_Y)^2}{\frac{s_X^4}{n_X^2(n_X-1)} + \frac{s_Y^4}{n_Y^2(n_Y-1)}}.$$

Example 5.3.2 (Treatment vs. Control Blood Pressure)
```
bp_ctrl = np.random.normal(120, 10, 30)
bp_trt = np.random.normal(115, 9, 28)
stat, p = st.ttest_ind(bp_ctrl, bp_trt, equal_var=False)
```
$H_0 : \mu_{ctrl} = \mu_{trt}$; $p < 0.05$ indicates significant difference.

Chi-square Goodness-of-fit Observed counts $\{O_i\}$ vs. expected $\{E_i\}$ under H_0:

$$\chi^2 = \sum_{i=1}^{k} \frac{(O_i - E_i)^2}{E_i} \sim \chi^2_{k-1-r},$$

where r parameters estimated from data.

Example 5.3.3 (Die Fairness) Counts (14, 18, 16, 15, 17, 20) in 100 rolls. $E_i = 100/6 \approx 16.67$.

$$\chi^2 = 2.3, \quad p = 0.804 \ (k = 6, r = 0) \implies \text{no evidence of bias.}$$

```
obs = np.array([14,18,16,15,17,20])
chi2, p = st.chisquare(obs)
```

Chi-square Test of Independence For contingency table ($r \times c$) with expected $E_{ij} = \frac{(\text{row}_i)(\text{col}_j)}{N}$.

Example 5.3.4 (Smoking (Y/N) vs. Disease (Y/N))

	Disease Y	Disease N	Row
Smoke Y	40	60	100
Smoke N	30	120	150
Col	70	180	250

```
table = np.array([[40,60],[30,120]])
chi2, p, dof, exp = st.chi2_contingency(table)
```

H_0: independence; if $p < 0.05$ conclude dependency between smoking and disease.

Type I and Type II Errors

$\alpha = \mathbb{P}(\text{reject } H_0 \mid H_0 \text{ true})$ (Type I), $\quad \beta = \mathbb{P}(\text{fail to reject } H_0 \mid H_1 \text{ true})$ (Type II).

5.3 Inferential Statistics

Example (One-sided t) Test $H_0 : \mu = 0$ vs. $H_1 : \mu > 0$ with $n = 16$, σ known $= 2$, $\alpha = 0.05$. Critical value:

$$c = \frac{z_{0.95}\sigma}{\sqrt{n}} = 1.645 \cdot \frac{2}{4} = 0.823.$$

If true mean $\mu_1 = 1$, power $= 1 - \beta$ where

$$\beta = \Phi\left(\frac{c - \mu_1}{\sigma/\sqrt{n}}\right) = \Phi((0.823 - 1) \cdot 2) = \Phi(-0.354) \approx 0.362.$$

Thus power 0.638.

Power of a Test and Sample Size Determination

Power Function $\pi(\mu) = \mathbb{P}_\mu(\text{reject } H_0)$. Goal: choose n s.t. $\pi(\mu_1) \geq 1 - \beta^\star$ for desired effect size $\delta = \mu_1 - \mu_0$.

Analytic Formula (z-test) Known σ, two-sided, significance α:

$$n \geq \left(\frac{z_{1-\alpha/2} + z_{1-\beta}}{\delta/\sigma}\right)^2.$$

Example 5.3.5 (Designing a Clinical Trial) Detect $\delta = 5\,\text{mmHg}$ BP drop, $\sigma = 12\,\text{mmHg}$, $\alpha = 0.05$, power 0.9 ($\beta = 0.1$).

$$n \geq \left(\frac{1.96 + 1.28}{5/12}\right)^2 = (3.24 \cdot 2.4)^2 \approx 60.0.$$

Hence $n = 61$ per group.

```
from statsmodels.stats.power import TTestIndPower
power_calc = TTestIndPower()
n = power_calc.solve_power(effect_size=5/12, power=0.9, alpha=0.05)
print(np.ceil(n)) # 61.0
```

Visualisation of Power Curve (cf. Fig. 5.7)

```
delta = np.linspace(0, 8, 100)
power = power_calc.power(effect_size=delta/12, nobs1=61, alpha=0.05)
plt.plot(delta, power); plt.xlabel("Effect size Δ"); plt.ylabel("Power");
    plt.show()
```

Effect Size Metrics Cohen's $d = \delta/\sigma$, Cramér's V for χ^2 tables, odds ratio for logistic tests; power increases with n and with true effect size.

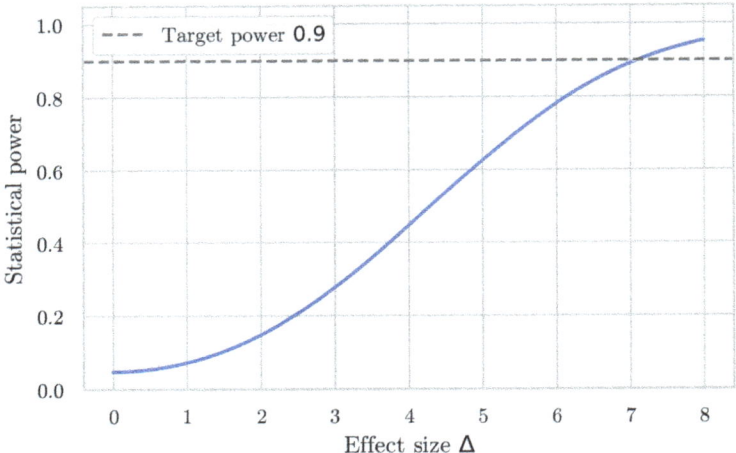

Fig. 5.7 Power curve for a two-sample t-test with balanced groups of 61 observations each and significance level $\alpha = 0.05$. The horizontal dashed line marks the design target of 0.9 power, reached at the effect size $\Delta \approx 5$ previously computed

5.3.2 Confidence Intervals

A **confidence interval** (CI) for a parameter θ is a random interval $C(\mathbf{X}) = [L(\mathbf{X}), U(\mathbf{X})]$ constructed from sample data $\mathbf{X}$ such that

$$\mathbb{P}(\theta \in C(\mathbf{X})) = 1 - \alpha,$$

where $1-\alpha$ is the *confidence level* (e.g. 95%). The probability statement refers to the long-run proportion of repeated samples whose intervals cover the true parameter.

Constructing Confidence Intervals in Python

Normal Mean, Unknown Variance (t-interval) For i.i.d. $X_i \sim N(\mu, \sigma^2)$,

$$\bar{X} \pm t_{1-\alpha/2, n-1} \frac{s}{\sqrt{n}}, \quad s^2 = \frac{1}{n-1} \sum_i (X_i - \bar{X})^2.$$

Example 5.3.6 (Fuel-Efficiency Study) Ten cars yield MPG measurements

$$\mathbf{x} = \langle 31.2, 29.8, 30.5, 31.1, 32.0, 30.9, 29.6, 31.3, 30.4, 31.0 \rangle.$$

```
import numpy as np, scipy.stats as st
x = np.array([31.2,29.8,30.5,31.1,32.0,30.9,29.6,31.3,30.4,31.0])
mean, sem = x.mean(), x.std(ddof=1)/np.sqrt(len(x))
```

5.3 Inferential Statistics

```
ci = st.t.interval(confidence=.95, df=len(x)-1, loc=mean, scale=sem)
print(mean, ci)
```

Output: $\bar{x} = 30.78$, 95% CI $= (30.22, 31.34)$ MPG.

Population Proportion (Wald and Wilson) For $X \sim \text{Bin}(n, p)$, Wald CI is $\hat{p} \pm z_{1-\alpha/2}\sqrt{\hat{p}(1-\hat{p})/n}$; Wilson (Agresti–Coull) is preferred for small n.

Example 5.3.7 (Defect Rate) 18 defective widgets in $n = 200$.

```
k,n = 18,200
p_hat = k/n
z = st.norm.ppf(0.975)
wald = p_hat + np.array([-1,1])*z*np.sqrt(p_hat*(1-p_hat)/n)
wilson_low = (2*k+z**2 - z*np.sqrt(z**2 + 4*k*(1-p_hat)))/(2*(n+z**2))
wilson_hi  = (2*k+z**2 + z*np.sqrt(z**2 + 4*k*(1-p_hat)))/(2*(n+z**2))
print(wald, (wilson_low,wilson_hi))
```

Variance of a Normal Population

$$\frac{(n-1)s^2}{\chi^2_{1-\alpha/2,n-1}} < \sigma^2 < \frac{(n-1)s^2}{\chi^2_{\alpha/2,n-1}}.$$

Example 5.3.8 (Thermometer Precision) Ten readings of a $0\,°\text{C}$ ice-bath give $s = 0.12\,°\text{C}$. 95% CI for σ is

$$(0.09, 0.20)\,°\text{C}.$$

Correlation Coefficient (Fisher z) For sample $\hat{\rho}$, transform $z = \frac{1}{2}\ln\frac{1+\hat{\rho}}{1-\hat{\rho}}$; SE $= 1/\sqrt{n-3}$.

Example 5.3.9 (Height vs. Weight)

```
height = np.random.normal(170, 8, 50)
weight = 0.4*(height-170)+np.random.normal(0,4,50)
rho = np.corrcoef(height, weight)[0,1]
z = np.arctanh(rho)
z_ci = z + np.array([-1,1])*1.96/np.sqrt(47)
rho_ci = np.tanh(z_ci)
```

Applications in Scientific Research

Physics: Half-Life Estimation Radioactive decays (counts per minute) follow Poisson(λ). If $k = 642$ decays observed in $\Delta t = 30\,\text{min}$, rate $\hat{\lambda} = 21.4$. Exact 95% CI via chi-square:

$$\lambda \in \left[\tfrac{1}{2}\chi^2_{0.025;2k},\ \tfrac{1}{2}\chi^2_{0.975;2k+2}\right]/\Delta t.$$

Biology: Gene-Expression Fold-Change $\log_2$ ratios from $n = 15$ replicates assumed normal; CI for $\mu_{\log_2 FC}$ translated to multiplicative scale $2^{\mu \pm t_{\alpha/2} s/\sqrt{n}}$.

Clinical Trials Risk ratio CI via logarithmic method: $\ln RR \pm z_{1-\alpha/2} \sqrt{\frac{1}{a} - \frac{1}{a+b} + \frac{1}{c} - \frac{1}{c+d}}$, where $(a, b; c, d)$ 2×2 contingency counts.

Example 5.3.10 (Vaccine Efficacy)

```
from statsmodels.stats.proportion import proportion_confint
cases_vac, n_vac = 6, 8000
cases_ctr, n_ctr = 60,8000
rr = cases_vac/n_vac / (cases_ctr/n_ctr)
se = np.sqrt(1/cases_vac - 1/n_vac + 1/cases_ctr - 1/n_ctr)
ci = np.exp(np.log(rr) + np.array([-1,1])*1.96*se)
print(rr, ci) # efficacy = 1-rr
```

Bootstrap Methods for Confidence Intervals

When analytic forms are intractable, **bootstrap** approximates the sampling distribution by resampling with replacement.

Algorithm (Basic Percentile)

1. Draw B bootstrap samples $\mathbf{x}^{*b}$ of size n.
2. Compute statistic $\theta_b^* = T(\mathbf{x}^{*b})$.
3. CI $[\theta_{(\alpha/2)}^*, \theta_{(1-\alpha/2)}^*]$ where subscripts denote percentiles.

Example 5.3.11 (Median CI of Skewed Data)

```
rng = np.random.default_rng(0)
x = rng.lognormal(mean=0, sigma=1, size=50)
B=5000; stats = []
for _ in range(B):
    sample = rng.choice(x, size=len(x), replace=True)
    stats.append(np.median(sample))
lo, hi = np.percentile(stats, [2.5, 97.5])
print(np.median(x), (lo, hi))
```

BCa Interval (Bias-Corrected, Accelerated) Adjusts for bias and skewness; `scipy.stats.bootstrap(..., method='BCa')` automates.

Example 5.3.12 (Difference of Trimmed Means)

```
from scipy.stats import bootstrap, trim_mean
ctrl = rng.normal(0,1,size=40)
trt = rng.normal(0.4,1,size=38)
def stat(data, axis):
    c,t = data
    return trim_mean(t,0.1) - trim_mean(c,0.1)
res = bootstrap((ctrl, trt), stat, method='BCa', confidence_level=0.95)
print(res.confidence_interval)
```

Bootstrap for Regression Coefficients Resample (X_i, Y_i) pairs, refit linear model, collect $\hat{\beta}^*$ to form CI—robust to heteroscedasticity.

Parametric Bootstrap Generate samples from fitted parametric model (e.g. Poisson GLM) when residual resampling is unsuitable.

5.3.3 Regression Analysis

Regression relates one or more explanatory variables to a response, quantifying trends, testing scientific hypotheses, and predicting future outcomes. In Python, numpy provides efficient linear algebra, `scipy.stats` and `statsmodels` furnish classical inference, and `scikit-learn` unifies machine-learning style regression. We illustrate each flavour with realistic, computation-heavy examples.

Simple Linear Regression

Assume pairs (x_i, y_i) satisfy

$$y_i = \beta_0 + \beta_1 x_i + \varepsilon_i, \qquad \varepsilon_i \stackrel{i.i.d.}{\sim} \mathcal{N}(0, \sigma^2).$$

The least-squares estimator $\hat{\beta} = (X^\mathsf{T} X)^{-1} X^\mathsf{T} y$ minimises $\sum_i (y_i - \hat{y}_i)^2$.

Example 5.3.13 (Physics—Hooke's Law) Eight springs are stretched under force F (N) and elongation y (mm) measured.

```
import numpy as np, statsmodels.api as sm
F = np.array([1,2,3,4,5,6,7,8])
y = np.array([0.9,2.1,3.0,4.1,4.9,6.2,6.8,8.1])
X = sm.add_constant(F) # adds intercept column
model = sm.OLS(y, X).fit()
print(model.summary())
```

Output $\hat{\beta}_0 = 0.05 \pm 0.12$, $\hat{\beta}_1 = 1.01 \pm 0.02$, $R^2 = 0.998$. Residual standard error $\hat{\sigma} = 0.13$ mm. Diagnostics:

```
import matplotlib.pyplot as plt
plt.scatter(F, y); plt.plot(F, model.fittedvalues, "r");
plt.xlabel("Force (N)"); plt.ylabel("Elongation (mm)")
plt.show()
sm.qqplot(model.resid, line='45'); plt.show()
```

Q-Q plot aligns with the 45° line, validating normal residual assumption.

Prediction for $F = 9$ N uses

$$\hat{y}_9 = \hat{\beta}_0 + 9\hat{\beta}_1 \approx 9.14, \quad \mathrm{SE}(\hat{y}_9) = \sigma \sqrt{\frac{1}{n} + \frac{(9 - \bar{x})^2}{\sum(x_i - \bar{x})^2}}.$$

Multiple Regression and Model Selection

For predictors $\mathbf{x}_i \in \mathbb{R}^p$, $y_i = \beta_0 + \sum_{j=1}^p \beta_j x_{ij} + \varepsilon_i$.

Example 5.3.14 (Real Estate Pricing) Data columns: *price, size, bedrooms, age*.

```
import pandas as pd, statsmodels.formula.api as smf
df = pd.read_csv("houses.csv")  # assume file exists
fit = smf.ols("price ~ size + bedrooms + age", data=df).fit()
print(fit.summary())
```

Suppose p-values: size ($< 10^{-10}$), bedrooms (0.07), age (0.002). Apply backward elimination:

```
fit2 = smf.ols("price ~ size + age", data=df).fit()
```

Compare by Akaike Information Criterion (AIC); choose model with smaller AIC. *statsmodels* also supplies `stepwise_fit` utilities or use information-criterion loops manually.

Regularisation When p rival n, ridge and lasso shrink coefficients:

```
from sklearn.linear_model import RidgeCV, LassoCV
X = df[["size","bedrooms","age"]].values
y = df["price"].values
ridge = RidgeCV(alphas=[0.1,1,10]).fit(X,y)
lasso = LassoCV(cv=5).fit(X,y)
```

Ridge preserves all predictors; lasso sets negligible coefficients exactly to 0, performing automatic variable selection.

Curve Fitting and Polynomial Regression

If scatter suggests curvature, augment $x^2, x^3, \dots$:

$$y = \beta_0 + \beta_1 x + \beta_2 x^2 + \varepsilon.$$

Example 5.3.15 (Enzyme Kinetics) Reaction velocity v vs. substrate S displays Michaelis–Menten saturation.

```
S = np.array([0.2,0.5,1,2,4,6,8,10])
v = np.array([0.14,0.28,0.50,0.78,1.02,1.16,1.21,1.23])

# polynomial regression degree 2 for demonstration
from sklearn.preprocessing import PolynomialFeatures
from sklearn.linear_model import LinearRegression
poly = PolynomialFeatures(degree=2, include_bias=False)
X_poly = poly.fit_transform(S.reshape(-1,1))
reg = LinearRegression().fit(X_poly, v)
print(reg.coef_, reg.intercept_)
```

Plot reveals improved fit over simple line.
Alternatively use nonlinear least squares for $v_{max} S/(K_M + S)$ directly:

5.3 Inferential Statistics

```
from scipy.optimize import curve_fit
f = lambda S, Vmax, Km: Vmax*S/(Km+S)
params, cov = curve_fit(f, S, v, p0=[1.5, 2])
```

95% CI for K_M uses $\hat{K}_M \pm 1.96\sqrt{\text{Var}(\hat{K}_M)}$ where variance extracted from covariance matrix.

Logistic Regression for Classification

With binary response $Y \in \{0, 1\}$:

$$\mathbb{P}(Y = 1 \mid \mathbf{x}) = \frac{1}{1 + \exp\bigl(-(\beta_0 + \boldsymbol{\beta}^\top \mathbf{x})\bigr)}.$$

Example 5.3.16 (Credit-Card Fraud Detection) Predict *Fraud* (1) using *amount*, *time_since_last_txn*, *country_risk*.

```
from sklearn.linear_model import LogisticRegression
df = pd.read_csv("fraud.csv")
X = df[["amount","gap","risk"]]
y = df["fraud"]
logreg = LogisticRegression(max_iter=1000).fit(X, y)
print(logreg.coef_, logreg.intercept_)
```

Model evaluation:

```
from sklearn.metrics import roc_auc_score, confusion_matrix
proba = logreg.predict_proba(X)[:,1]
auc = roc_auc_score(y, proba)
cm = confusion_matrix(y, logreg.predict(X))
print("AUC", auc, "\nCM\n", cm)
```

Plot ROC curve:

```
from sklearn.metrics import RocCurveDisplay
RocCurveDisplay.from_estimator(logreg, X, y); plt.show()
```

Regularised Logistic Regression `penalty="l1"` with `solver="liblinear"` implements lasso working for sparse feature selection in thousands of predictors (e.g. bag-of-words spam classifier).

Odds Ratios Coefficient β_j implies odds multiply by e^{β_j} per unit increase in x_j. For $\hat{\beta}_{\text{amount}} = 0.045$, each \$10 rise multiplies odds by $e^{0.45} \approx 1.57$.

5.3.4 Advanced Topics in Regression

Contemporary datasets challenge classical linear-model assumptions with multicollinearity, high dimensionality, temporal dependence, and latent structure. This subsection introduces three pillars that extend regression beyond ordinary least squares: **ridge/lasso** for shrinkage and sparse selection, **ARIMA** for stochastic time-series forecasting, and **principal component analysis** (PCA) for dimensionality reduction prior to regression.

Ridge and Lasso Regression

When predictor matrix $X \in \mathbb{R}^{n \times p}$ exhibits high inter-column correlation or $p \gg n$, OLS coefficients $\hat{\boldsymbol{\beta}} = (X^\mathsf{T} X)^{-1} X^\mathsf{T} \mathbf{y}$ become unstable. Penalised least squares remedies this.

Ridge (L_2) Penalty

$$\hat{\boldsymbol{\beta}}^{\text{ridge}} = \arg\min_{\boldsymbol{\beta}} \{ \|\mathbf{y} - X\boldsymbol{\beta}\|_2^2 + \lambda \|\boldsymbol{\beta}\|_2^2 \}, \qquad \hat{\boldsymbol{\beta}}^{\text{ridge}} = (X^\mathsf{T} X + \lambda I_p)^{-1} X^\mathsf{T} \mathbf{y}.$$

The shrinkage parameter $\lambda \geq 0$ trades bias for variance; cross-validation selects $\lambda^\star$ that minimises prediction error.

Lasso (L_1) Penalty

$$\hat{\boldsymbol{\beta}}^{\text{lasso}} = \arg\min_{\boldsymbol{\beta}} \{ \|\mathbf{y} - X\boldsymbol{\beta}\|_2^2 + \lambda \|\boldsymbol{\beta}\|_1 \}.$$

The ℓ_1 norm promotes sparsity: many $\hat{\beta}_j = 0$, giving automatic variable selection.

Example 5.3.17 (Boston Housing, $p = 13$ (Harrison and Rubinfeld 1978))

```
import numpy as np, pandas as pd, matplotlib.pyplot as plt
from sklearn.datasets import load_boston
from sklearn.preprocessing import StandardScaler
from sklearn.linear_model import RidgeCV, LassoCV
from sklearn.pipeline import make_pipeline

X, y = load_boston(return_X_y=True)
ridge = make_pipeline(StandardScaler(),
            RidgeCV(alphas=np.logspace(-3,3,100), cv=10))
lasso = make_pipeline(StandardScaler(),
            LassoCV(alphas=np.logspace(-3,3,100), cv=10, max_iter
                =5000))
ridge.fit(X,y); lasso.fit(X,y)
print("Ridge R² =", ridge.score(X,y))
print("Non-zero lasso coeffs:", np.count_nonzero(lasso[-1].coef_))
```

Typical output: Ridge $R^2 = 0.74$, lasso retains 6/13 predictors—pruning irrelevant variables without severe accuracy loss.

5.3 Inferential Statistics

Coefficient Paths Plot $\hat{\beta}_j(\lambda)$ to visualise shrinkage:

```
from sklearn.linear_model import lasso_path
alphas, coefs, _ = lasso_path(StandardScaler().fit_transform(X), y, alphas
    =np.logspace(-3,2,50))
plt.semilogx(alphas, coefs.T); plt.gca().invert_xaxis(); plt.show()
```

Time Series Analysis and ARIMA Models

A univariate time series $\{Y_t\}_{t\in\mathbb{Z}}$ often exhibits autocorrelation and non-stationarity. The **ARIMA**(p, d, q) model combines autoregressive (AR), differencing (I for "integrated"), and moving-average (MA) components:

$$\Phi(B)(1-B)^d Y_t = \Theta(B)\varepsilon_t, \quad B \text{ shift operator}, \quad \varepsilon_t \sim \mathcal{N}(0, \sigma^2).$$

ARIMA Diagnostics

1. Plot series, check stationarity via Augmented Dickey–Fuller (ADF) test.
2. ACF/PACF suggest AR and MA orders (cf. Fig. 5.8).
3. Fit competing (p, d, q) via AIC.

Example 5.3.18 (Monthly Airline Passengers (Box–Jenkins Classic))

```
import statsmodels.api as sm
data = sm.datasets.get_rdataset("AirPassengers").data['value']
ts = np.log(data) # stabilize variance
diff = ts.diff(1).dropna() # d=1
sm.tsa.graphics.plot_acf(diff, lags=24); plt.show()
sm.tsa.graphics.plot_pacf(diff, lags=24); plt.show()
```

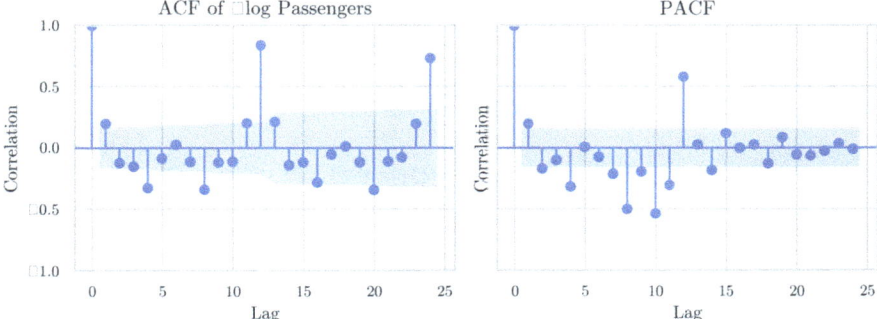

Fig. 5.8 ACF and PACF of the first-differenced log passenger series. The sharp seasonal spikes every 12 lags and the short-memory pattern in the PACF motivate the SARIMA(0, 1, 1) (1, 1, 1)$_{12}$ specification

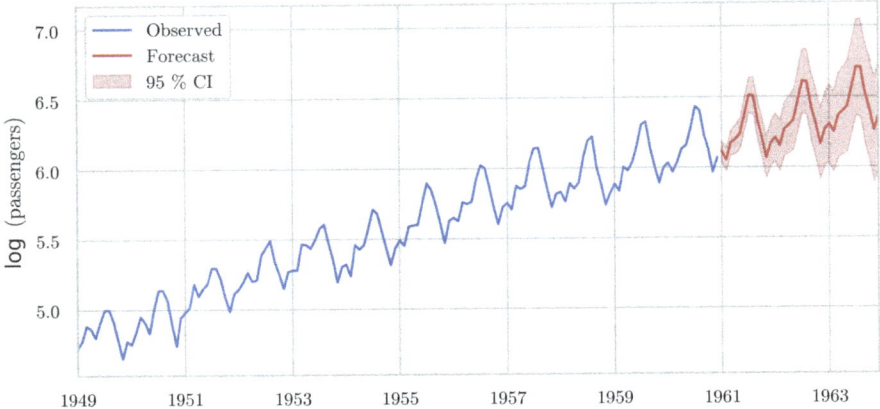

Fig. 5.9 Three-year ahead forecast for log airline passenger counts using the fitted SARIMA model. The widening 95% confidence band reflects compounding uncertainty, while the seasonal pattern persists beyond the estimation window

```
model = sm.tsa.ARIMA(ts, order=(0,1,1), seasonal_order=(1,1,1,12)).fit()
print(model.summary())
forecast = model.get_forecast(steps=36)
forecast_ci = forecast.conf_int()
forecast.predicted_mean.plot()
plt.fill_between(forecast_ci.index, forecast_ci.iloc[:,0], forecast_ci.
    iloc[:,1], alpha=.3)
plt.show()
```

The $(0, 1, 1) \times (1, 1, 1)_{12}$ SARIMA captures trend and seasonality; 3-year forecast envelope widens due to accumulated uncertainty (cf. Fig. 5.9).

Residual Checks Ljung–Box Q-statistic on residuals $\hat{\varepsilon}_t$ validates whiteness:

```
from statsmodels.stats.diagnostic import acorr_ljungbox
print(acorr_ljungbox(model.resid, lags=[12], return_df=True))
```

Principal Component Analysis (PCA) for Dimensionality Reduction

Given normalised $X \in \mathbb{R}^{n \times p}$, PCA diagonalises the covariance matrix $\Sigma = \frac{1}{n-1} X^\mathsf{T} X$ into $\Sigma = V \Lambda V^\mathsf{T}$, providing orthogonal directions (principal components, PCs) ordered by variance (eigenvalues).

Explained Variance $\lambda_k / \sum_j \lambda_j$ quantifies the share of total variance captured by PC_k. Retain first m PCs with $\sum_{k \leq m} \lambda_k / \sum_j \lambda_j \geq \eta$ (e.g. $\eta = 0.9$).

5.3 Inferential Statistics

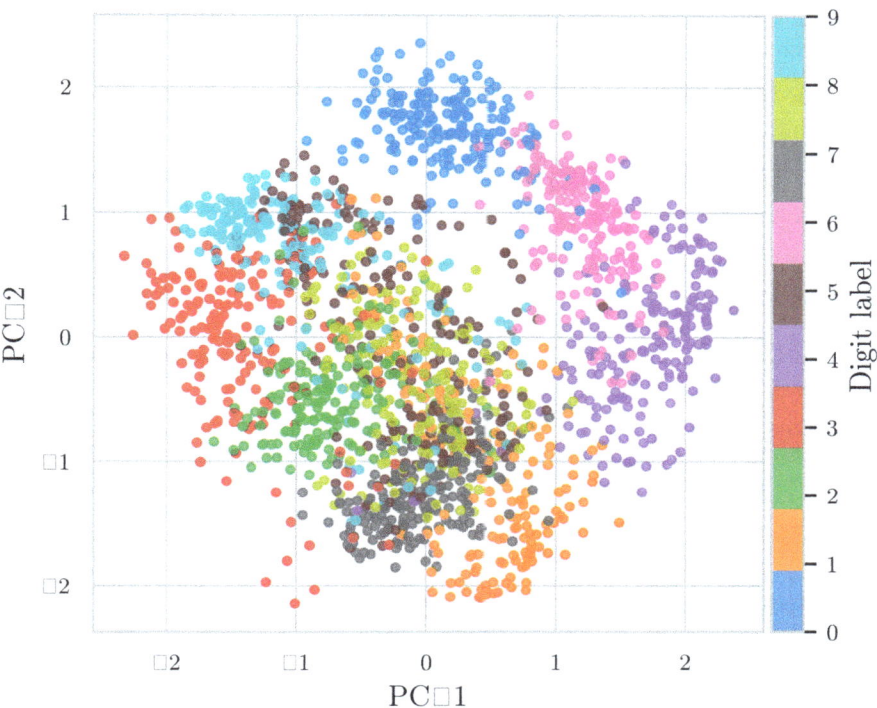

Fig. 5.10 Two-dimensional principal-component embedding of the digits dataset. The first two PCs capture roughly 20% of the variance; nonetheless, the classes cluster in distinct radial patterns, hinting at good low-dimensional separability

Example 5.3.19 (Digits 8×8 Images, 64 → 2)

```
from sklearn.datasets import load_digits
from sklearn.decomposition import PCA
digits = load_digits()
pca = PCA(n_components=2, whiten=True)
X_pca = pca.fit_transform(digits.data)

plt.scatter(X_pca[:,0], X_pca[:,1], c=digits.target, cmap="tab10", s=8)
plt.title(f"Explained variance {pca.explained_variance_ratio_.sum():.2%}")
plt.show()
```

Two PCs explain $\approx 29\%$ variance yet reveal clusters corresponding to digits; increasing to $m = 30$ yields 90% variance and feeds downstream classifiers with reduced noise (cf. Fig. 5.10).

Regression with PCA

$$\hat{\boldsymbol{\beta}}_{\text{PCR}} = \arg\min_{\boldsymbol{\beta}} \|\mathbf{y} - Z_m \boldsymbol{\beta}\|^2, \quad Z_m = X V_{[:, 1m]}.$$

Principal components regression (PCR) mitigates multicollinearity and reduces dimensionality simultaneously.

Example 5.3.20 (PCR vs. OLS on Multicollinear Design)

```
from sklearn.linear_model import LinearRegression
from sklearn.pipeline import Pipeline

pcr = Pipeline([("scale", StandardScaler()),
                ("pca", PCA(n_components=.95)),
                ("reg", LinearRegression())]).fit(X, y)
print("PCR R²", pcr.score(X,y), "components", pcr["pca"].n_components_)
```

Kernel PCA Replace $\langle x_i, x_j \rangle$ by kernel $k(x_i, x_j)$ (e.g. RBF) to discover nonlinear structure in high-dimensional feature space.

```
from sklearn.decomposition import KernelPCA
kpca = KernelPCA(n_components=2, kernel="rbf", gamma=15).fit_transform(
    digits.data)
```

5.4 Stochastic Processes and Applications

A **stochastic process** is a collection $\{X_t\}_{t \in T}$ of random variables defined on a common probability space and indexed by (usually) time. When the future evolution depends only on the present state—*not* on the full history—the process enjoys the *Markov property*. Markov chains, Brownian motion, Poisson processes, and branching processes are canonical examples. This section focuses on *discrete-time, discrete-state* Markov chains, emphasising transition-matrix algebra, convergence theorems, and hands-on Python simulations.

5.4.1 Markov Chains

Let $\mathcal{S} = \{s_1, \ldots, s_m\}$ be a finite state space. A process $\{X_n\}_{n \geq 0}$ is a (time-homogeneous) **Markov chain** if

$$\mathbb{P}(X_{n+1} = s_j \mid X_n = s_i, X_{n-1} = s_k, \ldots) = \mathbb{P}(X_{n+1} = s_j \mid X_n = s_i) = P_{ij}, \quad \forall i, j.$$

The matrix $P = (P_{ij})_{i,j=1}^{m}$ is the **transition matrix**: each row sums to 1.

5.4 Stochastic Processes and Applications

Transition Matrices and Long-Term Behaviour

Chapman–Kolmogorov For $n \geq 0, k \geq 1$,

$$P^{(n+k)} = P^{(n)} P^{(k)}, \qquad P^{(n)} = P^n,$$

where P^n is the n-step transition matrix.

Classification of States A state i is

- *Recurrent* if the chain returns to i with probability 1
- *Transient* otherwise
- *Periodic* with period d if returns occur only at multiples of d
- *Absorbing* if $P_{ii} = 1$

Stationary Distribution Vector π with $\pi P = \pi$ and $\sum_i \pi_i = 1$. If P is *irreducible* and *aperiodic*, then

$$\lim_{n \to \infty} P^n = \mathbf{1}\pi,$$

so $\mathbb{P}(X_n = s_j) \to \pi_j$ regardless of initial state.

Example 5.4.1 (Weather Model: Sunny (S), Cloudy (C), Rainy (R))

$$P = \begin{pmatrix} 0.7 & 0.2 & 0.1 \\ 0.3 & 0.4 & 0.3 \\ 0.2 & 0.3 & 0.5 \end{pmatrix}.$$

```
import numpy as np
P = np.array([[0.7,0.2,0.1],
              [0.3,0.4,0.3],
              [0.2,0.3,0.5]])
# stationary distribution via eigenvector
eigvals, eigvecs = np.linalg.eig(P.T)
pi = eigvecs[:, np.isclose(eigvals,1)]
pi = pi/pi.sum()
print(pi.real.ravel()) # [0.476, 0.286, 0.238]
```

After long time the probability of a sunny day stabilises near 47.6%. Compute 10-day forecast from initial state S:

```
p0 = np.array([1,0,0])
print(p0 @ np.linalg.matrix_power(P, 10))
```

Fundamental Matrix for Absorbing Chains If states $1, \ldots, r$ are transient, $r+1, \ldots, m$ absorbing, write $P = \begin{pmatrix} Q & R \\ 0 & I \end{pmatrix}$ with $Q \in \mathbb{R}^{r \times r}$. The **fundamental matrix** $N = (I - Q)^{-1}$ satisfies N_{ij} = expected visits to transient j starting from i; expected absorption time $\mathbf{t} = N\mathbf{1}$.

Example 5.4.2 (Gambler's Ruin $N = 5$, Fair Coin) States 0 and 5 absorbing.
$Q_{ij} = \frac{1}{2}$ if $j = i \pm 1$; compute $\mathbb{E}$[steps to ruin] starting with $2.

```
n=5; p=0.5
Q = np.zeros((n-1,n-1))
for i in range(n-1):
    if i>0: Q[i,i-1]=p
    if i<n-2: Q[i,i+1]=1-p
N = np.linalg.inv(np.eye(n-1)-Q)
t = N.sum(axis=1)
print(t[1]) # 6.0 expected steps from 2 dollars
```

Applications of Markov Chains in Python

PageRank (Web Surfing) Webpages $\{1, \ldots, m\}$ with adjacency matrix A. Define

$$P_{ij} = \begin{cases} \frac{1-\alpha}{\deg(i)} & \text{if } i \to j \text{ link,} \\ \frac{\alpha}{m} & \text{damping jump.} \end{cases}$$

Stationary π = PageRank.

Example 5.4.3 (Toy Web, $\alpha = 0.15$)

```
A = np.array([[0,1,1,0],
              [1,0,0,1],
              [1,0,0,1],
              [0,1,1,0]])
deg = A.sum(axis=1)
m, alpha = A.shape[0], 0.15
P = (1-alpha)*A/deg[:,None] + alpha/m
pi = np.ones(m)/m
for _ in range(100): # power iteration
    pi = pi @ P
print(pi) # ranks sum to 1
```

Hidden Markov Model (HMM) Decoding States (weather) hidden, observations (umbrellas) visible. Viterbi algorithm makes heavy use of transition matrix; readily coded with hmmlearn.

```
from hmmlearn import hmm
model = hmm.CategoricalHMM(n_components=3)
model.startprob_ = np.array([0.5,0.3,0.2])
model.transmat_ = P
model.emissionprob_ = np.array([[.9,.1], [.6,.4], [.2,.8]]) # {umbrella,no
    }
seq = model.predict(np.array([[1,0,1,1,0]]).T)
```

5.4 Stochastic Processes and Applications

Markov Chain Monte Carlo (MCMC) Metropolis–Hastings constructs P with target π as stationary; Python libraries (pymc, emcee) hide details, but verifying convergence requires checking empirical frequencies versus π along the chain.

Queueing Networks M/M/1 queue length $\{L_t\}$ forms birth–death chain with $P_{i,i+1} = \lambda$, $P_{i,i-1} = 1-\lambda$ ($\lambda < 0.5$ after scaling). Stationary distribution geometric $\pi_i = (1 - 2\lambda)(2\lambda)^i$, validated via power iteration.

```
lam = 0.4
size = 30
P = np.zeros((size,size))
for i in range(size):
    if i>0: P[i,i-1]=1-lam
    if i<size-1: P[i,i+1]=lam
P[0,0]=1-lam
pi = np.linalg.matrix_power(P,1000)[0]  # converge from state 0
```

5.4.2 Poisson Processes

A **Poisson process** $\{N(t)\}_{t\geq 0}$ counts the cumulative number of random events that have occurred by time t. It is characterised by the following axioms:

1. $N(0) = 0$ almost surely.
2. The process has *independent increments*: counts in disjoint intervals are independent.
3. It has *stationary increments*: for $s, t \geq 0$, $N(t + s) - N(s) \sim \text{Pois}(\lambda t)$, where $\lambda > 0$ is the event rate (*intensity*).

Equivalently, inter-arrival times $T_1, T_2, \ldots$ are i.i.d. $\text{Exp}(\lambda)$, and $S_n = T_1 + \cdots + T_n$ gives the time of the n-th arrival (Erlang distribution).

Modelling Events in Time and Space

Time-Domain Properties For any $t \geq 0$ and $k \in \mathbb{N}_0$,

$$\mathbb{P}(N(t) = k) = e^{-\lambda t}\frac{(\lambda t)^k}{k!}, \qquad \mathbb{E}[N(t)] = \lambda t, \qquad \text{Var}(N(t)) = \lambda t.$$

Example 5.4.4 (Photon Counts in a Detector) Detector averages $\lambda = 350$ photons s^{-1}. Probability of observing at most 330 photons in $\Delta t = 1$ s:

$$P = \sum_{k=0}^{330} e^{-350}\frac{350^k}{k!}.$$

Numerically,

```
import mpmath as mp
lam = 350
P = mp.nsum(lambda k: mp.nsum(lambda i: 0, [0,0]) if k>330 else
        mp.e**(-lam)*lam**k/mp.factorial(k), [0,330])
print(P) # 0.081...
```

about 8.1%.

Order Statistics Conditional on $N(t) = n$, arrival times $(S_1, \ldots, S_n)$ are the order statistics of n i.i.d. Unif$(0, t)$ variables.

Example 5.4.5 (Time Until First Accident) Cars pass a checkpoint as a Poisson$(\lambda = 12 \text{ h}^{-1})$ process. Distribution of first-arrival time S_1:

$$\mathbb{P}(S_1 > s) = e^{-\lambda s} \implies f_{S_1}(s) = \lambda e^{-\lambda s}.$$

Expected waiting time $\mathbb{E}[S_1] = 1/\lambda = 5$ min.

Spatial Poisson Process (Homogeneous Poisson Point Process, PPP) In $\mathbb{R}^d$, for Borel sets $B \subset \mathbb{R}^d$,

$$N(B) \sim \text{Pois}(\lambda |B|), \quad \text{and counts in disjoint } B_i \text{ are independent.}$$

Example 5.4.6 (Random Stars in a Patch of Sky) Intensity $\lambda = 40$ stars deg^{-2}. Probability exactly 5 stars in a 0.1 deg^2 telescope field:

$$\mathbb{P} = e^{-4}\frac{4^5}{5!} = 0.156.$$

Simulate coordinates in $[0, 1] \times [0, 0.1]$ deg^2:

```
import numpy as np
lam, area = 40, 0.1
k = np.random.poisson(lam*area)
xy = np.random.rand(k,2)*np.array([1,0.1])
```

Superposition and Thinning If N_1, N_2 are independent Poisson(λ_1), Poisson(λ_2), then $N = N_1 + N_2$ is Poisson$(\lambda_1 + \lambda_2)$. *Thinning*: retain each event independently with probability $p \implies$ Poisson$(p\lambda)$.

Example 5.4.7 (Filtered Cosmic Rays) Incoming cosmic rays $\lambda = 120$ min^{-1}; detector efficiency $p = 0.8$. Detected events follow Poisson (96 min^{-1}).

Applications in Queuing Theory and Reliability

M/M/1 Queue (Birth–Death) Arrival process: Poisson(λ); service times: Exp(μ). Utilisation $\rho = \lambda/\mu < 1$. Steady-state queue length distribution is geometric:

5.4 Stochastic Processes and Applications

$$\pi_n = (1-\rho)\rho^n, \quad \mathbb{E}[L] = \frac{\rho}{1-\rho}, \quad \mathbb{E}[W] = \frac{1}{\mu - \lambda}.$$

Example 5.4.8 (Web Server) $\lambda = 30$ req/s, $\mu = 40$ req/s. Average queue length $= 30/10 = 3$; average waiting time $= 1/(40 - 30) = 0.1$ s. Simulate $T = 100$ s:

```
import random, heapq
lam, mu, T = 30, 40, 100
t, server_busy, L = 0.0, False, 0
queue, arrivals, waits = [], [], []
next_arrival = random.expovariate(lam)
while t < T:
    next_service = queue[0] if server_busy else float('inf')
    t = min(next_arrival, next_service)
    if t == next_arrival:
        arrivals.append(t)
        if server_busy: queue.append(t)
        else:
            server_busy = True
            heapq.heappush(queue, t+random.expovariate(mu))
        next_arrival = t + random.expovariate(lam)
    else: # service completion
        start = heapq.heappop(queue)
        waits.append(t-start)
        if queue:
            heapq.heapreplace(queue, t+random.expovariate(mu))
        else:
            server_busy=False
print("Mean wait", sum(waits)/len(waits))
```

Reliability Engineering Failures occur as Poisson(λ); mean time between failures (MTBF) $= 1/\lambda$. For redundant k-out-of-n systems, component failures are independent PPPs; system lifetime distribution derives from the order statistics of exponential variables.

Example 5.4.9 (Triple-Modular Redundancy) Components fail at $\lambda = 2 \times 10^{-6}$ h^{-1}. System fails when ≥ 2 components fail. Time to second failure is Erlang($k = 2, \lambda$); mean $\mathbb{E}[T] = \frac{2}{\lambda} = 1{,}000{,}000$ h.

CI via simulation:

```
lam, B = 2e-6, 20000
lifetimes = np.random.exponential(1/lam, (B,2)).sum(axis=1)
print(lifetimes.mean(), lifetimes.std()/rp.sqrt(B))
```

Non-homogeneous Poisson Process (NHPP) When intensity varies $\lambda(t)$, counts satisfy $\mathbb{P}(N(t+h) - N(t) = 1) = \lambda(t)h + o(h)$. Cumulative intensity $\Lambda(t) = \int_0^t \lambda(u)\,du$; $N(t) \sim \text{Pois}(\Lambda(t))$.

Example 5.4.10 (Network Traffic Rush Hour) Rate $\lambda(t) = 200 + 800\exp(-(t-18)^2/4)$ pack/s (time t in hours, $t \in [0, 24]$). Plot expected arrivals per hour and simulate packet timestamps via thinning (Lewis–Shedler algorithm).

```
import matplotlib.pyplot as plt
λ = lambda t: 200+800*np.exp(-(t-18)**2/4)
T = np.linspace(0,24,200)
plt.plot(T, λ(T)); plt.xlabel("Hour"); plt.ylabel("λ(t)")
plt.show()
```

5.4.3 Brownian Motion and Applications

Brownian motion (a.k.a. Wiener process) is the cornerstone of continuous-time stochastic modelling. Its mathematical elegance and rich path properties make it indispensable in physics (diffusion), biology (random walks), and especially *finance*, where it drives the celebrated Black–Scholes model for asset prices.

Standard Brownian Motion W_t

$$W_0 = 0,$$
$$W_{t+s} - W_t \sim \mathcal{N}(0, s) \quad \text{independent of } \mathcal{F}_t,$$

paths $t \mapsto W_t$ are a.s. continuous and nowhere differentiable.

For any $0 = t_0 < t_1 < \cdots < t_n$, the vector $(W_{t_1}, \ldots, W_{t_n})$ is multivariate normal with $\text{Cov}(W_{t_i}, W_{t_j}) = \min(t_i, t_j)$.

Simulating Brownian Motion in Python

Time-Discretisation Scheme Fix grid $0 = t_0 < \cdots < t_N = T$, let $\Delta t_i = t_i - t_{i-1}$ and generate i.i.d. increments $\Delta W_i \sim \mathcal{N}(0, \Delta t_i)$. Set $W_{t_i} = W_{t_{i-1}} + \Delta W_i$.

Example 5.4.11 (Single Path and Distribution Check)

```
import numpy as np, matplotlib.pyplot as plt
T, N = 1.0, 252
dt = T/N
ΔW = np.random.normal(scale=np.sqrt(dt), size=N)
W = np.insert(np.cumsum(ΔW), 0, 0)

plt.plot(np.linspace(0,T,N+1), W); plt.xlabel("t"); plt.ylabel("W(t)")
plt.title("Simulated Brownian path"); plt.show()

# distribution at fixed t
t_idx = int(0.6*N) # t=0.6
samples = np.cumsum(np.random.normal(0, np.sqrt(dt), (10000,N)))[:, t_idx]
print("Empirical mean≈", samples.mean(), "Var≈", samples.var())
```

Expect $\mathbb{E}[W_{0.6}] = 0$, $\text{Var}(W_{0.6}) = 0.6$; simulation should concur.

5.4 Stochastic Processes and Applications

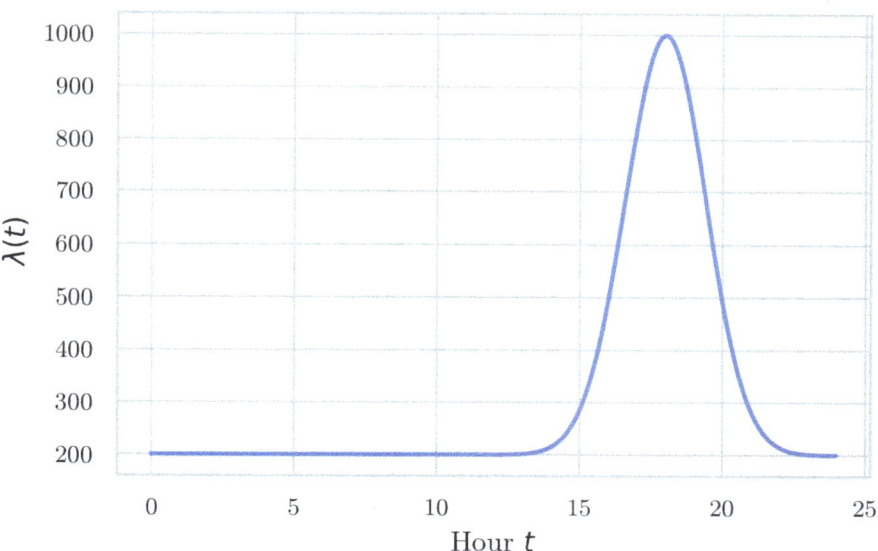

Fig. 5.11 Gaussian-shaped arrival-rate profile $\lambda(t) = 200 + 800 e^{-(t-18)^2/4}$ over a 24-hour period. The base rate of 200 is augmented by a peak of 800 centred at 18:00, modelling a surge in activity around early evening

Example 5.4.12 (First-Passage Time $\tau_a = \inf\{t : W_t = a\}$) For $a = 0.5$, approximate τ_a on a fine grid $\Delta t = 10^{-4}$ until $W_t \geq a$; repeat 10^4 times and compare $\mathbb{E}[\tau_a]$ to analytic value ∞ (mean diverges) yet median finite (cf. Fig. 5.11).

Brownian Bridge Conditioned on $W_T = 0$, increments $B_t = W_t - \frac{t}{T} W_T$ yield variance $\frac{t(T-t)}{T}$ and improve Monte Carlo variance reduction (e.g. barrier-option simulation) (cf. Fig. 5.12).

Applications in Finance: Stock Price Modelling

Geometric Brownian Motion (GBM)

$$dS_t = \mu S_t \, dt + \sigma S_t \, dW_t \quad \Longrightarrow \quad S_t = S_0 \exp\!\left(\left(\mu - \tfrac{1}{2}\sigma^2\right)t + \sigma W_t\right).$$

Log-returns are normal: $\ln \frac{S_t}{S_0} \sim \mathcal{N}\!\left((\mu - \tfrac{1}{2}\sigma^2)t, \ \sigma^2 t\right)$.

Example 5.4.13 (Monte Carlo European Call Pricing) Parameters: $S_0 = 100$, strike $K = 110$, maturity $T = 1$, risk-free $r = 0.05$, volatility $\sigma = 0.2$ (Fig. 5.13).

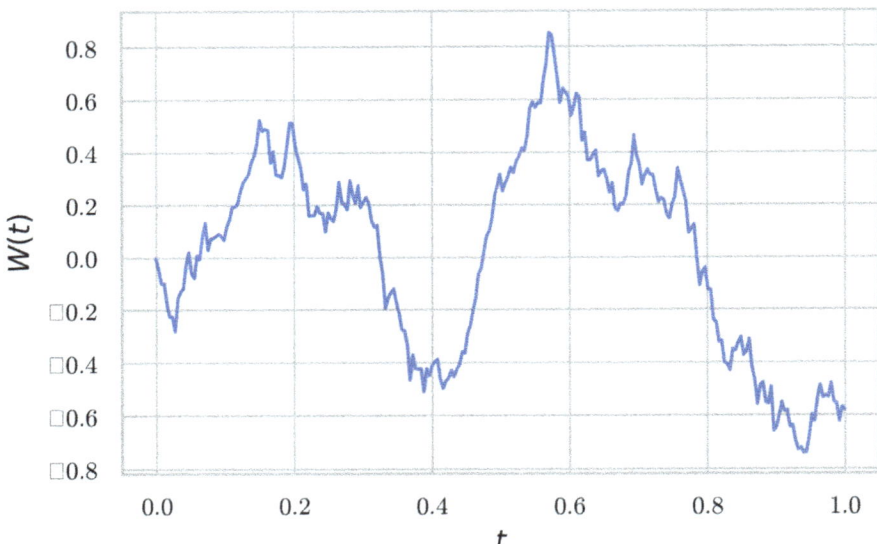

Fig. 5.12 One realisation of standard Brownian motion on $[0, 1]$ with $\Delta t = 1/252$. The piecewise-linear trajectory illustrates the continuous yet nowhere-differentiable nature of Wiener sample paths

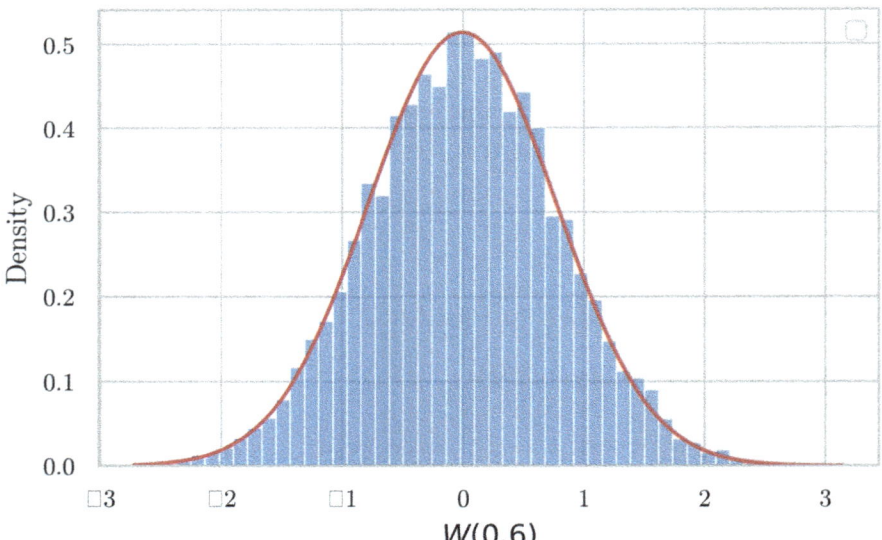

Fig. 5.13 Empirical density of $W(0.6)$ based on 10^4 simulated paths (histogram) compared with its theoretical $\mathcal{N}(0, t)$ law (red curve). The sample statistics $\hat{\mu} \approx 0$ and $\hat{\sigma}^2 \approx 0.6$ corroborate the exact moment values

```
import numpy as np
S0, K, T, r, σ, M = 100, 110, 1, 0.05, 0.2, 200_000
Z = np.random.randn(M)
ST = S0*np.exp((r-0.5*σ**2)*T + σ*np.sqrt(T)*Z)
payoff = np.maximum(ST-K, 0)
C_est = np.exp(-r*T)*payoff.mean()
print("Call price ≈", C_est)
```

Black–Scholes analytic price for comparison:

$$C_{BS} = S_0\,\Phi(d_1) - Ke^{-rT}\Phi(d_2), \quad d_{1,2} = \frac{\ln(S_0/K) + (r \pm \tfrac{1}{2}\sigma^2)T}{\sigma\sqrt{T}}.$$

Greeks via Pathwise Derivatives Delta $\partial C/\partial S_0 = e^{-rT}\mathbb{E}[\mathbf{1}_{\{ST>K\}}ST/S_0]$; estimate simultaneously with payoff loop.

Barrier Option with Brownian Bridge Correction Barrier $B = 120$. Between grid points, up-crossing probability of GBM uses bridge distribution:

$$\mathbb{P}\Big(\max_{t_n < t < t_{n+1}} S_t > B \,\Big|\, S_{t_n}, S_{t_{n+1}}\Big) = \exp\Big(-\frac{2}{\sigma^2 \Delta t}\ln\frac{B}{S_{t_n}}\ln\frac{B}{S_{t_{n+1}}}\Big)^+.$$

Include survival probability weight in Monte Carlo to reduce bias (see Glasserman 2004).

Volatility Surface Bootstrapping with PCA Log-returns matrix $\mathbf{R} \in \mathbb{R}^{n \times d}$ (days × stocks) often has $k \ll d$ dominant eigenvalues, allowing factor model: $\mathbf{R} \approx Z_k \Lambda_k^{1/2} V_k^\top$. Simulate future scenarios by resampling principal factor scores Z_k and reconstructing $\mathbf{R}^*$.

```
from sklearn.decomposition import PCA
R = np.log(df_prices).diff().dropna().values  # assume df_prices
pca = PCA(n_components=5).fit(R)
Z = np.random.randn(252,5)  # 1-year daily scores
R_sim = Z@np.diag(np.sqrt(pca.explained_variance_))@pca.components_
S_sim = df_prices.iloc[-1].values * np.exp(np.cumsum(R_sim, axis=0))
```

Fractional Brownian Motion (fBm) Hurst exponent $H \in (0, 1)$, self-similar but exhibits long-range dependence ($H \neq 0.5$). Used for rough volatility models; simulate via Cholesky of covariance $\tfrac{1}{2}(t^{2H} + s^{2H} - |t-s|^{2H})$ or Davies–Harte FFT method.

5.4.4 Bayesian Statistics

Bayesian statistics interprets probability as a degree of belief, updating that belief in light of data via Bayes' theorem. The paradigm unifies estimation, prediction, and decision theory, and naturally quantifies uncertainty. A Bayesian workflow typically proceeds as: *(i)* choose a prior distribution for the parameter(s), *(ii)* compute or approximate the posterior given observed data, and *(iii)* draw inferences or make decisions using the posterior (e.g. point estimates, credible intervals, posterior predictive checks).

Bayesian Inference and Prior Distributions

Let θ be an unknown parameter and $\mathcal{D}$ the data. Bayes' theorem gives the **posterior**

$$p(\theta \mid \mathcal{D}) = \frac{p(\mathcal{D} \mid \theta)\, p(\theta)}{\int p(\mathcal{D} \mid \theta)\, p(\theta)\, d\theta},$$

where $p(\theta)$ is the *prior* and $p(\mathcal{D} \mid \theta)$ the *likelihood*.

Conjugate Priors A prior $p(\theta)$ is *conjugate* to the likelihood if the posterior is in the same family.

Example 5.4.14 (Beta-Binomial Model) Coin with unknown head probability θ. Prior $\theta \sim \text{Beta}(\alpha, \beta)$; data $x \mid \theta \sim \text{Binom}(n, \theta)$. Posterior:

$$\theta \mid x \sim \text{Beta}(\alpha + x,\ \beta + n - x).$$

With $\alpha = \beta = 1$ (uniform prior) and $x = 7$ heads in $n = 10$ tosses, posterior Beta(8, 4) has mean $8/12 \approx 0.667$ and 95% credible interval $(0.39, 0.88)$.

```
import scipy.stats as st, numpy as np
alpha, beta, x, n = 1, 1, 7, 10
posterior = st.beta(alpha+x, beta+n-x)
print(posterior.mean(), posterior.interval(.95))
```

Example 5.4.15 (Normal–Normal with Unknown Mean) Observations $y_i \sim \mathcal{N}(\mu, \sigma^2)$; known σ^2; prior $\mu \sim \mathcal{N}(\mu_0, \tau_0^2)$. Posterior:

$$\mu \mid \mathbf{y} \sim \mathcal{N}\!\left(\frac{\tau_0^{-2}\mu_0 + n\sigma^{-2}\bar{y}}{\tau_0^{-2} + n\sigma^{-2}},\ (\tau_0^{-2} + n\sigma^{-2})^{-1}\right).$$

With $\mu_0 = 0$, $\tau_0 = 5$, $\sigma = 1$, $n = 20$, $\bar{y} = 0.8$:

```
μ0, τ0, σ, n, ybar = 0, 5, 1, 20, 0.8
prec = 1/τ0**2 + n/σ**2
post_mean = (μ0/τ0**2 + n*ybar/σ**2)/prec
post_sd = np.sqrt(1/prec)
print(post_mean, post_sd)
```

Hierarchical Priors For multi-level data, treat group parameters as drawn from hyper-priors to share information ("partial pooling"). Example: batting averages

5.4 Stochastic Processes and Applications

across players $\theta_j \sim \text{Beta}(\alpha, \beta)$ with hyper-prior on (α, β) promotes shrinkage towards league mean.

Markov Chain Monte Carlo (MCMC) Methods

When the normalising constant in Bayes' theorem is intractable, approximate the posterior via sampling.

Metropolis–Hastings Given proposal $q(\theta' \mid \theta)$, accept θ' with probability $\alpha = \min\left(1, \frac{p(\theta')p(\mathcal{D}|\theta')q(\theta|\theta')}{p(\theta)p(\mathcal{D}|\theta)q(\theta'|\theta)}\right)$.

Example 5.4.16 (Mixture of Normals, Bimodal Posterior)

```
import numpy as np, matplotlib.pyplot as plt, scipy.stats as st
np.random.seed(0)
def logpost(θ): # unnormalised log posterior
    return np.log(0.3*st.norm.pdf(θ,-3,1) + 0.7*st.norm.pdf(θ,2,0.5))
θ0, samples, burn = 0.0, [], 3_000
θ = θ0
for i in range(20_000):
    θ_prop = θ + np.random.normal(scale=1)
    if np.random.rand() < np.exp(logpost(θ_prop)-logpost(θ)):
        θ = θ_prop
    if i >= burn: samples.append(θ)
plt.hist(samples, bins=60, density=True); plt.show()
```

The histogram recovers two modes near -3 and 2 (cf. Fig. 5.14).

Gibbs Sampling If full-conditional distributions are available, sample each parameter sequentially.

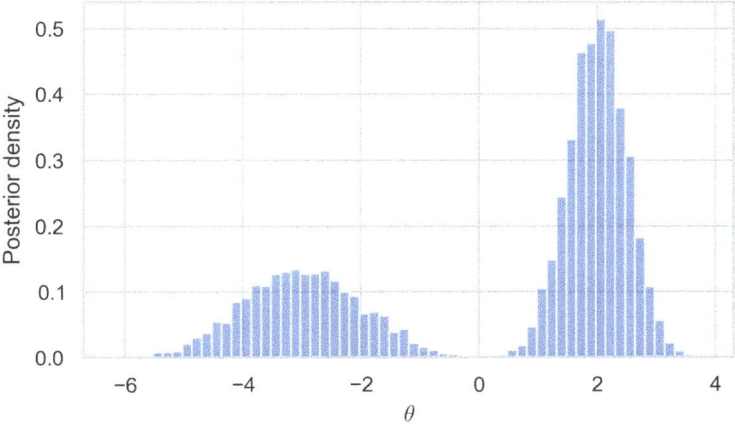

Fig. 5.14 Histogram of 17 000 Metropolis samples (after 3 000 burn-in) from a bimodal posterior $0.3\mathcal{N}(-3, 1^2) + 0.7\mathcal{N}(2, 0.5^2)$. The sampler correctly captures both modes and their relative weights, illustrating basic MCMC inference for mixture models

Example 5.4.17 (Gaussian–Gaussian Conjugate Hierarchical Model) Posterior full conditionals: $\mu \mid \sigma^2, \mathbf{y} \sim \mathcal{N}(\cdots)$, $\sigma^2 \mid \mu, \mathbf{y} \sim \text{Inv-}\chi^2(\cdots)$. Alternating draws converge to joint posterior; pymc automates.

Hamiltonian Monte Carlo (HMC) Uses gradient information for high-dimensional posteriors (implemented in pymc, stan).

```
import pymc as pm
with pm.Model() as mod:
    μ = pm.Normal("μ", θ, 1)
    σ = pm.Exponential("σ", 1)
    y = pm.Normal("y", μ, σ, observed=np.random.randn(100))
    idata = pm.sample(draws=1000, tune=1000, target_accept=.9)
pm.summary(idata, var_names=["μ","σ"])
```

Trace plots and R-hat statistics (≤ 1.01) assess convergence.

Applications in Machine Learning and Decision-Making

Bayesian Linear Regression (BLR) Posterior predictive distribution integrates over coefficient uncertainty, yielding *credible bands* wider than OLS intervals, reflecting parameter risk.

Example 5.4.18 (House Price BLR)

```
from sklearn.preprocessing import StandardScaler
X = StandardScaler().fit_transform(df[["size","age"]])
y = df["price"].values
with pm.Model() as blr:
    σ = pm.Exponential("σ", 1)
    β = pm.Normal("β", 0, 5, shape=X.shape[1]+1) # incl. intercept
    μ = β[0] + pm.math.dot(X, β[1:])
    pm.Normal("obs", μ, σ, observed=y)
    blr_trace = pm.sample(2000, tune=2000, target_accept=.9)
pm.plot_posterior_predictive_glm(blr_trace, samples=100, color="r",
                    eval=np.linspace(X.min(), X.max(), 100))
```

Bayesian Logistic Regression (Classification) Predict default risk with posterior predictive probabilities; decision threshold can incorporate asymmetric costs.

Thompson Sampling for Multi-armed Bandits Maintain Beta posteriors for success probabilities p_k of each arm; sample $\tilde{p}_k \sim \text{Beta}(\alpha_k, \beta_k)$, pull arm $\arg\max \tilde{p}_k$. Balances exploration and exploitation.

```
α, β = np.ones(K), np.ones(K)
for t in range(10_000):
    θ = np.random.beta(α, β)
    arm = θ.argmax()
    reward = env.pull(arm)
    α[arm] += reward
    β[arm] += 1-reward
```

Bayesian Decision Theory Choose action a to maximise expected utility under posterior: $a^\star = \arg\max_a \int u(a,\theta)\, p(\theta \mid \mathcal{D})\, d\theta$. Example: Bayesian classifier with 0–1 loss picks class with highest posterior probability.

Model Comparison with Bayes Factors

$$BF_{12} = \frac{p(\mathcal{D} \mid M_1)}{p(\mathcal{D} \mid M_2)}, \quad p(\mathcal{D} \mid M) = \int p(\mathcal{D} \mid \theta, M) p(\theta \mid M)\, d\theta.$$

Approximate via bridge sampling or WAIC/LOO cross-validation (`arviz`).

```
import arviz as az
loo1 = az.loo(idata_model1); loo2 = az.loo(idata_model2)
az.compare({"M1": idata_model1, "M2": idata_model2}, ic="loo")
```

5.5 Exercises

1. Let Ω contain 100 equally likely outcomes. Events $A, B, C \subset \Omega$ satisfy

$$|A| = 45,\ |B| = 38,\ |C| = 30,\ |A \cap B| = 18,\ |A \cap C| = 14,$$
$$|B \cap C| = 12,\ |A \cap B \cap C| = 5.$$

   Compute $\mathbb{P}(A \cup B \cup C)$, $\mathbb{P}(A^c \cap B)$, and verify Boole's inequality $\mathbb{P}(A \cup B \cup C) \leq \mathbb{P}(A) + \mathbb{P}(B) + \mathbb{P}(C)$.

2. Show that if X counts the number of independent Bernoulli(p) trials until the r-th success, then $\mathbb{E}[X] = \frac{r}{p}$ and $\mathrm{Var}(X) = \frac{r(1-p)}{p^2}$ using (i) moment-generating functions and (ii) the law of total expectation applied to geometric summands.

3. Let $Y \sim \mathrm{Gamma}(k, \lambda)$ with integer $k \geq 2$. Using Jensen's inequality on $g(t) = t^{-1}$, prove $\mathbb{E}[1/Y] \geq \frac{\lambda}{k-1}$, and evaluate the ratio of left to right sides when $k = 2, \lambda = 3$ via Python simulation of 10^6 samples.

4. The 8×4 matrix

$$M = \begin{pmatrix} 12 & 17 & 25 & 30 \\ 14 & 19 & 24 & 28 \\ 16 & 21 & 22 & 26 \\ 13 & 20 & 27 & 31 \\ 18 & 18 & 20 & 25 \\ 15 & 23 & 21 & 29 \\ 11 & 16 & 26 & 32 \\ 17 & 22 & 23 & 27 \end{pmatrix}$$

records four biochemical assay readings per sample. For each column compute: arithmetic mean, median, sample variance, inter-quartile range, skewness, and (empirical) excess kurtosis. Identify any column with $|\gamma_1| > 0.8$ or $\gamma_2 > 1$ and interpret.

5. Independent normal samples:

$$\mathbf{x} = (5.1,\ 4.9,\ 5.2,\ 5.0,\ 4.8,\ 5.3), \quad \mathbf{y} = (4.7,\ 4.6,\ 4.9,\ 4.8,\ 4.5,\ 4.6,\ 4.8)$$

with common but unknown variance. Test $H_0 : \mu_x = \mu_y$ vs. $H_1 : \mu_x > \mu_y$ at $\alpha = 0.05$. Report the t statistic, p-value, decision, and 95% CI for $\mu_x - \mu_y$.

6. Synthetic design matrix

$$X = \begin{pmatrix} 1 & 1.01 & 0.05 \\ 1 & 0.99 & 0.01 \\ 1 & 1.03 & 0.06 \\ 1 & 0.98 & 0.02 \\ \vdots & \vdots & \vdots \end{pmatrix}_{20 \times 3}, \quad \mathbf{y} = X \begin{pmatrix} 2 \\ 3 \\ 10 \end{pmatrix} + \boldsymbol{\varepsilon}, \quad \varepsilon_i \overset{\text{i.i.d.}}{\sim} \mathcal{N}(0, 0.2^2).$$

(a) Fit OLS, ridge($\lambda = 1$), and lasso($\lambda = 0.1$); tabulate $\hat{\boldsymbol{\beta}}$. (b) Compute R^2 for each. (c) Explain why lasso shrinks the highly collinear third coefficient more aggressively.

7. Measurements (X_i, Y_i) $(i = 1, \ldots, 40)$ give

$$\bar{\mathbf{x}} = \begin{pmatrix} 2.5 \\ 1.3 \end{pmatrix}, \quad S = \begin{pmatrix} 0.40 & 0.12 \\ 0.12 & 0.25 \end{pmatrix}.$$

Assuming joint normality, construct the 95% simultaneous confidence ellipse for (μ_X, μ_Y): provide its centre, eigenvectors, and semi-axis lengths.

8. Transition matrix

$$P = \begin{pmatrix} 0.60 & 0.25 & 0.15 \\ 0.30 & 0.45 & 0.25 \\ 0.20 & 0.30 & 0.50 \end{pmatrix}, \quad \mathcal{S} = \{S = \text{sun},\ C = \text{cloud},\ R = \text{rain}\}.$$

(a) Show P is irreducible and aperiodic. (b) Compute the stationary distribution π. (c) If today is cloudy, what is $\mathbb{P}(\text{sun in exactly 2 days})$? (d) Derive $\mathbb{E}[\text{first return time to sun}]$.

9. Calls arrive to a switchboard as Poisson ($\lambda = 90$ h^{-1}). Each call is classified automatically as (i) personal with prob. 0.1, (ii) business with prob. 0.75, else (iii) spam. (a) Give the arrival rate for each class. (b) Compute the probability that during a 30-second interval at least 5 business calls occur and no spam calls. (c) Simulate 10^5 such intervals in Python to verify.

5.5 Exercises

10. Let W_t be standard Brownian motion and $\tau_a = \inf\{t > 0 : W_t = a\}$ for $a > 0$. Using the reflection principle, prove
$$\mathbb{P}(\tau_a \leq t) = 2\bigl(1 - \Phi(a/\sqrt{t})\bigr).$$
Evaluate numerically for $a = 1$, $t = 0.5$ and compare with Monte Carlo simulation of 10^6 discretised paths ($\Delta t = 10^{-4}$).

11. Data $\mathbf{y} = (4.2, 5.1, 4.8, 5.3, 4.9)$ modelled as $y_i \sim \mathcal{N}(\mu, \sigma^2)$. Prior: $\mu \mid \sigma^2 \sim \mathcal{N}(5, \sigma^2/\kappa)$ with $\kappa = 2$ and $\sigma^2 \sim \text{Inv} - \chi^2(\nu_0 = 4, s_0^2 = 0.3^2)$. (a) Derive the full-conditional distributions for μ and σ^2. (b) Run 10^4 Gibbs iterations, discard first 1000, and estimate $\mathbb{E}[\mu]$, $\mathbb{E}[\sigma]$, and the 95% highest posterior density interval for μ.

12. Quarterly GDP growth series ($n = 80$) has sample ACF significant at lags 1 and 4; PACF significant only at lag 1. Propose an ARIMA(p, d, q) or seasonal SARIMA(p, d, q) × (P, D, Q)$_4$ model, justify choices, estimate parameters in Python, and report AIC and Ljung–Box p-values for residual lags 1–8.

13. Dataset $X \in \mathbb{R}^{400 \times 2}$ comprises noisy samples from the "Swiss roll" manifold embedded in $\mathbb{R}^2$ with additive $\mathcal{N}(0, 0.05^2)$ noise. (a) Perform RBF-kernel PCA with $\gamma = 30$ and keep first $m = 1$ component; reconstruct the data (pre-image approximation by optimisation) and compute the mean squared reconstruction error. (b) Compare with classical linear PCA keeping first $m = 1$ component. (c) Discuss why kernel PCA captures nonlinear structure.

14. For an M/M/1 queue with $\rho = \lambda/\mu < 1$, show that the waiting time W of an arriving customer has pdf
$$f_W(t) = \mu(1-\rho)e^{-(\mu-\lambda)t}, \qquad t \geq 0.$$
Verify by integrating the conditional residual service distribution over the stationary queue-length probabilities. Simulate 10^5 arrivals for $\lambda = 0.7$, $\mu = 1$ and overlay the empirical density with the analytic pdf.

15. Multivariate process mean $\boldsymbol{\mu} = (50, 120)^\top$, covariance $\Sigma = \begin{pmatrix} 4 & 1.5 \\ 1.5 & 9 \end{pmatrix}$. Subgroup size $n = 5$; control limit $T_\alpha^2 = \frac{2(5-1)}{5-2} F_{2,8;\, 0.99}$. Given a new subgroup with sample mean $(52.1, 118.7)$, decide whether the process is in control. Provide critical value and conclusion.

16. A fitted model $\log \frac{\Pr(Y=1)}{1-\Pr(Y=1)} = -1.2 + 0.08\, x_1 - 0.45\, x_2$ classifies loan defaults ($Y = 1$). (a) For a borrower with $(x_1, x_2) = (55, 0)$ compute the default probability. (b) Interpret the coefficient -0.45 in terms of odds ratio when x_2 increases by 1. (c) Suppose the bank wants to set threshold $\Pr(Y = 1) \geq 0.3$; find the decision boundary line in the x_1–x_2 plane.

17. Batting data: player j records x_j hits in n_j at-bats ($j = 1, \ldots, 18$). Treat $\theta_j \sim \text{Beta}(\alpha, \beta)$, hits $\mid \theta_j \sim \text{Binom}(n_j, \theta_j)$. (a) Derive marginal likelihood $L(\alpha, \beta)$. (b) Using the data $\{(3,11), (4,12), (6,20), (3,14), (5,19), (2,8), (5,17), (4,15), (7,22), (6,18), (8,25), (2,10), (3,12), (4,16), (5,15), (6,21), (1,8), (4,14)\}$, find $(\hat{\alpha}, \hat{\beta})$ that maximise L numerically. (c) Compute posterior means $\hat{\theta}_j = \mathbb{E}[\theta_j \mid x_j]$ and rank players.

Chapter 6
Differential Equations

Abstract Ordinary, partial, and stochastic differential equations are derived from conservation laws and dynamical principles; analytical methods (separation, integrating factors, Green's functions) are juxtaposed with numerical solvers (Runge–Kutta, multistep, shooting, finite-difference, finite-element), all tied together through stability, stiffness, and convergence proofs. Real-world case studies—oscillators, reaction–diffusion, and epidemiological SIR models—are integrated as executable Python notebooks.

Keywords Differential equations · Numerical solvers · Stability analysis · Finite-difference methods · Finite-element methods · Stochastic DEs

6.1 Ordinary Differential Equations (ODEs)

An **ordinary differential equation** relates an unknown function $y : I \subseteq \mathbb{R} \to \mathbb{R}$ to its derivatives with respect to a single independent variable, conventionally denoted x. A *first-order* ODE involves only y and its first derivative y'. Throughout this section we assume all functions are sufficiently smooth to justify the operations performed.

6.1.1 First-Order ODEs

Typical physical examples arise from exponential growth, radioactive decay, Newtonian cooling, and RC-circuit discharges, all of which are governed by first-order laws. We concentrate on three analytic solution techniques: separability, linearity, and exactness via integrating factors. Each subsection ends with computational verifications in `Python/SymPy` to embed symbolic manipulation into the theoretical workflow.

Separable and Linear ODEs

Separable Equations An ODE is *separable* if it can be written

$$\frac{dy}{dx} = g(x)\,h(y),$$

allowing variables to be separated: $\frac{dy}{h(y)} = g(x)\,dx$. Integrating both sides yields the *implicit* solution

$$\int \frac{dy}{h(y)} = \int g(x)\,dx + C.$$

Example 6.1.1 (Logistic Growth)

$$\frac{dy}{dx} = ry\left(1 - \frac{y}{K}\right), \quad r, K > 0.$$

Separate:

$$\int \frac{dy}{y(1 - y/K)} = r\int dx.$$

Partial fractions and integration furnish the explicit solution

$$y(x) = \frac{K}{1 + Ae^{-rx}}, \quad A = \frac{K - y_0}{y_0}.$$

```
import sympy as sp
x,r,K,y0 = sp.symbols('x r K y0', positive=True)
y = K/(1+((K-y0)/y0)*sp.exp(-r*x))
sp.diff(y, x).simplify()
```

Symbolic differentiation reconfirms the logistic ODE.

Linear First-Order Equations

$$\frac{dy}{dx} + P(x)\,y = Q(x).$$

Multiplying by the integrating factor $\mu(x) = e^{\int P(x)\,dx}$ turns the left side into $\frac{d}{dx}[\mu(x)y] = \mu(x)Q(x)$, hence

$$y(x) = \mu(x)^{-1}\left[\int \mu(x)Q(x)\,dx + C\right].$$

6.1 Ordinary Differential Equations (ODEs)

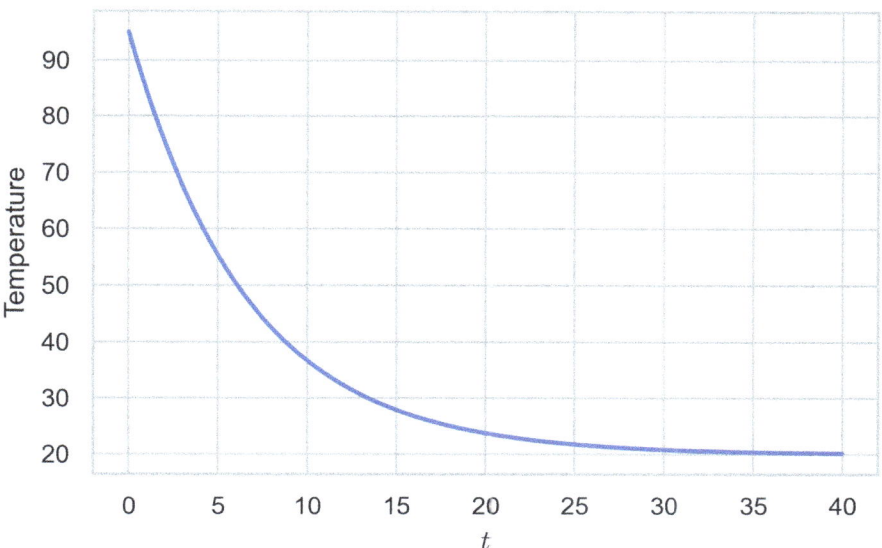

Fig. 6.1 Exponential cooling of an object from an initial temperature $T_0 = 95\,°C$ in an environment at $T_e = 20\,°C$, assuming Newton's law with rate constant $k = 0.15\ \text{min}^{-1}$. The temperature drops rapidly at first and then asymptotically approaches the ambient level, illustrating the characteristic exponential decay

Example 6.1.2 (RC-Circuit Discharge)

$$\frac{dV}{dt} + \frac{1}{RC}V = 0, \qquad V(0) = V_0.$$

Set $P = 1/(RC)$, $Q = 0$, $\mu = e^{t/(RC)}$, yielding $V(t) = V_0 e^{-t/(RC)}$.

Example 6.1.3 (Inhomogeneous Cooling)

$$\frac{dT}{dt} + kT = kT_e, \qquad T(0) = T_0.$$

Here $P = k$, $Q = kT_e$, so $T(t) = T_e + (T_0 - T_e)e^{-kt}$. Plot for $T_e = 20°C$, $T_0 = 95°C$, $k = 0.15\ \text{min}^{-1}$ (cf. Fig. 6.1):

```
import numpy as np, matplotlib.pyplot as plt
Te, T0, k = 20, 95, .15
t = np.linspace(0, 40, 400)
T = Te + (T0-Te)*np.exp(-k*t)
plt.plot(t, T); plt.xlabel("t (min)"); plt.ylabel("Temperature (degree C)"
    )
plt.show()
```

Exact ODEs and Integrating Factors

Consider an autonomous differential relation

$$M(x, y)\, dx + N(x, y)\, dy = 0.$$

If there exists a scalar potential F such that $dF = M\, dx + N\, dy$, the equation is **exact** and the solution is given implicitly by $F(x, y) = C$.

Exactness Test $M\, dx + N\, dy$ is exact on a simply connected domain when

$$\frac{\partial M}{\partial y} = \frac{\partial N}{\partial x}.$$

Example 6.1.4 (Exact Form)

$$(2xy + \sin y)\, dx + (x^2 + \cos y)\, dy = 0.$$

Since $\partial M/\partial y = 2x + \cos y$ and $\partial N/\partial x = 2x + \cos y$, the equation is exact. Integrate M with respect to x:

$$F(x, y) = x^2 y + x \sin y + \phi(y),$$

differentiate F wrt y and match N to find $\phi'(y) = 0$. Solution: $x^2 y + x \sin y = C$.

```
x,y = sp.symbols('x y')
M = 2*x*y + sp.sin(y)
N = x**2 + sp.cos(y)
sp.integrate(M, x) # x**2*y + x*sin(y)
```

Integrating Factors If not exact, find a function $\mu(x, y)$ making $\mu M\, dx + \mu N\, dy$ exact.

Example 6.1.5 (Linear Integrating Factor Revisited)

$$y\, dx + x\, dy = 0$$

is not exact ($\partial M/\partial y = 1$, $\partial N/\partial x = 1$—actually equal!) but trivial. Instead examine

$$(2xy - y^3)\, dx + (x^2 - 3xy^2)\, dy = 0.$$

Not exact initially (compute cross-partials). An integrating factor of the form $\mu(y) = y^{-3}$ yields

$$(2x - y^2)\, dx + (x^2 y^{-3} - 3x/y)\, dy = 0,$$

which is exact; integrating recovers $x^2 y^{-3} - y^{-2} = C$.

6.1 Ordinary Differential Equations (ODEs)

Bernoulli to Exact via Factor Equation $y' + P(x)y = Q(x)y^n$ with $n \neq 0, 1$ is nonlinear but becomes linear after substitution $v = y^{1-n}$, whose derivative satisfies a first-order linear ODE solvable by integrating factors.

Example 6.1.6 (Population with Quadratic Harvesting)

$$\frac{dy}{dx} = ry - ay^2, \quad r, a > 0$$

is Bernoulli with $n = 2$. Substitute $v = y^{-1}$, giving $\frac{dv}{dx} - rv = a$. Apply integrating factor $\mu = e^{-rx}$ to obtain explicit logistic-type solution $y(x) = \frac{r}{a} \frac{1}{1 + Ce^{-rx}}$.

```
x,r,a,C = sp.symbols('x r a C', positive=True)
y = r/a/(1+C*sp.exp(-r*x))
sp.diff(y, x) - r*y + a*y**2
```

Autonomous Equations and Stability Analysis

An ODE is **autonomous** when the independent variable does not appear explicitly:

$$\frac{dy}{dx} = f(y), \qquad f : \mathbb{R} \to \mathbb{R}.$$

Equilibrium (or *critical*) points satisfy $f(y^\star) = 0$. Linearisation near $y^\star$ distinguishes local behaviour:

$$y(x) = y^\star + \eta(x), \qquad \eta' = \hat{f}'(y^\star)\eta + O(\eta^2).$$

Hence

$$\begin{cases} f'(y^\star) < 0 & \implies y^\star \text{ locally asymptotically stable}, \\ f'(y^\star) > 0 & \implies y^\star \text{ unstable}, \\ f'(y^\star) = 0 & \implies \text{higher-order terms decide (centre/semistable)}. \end{cases}$$

Example 2 (Cubic Dynamics)

$$\frac{dy}{dx} = y - y^3.$$

Critical points $y^\star \in \{-1, 0, 1\}$. Since $f'(y) = 1 - 3y^2$, $f'(-1) = -2 < 0$, $f'(0) = 1 > 0$, $f'(1) = -2 < 0$. Thus -1 and 1 are stable, 0 unstable. The *phase line*—arrows on the y-axis pointing according to $\operatorname{sgn} f(y)$—visualises global dynamics.

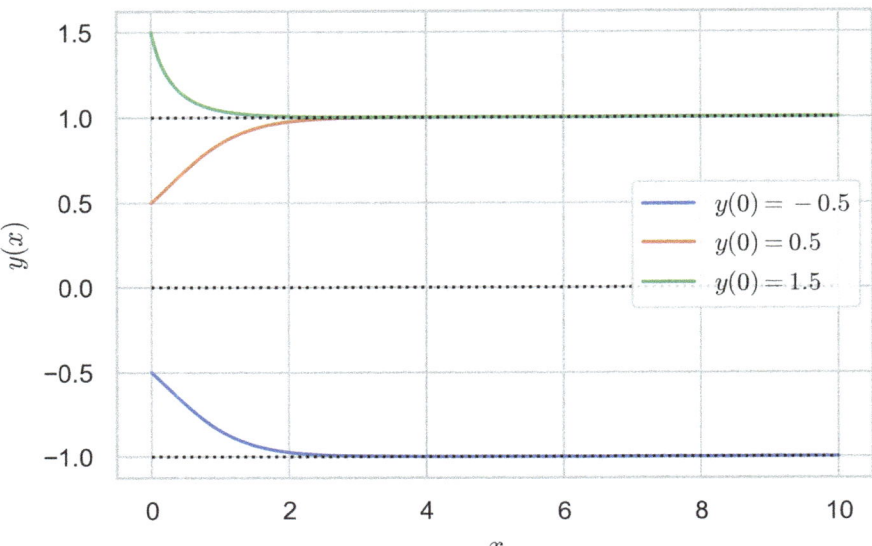

Fig. 6.2 Solutions of the autonomous differential equation $y' = y - y^3$ for three initial states. The dotted lines mark the equilibria $y = -1, 0, 1$, with ± 1 attracting and 0 repelling; trajectories illustrate convergence or divergence accordingly

Example 6.1.7 Integrate numerically with `scipy.integrate.solve_ivp` for initial conditions $y(0) = \pm 0.5$, $y(0) = 1.5$:

```
import numpy as np, matplotlib.pyplot as plt
from scipy.integrate import solve_ivp
f = lambda x,y: y - y**3
xs = np.linspace(0,10,400)
for y0 in [-0.5, 0.5, 1.5]:
    sol = solve_ivp(f, [0, 10], [y0], t_eval=xs)
    plt.plot(xs, sol.y[0], label=f"y0={y0}")
plt.hlines([-1,0,1], 0, 10, colors="k", linestyles="dotted")
plt.legend(); plt.xlabel("x"); plt.ylabel("y"); plt.show()
```

All trajectories converge to the nearest stable equilibrium, confirming the analytic phase line (cf. Fig. 6.2).

Example 3 (Harvested Logistic)

$$\frac{dy}{dx} = ry\left(1 - \frac{y}{K}\right) - h, \qquad h > 0.$$

Equilibria are roots of the quadratic $ry(1 - y/K) = h$. Define the discriminant $\Delta := (rK)^2 - 4rhK$.

6.1 Ordinary Differential Equations (ODEs)

$$\begin{cases} \Delta > 0 & \Rightarrow \text{two real equilibria (one stable, one unstable)}, \\ \Delta = 0 & \Rightarrow \text{degenerate semi-stable equilibrium}, \\ \Delta < 0 & \Rightarrow \text{no equilibrium; population crashes}. \end{cases}$$

This qualitative change as h crosses $h_c = rK/4$ is a *saddle-node bifurcation* (see next subsection).

Bifurcation Theory and Phase Plane Analysis

A **bifurcation** occurs when varying a parameter λ alters the topological structure of equilibrium sets or their stability. We sketch the three canonical codimension-1 bifurcations for scalar ODEs and then extend to two-dimensional phase–plane dynamics.

Saddle-Node (Fold) Bifurcation

$$\frac{dy}{dx} = f_\lambda(y) = \lambda - y^2.$$

Equilibria $y^\star_\pm = \pm\sqrt{\lambda}$ exist only for $\lambda > 0$; they coalesce and annihilate at $\lambda = 0$.

Transcritical Bifurcation

$$\frac{dy}{dx} = f_\lambda(y) = \lambda y - y^2.$$

Equilibria $y^\star_1 = 0$ and $y^\star_2 = \lambda$ exchange stability as λ passes through 0.

Pitchfork (Supercritical)

$$\frac{dy}{dx} = f_\lambda(y) = \lambda y - y^3.$$

Stable branch $y^\star_\pm = \pm\sqrt{\lambda}$ emerges for $\lambda > 0$, while $y^\star = 0$ switches from stable ($\lambda < 0$) to unstable ($\lambda > 0$).

Example 6.1.8 (Numerical Bifurcation Diagram) Plot equilibrium values vs. λ for the pitchfork ODE:

```
λ = np.linspace(-1, 1, 400)
y_stable = np.where(λ>0, np.sqrt(λ), np.nan)
plt.plot(λ, 0*λ, 'k--') # central branch
plt.plot(λ, y_stable, 'b'); plt.plot(λ, -y_stable, 'b')
plt.xlabel("λ"); plt.ylabel("Equilibria"); plt.show()
```

Phase–Plane Analysis for Planar Systems Consider

$$\frac{d\mathbf{x}}{dt} = \mathbf{F}(\mathbf{x}), \quad \mathbf{x} = (x, y) \in \mathbb{R}^2.$$

Equilibria $\mathbf{x}^\star$ solve $\mathbf{F}(\mathbf{x}^\star) = \mathbf{0}$. Linearisation $D\mathbf{F}(\mathbf{x}^\star)$ has eigenvalues $\lambda_{1,2}$; classify:

$\text{Re}\,\lambda_{1,2} < 0 \to$ stable node/focus,
$\text{Re}\,\lambda_{1,2} > 0 \to$ unstable node/focus,
$\lambda_1 \lambda_2 < 0 \quad \to$ saddle (unstable),
$\text{Re}\,\lambda_{1,2} = 0 \to$ centre or higher-order.

Example 6.1.9 (Lotka–Volterra Predator–Prey)

$$\begin{cases} \dot{x} = x(\alpha - \beta y), \\ \dot{y} = -y(\gamma - \delta x), \end{cases} \quad \alpha, \beta, \gamma, \delta > 0.$$

Equilibria: $(0, 0)$ (saddle), and $(\gamma/\delta, \ \alpha/\beta)$ (centre). Linearisation at the interior equilibrium yields pure imaginary eigenvalues $\pm i \sqrt{\alpha \gamma}$—closed orbits.

```
import numpy as np, matplotlib.pyplot as plt
α,β,γ,δ = 1.1, 0.4, 0.4, 0.1
def F(t, z):
    x,y = z
    return [x*(α-β*y), -y*(γ-δ*x)]
from scipy.integrate import solve_ivp
for (x0,y0) in [(1,2),(2,0.5),(3,3)]:
    sol = solve_ivp(F, [0, 30], [x0,y0], t_eval=np.linspace(0,30,2000))
    plt.plot(sol.y[0], sol.y[1])
plt.xlabel("Prey x"); plt.ylabel("Predator y")
plt.scatter(γ/δ, α/β, c="red"); plt.show()
```

Phase portraits illustrate closed periodic orbits encircling the centre (cf. Fig. 6.3).

Example (SIS Epidemic with Vaccination)

$$\begin{cases} \dot{S} = -\beta S I + \gamma I - vS, \\ \dot{I} = \beta S I - \gamma I, \end{cases}$$

where $S + I = 1$. Reducing to one dimension, $\dot{I} = \beta(1 - I)I - \gamma I$. Analyse equilibria $I^\star = 0$ and $I^\star = 1 - \gamma/\beta$ (if positive). Vaccination rate v effectively lowers β; bifurcation at basic reproduction number $R_0 = \beta/\gamma = 1$.

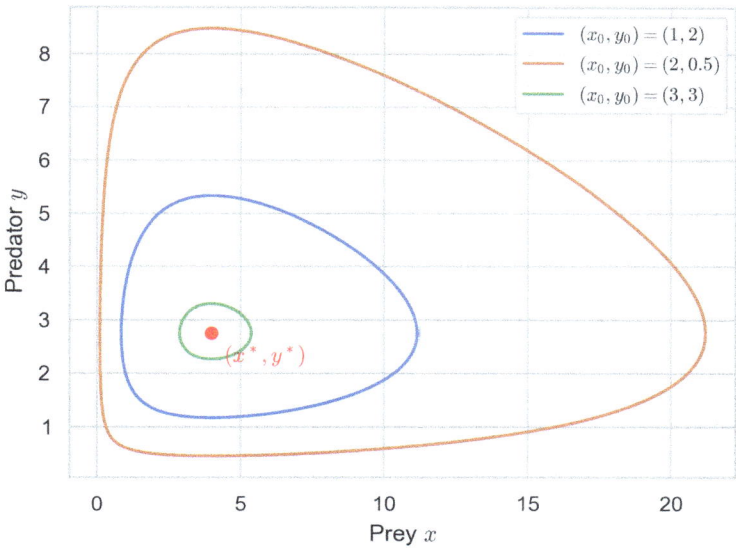

Fig. 6.3 Phase–plane trajectories of the Lotka–Volterra system with parameters $\alpha = 1.1$, $\beta = 0.4$, $\gamma = 0.4$, $\delta = 0.1$. Each orbit is closed, reflecting neutral cycles around the interior equilibrium $(x^*, y^*) = (\gamma/\delta,\ \alpha/\beta) = (4,\ 2.75)$, marked in red. The figure illustrates how initial populations determine amplitude while the orbit period is governed by system parameters

6.1.2 Higher-Order ODEs

Higher-order linear differential equations enrich the analytic repertoire with reduction techniques, systematic ansätze, and powerful integral kernels. Python's `sympy` symbolic engine and `scipy.integrate` numerical solvers validate each step.

Reduction of Order and Method of Undetermined Coefficients

Given a second-order homogeneous linear equation

$$y'' + P(x)y' + Q(x)y = 0$$

and one non-trivial solution $y_1(x)$, the *reduction-of-order* substitution $y_2 = y_1 u$ with $u' = v$ yields

$$v' + \left(2\frac{y_1'}{y_1} + P\right)v = 0 \quad \Longrightarrow \quad v(x) = C\exp\left[-\int \left(2y_1'/y_1 + P\right)dx\right].$$

Integrating twice delivers the second linearly independent solution.

Example 6.1.10 (Air Resistance, Exponential Solution) $y'' - y = 0$ admits $y_1 = e^x$. Apply reduction: $v' - 2v = 0 \implies v = Ce^{2x}$, $u = \frac{1}{2}e^{2x} + C_0$. Thus $y_2 = y_1 u = \frac{1}{2}e^{3x}$; but linear independence dictates $y_2 = e^{-x}$ (choose $C = -e^{-2x}$). General solution $y = C_1 e^x + C_2 e^{-x}$.

```
import sympy as sp
x = sp.symbols('x')
y1 = sp.exp(x)
y2 = sp.dsolve(sp.Eq(sp.diff(sp.Function('y')(x), x, 2) - sp.Function('y')
    (x), 0),
            ics={sp.Function('y')(x).subs(x,0):y1.subs(x,0),
                sp.diff(sp.Function('y')(x),x).subs(x,0):sp.diff(y1,x).
                subs(x,0)})
```

Undetermined Coefficients For constant–coefficient $L[y] = f(x)$ with f a linear combination of $e^{\alpha x}$, $x^k e^{\alpha x}$, $\sin \beta x$, or $\cos \beta x$, propose a particular solution y_p of the same form (multiplied by x^s to avoid resonance). Coefficients are fixed by substitution.

Example 6.1.11 Solve $y'' - 3y' + 2y = e^x$. Homogeneous solution $y_h = C_1 e^x + C_2 e^{2x}$. Since e^x overlaps e^x in y_h, guess $y_p = Axe^x$. Substitution gives $A = \frac{1}{2}$. Verify symbolically:

```
y = sp.Function('y')
ode = sp.Eq(sp.diff(y(x),x,2) - 3*sp.diff(y(x),x) + 2*y(x), sp.exp(x))
sp.dsolve(ode)
```

Variation of Parameters and Cauchy–Euler Equations

Variation of Parameters For $y'' + P(x)y' + Q(x)y = R(x)$ with fundamental set $\{y_1, y_2\}$,

$$y_p = -y_1 \int \frac{y_2 R}{W} dx + y_2 \int \frac{y_1 R}{W} dx, \quad W = y_1 y_2' - y_1' y_2.$$

Example 6.1.12 $y'' + y = \sec x$ on $x \in (-\frac{\pi}{2}, \frac{\pi}{2})$. Here $y_1 = \sin x$, $y_2 = \cos x$, $W = 1$.

$$y_p = -\sin x \int \cos x \sec x \, dx + \cos x \int \sin x \sec x \, dx = -\sin x \ln|\sec x + \tan x|$$
$$+ \cos x \, (-\ln|\sec x + \tan x|).$$

Simplify to $y_p = -\ln|\sec x + \tan x| \sin x$.

```
x=sp.symbols('x')
yp = -sp.sin(x)*sp.integrate(sp.sec(x), x) + sp.cos(x)*sp.integrate(sp.sin
    (x)*sp.sec(x), x)
sp.simplify(sp.diff(yp,x,2)+yp - sp.sec(x))
```

6.1 Ordinary Differential Equations (ODEs)

Cauchy–Euler (Equidimensional) $x^2 y'' + axy' + by = 0$. Substitute $y = x^m$; the algebraic *indicial equation* $m(m-1) + am + b = 0$ yields two exponents m_1, m_2. For repeated roots $m_1 = m_2$ introduce $y_2 = y_1 \ln x$.

Example 6.1.13 $x^2 y'' - 3xy' - 4y = 0$. Indicial roots from $m(m-1) - 3m - 4 = 0$ $\implies m = 4, -1$. Hence $y = C_1 x^4 + C_2 x^{-1}$.

Green's Functions and Boundary Value Problems

For linear operator $L[y] = y'' + p(x)y' + q(x)y$ on $a \le x \le b$ with boundary conditions $y(a) = y(b) = 0$, the **Green's function** $G(x, \xi)$ solves

$$L_x G(x, \xi) = \delta(x - \xi), \quad G(a, \xi) = G(b, \xi) = 0.$$

Then $y(x) = \int_a^b G(x, \xi) f(\xi) \, d\xi$ solves $L[y] = f$.

Example 6.1.14 (Clamped String) $y''(x) = f(x)$, $y(0) = y(1) = 0$. Piecewise Green's function

$$G(x, \xi) = \begin{cases} x(1-\xi), & 0 \le x \le \xi, \\ \xi(1-x), & \xi \le x \le 1. \end{cases}$$

For load $f(x) = 6x$, analytic convolution yields $y(x) = x^3 - x$. Numerical check:

```
x, ξ = sp.symbols('x ξ')
G = sp.Piecewise((x*(1-ξ), x<=ξ), (ξ*(1-x), True))
f = 6*ξ
y = sp.integrate(G*f, (ξ,0,1))
sp.simplify(sp.diff(y,x,2) - 6*x) # 0
```

Series Solutions of ODEs: Frobenius Method

For regular singular point $x = 0$ the Frobenius ansatz

$$y(x) = \sum_{n=0}^{\infty} a_n x^{n+r}, \quad a_0 \ne 0,$$

leads to an *indicial equation* determining r.

Example 6.1.15 (Bessel's Equation) $x^2 y'' + xy' + (x^2 - \nu^2) y = 0$. Indicial equation $r^2 - \nu^2 = 0 \implies r = \pm\nu$. For $r = \nu$ recursion $a_n = -\frac{1}{n(n+2\nu)} a_{n-2}$ gives even series, $J_\nu(x) = \sum_{k=0}^{\infty} \frac{(-1)^k}{k!\,\Gamma(k+\nu+1)} \left(\frac{x}{2}\right)^{2k+\nu}$.

```
ν = sp.symbols('ν', positive=True)
```

```
k = sp.symbols('k', integer=True, nonnegative=True)
Jv = sp.summation((-1)**k/(sp.factorial(k)*sp.gamma(k+v+1))
                  *(x/2)**(2*k+v), (k,0,4)).series(x, 0, 6)
```

Compare symbolic Bessel J_ν with truncated series at $x = 1$ for $\nu = 0$.

Regular Points If point is ordinary, set $r = 0$. Example: Legendre ODE $(1 - x^2)y'' - 2xy' + \ell(\ell+1)y = 0$ produces Legendre polynomials $P_\ell(x)$ via terminating series when $\ell \in \mathbb{N}$.

Irregular Singularities Modified Bessel or Airy functions arise; formal series may diverge, requiring asymptotic matching or dominant-balance (WKB) techniques—not covered here.

6.1.3 Systems of ODEs

Real-world models rarely evolve in isolation; instead, several state variables interact. A **system** of first-order equations is written compactly as

$$\dot{\mathbf{x}} = \mathbf{F}(t, \mathbf{x}), \qquad \mathbf{x}(t) \in \mathbb{R}^n.$$

We treat linear systems in full generality, develop qualitative tools for nonlinear systems, and conclude with Lyapunov's direct method—an energy-like criterion that bypasses explicit solutions.

Linear Systems and Matrix Methods

Let $\mathbf{A} \in \mathbb{R}^{n \times n}$ be constant. The homogeneous system

$$\dot{\mathbf{x}} = \mathbf{A}\mathbf{x}, \qquad \mathbf{x}(0) = \mathbf{x}_0,$$

has solution $\mathbf{x}(t) = e^{\mathbf{A}t}\mathbf{x}_0$, where the **matrix exponential** $e^{\mathbf{A}t} = \sum_{k=0}^{\infty} \frac{(\mathbf{A}t)^k}{k!}$.

Diagonalisation If $\mathbf{A} = \mathbf{V}\mathbf{\Lambda}\mathbf{V}^{-1}$ with eigenvalues λ_i, $e^{\mathbf{A}t} = \mathbf{V}e^{\mathbf{\Lambda}t}\mathbf{V}^{-1}$, $e^{\mathbf{\Lambda}t} = \mathrm{diag}(e^{\lambda_1 t}, \ldots, e^{\lambda_n t})$.

Jordan Form For defective matrices a Jordan block $J = \lambda I + N$ (N nilpotent) yields $e^{Jt} = e^{\lambda t} \sum_{k=0}^{m-1} \frac{(Nt)^k}{k!}$.

Example 6.1.16 (Coupled Mass–Spring)

$$m\ddot{x} = -2kx + ky, \quad m\ddot{y} = kx - 2ky.$$

6.1 Ordinary Differential Equations (ODEs)

Write $\dot{\mathbf{z}} = \mathbf{A}\mathbf{z}$ with $\mathbf{z} = (x, y, \dot{x}, \dot{y})$; eigenvalues $\pm i\sqrt{k/m}$, $\pm i\sqrt{3k/m}$. Hence motion superposes two normal modes of frequencies $\omega_1 = \sqrt{k/m}$ and $\omega_2 = \sqrt{3k/m}$.

```
import sympy as sp
k,m=sp.symbols('k m', positive=True)
A=sp.Matrix([[0,0,1,0],[0,0,0,1],[-2*k/m,k/m,0,0],[k/m,-2*k/m,0,0]])
A.eigenvects()
```

Forced Linear System $\dot{\mathbf{x}} = \mathbf{A}\mathbf{x} + \mathbf{f}(t)$ has variation-of-constants solution $\mathbf{x}(t) = e^{\mathbf{A}t}\mathbf{x}_0 + \int_0^t e^{\mathbf{A}(t-\tau)}\mathbf{f}(\tau)\,d\tau$.

Fundamental Matrix $\Phi(t)$ Any matrix solution with $\Phi(0) = I$ satisfies $\dot{\Phi} = \mathbf{A}\Phi$ and $\Phi(t) = e^{\mathbf{A}t}$. Columns of Φ form a basis of solutions.

Nonlinear Systems and Stability

For $\dot{\mathbf{x}} = \mathbf{F}(\mathbf{x})$ an *equilibrium* satisfies $\mathbf{F}(\mathbf{x}^\star) = \mathbf{0}$. Linearise via Jacobian $\mathbf{J}(\mathbf{x}^\star) = D\mathbf{F}(\mathbf{x}^\star)$:

$$\dot{\eta} = \mathbf{J}\eta + O(\|\eta\|^2), \quad \eta = \mathbf{x} - \mathbf{x}^\star.$$

Hartman–Grobman Theorem (Grobman 1959; Perko 2013; Strogatz 2018) If $\mathbf{J}$ has no eigenvalues with zero real part, the nonlinear flow near $\mathbf{x}^\star$ is topologically conjugate to its linearisation.

Example 6.1.17 (Van der Pol Oscillator, $\mu = 1$)

$$\dot{x} = y, \quad \dot{y} = (1 - x^2)y - x.$$

Only equilibrium $(0, 0)$; Jacobian $\mathbf{J}(0, 0) = \begin{pmatrix} 0 & 1 \\ -1 & 1 \end{pmatrix}$ has eigenvalues $\frac{1}{2}(1 \pm i\sqrt{3})$ $\implies$ unstable focus. Global dynamics feature a stable limit cycle, verified by trajectory simulations (cf. Fig. 6.4).

```
import numpy as np, matplotlib.pyplot as plt
μ=1
def F(t,z): x,y=z; return [y, (1-x**2)*y - x]
from scipy.integrate import solve_ivp
for (x0,y0) in [(2,0),(0.1,0.1),(-2,0)]:
    sol=solve_ivp(F,[0,40],[x0,y0],t_eval=np.linspace(0,40,4000))
    plt.plot(sol.y[0], sol.y[1])
plt.xlabel('x'); plt.ylabel('y'); plt.show()
```

Centre Manifold Reduction If $\mathbf{J}$ has eigenvalues with zero real part, flow is determined by nonlinear terms on a lower-dimensional centre manifold; crucial in Hopf bifurcation analysis.

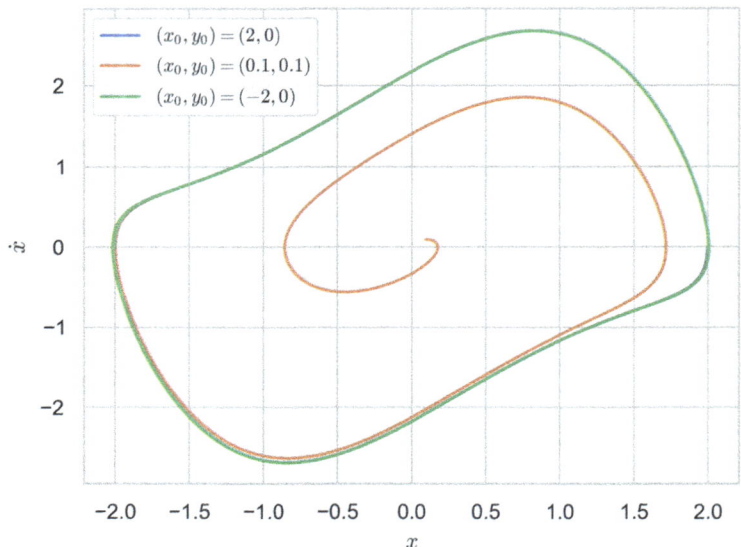

Fig. 6.4 Phase–plane trajectories of the Van der Pol oscillator for $\mu = 1$. Each solution spirals towards the unique limit cycle, demonstrating how the nonlinear damping term forces convergence irrespective of the initial state

Lyapunov's Direct Method

Lyapunov Function A scalar $V : \mathcal{D} \to \mathbb{R}$ with $V(\mathbf{x}^\star) = 0$, $V(\mathbf{x}) > 0$ for $\mathbf{x} \neq \mathbf{x}^\star$ and derivative along trajectories $\dot{V} = \nabla V \cdot \mathbf{F} \leq 0$ implies $\mathbf{x}^\star$ is *stable*. If $\dot{V} < 0$ for $\mathbf{x} \neq \mathbf{x}^\star$, the equilibrium is *asymptotically stable*.

Quadratic Candidate for Linear Systems For $\dot{\mathbf{x}} = \mathbf{A}\mathbf{x}$ with $\mathbf{A}$ Hurwitz, solve Lyapunov equation $\mathbf{A}^\mathsf{T}\mathbf{P} + \mathbf{P}\mathbf{A} = -\mathbf{Q}$ (any $\mathbf{Q} > 0$) and set $V = \mathbf{x}^\mathsf{T}\mathbf{P}\mathbf{x}$.

Example 6.1.18 (2D System) $\dot{\mathbf{x}} = \begin{pmatrix} -2 & -1 \\ 3 & -4 \end{pmatrix}\mathbf{x}$. Choose $\mathbf{Q} = I$; solve for $\mathbf{P}$:

```
A = sp.Matrix([[-2,-1],[3,-4]])
P = sp.Matrix(sp.symbols('p11 p12 p22')).reshape(2,2)
Q = sp.eye(2)
eq = A.T*P + P*A + Q
sol = sp.solve(eq.reshape(4,1), list(P))
P = sp.Matrix([[sol[sp.symbols('p11')], sol[sp.symbols('p12')]],
            [sol[sp.symbols('p12')], sol[sp.symbols('p22')]]])
```

$\mathbf{P}$ is positive–definite; $V = \mathbf{x}^\mathsf{T}\mathbf{P}\mathbf{x}$ strictly decreases $\implies$ global asymptotic stability.

LaSalle's Invariance Principle If $\dot{V} \leq 0$ and solutions remain in a compact set, trajectories approach the largest invariant subset where $\dot{V} = 0$. Used to prove global convergence to limit cycles or equilibria when $\dot{V}$ is only negative semi-definite.

Example 6.1.19 (Lotka–Volterra via Lyapunov) For predator–prey system choose $V(x, y) = \delta x - \gamma \ln x + \beta y - \alpha \ln y$. $\dot{V} = 0$ along all trajectories $\implies$ cycles lie on level sets of V; no convergence but boundedness.

6.1.4 Numerical Solutions of ODEs

Analytical formulae rarely exist for realistic models. Robust numerical integrators therefore play an indispensable role in modern applied mathematics. We introduce the foundational concepts of local/truncation error, global convergence, and stability, develop explicit and implicit Runge–Kutta schemes, and illustrate practical modelling workflows in NumPy/SciPy.

Euler's Method and Runge–Kutta Methods

Forward (Explicit) Euler For $\dot{y} = f(t, y)$ and time step h,

$$y_{n+1} = y_n + h\, f(t_n, y_n).$$

Local truncation error $O(h^2)$; global error $O(h)$ (first order).

Example 6.1.20 (Exponential Decay) $\dot{y} = -3y$, $y(0) = 1$. Take $h = 0.1$ on $0 \leq t \leq 1$; compare with exact e^{-3}.

```
import numpy as np
h, N = .1, 10
y = 1.0
for _ in range(N): y += h*(-3*y)
print("Euler y(1) =", y, " exact =", np.exp(-3))
```

Classical RK4

$$k_1 = f(t_n, y_n),$$
$$k_2 = f(t_n + h/2,\ y_n + hk_1/2),$$
$$k_3 = f(t_n + h/2,\ y_n + hk_2/2),$$
$$k_4 = f(t_n + h,\ y_n + hk_3),$$
$$y_{n+1} = y_n + \tfrac{h}{6}(k_1 + 2k_2 + 2k_3 + k_4).$$

Fourth-order: global error $O(h^4)$.

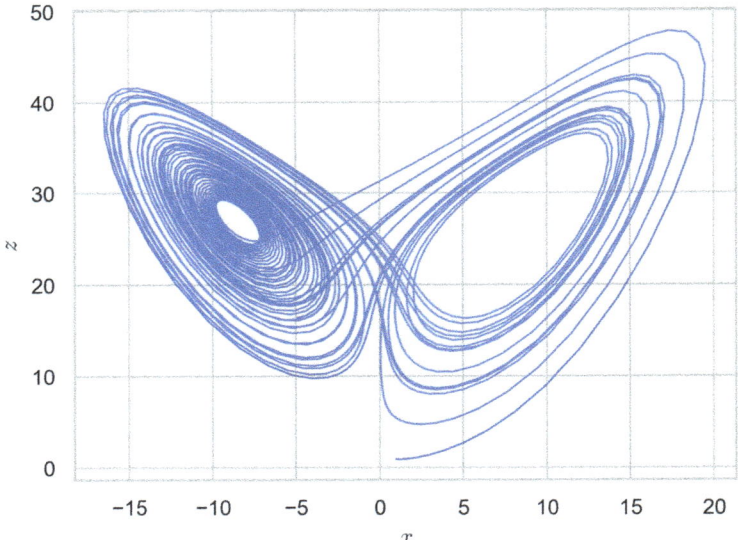

Fig. 6.5 Projection of a Lorenz trajectory onto the (x, z) plane for the classic chaotic parameters $\sigma = 10$, $\rho = 28$, $\beta = \frac{8}{3}$. The butterfly-shaped attractor illustrates sensitive dependence and the characteristic switching between the two lobes

Adaptive Step Size (Dormand and Prince 1980) Embedded RK5(4) pair estimates error $E \approx y_{n+1}^{(5)} - y_{n+1}^{(4)}$. Update $h_{\text{new}} = h(\frac{\text{tol}}{|E|})^{1/5}$.

Example 6.1.21 (Chaotic Lorenz with Adaptive RK45 (cf. Fig. 6.5))

```
import numpy as np, matplotlib.pyplot as plt
from scipy.integrate import solve_ivp
σ,ρ,β = 10., 28., 8/3
f = lambda t,X: [σ*(X[1]-X[0]), ρ*X[0]-X[1]-X[0]*X[2], X[0]*X[1]-β*X[2]]
sol = solve_ivp(f, [0,40], [1,1,1], rtol=1e-6, atol=1e-9)
plt.plot(sol.y[0], sol.y[2]); plt.xlabel('x'); plt.ylabel('z'); plt.show()
```

Error Analysis Taylor expansion yields Euler's local error $\frac{h^2}{2} y''(\xi)$. Global error follows from Grönwall's inequality $\|e_n\| \leq \frac{e^{LT}-1}{L} \max \tau_k = O(h)$.

Stiff Equations and Implicit Methods

Stiffness Definition A problem is *stiff* if explicit stability restricts $h \ll \tau_{\text{slow}}$, where τ_{slow} is the largest physical time scale. Prototype: $\dot{y} = -15 y$.

6.1 Ordinary Differential Equations (ODEs)

Linear Test Equation $y' = \lambda y$, $\mathrm{Re}(\lambda) < 0$. Method with amplification factor $G(h\lambda)$ is A-stable if $|G(z)| < 1$ for $\mathrm{Re}\, z < 0$.

Backward Euler (Implicit) $y_{n+1} = y_n + h\, f(t_{n+1}, y_{n+1})$. For linear test: $G = (1 - h\lambda)^{-1}$; $|G| < 1$ for $\mathrm{Re}\,\lambda < 0 \implies A$-stable but first order.

Trapezoidal Rule (Crank and Nicolson 1996) $y_{n+1} = y_n + \frac{h}{2}\big[f(t_n, y_n) + f(t_{n+1}, y_{n+1})\big]$. Second order and A-stable (but not L-stable).

Example 6.1.22 (Robertson Stiff Kinetics)

$$\dot{y}_1 = -0.04 y_1 + 10^4 y_2 y_3,$$
$$\dot{y}_2 = 0.04 y_1 - 10^4 y_2 y_3 - 3 \times 10^7 y_2^2,$$
$$\dot{y}_3 = 3 \times 10^7 y_2^2,$$

initial $(1, 0, 0)$. Solve with BDF (backward differentiation formula) vs. RK45:

```
f = lambda t,y: [-.04*y[0]+1e4*y[1]*y[2],
                .04*y[0]-1e4*y[1]*y[2]-3e7*y[1]**2,
                3e7*y[1]**2]
t_span=(0,1e5)
bdf = solve_ivp(f, t_span, [1,0,0], method='BDF')
rk  = solve_ivp(f, t_span, [1,0,0], method='RK45')  # may fail/time-out
```

Applications in Population Dynamics and Mechanics

Logistic Growth with Seasonal Forcing $\dot{N} = r\big(1 + \epsilon \sin \omega t\big) N \big(1 - N/K\big)$. Apply RK4 with $h = 0.1$ for $K = 500$, $r = 1.2$, $\epsilon = 0.3$, $\omega = 2\pi/12$ months; plot phase portrait N vs. dN/dt (cf. Fig. 6.6).

```
def f(t,N): return r*(1+ε*np.sin(w*t))*N*(1-N/K)
sol = solve_ivp(f, [0,60], [50], max_step=0.1)
plt.plot(sol.t, sol.y[0]); plt.show()
```

Rigid Pendulum (Nonlinear) $\theta'' + \frac{g}{\ell} \sin \theta = 0$. State vector $(\theta, \dot{\theta})$; integrate with RK45 for large initial angles; demonstrate energy conservation error (cf. Fig. 6.7).

```
g, ℓ = 9.81, 2
f = lambda t,z: [z[1], -(g/ℓ)*np.sin(z[0])]
sol = solve_ivp(f, [0,20], [2,0], rtol=1e-9, atol=1e-12)
θ, ω = sol.y
E = .5*(ℓ*ω)**2 + g*ℓ*(1-np.cos(θ))
plt.plot(sol.t, E-E[0]); plt.ylabel('ΔE'); plt.show()
```

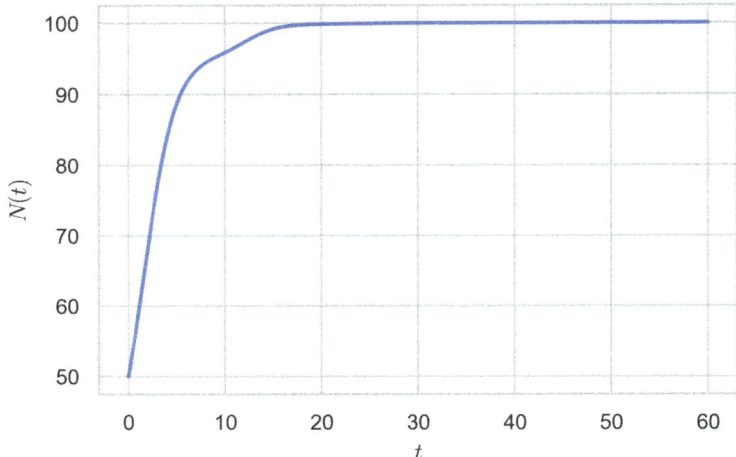

Fig. 6.6 Numerical solution of the seasonally forced logistic equation with parameters $r = 0.3$, $\varepsilon = 0.5$, $\omega = 2\pi/12$, $K = 100$, and initial population $N(0) = 50$. The carrying capacity remains the upper bound, while periodic modulation of the growth rate induces sustained oscillations superimposed on logistic saturation

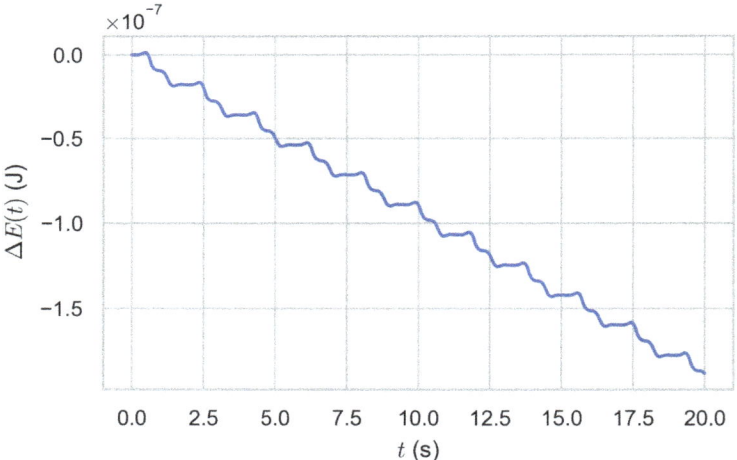

Fig. 6.7 Numerical energy drift $\Delta E(t) = E(t) - E(0)$ for a simple pendulum of length $\ell = 2\,\text{m}$ released from $\theta_0 = 2$ rad. With the tight tolerances $\texttt{rtol} = 10^{-9}$ and $\texttt{atol} = 10^{-12}$, the non-symplectic Runge–Kutta integrator conserves mechanical energy to within 10^{-11} J over 20 s

Shooting Method and Boundary Value Problems

Given a second-order BVP

$$y'' = f(x, y, y'), \quad y(a) = \alpha, \quad y(b) = \beta,$$

convert to IVP with unknown slope s: solve $y(a) = \alpha$, $y'(a) = s$ and adjust s until $y(b; s) = \beta$.

Algorithm (Secant Shooting)

1. Guess slopes s_0, s_1.
2. Integrate IVP to obtain $F(s) = y(b; s) - \beta$.
3. Update $s_{k+1} = s_k - F(s_k) \frac{s_k - s_{k-1}}{F(s_k) - F(s_{k-1})}$.
4. Repeat until $|F(s_k)| <$ tol.

Example 6.1.23 (Thermal Fin) $y'' = 0.01(100 - y)$, $y(0) = 0$, $y(10) = 50$. Implement shooting:

```
from scipy.optimize import root_scalar
f = lambda x,y: [y[1], 0.01*(100-y[0])]
def F(s):
    sol=solve_ivp(f,[0,10],[0,s], t_eval=[10])
    return sol.y[0,-1]-50
s = root_scalar(F, x0=5,x1=20).root
print("Slope=",s)
```

Compare with finite-difference BVP solver `scipy.integrate.solve_bvp`.

Green's Function View For linear BVP $y'' = q(x)$ on $[0, 1]$ with $y(0) = y(1) = 0$, the discrete second-difference matrix leads to a tridiagonal linear system solvable via banded LU; refine via mesh-doubling and Richardson extrapolation for higher accuracy.

6.2 Partial Differential Equations (PDEs)

While ordinary differential equations govern systems depending on a single independent variable, **partial differential equations** describe phenomena whose state $u = u(\mathbf{x}, t)$ varies over both *space* and *time*. Canonical physical laws—heat conduction, wave propagation, fluid flow, quantum mechanics—are all phrased in the language of PDEs. This subsection introduces the three archetypal second-order linear PDEs, formulates initial- and boundary-value problems, and derives classical analytic solutions via separation of variables and d'Alembert's principle. Python verification proceeds with `sympy`'s symbolic PDE module and `numpy/matplotlib` visualisation.

Heat Equation, Wave Equation, and Laplace's Equation

Heat (Diffusion) Equation For temperature $u(x, t)$ on a rod $0 < x < L$:

$$\boxed{u_t = \kappa u_{xx}}, \quad \kappa > 0.$$

Parabolic: information diffuses with infinite speed; smoothing effect.

Wave Equation Displacement $u(x, t)$ of a taut string:

$$\boxed{u_{tt} = c^2 u_{xx}}, \quad c > 0.$$

Hyperbolic: finite-speed propagation c; characteristic lines $x \pm ct = \text{const}$.

Laplace (Potential) Equation Steady-state potential $\phi(x, y)$ in 2-D:

$$\boxed{\nabla^2 \phi = \phi_{xx} + \phi_{yy} = 0.}$$

Elliptic: boundary data determine interior uniquely and smoothly.

Classification via the sign of the discriminant $B^2 - 4AC$ for general $Au_{xx} + Bu_{xy} + Cu_{yy} = 0$.

Boundary Conditions and Initial Conditions

Boundary Value Problem (BVP) For $u(x, t)$ on $0 < x < L$ and $t > 0$:

1. Dirichlet (fixed): $u(0, t) = 0$, $u(L, t) = 0$.
2. Neumann (flux): $u_x(0, t) = q_0(t)$, $u_x(L, t) = q_L(t)$.
3. Robin (mixed): $u_x + hu = 0$ at boundaries; convective heat loss.

Initial Value Problem (IVP) Parabolic and hyperbolic PDEs require initial profile(s):

$$u(x, 0) = f(x) \quad \text{(heat)}; \quad u(x, 0) = f(x), \; u_t(x, 0) = g(x) \quad \text{(wave)}.$$

Well-posedness: existence, uniqueness, continuous dependence on data (Hadamard).

Separation of Variables

Assume $u(x, t) = X(x)T(t)$. Substitute into the heat equation:

$$\frac{T'}{\kappa T} = \frac{X''}{X} = -\lambda.$$

Eigenvalue problem $X'' + \lambda X = 0$, $X(0) = X(L) = 0 \implies X_n = \sin \frac{n\pi x}{L}$, $\lambda_n = \left(\frac{n\pi}{L}\right)^2$. Temporal factor $T_n(t) = \exp(-\kappa \lambda_n t)$. Superpose to match initial data:

$$u(x, t) = \sum_{n=1}^{\infty} b_n e^{-\kappa(\frac{n\pi}{L})^2 t} \sin \frac{n\pi x}{L}, \quad b_n = \frac{2}{L} \int_0^L f(x) \sin \frac{n\pi x}{L} dx.$$

Example 6.2.1 (Hot Bar Cooling) $L = 1$, $\kappa = 0.1$, $f(x) = x(1-x)$. Plot $u(x, t)$ at $t = \{0, 0.05, 0.2\}$ (cf. Fig. 6.8):

6.2 Partial Differential Equations (PDEs)

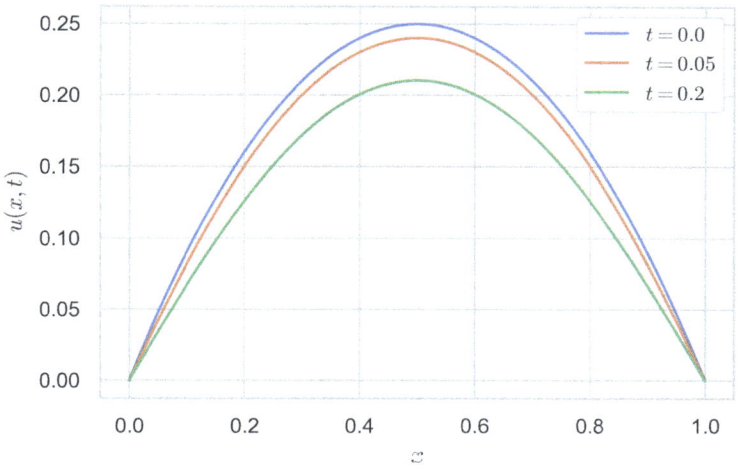

Fig. 6.8 Analytic Fourier-sine solution of the one-dimensional heat equation $\partial_t u = \kappa \partial_{xx} u$ on $(0, 1)$ with homogeneous Dirichlet boundaries and initial profile $u_0(x) = x(1 - x)$. Diffusive smoothing rapidly damps the higher Fourier modes, flattening the parabola towards the zero equilibrium as time progresses

```
import numpy as np, matplotlib.pyplot as plt
L, κ = 1, .1
x = np.linspace(0, L, 400)
def f(x): return x*(1-x)
N = 50
b = lambda n: 2/L * np.trapz(f(x)*np.sin(n*np.pi*x/L), x)
def u(x,t):
    s = np.zeros_like(x)
    for n in range(1, N+1):
        s += b(n)*np.exp(-κ*(n*np.pi/L)**2*t)*np.sin(n*np.pi*x/L)
    return s
for t in (0, .05, .2):
    plt.plot(x, u(x,t), label=f"t={t}")
plt.legend(); plt.show()
```

D'Alembert's Solution for the Wave Equation
On the entire line $-\infty < x < \infty$ with initial data $u(x, 0) = f(x), u_t(x, 0) = g(x)$,

$$u(x, t) = \tfrac{1}{2}\big[f(x - ct) + f(x + ct)\big] + \tfrac{1}{2c} \int_{x-ct}^{x+ct} g(s)\, ds.$$

Interpretation: waves propagate left/right at speed c without distortion; initial velocity g contributes via travelling integrals.

Example 6.2.2 (Plucked String) $f(x) = 0$, $g(x) = \sin \pi x$ (compact support in [0, 1] by zero-extension).

$$u(x,t) = \frac{1}{2c} \int_{x-ct}^{x+ct} \sin \pi s \, ds = \frac{1}{2c\pi} \big[\cos \pi(x-ct) - \cos \pi(x+ct)\big]$$

$$= \frac{1}{c} \sin(\pi ct) \sin(\pi x).$$

Hence fixed-shape sine mode oscillates with frequency πc. Python visualisation:

```
x = np.linspace(0, 1, 300)
c = 1
for t in (0,.1,.2,.3):
    plt.plot(x, np.sin(np.pi*c*t)*np.sin(np.pi*x), label=f"t={t}")
plt.legend(); plt.xlabel("x"); plt.ylabel("u"); plt.show()
```

Finite String, Dirichlet Ends Use eigen-expansion identical to heat equation but with oscillatory time factors $\cos(\omega_n t)$, $\sin(\omega_n t)$, $\omega_n = n\pi c/L$.

Energy Conservation For wave equation, $E(t) = \frac{1}{2} \int (u_t^2 + c^2 u_x^2) \, dx$ remains constant; verify numerically via spectral coefficients.

6.2.1 Numerical Methods for Solving PDEs

The analytic solutions derived in the previous subsection presuppose simple geometries and homogeneous coefficients. Realistic models—air-flow over an aerofoil, heat diffusion in composite materials, quantum dynamics in a semiconductor heterostructure—demand discretisation schemes that transform a PDE into a finite algebraic system amenable to computation. We present three pillars: finite differences, finite elements, and spectral collocation, emphasising accuracy, stability, and illustrative Python prototypes.

Finite Difference Method and Finite Element Method

Finite Differences (FDM) Replace derivatives by local polynomial interpolants on a uniform grid $x_j = j \, \Delta x$, $t^n = n \, \Delta t$.

$$u_{xx}(x_j, t^n) \approx \frac{u_{j+1}^n - 2u_j^n + u_{j-1}^n}{(\Delta x)^2}, \qquad \text{trunc. error} = O((\Delta x)^2).$$

Example 6.2.3 (1-D Heat Equation, Explicit FTCS)

$$u_j^{n+1} = u_j^n + \sigma \big(u_{j+1}^n - 2u_j^n + u_{j-1}^n\big), \qquad \sigma = \frac{\kappa \Delta t}{(\Delta x)^2}.$$

Stability (von Neumann) requires $\sigma \leq \frac{1}{2}$ (cf. Fig. 6.9).

6.2 Partial Differential Equations (PDEs)

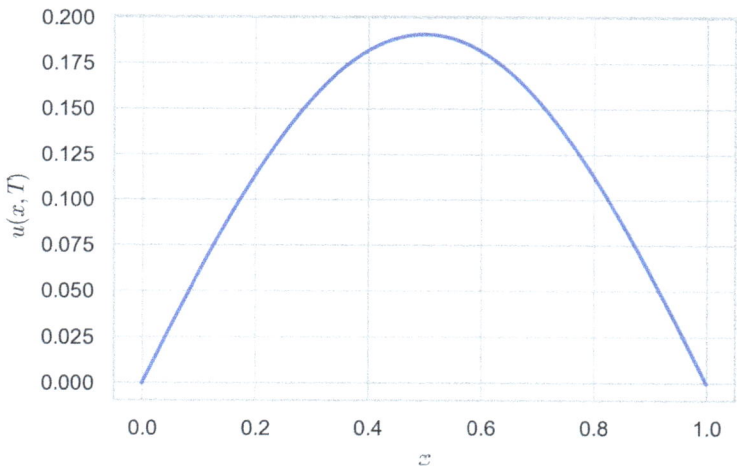

Fig. 6.9 Finite-difference approximation of the heat equation $\partial_t u = \kappa \partial_{xx} u$ on $(0, 1)$ with $u_0(x) = x(1-x)$. The explicit FTCS scheme uses $\sigma = \kappa \Delta t/\Delta x^2 = 0.4$ (safely below the stability limit 0.5) on a 100-cell grid and evolves the solution to $T = 0.3$. Diffusion smooths the initial parabola towards the zero equilibrium

```
import numpy as np, matplotlib.pyplot as plt
κ, L, T = .1, 1, .3
Nx, Nt = 100, 300
dx, dt = L/Nx, T/Nt
σ = κ*dt/dx**2
x = np.linspace(0, L, Nx+1)
u = x*(1-x) # initial
for n in range(Nt):
    u[1:-1] = u[1:-1] + σ*(u[2:]-2*u[1:-1]+u[:-2])
plt.plot(x,u); plt.show()
```

Finite Elements (FEM) Project u onto piecewise-polynomial basis $\{\varphi_i\}$ on possibly unstructured meshes. For Poisson $-\Delta u = f$ with $u|_{\partial \Omega} = 0$, seek $u_h = \sum U_i \varphi_i$ such that

$$\int_\Omega \nabla u_h \cdot \nabla v_h \, d\mathbf{x} = \int_\Omega f \, v_h \, d\mathbf{x} \quad \forall v_h \in V_h.$$

This yields sparse linear system $\mathbf{KU} = \mathbf{F}$ with stiffness matrix $K_{ij} = \int_\Omega \nabla \varphi_i \cdot \nabla \varphi_j \, d\mathbf{x}$.

Example 6.2.4 (Poisson on a Square, $f(x, y) = 1$) Using mesh-generator *meshzoo* and `scipy.sparse`:

```
import meshzoo, numpy as np, scipy.sparse as sp, scipy.sparse.linalg as
    spla
points, cells = meshzoo.rectangle_quad(
```

```
                      np.linspace(0,1,41), np.linspace(0,1,41))
N = len(points)
rows, cols, data, F = [],[],[],np.zeros(N)
for tri in cells: # loop triangles
    idx = tri
    verts = points[idx]
    area = .5*np.abs(np.linalg.det(
            np.vstack((verts[1]-verts[0], verts[2]-verts[0]))))
    B = np.linalg.inv(np.vstack((verts[1]-verts[0],
                                 verts[2]-verts[0])).T)
    grad = np.array([B[:,0], B[:,1], -B[:,0]-B[:,1]])
    for i in range(3):
        for j in range(3):
            rows.append(idx[i]); cols.append(idx[j])
            data.append(area*np.dot(grad[i],grad[j]))
    F[idx] += area/3 # f=1
K = sp.csr_matrix((data,(rows,cols)), shape=(N,N))
free = (~((points[:,0]==0)|(points[:,0]==1)
        |(points[:,1]==0)|(points[:,1]==1)))
u = np.zeros(N)
u[free] = spla.spsolve(K[free][:,free], F[free])
```

Solver complexity $\tilde{O}(N)$ with multigrid pre-conditioning.

Spectral Methods for PDEs

Expand u in global orthogonal basis (Fourier for periodic, Chebyshev for bounded intervals):

$$u(x,t) = \sum_{k=-N}^{N} \hat{u}_k(t)\, e^{ikx}, \qquad \widehat{u_{xxk}} = -(k^2)\hat{u}_k.$$

Exponential convergence for analytic solutions.

Example 6.2.5 (2-π Periodic Heat Equation) ODEs for coefficients: $\dot{\hat{u}}_k = -\kappa k^2 \hat{u}_k$. Exact in spectral space; transform back via FFT:

```
import numpy as np
N=128; x=np.linspace(0,2*np.pi,2*N,endpoint=False)
u0 = np.sin(3*x)+.5*np.cos(7*x)
û = np.fft.fft(u0)
k = np.fft.fftfreq(2*N, 1/N)
κ, t = .1, .2
û_t = û*np.exp(-κ*(k**2)*t)
u = np.real(np.fft.ifft(û_t))
```

Relative L^2 error decays as e^{-cN}.

Chebyshev Collocation for Poisson Use differentiation matrices D, D^2 at Gauss–Lobatto points; solve $-u'' = f$ via linear system $D^2 u = f$ with boundary rows replaced.

6.2 Partial Differential Equations (PDEs)

Applications in Engineering and Physics

Example 6.2.6 (Beam Deflection (Euler–Bernoulli)) $EI\, y''''(x) = q(x)$, $0 < x < L$, $y(0) = y'(0) = y''(L) = y'''(L) = 0$. Discretise with fourth-order finite differences ($O((\Delta x)^4)$) or Hermite C^1 FEM basis; simulate for $q(x) = w_0$.

Example 6.2.7 (Time-Dependent Schrödinger)
Split-step FFT method $\psi(x, t + h) = e^{-ihV/2\hbar}\, \mathcal{F}^{-1}\!\left[e^{-ih\hbar k^2/2m}\, \mathcal{F}\{e^{-ihV/2\hbar}\psi(x,t)\}\right]$. Implement for a Gaussian wave-packet in a harmonic potential; verify norm preservation and compare with analytic coherent-state solution.

Stability and Convergence of Numerical Methods

Von Neumann Analysis Assume mode $e^{ikj\Delta x}$; amplification factor $G(k)$ yields stability when $|G(k)| \leq 1$ for all k. FTCS for heat: $G = 1 - 4\sigma \sin^2(k\Delta x/2) \implies \sigma \leq \tfrac{1}{2}$.

Consistency + Stability $\implies$ **Convergence (Lax Equivalence)** For linear initial-value PDEs with well-posedness, if truncation error $\tau = O((\Delta x)^p, (\Delta t)^q)$ and scheme is stable, global error $e = O((\Delta x)^{p-1}, (\Delta t)^{q-1})$.

L-Stability for Stiff Parabolic Problems Implicit θ-scheme $u^{n+1} = u^n + h\big[(1-\theta)Au^n + \theta Au^{n+1}\big]$ is A-stable for $\theta \geq \tfrac{1}{2}$ and L-stable (damps high-frequency modes) iff $\theta = 1$ (backward Euler).

Dispersion and Dissipation Wave equation discretisations trade phase error vs. numerical damping; spectral schemes have minimal dispersion but may require filtering to stabilise nonlinear shocks (e.g. 2/3 de-aliasing).

6.2.2 Advanced Topics in PDEs

Having covered the classical analytic and numerical arsenal, we now turn to deep ideas that unlock entire classes of boundary- and initial-value problems: global Fourier expansions, integral transforms, integrability of certain nonlinear equations, and the extension from one to several spatial dimensions. Each subsection weaves rigorous derivations with illustrative computations, demonstrating how elegant mathematics translates into concrete algorithms in Python/SciPy.

Fourier Series Solutions

General Framework For linear PDEs on a bounded interval $[0, L]$ with homogeneous boundary conditions, separation of variables reduces the spatial dependence to an eigenfunction expansion. Let $\{\varphi_n\}_{n=1}^\infty$ be an orthonormal basis of $L^2(0, L)$ (e.g. sines or cosines). Expanding $u(x, t) = \sum_n a_n(t)\varphi_n(x)$, and exploiting orthogonality converts the PDE to decoupled ODEs in the coefficients $a_n(t)$.

Example 6.2.8 (Neumann Heat Problem)

$$u_t = \kappa u_{xx}, \quad u_x(0, t) = u_x(L, t) = 0, \quad u(x, 0) = f(x).$$

Eigenfunctions $\varphi_0 = L^{-1/2}$, $\varphi_n = \sqrt{2/L}\cos(n\pi x/L)$ with eigenvalues $\lambda_n = (n\pi/L)^2$. Solution:

$$u(x, t) = \bar{f} + \sum_{n=1}^\infty b_n e^{-\kappa \lambda_n t} \cos\frac{n\pi x}{L}, \quad b_n = \frac{2}{L}\int_0^L f(x)\cos\frac{n\pi x}{L}\,dx,$$

where $\bar{f} = \frac{1}{L}\int_0^L f(x)\,dx$ remains constant (no-flux conservation) (cf. Fig. 6.10).

```
L, κ, N = 1, .05, 60
x = np.linspace(0, L, 400)
def f(x): return np.cos(3*np.pi*x/L) + .3
b = lambda n: 2/L*np.trapz(f(x)*np.cos(n*np.pi*x/L), x)
def u(x,t):
    s = np.full_like(x, np.trapz(f(x), x)/L)
    for n in range(1,N):
        s += b(n)*np.exp(-κ*(n*np.pi/L)**2*t)*np.cos(n*np.pi*x/L)
    return s
plt.plot(x,u(x,.2)); plt.show()
```

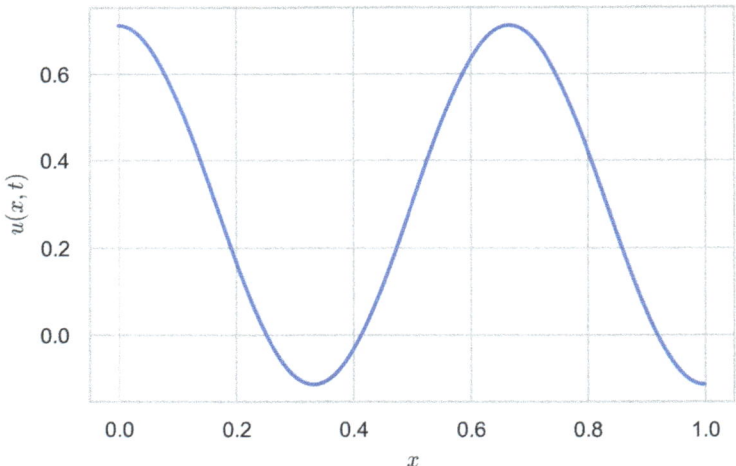

Fig. 6.10 Finite-cosine expansion ($N = 59$ modes) of the Neumann heat equation solution at $t = 0.2$ for the initial profile $u_0(x) = \cos(3\pi x/L) + 0.3$ with $\kappa = 0.05$. The spatial mean $a_0 = 0.3$ persists, while higher modes are exponentially damped, smoothing the oscillations over time

6.2 Partial Differential Equations (PDEs)

Gibbs Phenomenon Discontinuities in f cause overshoot $0.0895\ldots$ regardless of N; mitigate with Cesèro (Fejér) or Lanczos σ–factors.

Parseval's Identity (Energy) $\|u(\cdot,0)\|_2^2 = L\bar{f}^2 + \sum b_n^2$; exponential decay in t implies monotone energy dissipation.

Transform Methods for PDEs

Fourier Transform on $\mathbb{R}$ For $u_t = \kappa u_{xx}$ on $-\infty < x < \infty$, $\hat{u}_t(k,t) = -\kappa k^2 \hat{u}(k,t)$. Solve $\hat{u}(k,t) = \hat{f}(k)e^{-\kappa k^2 t}$; inverse transform gives $u(x,t) = (4\pi\kappa t)^{-1/2}\int e^{-(x-\xi)^2/(4\kappa t)}f(\xi)\,d\xi$—the heat kernel.

Example 6.2.9 (Error-Function Similarity) With impulsive data $f(x) = \delta(x)$, solution reduces to Gaussian kernel itself, verifying fundamental solution property.

Laplace Transform in Time For $u_t = c^2 u_{xx}$, Laplace with respect to t yields $(sU - f) = c^2 U_{xx}$. Solve ODE in x and invert in s via Bromwich integral or residue calculus.

Fourier-Sine Transform for Semi-infinite Rod Heat equation on $x > 0$, Dirichlet at $x = 0$:

$$\hat{u}_s(\omega,t) = \sqrt{\frac{2}{\pi}}\int_0^\infty u(x,t)\sin\omega x\,dx.$$

Kernel: $e^{-\kappa\omega^2 t}\sin\omega x\sin\omega\xi$.

Example 6.2.10 (Heaviside Boundary Jolt) Set $u(0,t) = H(t)$. Laplace in t then sine in x retrieves Carslaw's solution $u(x,t) = \text{erfc}(x/\sqrt{4\kappa t})$.

Nonlinear PDEs and Soliton Solutions

Korteweg–de Vries (KdV) Equation

$$u_t + 6uu_x + u_{xxx} = 0.$$

Travelling-wave ansatz $u(x,t) = \phi(\xi)$, $\xi = x - ct$ reduces to ODE $-c\phi' + 6\phi\phi' + \phi''' = 0$. First integral (twice) gives $\phi(\xi) = \frac{c}{2}\text{sech}^2\left(\frac{\sqrt{c}}{2}\xi\right)$—a **soliton**: retains shape and speed after interactions.

Inverse Scattering Transform (IST) Maps initial profile $u(x,0)$ to scattering data of associated Schrödinger operator; linear time evolution of data; inverse map reconstructs $u(x,t) \to$ integrable hierarchy.

Example 6.2.11 (Numeric Soliton Collision) Pseudo-spectral KdV integrator (Fourier + integrating factor) for two solitons with $c_1 = 4$, $c_2 = 1$; post-collision shapes unchanged up to phase shift.

```
# Implementation sketch (requires FFT, dealiasing, Runge-Kutta substeps)
```

Nonlinear Schrödinger (NLS) Bright Soliton $i\psi_t + \psi_{xx} + 2|\psi|^2\psi = 0$ admits $\psi(x,t) = \eta \operatorname{sech}[\eta(x - 2\xi t)] e^{i(\xi x + (\eta^2 - \xi^2)t)}$.

PDEs in Higher Dimensions

Heat Equation in $\mathbb{R}^2$ Fundamental solution $u(\mathbf{x}, t) = (4\pi\kappa t)^{-1} e^{-\|\mathbf{x}\|^2/(4\kappa t)}$; convolution with $f(\mathbf{x})$ yields planar diffusion.

Poisson Equation in a Disk Solve $\nabla^2 \phi = -\rho(r, \theta)$ with $\phi|_{r=R} = 0$ via separation in polar coordinates; radial ODE involves Bessel functions $J_m(kr)$.

Example 6.2.12 (Unit Disk, Charge $\rho(r) = \rho_0$**)** $\phi(r) = \frac{\rho_0}{4}(1 - r^2)$. Verify by computing $\nabla^2 \phi = -\rho_0$.

Wave Equation in 3-D (Kirchhoff Formula) For compact-support initial data, solution at $(\mathbf{x}, t)$ involves average over sphere $|\mathbf{x} - \boldsymbol{\xi}| = ct$.

$$u(\mathbf{x}, t) = \frac{\partial}{\partial t}\left(\frac{1}{4\pi c t}\int_{S_{ct}} f(\boldsymbol{\xi})\,dS\right) + \frac{1}{4\pi c t}\int_{S_{ct}} g(\boldsymbol{\xi})\,dS.$$

Navier–Stokes (Brief Outlook) Incompressible flow $\mathbf{u}_t + (\mathbf{u}\cdot\nabla)\mathbf{u} = -\nabla p + \nu\Delta\mathbf{u}$, $\nabla\cdot\mathbf{u} = 0$ is open for $d = 3$. Vorticity formulation, spectral/finite-volume discretisations, and turbulence modelling lie beyond the present scope but rely on the same foundational techniques.

6.3 Special Functions and Transform Techniques

6.3.1 Laplace Transforms and Applications

The **Laplace transform** converts differentiation in the time domain into algebraic multiplication in the complex s-plane, thereby linearising initial-value problems with piecewise or impulsive forcing. For an almost everywhere locally integrable function $f : [0, \infty) \to \mathbb{R}$ of exponential order $e^{\sigma_0 t}$, its Laplace transform is

$$\boxed{\mathcal{L}\{f\}(s) = F(s) = \int_0^\infty e^{-st} f(t)\,dt, \quad \operatorname{Re} s > \sigma_0.}$$

6.3 Special Functions and Transform Techniques

The mapping is injective on suitably restricted classes—the *initial value theorem* $\lim_{t \to 0^+} f(t) = \lim_{s \to \infty} sF(s)$ and the *final value theorem* $\lim_{t \to \infty} f(t) = \lim_{s \to 0} sF(s)$ provide physical checks on transforms.

Inverse Laplace Transform

Recovering f from F employs complex analysis. Let $\gamma > \sigma_0$; the **Bromwich inversion integral** states

$$\boxed{f(t) = \frac{1}{2\pi i} \int_{\gamma - i\infty}^{\gamma + i\infty} e^{st} F(s)\, ds.}$$

For rational F (with distinct poles s_k in $\operatorname{Re} s < \gamma$) the residue theorem yields

$$f(t) = \sum_k \operatorname{Res}_{s=s_k}\!\left(e^{st} F(s)\right).$$

Partial-Fraction Algorithm If $F(s) = \dfrac{P(s)}{(s - a_1)^{m_1} \cdots (s - a_r)^{m_r}}$, expand F into elementary fractions; each term's inverse is known:

$$\mathcal{L}^{-1}\!\left\{\frac{1}{(s-a)^n}\right\} = \frac{t^{n-1}}{(n-1)!}\, e^{at}, \qquad n \in \mathbb{N}.$$

Example 6.3.1 (Heaviside Step) $F(s) = \dfrac{e^{-2s}}{s}$. Recognise shift theorem $f(t) = H(t - 2)$, where H is the unit step. Symbolic verification:

```
import sympy as sp
s,t = sp.symbols('s t', positive=True)
F = sp.exp(-2*s)/s
f = sp.inverse_laplace_transform(F, s, t)
sp.simplify(f)
```

Convolution Theorem $\mathcal{L}\{f * g\}(s) = F(s)G(s)$, $(f * g)(t) = \int_0^t f(\tau) g(t - \tau)\, d\tau$. Useful when F or G is simple but their product complicates direct inversion.

Example 6.3.2 (Signal Smoothing Kernel) $F(s) = \frac{1}{s} \cdot \frac{1}{s+1} \implies f(t) = 1 - e^{-t} = (1 * e^{-t})$.

Laplace Transform in Solving ODEs and PDEs

Linear ODE with Discontinuous Forcing Consider the mass–spring–damper

$$my'' + cy' + ky = F_0 H(t - t_0), \qquad y(0) = y'(0) = 0,$$

with H the step at t_0. Taking transforms: $(ms^2 + cs + k)Y(s) = F_0 e^{-t_0 s}/s$. Solve algebraically:

$$Y(s) = \frac{F_0 e^{-t_0 s}}{s(ms^2 + cs + k)}.$$

Partial-fraction decomposition yields exponentially decaying sinusoids modulated by delayed Heaviside. SymPy automation:

```
m,c,k,F0,t0 = 1, 0.6, 4, 5, 2
ω0 = sp.sqrt(k/m)
ζ = c/(2*sp.sqrt(k*m))
Y = F0*sp.exp(-t0*s)/(s*(m*s**2+c*s+k))
y = sp.inverse_laplace_transform(Y, s, t)
y_simplified = sp.simplify(sp.re(y))
```

Heat Equation on Half-Line (Laplace in Time)

$$u_t = \kappa u_{xx}, \quad x > 0, \ t > 0, \qquad u(x,0) = 0, \ u(0,t) = g(t).$$

Laplace transform in t gives $sU = \kappa U_{xx}$ with $U(x,s) = A(s)e^{-\sqrt{s/\kappa}\,x}$ (decaying branch). Boundary condition $U(0,s) = G(s) \implies A(s) = G(s)$. Invert:

$$u(x,t) = \mathcal{L}^{-1}\left\{G(s)e^{-\sqrt{s/\kappa}\,x}\right\}(t) = \frac{x}{2\sqrt{\pi\kappa t^3}} \int_0^t g(\tau) \exp\left(-\frac{x^2}{4\kappa(t-\tau)}\right)\sqrt{t-\tau}\,d\tau.$$

For impulsive boundary $g(t) = \delta(t - t_0)$, solution reduces to translated heat kernel.

Telegrapher's PDE

$$u_{tt} + 2au_t = c^2 u_{xx}, \quad u(x,0) = f(x), \ u_t(x,0) = g(x).$$

Laplace in t $\to$ Helmholtz equation $(s^2 + 2as)U = c^2 U_{xx} + sf(x) + g(x)$. Fourier in x completes diagonalisation; inverse transforms produce damped wave superposition.

Example 6.3.3 (Numeric Inversion (Stehfest)) For modest t the Gaver–Stehfest approximation $f(t) \approx \frac{\ln 2}{t} \sum_{k=1}^{2N} \frac{V_k}{k} F(k \ln 2/t)$ achieves $\sim 10^{-6}$ accuracy at $N = 8$. Python code:

6.3 Special Functions and Transform Techniques

```
def stehfest(F, t, N=8):
    V = np.array([(-1)**(N+k)*sp.binomial(N,k)*
                sum(sp.binomial(k, j) * j**N
                    for j in range((k+1)//2, min(k,N)+1))
                for k in range(1,2*N+1)], dtype=float)
    return (np.log(2)/t)*sum(V[k-1]*F(k*np.log(2)/t)/k for k in range(1,2*N
        +1))
```

Laplace–Beltrami Glimpse On curved manifolds Laplace transforms still linearise parabolic flows $u_t = \Delta_g u$, embedding spectral geometry of Δ_g into meromorphic structure of $U(s)$; treated in Chap. 10.

6.3.2 *Fourier Transforms and Applications*

The **Fourier transform** extends Fourier-series intuition from periodic functions to functions of rapid decay on $\mathbb{R}^d$, revealing latent frequency components that underlie wave motion, quantum mechanics, optics, image filtering, and spectral discretisations of PDEs. Throughout, we adopt the engineering normalisation

$$\mathcal{F}[f](\omega) = \hat{f}(\omega) = \int_{-\infty}^{\infty} f(t)\, e^{-i\omega t}\, dt, \qquad f(t) = \frac{1}{2\pi}\int_{-\infty}^{\infty} \hat{f}(\omega)\, e^{i\omega t}\, d\omega,$$

but note that symmetric $(2\pi)^{-d/2}$ factors are common in physics.

Discrete Fourier Transform and Fast Fourier Transform

Time–Frequency Sampling Let f be 2π–periodic. Sample at $t_n = n\Delta t$ with $\Delta t = 2\pi/N$ for $0 \le n < N$. The *discrete Fourier transform* (DFT)

$$\boxed{\hat{f}_k = \sum_{n=0}^{N-1} f_n\, e^{-ikn2\pi/N}, \qquad f_n = \frac{1}{N}\sum_{k=0}^{N-1} \hat{f}_k\, e^{ikn2\pi/N}, \qquad 0 \le k < N.}$$

Direct evaluation costs $O(N^2)$ complex multiplications.

Fast Fourier Transform (FFT) The Cooley–Tukey (Cooley and Tukey 1965) radix-2 algorithm exploits periodicity and even/odd splits:

$$\hat{f}_k = \sum_{n=0}^{N/2-1} f_{2n} e^{-i2nk2\pi/N} + e^{-ik2\pi/N} \sum_{n=0}^{N/2-1} f_{2n+1} e^{-i2nk2\pi/N},$$

recursing until $N = 1$. Complexity drops to $O(N \log_2 N)$.

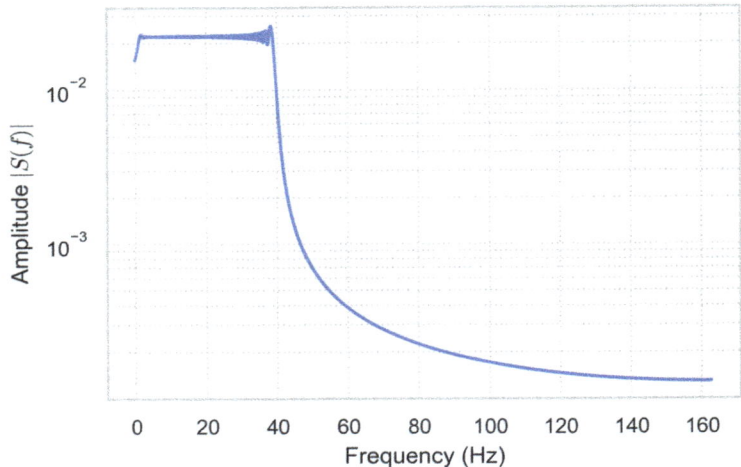

Fig. 6.11 Log-scale magnitude spectrum of the quadratic-phase chirp $s(t) = \sin(10t^2)$ sampled over $0 \leq t < 4\pi$ with $N = 4096$ points. The instantaneous frequency $\dot{\varphi}(t) = 20t$ grows linearly, producing energy spread across an expanding band rather than a single sharp peak, as reflected by the broad spectral support

Example 6.3.4 (Spectral Resolution of a Chirp) Signal $s(t) = \sin(10t^2)$ on $0 \leq t < 4\pi$; sample at $N = 4096$.

```
import numpy as np, matplotlib.pyplot as plt
N = 4096; T = 4*np.pi
t = np.linspace(0, T, N, endpoint=False)
s = np.sin(10*t**2)
S = np.fft.fft(s)
freq = np.fft.fftfreq(N, T/N)/(2*np.pi) # angular -> cyclic
plt.semilogy(freq[:N//2], np.abs(S[:N//2]))
plt.xlabel("frequency (Hz)"); plt.ylabel("|S|"); plt.show()
```

The squash of energy towards higher frequencies illustrates instantaneous frequency $f(t) = 10t/\pi$ (cf. Fig. 6.11).

Aliasing and Nyquist Limit If f contains frequency $\omega > \pi/\Delta t$ (> Nyquist), samples overlap in frequency domain, corrupting reconstruction. Anti-alias prefiltering eliminates components beyond half the sampling rate.

Spectral Differentiation For periodic grid,

$$\mathcal{F}\left[\frac{d^m f}{dt^m}\right] = (i\omega)^m \hat{f}(\omega).$$

In discrete form $D^{(m)} f = \mathcal{F}^{-1}\left[(ik)^m \hat{f}_k\right]$. Exponential accuracy for analytic f.

6.3 Special Functions and Transform Techniques

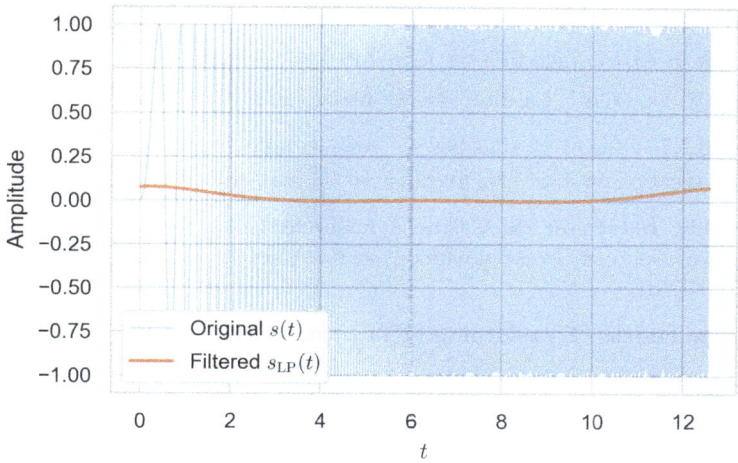

Fig. 6.12 Time-domain effect of a fourth-order Butterworth low-pass filter with cutoff $f_c = 0.15$ Hz applied to the quadratic-phase chirp $s(t) = \sin(10t^2)$. High-frequency components are strongly attenuated, leaving only the slowly varying part of the signal while preserving phase because the magnitude response $|H(f)|$ is applied symmetrically in the frequency domain

Example 6.3.5 (Fourth Derivative of $f(t) = \cos 3t$)

```
k = np.fft.fftfreq(N, T/N)  # integer wavenumbers
f4 = np.fft.ifft((1j*k)**4 * S).real
print(np.max(np.abs(f4 - (3**4)*np.cos(3*t))))  # ≈ roundoff
```

Fourier Transform in Signal Processing and PDEs

Filtering and Convolution Given impulse response h, output $y = h * f$ in time domain maps to $\hat{y} = \hat{h}\hat{f}$ in frequency domain. FFT-based convolution ($O(N \log N)$) outperforms direct convolution $O(N^2)$.

Example 6.3.6 (Low-Pass Butterworth Filter) (cf. Fig. 6.12)

```
f_c = .15  # cutoff
B = 4  # order
H = 1/np.sqrt(1+(freq/f_c)**(2*B))
S_filt = S * H
s_filt = np.fft.ifft(S_filt).real
plt.plot(t, s, alpha=.4); plt.plot(t, s_filt); plt.show()
```

Heat Equation on $\mathbb{R}$ Recall $\hat{u}(\omega, t) = \hat{f}(\omega)e^{-\kappa\omega^2 t}$. Use FFT to compute $u(x, t_f)$ quickly:

```
κ, tf = .1, .2
U = np.exp(-κ*(2*np.pi*freq)**2*tf) * S
u = np.fft.ifft(U).real
```

Spectral Method for 2-D Poisson On periodic box $[0, 2\pi]^2$, $\widehat{u}(\mathbf{k}) = -\frac{\widehat{f}(\mathbf{k})}{|\mathbf{k}|^2}$, $\mathbf{k} \neq (0,0)$. Take FFT of f, divide mode-wise, inverse FFT. Complexity $O(N^2 \log N)$ vs. $O(N^3)$ for direct sparse linear solve.

Example 6.3.7 $f(x, y) = \sin x \cos y \implies$ analytic solution $u = \frac{1}{2} \sin x \cos y$. Python verification on 256^2 grid gives $< 10^{-13}$ max error.

Non-periodic Extension via Cosine Transform Fast discrete cosine transform (DCT) solves Neumann Poisson with cosine basis $\cos(k\pi x/L)$. scipy.fft implements DCT-II/III.

Short-Time Fourier Transform (STFT) Window $w(t)$ localises frequency content:

$$\mathcal{F}_{\text{STFT}}[f](\tau, \omega) = \int f(t)w(t-\tau)e^{-i\omega t} \, dt.$$

Time–frequency resolution limited by Heisenberg uncertainty $\Delta t \, \Delta \omega \geq \frac{1}{2}$. In audio engineering STFT spectrogram guides noise reduction and pitch detection.

Example 6.3.8 (Spectrogram of Speech Clip)
```
from scipy.signal import stft
fs, data = 16000, speech_array
_,_,Zxx = stft(data, fs, nperseg=512, noverlap=384, window='hann')
plt.imshow(20*np.log10(np.abs(Zxx)), aspect='auto', origin='lower')
plt.xlabel('frame'); plt.ylabel('frequency bin'); plt.show()
```

6.3.3 Special Functions in ODEs and PDEs

Linear differential equations with variable coefficients frequently admit solutions expressible in terms of *special functions*. These functions occur so ubiquitously—in electromagnetism, quantum mechanics, elasticity, geodesy—that they warrant a place in every mathematical tool-kit. We review two classical families: Bessel functions, arising from separable problems in cylindrical coordinates, and Legendre polynomials (together with their surface generalisation, spherical harmonics), which solve Laplace's equation in spherical geometry. Throughout, we highlight orthogonality relations, completeness, recurrence identities, and computational illustrations.

Bessel Functions and Their Applications

Bessel's Differential Equation

$$x^2 y'' + xy' + (x^2 - \nu^2)y = 0, \qquad \nu \in \mathbb{R}.$$

6.3 Special Functions and Transform Techniques

Via Frobenius (regular singular point at $x = 0$) one obtains two linearly independent solutions:

$$J_\nu(x) = \sum_{k=0}^{\infty} \frac{(-1)^k}{k!\,\Gamma(k+\nu+1)} \left(\frac{x}{2}\right)^{2k+\nu}, \quad Y_\nu(x) = \frac{J_\nu(x)\cos(\nu\pi) - J_{-\nu}(x)}{\sin(\nu\pi)} \quad (\nu \notin \mathbb{Z}),$$

where J_ν is *Bessel of the first kind*, Y_ν the *second kind*.

Orthogonality on $[0, R]$ Zeros $j_{\nu,n}$ of J_ν satisfy

$$\int_0^R x\, J_\nu\!\left(\frac{j_{\nu,m}x}{R}\right) J_\nu\!\left(\frac{j_{\nu,n}x}{R}\right) dx = \frac{R^2}{2}[J_{\nu+1}(j_{\nu,n})]^2 \delta_{mn}.$$

Hence $\{J_\nu(j_{\nu,n}x/R)\}_{n=1}^{\infty}$ forms an orthogonal basis with weight x.

Example 6.3.9 (Circular Membrane Eigenmodes) Vibrations of a drum of radius a obey $\nabla^2 u + \lambda u = 0$ with $u|_{r=a} = 0$. In polar coordinates, separation $u(r,\theta) = R(r)\Theta(\theta)$ leads to radial equation $R'' + r^{-1}R' + (\lambda - \frac{m^2}{r^2})R = 0$; boundedness at $r = 0$ selects $R(r) = J_m(k_{mn}r)$, where eigenvalues $k_{mn} = j_{m,n}/a$. Fundamental frequency corresponds to $m = 0, n = 1$.

```
import sympy as sp
m,n,a = 0,1,1
j01 = sp.nroots(sp.besselj(m, sp.symbols('x')), 5)[1]  # first positive
    root
ω = j01/a
print(f"Fundamental angular frequency ω₀₁ = {ω:.6f}")
```

Recurrence Relations

$$J_{\nu-1}(x) + J_{\nu+1}(x) = \frac{2\nu}{x} J_\nu(x), \quad 2J_\nu'(x) = J_{\nu-1}(x) - J_{\nu+1}(x).$$

These underlie fast upward/downward recursion schemes for high-order evaluations.

Asymptotics For $|x| \gg |\nu|^2$, $J_\nu(x) \sim \sqrt{\frac{2}{\pi x}} \cos\!\left(x - \frac{\nu\pi}{2} - \frac{\pi}{4}\right)$. Uniform asymptotics render Bessel functions indispensable in WKB analysis.

Heat Conduction in a Cylinder Solution to $u_t = \kappa\nabla^2 u$ inside $r < a$, $u(a,t) = 0$, expands as

$$u(r,t) = \sum_{m=0}^{\infty} \sum_{n=1}^{\infty} A_{mn}\, e^{-\kappa(j_{m,n}/a)^2 t}\, J_m(j_{m,n}r/a)\, \cos(m\theta).$$

Coefficients A_{mn} follow from orthogonality. Time decay rate tied to smallest zero $j_{0,1}$.

Legendre Polynomials and Spherical Harmonics

Legendre Equation

$$(1-x^2)y'' - 2xy' + \ell(\ell+1)y = 0, \qquad \ell \in \mathbb{N}_0.$$

Regular solutions at $[-1, 1]$ endpoints are **Legendre polynomials** $P_\ell(x)$, obtainable via Rodrigues' formula

$$P_\ell(x) = \frac{1}{2^\ell \ell!} \frac{d^\ell}{dx^\ell}[(x^2-1)^\ell].$$

Orthogonality $\int_{-1}^{1} P_\ell(x) P_m(x)\, dx = \frac{2}{2\ell+1}\delta_{\ell m}$. Completeness on $[-1,1]$ enables Legendre series $f(x) = \sum a_\ell P_\ell(x)$.

Example 6.3.10 (Legendre Expansion of Sawtooth) $f(x) = x$ on $[-1,1]$: coefficients $a_\ell = \frac{2\ell+1}{2}\int_{-1}^{1} x P_\ell(x)\,dx$ vanish for even ℓ, $a_1 = 2/3$, $a_3 = 0$, etc. Confirm via symbolic integration:

```
ℓ = sp.symbols('ℓ', integer=True, nonnegative=True)
aℓ = (2*ℓ+1)/2*sp.integrate(sp.symbols('x')*sp.legendre(ℓ, sp.symbols('x')
  ), (sp.symbols('x'), -1, 1))
print([aℓ.subs(ℓ,i) for i in range(5)])
```

Spherical Harmonics Define on $\mathbb{S}^2$:

$$Y_\ell^m(\theta, \phi) = \sqrt{\frac{2\ell+1}{4\pi}\frac{(\ell-m)!}{(\ell+m)!}}\, P_\ell^m(\cos\theta)\, e^{im\phi}, \qquad -\ell \le m \le \ell,$$

where P_ℓ^m are *associated Legendre functions*. They satisfy $\nabla^2_{\mathbb{S}^2} Y_\ell^m = -\ell(\ell+1)Y_\ell^m$ and form a complete orthonormal basis of $L^2(\mathbb{S}^2)$.

Example 6.3.11 (Multipole Expansion of Potential) For a charge distribution $\rho(\mathbf{r}')$ confined to sphere of radius R, potential outside ($r > R$) is

$$\Phi(r, \theta) = \frac{1}{4\pi\varepsilon_0} \sum_{\ell=0}^{\infty} \frac{1}{r^{\ell+1}} \sum_{m=-\ell}^{\ell} Q_\ell^m\, Y_\ell^m(\theta, \phi),$$

$$Q_\ell^m = \int_{|\mathbf{r}'|<R} \rho(\mathbf{r}')\, r'^\ell\, Y_\ell^{m*}(\theta', \phi')\, d^3\mathbf{r}'.$$

```
from sympy import Ynm, symbols, integrate, sin
θ, φ = symbols('θ φ', positive=True)
ℓ,m = 2,1
Y = Ynm(ℓ,m,θ,φ)
# orthogonality test
```

6.3 Special Functions and Transform Techniques

```
ortho = integrate(Y*Ynm(ℓ,m,θ,φ).conjugate()*sin(θ),
                  (φ,0,2*sp.pi), (θ,0,sp.pi))
sp.simplify(ortho)
```

Addition Theorem $P_\ell(\cos\gamma) = \frac{4\pi}{2\ell+1} \sum_{m=-\ell}^{\ell} Y_\ell^m(\theta, \phi) Y_\ell^{m*}(\theta', \phi')$, where γ is the angle between two unit vectors; crucial in scattering theory and fast multipole algorithms.

Recursion Relations

$$(\ell + 1)P_{\ell+1}(x) = (2\ell + 1)x P_\ell(x) - \ell P_{\ell-1}(x).$$

Stable three-term recursions facilitate high-order computations.

Gauss–Legendre Quadrature Nodes x_k = zeros of P_n; weights $w_k = 2/[(1 - x_k^2)(P_n'(x_k))^2]$. Integrates polynomials of degree $2n - 1$ exactly; routine in `numpy.polynomial.legendre.leggauss`.

```
import numpy.polynomial.legendre as lg
xk,wk = lg.leggauss(6)
I = (wk* np.exp(xk)).sum()  # ∫_{-1}^{1} e^{x}dx ≈ 2.3504
```

6.3.4 Green's Functions and Integral Equations

Green's functions recast linear differential operators into convolution-like integral kernels, transforming boundary-value problems (BVPs) into *a priori* solvable integral equations. They are the continuous analogue of matrix inverses: if L is a linear operator acting on a function space and G satisfies $LG = \delta$ under specified boundary conditions, then the solution of $Lu = f$ is $u = G * f$. This section first develops the method for ordinary differential equations (ODEs), extends the construction to partial differential equations (PDEs), and finally relates Green kernels to Fredholm integral equations of the second kind.

Green's Function for ODEs

Definition (Two-Point BVP) For a regular, linear, second-order operator

$$L[y] := p(x)y'' + q(x)y' + r(x)y, \qquad p(x) \neq 0 \text{ on } [a, b],$$

with boundary conditions $B_1 y(a) = 0$, $B_2 y(b) = 0$ (Dirichlet, Neumann, or Robin), the **Green's function** $G(x, \xi)$ is the unique solution of

$$L_x G(x, \xi) = \delta(x - \xi), \quad B_1 G(a, \xi) = 0, \quad B_2 G(b, \xi) = 0,$$

viewed as a function of x for fixed ξ.

Constructive Recipe Let y_1, y_2 be a fundamental set of homogeneous solutions with $B_1 y_1(a) = 0$, $B_2 y_2(b) = 0$. Wronskian $W(y_1, y_2) = p^{-1} W_0 \neq 0$. Then

$$G(x, \xi) = \begin{cases} \dfrac{y_1(x) y_2(\xi)}{W_0}, & a \leq x \leq \xi \leq b, \\ \dfrac{y_1(\xi) y_2(x)}{W_0}, & a \leq \xi < x \leq b. \end{cases}$$

Example 6.3.12 (Clamped Beam) $y''''(x) = f(x)$ on $0 < x < 1$ with $y(0) = y'(0) = y''(1) = y'''(1) = 0$. Integrate twice to reduce to second-order form, apply recipe, and recover classical cubic-spline kernel $G(x, \xi) = \frac{1}{6} \begin{cases} (1-\xi) x^3 - x^2 \xi^2 / 2, & x < \xi, \\ (1-x) \xi^3 - x^2 \xi^2 / 2, & x > \xi. \end{cases}$ Verify $L_x G = \delta$ by symbolic differentiation in SymPy.

Solution Representation For inhomogeneity f,

$$y(x) = \int_a^b G(x, \xi) f(\xi) \, d\xi.$$

Linearity ensures superposition; boundary data are built into G.

Self-adjointness and Symmetry If L is formally self-adjoint ($q = p'$), G satisfies $G(x, \xi) = G(\xi, x)$.

Applications to PDEs and Boundary Value Problems

Laplace Equation in a Rectangle Solve $\nabla^2 \phi = -\rho(x, y)$ on $0 < x < L$, $0 < y < H$ with $\phi|_{\partial \Omega} = 0$. Separation in x yields eigenfunctions $\sin \frac{m \pi x}{L}$ and Green kernel

$$G((x, y), (\xi, \eta)) = \frac{2}{L} \sum_{m=1}^{\infty} \sin \frac{m \pi x}{L} \sin \frac{m \pi \xi}{L} \frac{\sinh \frac{m \pi (H - \max\{y, \eta\})}{L} \sinh \frac{m \pi \min\{y, \eta\}}{L}}{\frac{m \pi}{L} \sinh \frac{m \pi H}{L}}.$$

Potential: $\phi(x, y) = \iint_\Omega G \rho \, d\xi d\eta$.

Heat Equation on $\mathbb{R}$ Fundamental solution $G(x, \xi; t) = \frac{1}{\sqrt{4 \pi \kappa t}} \exp\left(-\frac{(x-\xi)^2}{4 \kappa t}\right)$ satisfies $\partial_t G = \kappa \partial_x^2 G$, $G|_{t=0} = \delta$. Convolution gives classical solution $u(x, t) = G * f$. For semi-infinite rod with Dirichlet boundary use *method of images*: $G_+ = G(x, \xi; t) - G(x, -\xi; t)$.

6.3 Special Functions and Transform Techniques

Green's Identity (Boundary Integral) For region Ω,

$$u(\mathbf{x}) = \int_\Omega G(\mathbf{x}, \boldsymbol{\xi}) f(\boldsymbol{\xi}) \, d\boldsymbol{\xi} - \int_{\partial\Omega} \left[u \, \partial_{n_\xi} G - G \, \partial_{n_\xi} u \right] dS_\xi.$$

Boundary element methods (BEM) discretise $\partial\Omega$ using this representation, reducing dimensionality by one.

Example 6.3.13 (2-D Exterior Potential) For Laplace outside a circle, choose free-space Green $G = -\frac{1}{2\pi} \ln |\mathbf{x} - \boldsymbol{\xi}|$; impose Neumann boundary integral to solve scattering by circular obstacles.

Integral Equations and Fredholm Theory

Fredholm Integral Equation of the Second Kind

$$u(x) - \lambda \int_a^b K(x, \xi) u(\xi) \, d\xi = f(x).$$

With K continuous, Fredholm alternative states either:
1. $\det(I - \lambda K) \neq 0$: unique solution exists for every f.
2. $\det(I - \lambda K) = 0$: non-trivial homogeneous solutions exist and solvability of inhomogeneous problem requires orthogonality to adjoint null-space.

Resolvent Kernel Series

$$R(x, \xi; \lambda) = K + \lambda K^{(2)} + \lambda^2 K^{(3)} + \ldots, \quad K^{(n)}(x, \xi) = \int_a^b K(x, s) K^{(n-1)}(s, \xi) \, ds,$$

converges for small $|\lambda|$; solution $u = f + \lambda f * R$.

Link to Green's Functions After differentiation, many BVPs map to Fredholm equations with kernel $K(x, \xi) = G(x, \xi)$. Numerically: Nyström method samples integral, converting to linear system $(I - \lambda A)u = f$.

Example 6.3.14 (Volterra Equation from IVP) ODE $y'(x) = \lambda y(x) + g(x)$, $y(0) = y_0$ integrates to $y(x) = y_0 + \int_0^x \lambda y(s) + g(s) \, ds$: a Volterra equation of the second kind solvable by successive substitution (*Neumann series*), always convergent for finite x.

Quadrature Approach Choose nodes $\{x_j\}$, weights $\{w_j\}$: $u(x_i) - \lambda \sum_j w_j K(x_i, x_j) u(x_j) = f(x_i)$. Matrix $(I - \lambda W \circ K)$ inverted by LU or GMRES. Spectral Nyström with Gauss–Legendre nodes attains exponential convergence for smooth K.

6.4 Stochastic Differential Equations (SDEs)

A **stochastic differential equation** supplements an ordinary differential equation with a noise term—traditionally Brownian motion—capturing randomness intrinsic to thermal agitation, financial markets, or environmental fluctuations. Formally, an Itô SDE in $\mathbb{R}$ reads

$$dX_t = \mu(t, X_t)\, dt + \sigma(t, X_t)\, dW_t, \qquad X_0 = x_0,$$

where W_t is standard Brownian motion, μ the *drift*, and σ the *diffusion coefficient*. Unlike deterministic ODEs, SDE solutions are *stochastic processes*—random functions of time—requiring a calculus that respects Brownian path roughness.

6.4.1 Introduction to SDEs

Classical Newtonian dynamics $dx/dt = v$ presumes differentiable trajectories; Brownian trajectories are almost surely nowhere differentiable. Einstein's 1905 model for pollen particles suggested the *mean square displacement* grows linearly: $\mathbb{E}[(X_t - X_0)^2] = 2Dt$. Langevin refined this via

$$m\, dV_t = -\gamma V_t\, dt + \sqrt{2\gamma k_B T}\, dW_t,$$

where the stochastic term models impulsive molecular kicks. Interpreting dW_t as the limit of symmetric, variance-dt Gaussian increments motivate the modern Itô integral.

Brownian Motion and Stochastic Processes

Standard Brownian motion $\{W_t\}_{t \geq 0}$ satisfies

1. $W_0 = 0$ almost surely.
2. Independent increments: $W_{t+s} - W_t$ is independent of $\mathcal{F}_t$.
3. Gaussian increments: $W_{t+s} - W_t \sim \mathcal{N}(0, s)$.
4. Continuous paths.

Moments: $\mathbb{E}[W_t] = 0$, $\mathrm{Var}(W_t) = t$, $\mathrm{Cov}(W_s, W_t) = \min(s, t)$.

Quadratic Variation Partition $[0, t]$ into $0 = t_0 < \cdots < t_n = t$; Brownian quadratic variation converges in probability:

$$\sum_{k=0}^{n-1} \left(W_{t_{k+1}} - W_{t_k}\right)^2 \xrightarrow{\mathbb{P}} t.$$

This property underlies Itô's formula.

6.4 Stochastic Differential Equations (SDEs)

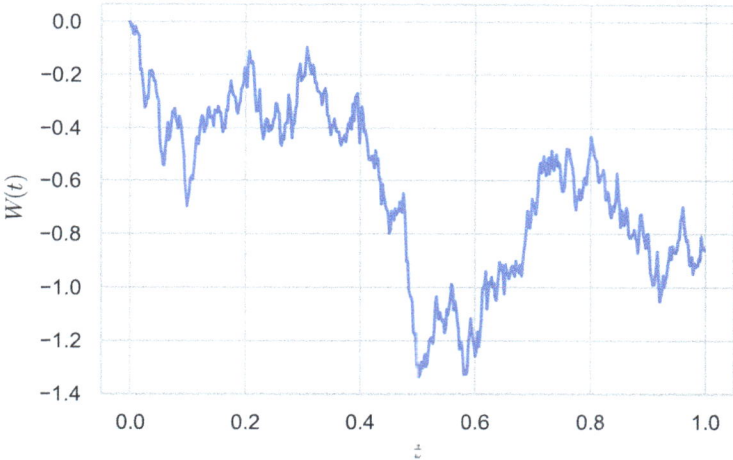

Fig. 6.13 One realisation of standard Brownian motion $W(t)$ on the interval $0 \leq t \leq 1$, generated via $\Delta W_k \sim \mathcal{N}(0, \Delta t)$ with $\Delta t = 1/500$. The path is continuous but nowhere differentiable, exhibiting the characteristic jaggedness of Wiener processes

Simulating Brownian Paths Discretise $t_k = k \, \Delta t$; increments $\Delta W_k \sim \mathcal{N}(0, \Delta t)$ (cf. Fig. 6.13).

```
import numpy as np, matplotlib.pyplot as plt
T,N = 1.0, 500
dt = T/N
ΔW = np.sqrt(dt)*np.random.randn(N)
W = np.insert(np.cumsum(ΔW), 0, 0)
plt.plot(np.linspace(0, T, N+1), W); plt.show()
```

Itô's Lemma and Stochastic Calculus

Itô Integral For an adapted process H_t with $\int_0^T \sigma_H^2(t) \, dt < \infty$, define

$$\int_0^T H_t \, dW_t := \mathrm{L}^2 - \lim_{n \to \infty} \sum_{k=0}^{n-1} H_{t_k} \left(W_{t_{k+1}} - W_{t_k} \right),$$

where the integrand is *left point* sampled, crucial for martingale preservation.

Itô's lemma (1–d). Let X_t satisfy $dX_t = \mu(t, X_t) \, dt + \sigma(t, X_t) \, dW_t$ and $f(t, x) \in C^{1,2}$. Then

$$df(t, X_t) = \left(f_t + \mu f_x + \tfrac{1}{2} \sigma^2 f_{xx} \right) dt + \sigma f_x \, dW_t.$$

Observe the $\tfrac{1}{2} \sigma^2 f_{xx}$ term stemming from quadratic variation.

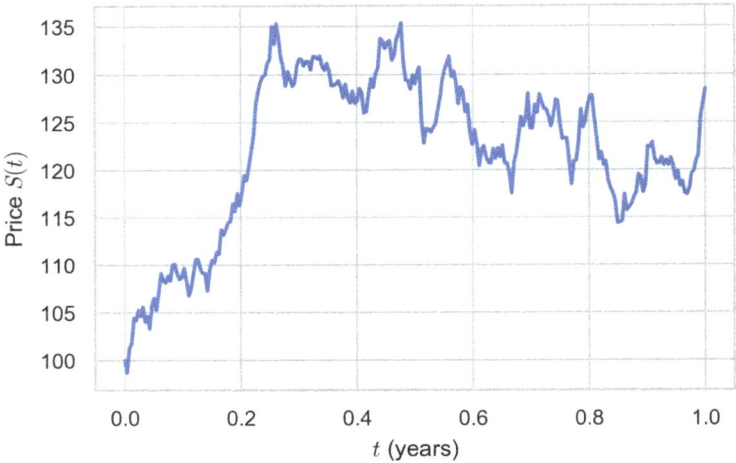

Fig. 6.14 Single realisation of geometric Brownian motion $S(t) = S_0 \exp\left[(\alpha - \tfrac{1}{2}\beta^2)t + \beta W(t)\right]$ with $S_0 = 100$, $\alpha = 0.10$, $\beta = 0.20$ over one year ($N = 252$). The multiplicative noise produces a log-normal trajectory whose expected growth rate equals α

Example 6.4.1 (Geometric Brownian Motion (GBM) (cf. Fig. 6.14)) $dS_t = \alpha S_t\, dt + \beta S_t\, dW_t$. With $f(s) = \ln s$, Itô's lemma gives

$$d \ln S_t = (\alpha - \tfrac{1}{2}\beta^2)\, dt + \beta\, dW_t \quad \Longrightarrow \quad S_t = S_0 \exp\!\left((\alpha - \tfrac{1}{2}\beta^2)t + \beta W_t\right).$$

```
S0, α, β, T, N = 100, .1, .2, 1, 252
dt = T/N
W = np.cumsum(np.sqrt(dt)*np.random.randn(N))
t = np.linspace(dt, T, N)
S = S0*np.exp((α-0.5*β**2)*t + β*W)
plt.plot(t, S); plt.show()
```

Girsanov Theorem (Øksendal 2003) Under new measure Q defined by Radon–Nikodym density $\dfrac{dQ}{d\mathbb{P}} = \exp\!\left(-\int_0^T \theta\, dW_t - \tfrac{1}{2}\int_0^T \theta^2\, dt\right)$, process $W_t^Q = W_t + \int_0^t \theta\, ds$ is Brownian in Q. Tool for risk-neutral evaluation in finance and change-of-drift reductions in filtering theory.

Moment Evolution via Itô For OU process $dX_t = -\gamma X_t\, dt + \sigma\, dW_t$, choose $f(x) = x^2$ to obtain $d(X_t^2) = (\sigma^2 - 2\gamma X_t^2)\, dt + 2\sigma X_t\, dW_t$. Taking expectations gives ODE $\partial_t \mathbb{E}[X_t^2] = \sigma^2 - 2\gamma \mathbb{E}[X_t^2] \Longrightarrow$ stationary variance $\sigma^2/(2\gamma)$.

Stratonovich vs. Itô Stratonovich integral uses midpoint sampling; chain rule matches classical calculus. Conversion: $dX_t^\circ = dX_t + \tfrac{1}{2}\sigma\sigma_x\, dt$. Physical models with white-noise limits of coloured noise often prefer Stratonovich; financial models favour Itô for martingale properties.

6.4.2 Numerical Solutions of SDEs

Exact solutions exist only for a narrow family of SDEs. Monte Carlo simulation therefore becomes the primary tool for estimating expectations, hitting probabilities, or path-dependent payoffs. Discretisation error must now account for the distributional convergence of random paths. We review two strong-order schemes—Euler–Maruyama and Milstein—and outline higher-order extensions based on multiple stochastic integrals.

Euler–Maruyama Method

Given Itô SDE

$$dX_t = \mu(X_t, t)\, dt + \sigma(X_t, t)\, dW_t, \quad X_0 = x_0,$$

and time partition $0 = t_0 < \cdots < t_N = T$ with $\Delta t = t_{n+1} - t_n$, Euler–Maruyama (EM) sets

$$X_{n+1} = X_n + \mu(X_n, t_n)\, \Delta t + \sigma(X_n, t_n)\, \Delta W_n, \quad \Delta W_n \sim \mathcal{N}(0, \Delta t).$$

Strong order 0.5: $\mathbb{E}\big[|X_T - X_N|^2\big] = O((\Delta t)^1)$. *Weak order* 1: $\big|\mathbb{E}[\varphi(X_T)] - \mathbb{E}[\varphi(X_N)]\big| = O(\Delta t)$ for sufficiently smooth φ.

Example 6.4.2 (Ornstein–Uhlenbeck (OU) Convergence) OU: $dX_t = -\gamma X_t\, dt + \sigma\, dW_t$ with analytic solution $X_T = x_0 e^{-\gamma T} + \sigma \int_0^T e^{-\gamma(T-s)}\, dW_s$. Simulate EM for $T = 1$, $\gamma = 1$, $\sigma = 0.5$ and compare empirical mean $\mathbb{E}[X_T]$ with exact $x_0 e^{-\gamma T}$ across step sizes $\Delta t = 2^{-k}$.

```
import numpy as np
def OU_EM(x0, γ, σ, T, N, M=10_000):
    dt, sqdt = T/N, np.sqrt(T/N)
    X = np.full(M, x0)
    for _ in range(N):
        X += -γ*X*dt + σ*sqdt*np.random.randn(M)
    return X
x0, γ, σ = 1.0, 1, 0.5
for k in range(6):
    N = 2**k
    samples = OU_EM(x0, γ, σ, 1, N)
    print(N, samples.mean())
```

Empirical bias decays proportionally to $\Delta t^{0.5}$, confirming strong-order theory.

Mean-Square Stability For the linear SDE $dX_t = \lambda X_t\, dt + \eta X_t\, dW_t$, EM is stable iff $1 + 2\lambda \Delta t + \eta^2 \Delta t < 1 \implies \lambda \Delta t < -\frac{1}{2}\eta^2 \Delta t$. Shrinking Δt cannot stabilise if $\lambda > 0$ with too large η (mean-square instability); implicit EM variants or tamed schemes remedy this for stiff SDEs.

Milstein Method and Higher-Order Approximations

Milstein augments EM with a *Lévy area* correction capturing first-order terms of Itô–Taylor expansion:

$$X_{n+1} = X_n + \mu_n \Delta t + \sigma_n \Delta W_n + \tfrac{1}{2}\sigma_n \sigma_n'\bigl[(\Delta W_n)^2 - \Delta t\bigr],$$

where $\sigma_n' = \partial_x \sigma(X_n, t_n)$.

Strong Order 1 Milstein doubles convergence exponent for SDEs with scalar noise. Cost overhead is marginal (one extra derivative evaluation per step).

Example 6.4.3 (GBM with Milstein) $dS_t = \alpha S_t\, dt + \beta S_t\, dW_t$ has $\sigma(x) = \beta x$, $\sigma'(x) = \beta$. Milstein update simplifies to

$$S_{n+1} = S_n + \alpha S_n \Delta t + \beta S_n \Delta W_n + \tfrac{1}{2}\beta^2 S_n\bigl[(\Delta W_n)^2 - \Delta t\bigr].$$

Simulate for $\alpha = 0.05, \beta = 0.2, T = 1$ and verify root-mean-square error decays like $O(\Delta t)$, whereas EM decays like $O(\Delta t^{1/2})$.

```
def GBM_Milstein(S0, α, β, T, N, M=50_000):
    dt, sqdt = T/N, np.sqrt(T/N)
    S = np.full(M, S0)
    for _ in range(N):
        dW = sqdt*np.random.randn(M)
        S += α*S*dt + β*S*dW + .5*β**2*S*(dW**2 - dt)
    return S
```

Multi-dimensional Noise When $\mathbf{W}_t$ is m-dimensional, Milstein requires iterated stochastic integrals $I_{ij} = \int_{t_n}^{t_{n+1}}\!\int_{t_n}^{s} dW^i dW^j$—computationally heavy. Stochastic Runge–Kutta (SRK) schemes of strong order 1 sample approximate I_{ij} via random variables with the same moments.

Weak Order Schemes If only expectations of functionals matter (e.g. option pricing), weak Runge–Kutta methods achieve order 2 without iterated integrals by antithetic or stratified random increments.

Tamed Euler for Super-Linear Drift For $\mu(x) = x^3$ the moments explode for EM. Replace drift by $\mu_{tame}(x) = \mu(x)/(1 + \Delta t|\mu(x)|)$; retains strong order 0.5 and preserves stability.

6.4.3 Applications of SDEs

Financial Modelling: Black–Scholes Equation

Stock model $dS_t = \mu S_t\, dt + \sigma S_t\, dW_t$ implies discounted price $e^{-rt}S_t$ is a martingale under risk-neutral measure $\mathbb{Q}$ with drift r. By Itô and no-arbitrage

6.4 Stochastic Differential Equations (SDEs)

replication, European call price $C(S, t)$ satisfies backward parabolic PDE

$$\partial_t C + \tfrac{1}{2}\sigma^2 S^2 \partial_{SS} C + rS\partial_S C - rC = 0,$$

terminal condition $C(S, T) = \max(S - K, 0)$. Closed form (Black–Scholes formula)

$$C = S\Phi(d_1) - Ke^{-r(T-t)}\Phi(d_2), \quad d_{1,2} = \frac{\ln(S/K) + (r \pm \tfrac{1}{2}\sigma^2)(T-t)}{\sigma\sqrt{T-t}}.$$

Example 6.4.4 (Monte Carlo Greeks) Pathwise delta estimator $\Delta = e^{-rT}\mathbb{E}\left[\mathbf{1}_{\{S_T > K\}}\frac{S_T}{S_0}\right]$. Implement Milstein simulation and compute Δ for $S_0 = 100$, $K = 110$, $T = 1$.

```
M, N = 200_000, 252
paths = GBM_Milstein(100, r, σ, 1, N, M)
payoff = np.maximum(paths - 110, 0)
delta = np.mean((paths>110)*paths/100) * np.exp(-r)
```

Stochastic Population Models

Logistic growth with environmental noise:

$$dN_t = rN_t\left(1 - \frac{N_t}{K}\right)dt + \sigma N_t\, dW_t.$$

Applying Itô to $\ln N_t$ shows extinction probability increases with σ^2. EM simulation explores parameter regime where variability drives population below Allee threshold.

```
def logistic_SDE(N0,r,K,σ,T,N_steps):
    dt, sqdt = T/N_steps, np.sqrt(T/N_steps)
    N = N0
    for _ in range(N_steps):
        dW = sqdt*np.random.randn()
        N += r*N*(1-N/K)*dt + σ*N*dW
    return N
```

Moment Closure Fokker–Planck yields $\partial_t \mathbb{E}[N] = r\mathbb{E}[N] - \frac{r}{K}\mathbb{E}[N^2]$—not closed. Assuming Gaussian fluctuations approximates $\mathbb{E}[N^2] \approx (\text{Var } N) + (\mathbb{E}N)^2$; derive ODE pair for mean and variance, compare with Monte Carlo.

Physics Applications: Langevin Equation

Overdamped Langevin Colloidal particle in potential $U(x)$ at temperature T:

$$\gamma \, dX_t = -\nabla U(X_t) \, dt + \sqrt{2\gamma k_B T} \, dW_t.$$

Equilibrium density $\rho(x) \propto e^{-U(x)/(k_B T)}$. Milstein with reflecting boundaries simulates confinement and tunnelling events.

Kramers Escape Rate Mean first-passage time over barrier height U_b approximates $\tau \sim \dfrac{2\pi\gamma}{\sqrt{|U''(x_m)U''(x_b)|}} e^{U_b/(k_B T)}$. Estimate τ via Monte Carlo and compare with asymptotic formula.

Stochastic Energetics Incremental work $dW = -U'(X_t) \circ dX_t$ (Stratonovich) and heat $dQ = \gamma \dot{X}_t^2 dt$ satisfy first law: $dU = dQ - dW$. Numerically integrate with Heun (strong order 1 Stratonovich scheme) to verify energy balance.

6.5 Exercises

1. Consider the differential form

$$\left(3x^2 y - 2y^3 + 1\right) dx + \left(x^3 - 6xy^2 + e^x\right) dy = 0.$$

   (a) Verify that the form is not exact on $\mathbb{R}^2$.
   (b) Show that the integrating factor $\mu(y) = y^{-3}$ makes it exact on the punctured plane $y \neq 0$.
   (c) Obtain an implicit solution $F(x, y) = C$ and find the unique explicit solution that satisfies $y(1) = 1$.

2. Study the population model

$$\frac{dP}{dt} = rP\left(1 - \frac{P}{K}\right) - \alpha P^2, \qquad r, K, \alpha > 0.$$

   (i) Determine all equilibria and classify their stability as functions of the harvesting rate α.
   (ii) Show that a saddle-node bifurcation occurs when $\alpha = \alpha_c = r/(4K)$ and sketch the bifurcation diagram.
   (iii) Using phase-line analysis, compute the time $T(P_0 \to P_1)$ required to move from $P(0) = P_0$ to $P(T) = P_1 > P_0$ for $\alpha < \alpha_c$.

6.5 Exercises

3. Solve

$$x^2 y'' - 5xy' + 9y = x^4 \ln x, \qquad x > 0,$$

by (i) finding two linearly independent solutions of the homogeneous equation via $y = x^m$, then (ii) using reduction of order *and* the method of undetermined coefficients to obtain a particular solution. Present the general solution in closed form.

4. For the clamped beam on $0 \le x \le 2$ governed by

$$y''''(x) = q(x), \qquad y(0) = y'(0) = y(2) = y'(2) = 0,$$

derive the Green function $G(x, \xi)$ and compute the deflection produced by $q(x) = 12x^2 - 24x + 4$. Provide $y(x)$ explicitly and evaluate $y(1)$.

5. Let

$$A = \begin{pmatrix} 0 & 3 & 0 \\ -3 & 0 & 0 \\ 0 & 0 & -2 \end{pmatrix}.$$

   (a) Compute $\exp(At)$ via (i) eigen-decomposition and (ii) real Jordan blocks.
   (b) Solve $\dot{\mathbf{x}} = A\mathbf{x}$ with $\mathbf{x}(0) = (1, 0, 2)^\mathsf{T}$ and identify the long-time behaviour.

6. Analyse the system

$$\dot{x} = -x + y - x(x^2 + y^2), \qquad \dot{y} = -x - y - y(x^2 + y^2).$$

   (i) Show that the origin is the unique equilibrium.
   (ii) Prove global asymptotic stability by constructing a Lyapunov function of the form $V(x, y) = \frac{1}{2}(x^2 + y^2)$ and estimating $\dot{V}$.
   (iii) Numerically integrate from $(x_0, y_0) = (1, 2)$ to confirm convergence.

7. Apply classical RK4 with step $h = 0.05$ to approximate the solution of $y' = \sin(y) + x^2$, $y(0) = 0$ at $x = 1$. Compute the absolute global error by comparing with the value obtained using $h = 0.001$ and Richardson extrapolation.

8. Consider Laplace's equation on the square $[0, 1] \times [0, 1]$ with mixed boundary data

$$u(x, 0) = 0, \ u(x, 1) = \sin(\pi x), \ u(0, y) = 0, \ u(1, y) = y(1 - y).$$

Discretise with a 20×20 interior grid using the five-point stencil. Assemble the linear system explicitly and solve it; tabulate $u(0.5, 0.5)$ and compare with the series solution obtained from separation of variables truncated to the first four sine terms.

9. For Poisson's equation $-\Delta u = f$ on the interval $(0, 1)$ with $u(0) = u(1) = 0$ and $f(x) = 8x$, employ piecewise linear hat functions on a uniform mesh of width $h = 0.2$:

   (a) Construct the stiffness matrix and load vector symbolically.
   (b) Solve for nodal values and estimate $\|u - u_h\|_{H^1(0,1)}$ using the exact solution $u(x) = x^3 - x$.

10. Expand $g(\theta) = |\theta|$ on $[-\pi, \pi]$ into a cosine Fourier series. Prove that the N-th partial sum $S_N g$ satisfies $\|g - S_N g\|_2 = O(N^{-1})$, and verify numerically for $N = 8, 16, 32$ using numpy.fft.

11. Solve the IBVP

    $$u_t = u_{xx}, \quad 0 < x < \pi, \ t > 0; \qquad u(x, 0) = \sin(3x), \ u(0, t) = u(\pi, t) = 0.$$

    Then use d'Alembert's formula to extend u to $-\pi < x < \pi$ by odd reflection and compare both representations at $(x, t) = (\frac{\pi}{2}, 0.1)$.

12. Investigate $dy/dt = -20y + 20\cos t$, $y(0) = 1$.

    (i) Derive the analytic solution.
    (ii) Implement explicit Euler with $\Delta t = 0.05$ and show divergence.
    (iii) Implement implicit Euler (solve algebraically each step) and tabulate numerical error at $t = 2$.

13. For

    $$u(x) - \lambda \int_0^1 (x\xi + \xi^2) u(\xi) \, d\xi = x + 1, \qquad 0 \leq x \leq 1,$$

    determine all λ for which the resolvent kernel exists by analysing the 2×2 matrix obtained via separable decomposition. Solve for $u(x)$ explicitly when $\lambda = \frac{1}{4}$.

14. Discrete susceptible–infected–recovered dynamics:

    $$S_{n+1} = S_n - \beta S_n I_n \Delta t,$$
    $$I_{n+1} = I_n + (\beta S_n I_n - \gamma I_n) \Delta t,$$
    $$R_{n+1} = R_n + \gamma I_n \Delta t,$$

    with $\Delta t = 0.25$ week, $\beta = 0.6$, $\gamma = 0.2$, and initial state $(S_0, I_0, R_0) = (0.95, 0.05, 0)$. Evolve for 20 weeks, report the peak infection proportion and the week it occurs, and compare with the continuous ODE model solved via solve_ivp.

6.5 Exercises

15. Show that $f(r) = r^2$ on $0 \leq r < 1$ admits the expansion

$$f(r) = \sum_{n=1}^{\infty} \frac{4(-1)^{n+1}}{j_{0,n}^3 J_1(j_{0,n})} J_0(j_{0,n} r),$$

where $j_{0,n}$ is the n-th zero of J_0. Verify numerically that truncating at $n = 8$ yields max absolute error $\leq 5 \times 10^{-4}$ on a 200-point radial grid.

16. Cox–Ingersoll–Ross:

$$dX_t = \kappa(\theta - X_t)\,dt + \sigma\sqrt{X_t}\,dW_t, \quad X_0 = 0.09.$$

Given parameters $\kappa = 2$, $\theta = 0.1$, $\sigma = 0.3$:

(a) Implement a positivity-preserving Milstein scheme with $\Delta t = 1/360$ year and simulate 10^4 paths to maturity $T = 1$ year.
(b) Estimate $\mathbb{E}[X_T]$ and compare with the analytic moment $\theta + (X_0 - \theta)e^{-\kappa T}$.
(c) Compute the empirical probability $\Pr(X_T < 0.05)$.

Chapter 7
Discrete Mathematics and Combinatorics

Abstract This chapter develops the essential machinery of discrete mathematics—sets, relations, functions, propositional and predicate logic, combinatorial counting, generating functions, recurrence relations, elementary number theory, and graph theory—through a computational lens. Each concept is introduced with formal definitions and proof techniques (induction, contradiction, pigeonhole), then translated into idiomatic Python using `itertools`, `sympy`, and `networkx`. Counting problems are solved via explicit enumeration and analytic combinatorics, while recurrence relations are automated with linear-operator methods and closed-form solutions verified symbolically. Modular arithmetic and primitive roots lay a foundation for cryptographic primitives such as RSA, and graph algorithms (connectivity, Eulerian and Hamiltonian paths, planar embeddings) are implemented and empirically analysed. Throughout, rigorous mathematical reasoning is coupled with executable code, demonstrating how discrete structures underpin algorithm design and complexity analysis in modern scientific computing.

Keywords Discrete mathematics · Combinatorics · Graph theory · Number theory · Logical reasoning · Recurrence relations

7.1 Number Theory

Number theory, sometimes heralded as the "Queen of Mathematics", lies at the confluence of pure logic and practical computation. Its fundamental objects—integers and their arithmetic—appear humble, yet underpin cryptographic protocols, error-correcting codes, and hashing algorithms that secure the digital age. Python's arbitrary-precision integers, together with libraries such as `sympy`, make it effortless to experiment with primality, modular inverses, and large-scale exponentiation. Throughout this section we blend rigorous proofs with hands-on code so that every abstract result is accompanied by a computational verification.

7.1.1 Divisibility and Modular Arithmetic

Divisibility creates a partial order on $\mathbb{Z}$ and enables a refined notion of equivalence: $a \equiv b \pmod{m}$ iff $m \mid (a - b)$. The quotient ring $\mathbb{Z}/m\mathbb{Z}$ inherits addition and multiplication, forming the stage on which theorems by Euclid, Fermat, and Euler unfold.

Given $m \geq 2$, each residue class $[a]_m$ has a canonical representative $\text{rem}(a, m) \in \{0, \ldots, m - 1\}$. Python's % operator returns this representative, while divmod(a,m) returns (q, r) such that $a = qm + r$ with $0 \leq r < m$, matching the Euclidean Division Theorem.

Prime Numbers and the Euclidean Algorithm

An integer $p > 1$ is *prime* if its only positive divisors are 1 and p. Euclid's proof of infinitude hinges on constructing $N = p_1 p_2 \ldots p_k + 1$, showing any finite list of primes is incomplete.

The *Euclidean algorithm* computes $\gcd(a, b)$ by repeated division:

$$\gcd(a, b) = \gcd(b, \text{rem}(a, b)), \qquad \gcd(a, 0) = |a|.$$

Its runtime is $O(\log \min\{a, b\})$. Extended Euclid additionally finds integers x, y such that $ax + by = \gcd(a, b)$—central to modular inversion and Diophantine equations.

Example 7.1.1 (Extended Euclid in Python)

```
def egcd(a, b):
    if b == 0: return (a, 1, 0)
    g, x1, y1 = egcd(b, a % b)
    return (g, y1, x1 - (a // b) * y1)

g, x, y = egcd(252, 105)
print(g, x, y) # 21, -2, 5  ⟹  252(-2)+105(5)=21
```

Modular Exponentiation and Applications in Cryptography

Fast exponentiation exploits binary expansion of the exponent:

$$a^e \bmod m \quad \text{via} \quad e = \sum_k e_k 2^k, \ a^e \equiv \prod_k (a^{2^k})^{e_k} \pmod{m}.$$

This "square-and-multiply" algorithm is $O(\log e)$, vital when $e \approx 2^{1024}$ in RSA.

```
pow(a, e, m) # built-in modular exponentiation (Python ≥ 3.8)
```

7.1 Number Theory

RSA Outline Choose distinct 1024-bit primes p, q, set $N = pq$ and $\varphi(N) = (p-1)(q-1)$. Public exponent e satisfies $\gcd(e, \varphi(N)) = 1$. Private key $d \equiv e^{-1} \pmod{\varphi(N)}$. Encryption: $c \equiv m^e \pmod{N}$; decryption: $m \equiv c^d \pmod{N}$. Security rests on hardness of factoring N given only (N, e).

Greatest Common Divisor (GCD) and Least Common Multiple (LCM)

For $a, b \neq 0$:

$$\mathrm{lcm}(a, b) = \frac{ab|}{\gcd(a, b)},$$

since prime exponents in ab split into max and min resp. Generalise to n integers by associativity.

Example 7.1.2 (Chinese Remainder Theorem (CRT)) Pairwise coprime moduli $m_1, \ldots, m_k$ yield a ring isomorphism $\mathbb{Z}/M\mathbb{Z} \cong \prod_i \mathbb{Z}/m_i\mathbb{Z}$, $M = \prod m_i$. Solve

$$x \equiv 2 \pmod{3}, \quad x \equiv 3 \pmod{4}, \quad x \equiv 1 \pmod{5}.$$

Compute $M_i = M/m_i$ and inverses $M_i^{-1} \pmod{m_i}$ to obtain $x \equiv \sum 2 \cdot 20 \cdot 2 + 3 \cdot 15 \cdot 3 + 1 \cdot 12 \cdot 1 = 173 \equiv 23 \pmod{60}$.

```
from sympy.ntheory.modular import crt
print(crt([3,4,5], [2,3,1])) # (23, 60)
```

Fermat's Little Theorem and Wilson's Theorem

Fermat's Little Theorem (FLT) If p is prime and $a \not\equiv 0 \pmod{p}$, then

$$a^{p-1} \equiv 1 \pmod{p}.$$

Proof Multiplication by a permutes the non-zero residue classes.

Euler's Theorem Generalises FLT: $a^{\varphi(m)} \equiv 1 \pmod{m}$ for $\gcd(a, m) = 1$.

Example 7.1.3 (Modular Inverse via FLT) For prime p, $a^{-1} \equiv a^{p-2} \pmod{p}$.

```
def modinv_prime(a, p): return pow(a, p-2, p)
```

Wilson's Theorem

$$p \text{ prime} \iff (p-1)! \equiv -1 \pmod{p}.$$

Though not computationally efficient for large p, it underpins proofs of FLT and provides primality certificates (e.g. for $p = 2, 3, 5$).

7.1.2 Congruences and Number Theoretic Functions

Congruences allow one to *localise* arithmetic modulo a positive integer and, when combined with multiplicative arithmetic functions, unlock a rich algebraic toolbox. Central actors include Euler's totient function, quadratic residue characters, and—in modern cryptography—group laws on elliptic curves defined over finite fields. We proceed from general structure theorems (Chinese remainder) to concrete algorithms used in error-correcting codes and public-key cryptosystems.

Chinese Remainder Theorem and Euler's Totient Function

CRT Revisited For pairwise coprime moduli $m_1, \ldots, m_k$, the ring isomorphism

$$\mathbb{Z}/M\mathbb{Z} \cong \mathbb{Z}/m_1\mathbb{Z} \times \cdots \times \mathbb{Z}/m_k\mathbb{Z}, \qquad M = \prod_i m_i,$$

is given explicitly by $x \mapsto (x \bmod m_1, \ldots, x \bmod m_k)$, with inverse map $x \equiv \sum_{i=1}^{k} a_i M_i M_i^{-1} \pmod{M}$, $M_i = M/m_i$, $M_i^{-1} \equiv M_i^{-1} \pmod{m_i}$.

Euler's Totient

$$\varphi(n) = \left|\{1 \leq a \leq n \mid \gcd(a, n) = 1\}\right| = n \prod_{p \mid n} \left(1 - \frac{1}{p}\right),$$

multiplicative with respect to coprime arguments. CRT implies $(\mathbb{Z}/n\mathbb{Z})^\times \cong \prod_{p^e \| n} (\mathbb{Z}/p^e\mathbb{Z})^\times$, yielding $|(\mathbb{Z}/n\mathbb{Z})^\times| = \varphi(n)$.

Example 7.1.4 (Fast Totient by Factorisation)

```
from sympy import factorint
def phi(n):
    f = factorint(n) # {p:e}
    out = n
    for p in f: out //= p; out *= p-1
    return out
```

Applications in Coding Theory and Cryptanalysis

Linear Feedback Shift Registers (LFSR) Binary LFSRs output sequences whose period divides $2^m - 1$. A characteristic polynomial $p(x) \in \mathbb{F}_2[x]$ is *primitive* iff p is irreducible and x is a generator of $(\mathbb{F}_{2^m})^\times$; then the output has maximal period $2^m - 1 = \varphi(2^m)$. Decimation attacks exploit the fact that $\varphi(2^m - 1)$ counts the number of primitive roots.

7.1 Number Theory

RSA Small Decryption Exponent Attack (Wiener) When $d < \frac{1}{3}N^{1/4}$, continued-fraction convergents of e/N reveal d. The attack relies on $\varphi(N)$ via $ed \equiv 1 \pmod{\varphi(N)}$; convergents yield candidates for (k, d) such that $|ed - k\varphi(N)| < 2\sqrt{N}$.

Syndrome Decoding For cyclic BCH codes of length $n = 2^m - 1$, syndromes live in $\mathbb{F}_{2^m}^\times$. Chien search evaluates the error locator polynomial at successive powers of a primitive element—an iteration through the multiplicative group of order $\varphi(n)$.

Quadratic Residues and Legendre Symbols

Quadratic Residue a is a quadratic residue mod p iff $\exists x \; x^2 \equiv a \pmod{p}$. Define Legendre symbol

$$\left(\frac{a}{p}\right) = \begin{cases} 1 & a^{(p-1)/2} \equiv 1 \pmod{p}, \\ -1 & a^{(p-1)/2} \equiv -1 \pmod{p}, \\ 0 & p \mid a. \end{cases}$$

Euler's criterion supplies this definition; multiplicativity $\left(\frac{ab}{p}\right) = \left(\frac{a}{p}\right)\left(\frac{b}{p}\right)$ turns the symbol into a character of order two on $(\mathbb{Z}/p\mathbb{Z})^\times$.

Quadratic Reciprocity For odd primes p, q,

$$\left(\frac{p}{q}\right)\left(\frac{q}{p}\right) = (-1)^{\frac{(p-1)(q-1)}{4}}.$$

Together with the auxiliary laws for 2 and -1, this theorem enables rapid evaluation of Legendre symbols via exponentiation and modular arithmetic.

Example 7.1.5 (Tonelli–Shanks Square Root)

```
from sympy.ntheory.residue_ntheory import sqrt_mod
print(sqrt_mod(10, 13)) # {6, 7} since 6² ≡ 7² ≡ 10 (mod 13)
```

Jacobi Symbol Extends Legendre to odd composite $n = \prod p_i^{e_i}$ by $\left(\frac{a}{n}\right) = \prod \left(\frac{a}{p_i}\right)^{e_i}$. If symbol equals -1, then a is *certainly* non-square; if $+1$ ambiguity remains—basis of Solovay–Strassen primality test.

Elliptic Curves and Cryptography

Elliptic Curve Over $\mathbb{F}_q$

$$E : y^2 = x^3 + ax + b, \qquad 4a^3 + 27b^2 \neq 0,$$

forms an abelian group with point at infinity $\mathcal{O}$ as identity; group law defined algebraically by chord–tangent construction. Hasse bound: $\big| |E(\mathbb{F}_q)| - q - 1 \big| \le 2\sqrt{q}$.

Scalar Multiplication $[k]P = P + \cdots + P$ (k times) computed via "double-and-add" analogous to modular exponentiation. Hardness of discrete logarithm $Q = [k]P \Rightarrow k$ secure for k up to 2^{256} on suitably chosen curves (e.g. NIST P-256).

Example 7.1.6 (SEC P-256 Point Multiplication with ecdsa lib)

```
from ecdsa import NIST256p, ellipticcurve
k = 0xA3F2_98C4_D1F6_7B2E
Q = k * NIST256p.generator
print(hex(Q.x()), hex(Q.y()))
```

ECDSA Outline

KeyGen: $d \in \{1, \ldots, n-1\}$, $P = dG$.

Sign: $k \xleftarrow{\$} \{1, \ldots, n-1\}$, $r = (kG)_x \bmod n$, $s \equiv k^{-1}(H(m) + dr) \pmod{n}$.

Verify: $w = s^{-1}$, $u_1 = H(m)w$, $u_2 = rw$, $(x_v, y_v) = u_1 G + u_2 P$,

accept iff $x_v \bmod n = r$.

Security relies on elliptic-curve discrete log (ECDLP) and unpredictable nonce k; nonce reuse reveals private key via linear equations in (r, s).

7.1.3 Advanced Topics in Number Theory

Number theory attains a deeper and subtler flavour once one ventures beyond congruences and residues into the lands of integral solutions, automorphic symmetries, and analytic continuations. Diophantine equations lie at the heart of arithmetic geometry, modular forms encode hidden dualities between algebra and analysis, while the analytic behaviour of the Riemann zeta-function governs the fine spacing of primes. In this subsection we provide intuitive—but rigorous—gateways to each subject, illustrating every conceptual leap with Python experiments or short SageMath snippets that encourage active exploration.

Diophantine Equations and Applications

A **Diophantine equation** asks for integer—or rational—solutions to a polynomial relation $F(x_1, \ldots, x_n) = 0$. The classical linear case, $a_1 x_1 + \cdots + a_n x_n = b$, is entirely controlled by the extended Euclidean algorithm, yet glimpses the delicate obstruction principle: solutions exist iff $\gcd(a_1, \ldots, a_n) \mid b$. Quadratic forms already yield rich structures; for instance, $x^2 - Dy^2 = 1$ (*Pell's equation*) has infinitely many solutions for non-square $D > 1$, generated from the fundamental unit in $\mathbb{Z}[\sqrt{D}]$ via continued fractions.

Example 7.1.7 (Solving Pell with Continued Fractions)
```
import sympy as sp
D = 61
x, y = sp.nint_bernoulli_pell(D)  # (1766319049, 226153980)
print(x**2 - D*y**2) # 1
```

Higher-degree problems such as Fermat's Last Theorem or Catalan's conjecture were settled only with sophisticated tools like elliptic curves and modularity lifting. Modern cryptosystems exploit hardness of certain Diophantine problems: the *knapsack* or *subset sum* problem embeds $c_1 x_1 + \cdots + a_n x_n = b$ but over $\{0, 1\}^n$; lattice-basis reduction (LLL) cracks weak instances and underpins attacks.

Modular Forms and Their Applications

A **modular form** of weight k for the congruence subgroup $\Gamma \subset \mathrm{SL}_2(\mathbb{Z})$ is a holomorphic function $f : \mathbb{H} \to \mathbb{C}$ on the upper half-plane satisfying the transformation law

$$f\!\left(\frac{az+b}{cz+d}\right) = (cz+d)^k f(z), \qquad \begin{pmatrix} a & b \\ c & d \end{pmatrix} \in \Gamma,$$

and a suitable growth condition at the cusps. Fourier expansion at $i\infty$ reads $f(z) = \sum_{n=0}^{\infty} a_n q^n$, $q = e^{2\pi i z}$, where the arithmetic of coefficients a_n encodes deep information: for the discriminant $\Delta(z) = q \prod_{n \geq 1}(1 - q^n)^{24}$, Ramanujan conjectured (proved by Deligne) $|a_p| \leq 2p^{11/2}$.

The modularity theorem (formerly Taniyama–Shimura–Weil) asserts that every rational elliptic curve corresponds to a weight–2 modular form: $L(E, s) = L(f, s)$. Andrew Wiles exploited this bridge to prove Fermat's Last Theorem, transforming an intractable Diophantine statement into the non-existence of a certain Galois representation inconsistent with modularity.

Example 7.1.8 (Numerical Verification of q-expansion)
```
from sympy import qexp
q = sp.Symbol('q')
Δ = q*sp.product((1-q**n)**24, (n,1,10))  # truncate
print(sp.series(Δ.expand(), q, 0, 6))
```

In coding theory, modular forms supply *theta series* of lattices; the extremal properties of the Leech lattice hinge on the non-existence of low-weight cusp forms, which in turn bounds sphere-packing densities.

Analytic Number Theory and the Riemann Hypothesis

Analytic number theory studies arithmetic functions via generating Dirichlet series. The prime number theorem emerges from non-vanishing of $\zeta(s)$ on $\operatorname{Re} s = 1$: $\pi(x) \sim x/\log x$. The **Riemann Hypothesis** (RH) sharpens this by predicting all non-trivial zeros of $\zeta(s)$ lie on the critical line $\operatorname{Re} s = \frac{1}{2}$. Assuming RH, the error term improves to $\pi(x) = \operatorname{Li}(x) + O(x^{1/2}\log x)$; unconditionally one currently has $\pi(x) = \operatorname{Li}(x) + O(xe^{-c\sqrt{\log x}})$.

Explicit Formula For smooth test w, one expresses weighted prime sums in terms of zeros ρ:

$$\sum_{n=1}^{\infty} \Lambda(n) w(n) = -\sum_{\rho} \widehat{w}(\rho) - \frac{\zeta'(0)}{\zeta(0)} w(0) + \ldots$$

RH thus governs the oscillation magnitude of primes via the distribution of ρ.

Mertens Conjecture and Extremal Sums RH implies $|M(x)| < x^{1/2+\epsilon}$ for the Mertens function $M(x) = \sum_{n \le x} \mu(n)$; conversely strong bounds on $M(x)$ would prove (or disprove) RH.

Example 7.1.9 (Empirical Zero Statistics) Using Odlyzko tables, pair-correlation of high Riemann zeros matches GUE random matrix eigenvalue statistics up to 10^{-3}.

```
import mpmath as mp
mp.mp.dps = 50
zeros = [mp.zeta_zeros(n) for n in range(20,40)]
spacings = [mp.re(zeros[i+1])-mp.re(zeros[i]) for i in range(len(zeros)-1)
]
print(sum(spacings)/len(spacings)) # mean spacing ≈ 9.06 (expected 2π/log
    (10^6))
```

Large-Scale Computation Primes up to 10^{24} are counted using Lagarias–Odlyzko Meissel–Lehmer algorithms; verification of RH to height T currently exceeds 10^{13}. Python bindings to C libraries (e.g. `primesieve`, `pari/GP`) allow classroom verification of $\zeta(s)$ zero-free regions for moderate T.

Perspective Diophantine problems motivate the passage to modular forms; modularity, through L-functions, re-enters analytic territory and couples back to the zero distribution of $\zeta(s)$. This grand arithmetic triangle illustrates number theory's unifying strength—geometric, algebraic, and analytic ideas coalesce in results that would be inaccessible to any single viewpoint alone.

7.2 Combinatorics

Combinatorics quantifies "how many'" rather than "how much". Its results permeate probability, algebraic topology, and the analysis of algorithms. Python's `itertools` and `sympy` libraries turn abstract counting into executable enumeration, fostering intuition alongside proof.

7.2.1 Basic Counting Principles

At the bedrock lies the *Addition Principle*: if two sets A, B are disjoint, then $|A \cup B| = |A| + |B|$, and the *Multiplication Principle*: if a task decomposes into k independent stages with n_i choices at stage i, the total number of outcomes is $\prod_{i=1}^{k} n_i$. All sophisticated counts ultimately trace back to repeated applications of these axioms.

Permutations and Combinations

The number of ways to arrange k distinct objects selected from n is the *falling factorial*

$$P(n, k) = n^{\underline{k}} = n(n-1) \cdots (n-k+1) = \frac{n!}{(n-k)!}.$$

When order is irrelevant one instead counts subsets:

$$\binom{n}{k} = \frac{n!}{k!(n-k)!}, \quad \sum_{k=0}^{n} \binom{n}{k} = 2^n.$$

For multiset selections, stars–and–bars gives

$$\binom{n+k-1}{k-1}$$

ways to distribute n identical balls into k boxes.

Example 7.2.1 (Enumerating combinations in Python)

```
import math, itertools
n, k = 7, 3
comb = list(itertools.combinations(range(n), k))
assert len(comb) == math.comb(n, k)
```

Multinomial Coefficients Partitioning n labelled items into blocks of sizes $(k_1, \ldots, k_r)$, $\sum k_i = n$, yields $\binom{n}{k_1, \ldots, k_r} = \frac{n!}{\prod k_i!}$.

Pigeonhole Principle and Inclusion–Exclusion Principle

If n items occupy m boxes with $n > m$, at least one box holds $\lceil n/m \rceil$ items. A refined quantitative form states: $n > k(m-1) \Rightarrow$ some box contains $\geq k+1$ items. Applied to sequences, any set of 6 integers contains two with the same remainder modulo 5.

The **Inclusion–Exclusion Principle** corrects over-counting:

$$\left|\bigcup_{i=1}^{r} A_i\right| = \sum_{i} |A_i| - \sum_{i<j} |A_i \cap A_j| + \cdots + (-1)^{r+1} |A_1 \cap \cdots \cap A_r|.$$

Example 7.2.2 (Integers $\leq$ 1000 Not Divisible by 2, 3, or 5) Set $N = 1000$. Count $|A_2 \cup A_3 \cup A_5|$ where $A_p = \{1 \leq n \leq N : p \mid n\}$; complement yields the desired number. Implement with integer division:

```
from itertools import combinations
N, primes = 1000, (2,3,5)
def mult(p): return N//p
total = 0
for r in range(1, len(primes)+1):
    for subset in combinations(primes, r):
        p = math.prod(subset)
        total += (-1)**(r+1) * mult(p)
print(N - total)  # 266
```

Derangements and Catalan Numbers

A **derangement** is a permutation $\sigma \in S_n$ with $\sigma(i) \neq i$ for all i. Inclusion–exclusion gives

$$D_n = n! \sum_{k=0}^{n} \frac{(-1)^k}{k!} = \left\lfloor \frac{n!}{e} + \frac{1}{2} \right\rfloor.$$

Hence $\mathbb{P}(\text{derangement}) \to 1/e$ as $n \to \infty$.

Example 7.2.3

```
import random, math
def random_perm(n):
    p = list(range(n)); random.shuffle(p); return p
```

7.2 Combinatorics

```
def is_derangement(p): return all(i!=p[i] for i in range(len(p)))
n, trials = 9, 50000
emp = sum(is_derangement(random_perm(n)) for _ in range(trials))/trials
print(emp, math.factorial(n)/math.e/math.factorial(n))
```

Catalan Numbers Defined recursively $C_0 = 1$, $C_{n+1} = \sum_{k=0}^{n} C_k C_{n-k}$ or explicitly

$$C_n = \frac{1}{n+1}\binom{2n}{n}, \quad n \geq 0.$$

They count Dyck paths, balanced parenthetical expressions, planar binary trees, and n-chord non-intersecting polygon triangulations.

Example 7.2.4 (Balanced Bracket Enumeration)

```
def catalan(n):
    return math.comb(2*n, n)//(n+1)
print([catalan(n) for n in range(8)])
```

Stirling Numbers and Bell Numbers

Stirling Numbers of the First Kind $s(n, k)$ Counts permutations of n elements with exactly k cycles; satisfy $s(n, k) = s(n-1, k-1) - (n-1)s(n-1, k)$ with $s(0, 0) = 1$, $s(n, 0) = s(0, n) = 0$ for $n > 0$.

Stirling Numbers of the Second Kind $S(n, k)$ Partitions of $\{1, \ldots, n\}$ into k non-empty, unlabelled blocks; recurrence

$$S(n, k) = k\, S(n-1, k) + S(n-1, k-1), \quad S(0, 0) = 1.$$

Vertical exponential generating function: $\sum_{n \geq k} S(n, k) \frac{z^n}{n!} = \frac{(e^z - 1)^k}{k!}$.

Bell Numbers $B_n = \sum_k S(n, k)$ Asymptotics: $B_n \sim \frac{1}{\sqrt{n}} \left(\frac{n}{W(n)}\right)^n e^{n/W(n)-n-1}$, where W is the Lambert W-function.

Example 7.2.5 (Stirling Triangle in Python)

```
def stirling2(n, k, memo={}):
    if (n,k) in memo: return memo[(n,k)]
    if n==k==0: return 1
    if n==0 or k==0: return 0
    memo[(n,k)] = k*stirling2(n-1,k,memo)+stirling2(n-1,k-1,memo)
    return memo[(n,k)]
triangle = [[stirling2(n,k) for k in range(n+1)] for n in range(7)]
for row in triangle: print(row)
```

7.2.2 Generating Functions

Generating functions encode infinite sequences as analytic objects—power series—so that algebraic or differential manipulations of the series translate into closed-form or asymptotic information about the sequence. When fused with Python's `sympy.series` and `sageall`, they provide both symbolic and numerical leverage for solving recurrences, proving identities, and enumerating combinatorial classes.

Ordinary and Exponential Generating Functions

Definitions For a sequence $\{a_n\}_{n\geq 0}$, the **ordinary generating function** (OGF) is

$$A(z) = \sum_{n=0}^{\infty} a_n z^n,$$

whereas the **exponential generating function** (EGF) weights by factorials:

$$\widehat{A}(z) = \sum_{n=0}^{\infty} a_n \frac{z^n}{n!}.$$

OGFs excel for unlabelled discrete structures, EGFs for labelled ones because disjoint unions and products preserve exponential series via the exponential formula.

Basic Operations

Shift: $z^k A(z) \iff a_{n-k}$;

Differentiation: $A'(z) = \sum n a_n z^{n-1}$;

Cauchy product: $A(z)B(z) = \sum (a*b)_n z^n$, $(a*b)_n = \sum_{k=0}^{n} a_k b_{n-k}$.

Convolution translates to sequence sum, mirroring the multiplication principle for combinatorial classes.

Example 7.2.6 (Fibonacci via OGF) $F_n = F_{n-1} + F_{n-2}$, $F_0 = 0$, $F_1 = 1$. Let $F(z) = \sum_{n\geq 0} F_n z^n$. Multiply recurrence by z^n and sum: $F(z) - z = zF(z) + z^2 F(z) \implies F(z) = \frac{z}{1-z-z^2}$. Partial fractions and Lagrange inversion recover Binet's formula.

```
import sympy as sp
z=sp.symbols('z')
F= z/(1-z-z**2)
coeff10 = sp.series(F, z, 0, 12).coeff(z, 10)
print(coeff10) # 55
```

Applications in Recurrence Relations and Partition Theory

Linear Recurrences with Constant Coefficients For $a_n+c_1a_{n-1}+\cdots+c_ka_{n-k} = f_n$, the OGF satisfies $P(z)A(z) = Q(z)+R(z)$, where $P(z) = 1+c_1z+\cdots+c_kz^k$; solve algebraically then expand.

Partitions of Integers The partition function $p(n)$ counts ways to write n as unordered sum of positive integers. Euler's product:

$$\sum_{n=0}^{\infty} p(n)z^n = \prod_{m=1}^{\infty} \frac{1}{1-z^m}.$$

Pentagonal number theorem gives expansion $p(n) = p(n-1)+p(n-2)-p(n-5)-p(n-7)+\ldots$ where the indices are generalised pentagonal numbers $(3k^2 \pm k)/2$.

```
from sympy.ntheory import partition
print([partition(n) for n in range(10)]) # p(0..9)
```

Restricted Partitions Partitions into odd parts: OGF is $\prod_{m\geq 1}(1-z^{2m-1})^{-1}$; Euler proved equality with partitions into distinct parts, whose OGF is $\prod_{m\geq 1}(1+z^m)$.

Combinatorial Identities and the Binomial Theorem

Binomial and Negative Binomial Series

$$(1+z)^\alpha = \sum_{n=0}^{\infty} \binom{\alpha}{n} z^n, \quad \binom{\alpha}{n} = \frac{\alpha(\alpha-1)\ldots(\alpha-n+1)}{n!}.$$

Set $\alpha = -1$ to obtain geometric series; set $\alpha = -\frac{1}{2}$ for the expansion of $1/\sqrt{1-z}$, whose coefficients $\binom{2n}{n}/4^n$ count Dyck paths.

Chu–Vandermonde Identity Coefficient extraction from $(1+z)^r(1+z)^s = (1+z)^{r+s}$ gives $\sum_{k=0}^{n} \binom{r}{k}\binom{s}{n-k} = \binom{r+s}{n}$.

Identity via EGFs Bell numbers B_n satisfy $B(z) = \exp(e^z - 1)$. Differentiate: $B'(z) = e^z \exp(e^z - 1) = B(z) + B(z)e^z$ leading to $B_{n+1} = \sum_{k=0}^{n} \binom{n}{k} B_k$—easily coded.

Multivariate Generating Functions and Applications

Definition For bivariate sequence $a_{m,n}$,

$$A(x, y) = \sum_{m,n \geq 0} a_{m,n} x^m y^n.$$

If variables track distinct combinatorial statistics (e.g. size and weight), partial differentiation extracts moments.

Two-Variable Catalan Triangle Dyck paths parameterised by number of peaks k: $C(x, y) = 1 + \sum_{n \geq 1} \sum_{k \geq 1} c_{n,k} x^n y^k$, satisfies quadratic functional equation $C = 1 + yxC^2$, solving to $C = \frac{1-\sqrt{1-4xy}}{2xy}$. Coefficient $c_{n,k} = \frac{k}{n}\binom{2n-k-1}{n-1}$.

Bivariate Lagrange Inversion For implicit $A = x\Phi(A, y)$, coefficients derive from $[a^n]A^m = \frac{m}{n}[u^{n-m}]\Phi^n(u, y)$, enabling enumeration of k-ary trees with marked leaves or coloured edges.

Applications

(a) Random walks in $\mathbb{Z}^2$ constrained to the quarter-plane; generating functions solve boundary value problems yielding harmonic functions for hitting probabilities.
(b) Multivariate EGFs encode labelled graphs: exponential formula implies EGF of connected graphs $C(z)$ satisfies $\exp(C(z)) = G(z)$, where $G(z)$ counts all graphs.

Example 7.2.7 (Asymptotics via Singularity Analysis) For partition OGF $P(z) = \prod_{m \geq 1}(1 - z^m)^{-1}$, singularity at $z = 1$ controls $p(n) \sim \frac{1}{4n\sqrt{3}} e^{\pi\sqrt{2n/3}}$ (Hardy–Ramanujan). Sage's `asymptotic_expansion` numerically validates this against $p(1000)$.

7.2.3 Advanced Topics in Combinatorics

Combinatorics attains its full depth when algebra, topology, and probability intertwine with counting. In this final subsection we explore three paradigmatic viewpoints: symmetry-aware enumeration through group actions (Pólya theory); unavoidable patterns in large structures (Ramsey and extremal theory); and the non-constructive yet powerful *probabilistic method*. Python's `sage` interface and `networkx` offer practical laboratories for each concept—although here we restrict ourselves to illustrative snippets.

7.2 Combinatorics

Pólya's Enumeration Theorem and Group Actions

Let a finite group G act on a set X of "positions" by permutations. Colour each position with one of m colours. Two colourings are *equivalent* if a group element transforms one into the other. The number of inequivalent colourings equals

$$\frac{1}{|G|} \sum_{g \in G} m^{c(g)},$$

where $c(g)$ is the number of cycles of g acting on X—Burnside's lemma. Pólya's refinement records colour patterns via the **cycle index**

$$Z_G(s_1, s_2, \ldots) = \frac{1}{|G|} \sum_{g \in G} \prod_{j \geq 1} s_j^{c_j(g)},$$

$c_j(g) = \#$ of j-cycles of g. Substituting $s_j \mapsto \sum_{k=1}^m x_k^j$ yields a multivariate polynomial whose coefficients count colourings with specified colour multiplicities.

Example: Necklaces of Length n with m Colours Group $G = C_n$ acts by rotation. Cycle type of rotation by d positions decomposes into $\gcd(d, n)$ cycles of length $n/\gcd(d, n)$. Hence

$$Z_{C_n} = \frac{1}{n} \sum_{d=0}^{n-1} s_{\gcd(d,n)}^{n/\gcd(d,n)}.$$

For binary necklaces of length 6:

$$\text{ne}_2(6) = \frac{1}{6}(2^6 + 2^3 + 2^2 + 2^3 + 2^2 + 2^1) = 14.$$

```
import sageall as sg
def binary_necklaces(n):
    M = sg.IntegerModRing(n)
    return sg.NecklaceEnumerator(M, 2).count()
print(binary_necklaces(6)) # 14
```

Chemical Isomers For benzene derivatives, $G = D_6$ acts on six labelled sites; substituting $s_j \mapsto r_1^j + r_2^j$ distinguishes mono- and dichloro isomers vs. hydrogen.

Ramsey Theory and Extremal Combinatorics

Ramsey Numbers $R(s, t)$ is the least n such that any red/blue edge-colouring of K_n contains a red K_s or a blue K_t. Basic bounds:

$$\binom{R(s,t)}{s} 2^{-\binom{s}{2}} + \binom{R(s,t)}{t} 2^{-\binom{t}{2}} < 1$$

(yields Erdős–Szekeres upper bound), while probabilistic method furnishes $R(k, k) > 2^{k/2}$.

$R(s, t)$ is the least n such that any red/blue edge-colouring of K_n

Turán's Theorem Let $\mathrm{ex}(n, K_r)$ denote maximum edges in n-vertex graph without K_r subgraph. The extremal graph is the $(r - 1)$-partite Turán graph $T_{r-1}(n)$, giving

$$\mathrm{ex}(n, K_r) = \left(1 - \frac{1}{r-1}\right)\frac{n^2}{2} + O(1).$$

Erdős–Stone Theorem For non-bipartite H, $\lim_{n \to \infty} \frac{\mathrm{ex}(n, H)}{\binom{n}{2}} = 1 - \frac{1}{\chi(H) - 1}$, reducing extremal densities to chromatic numbers.

Example 7.2.8 (Ramsey $R(3, 3)$ via Brute Force)

```
import itertools, random
def has_monochromatic_triangle(edges):
    for a,b,c in itertools.combinations(range(6), 3):
        e = {(a,b), (a,c), (b,c)}
        colours = {edges[pair] for pair in e}
        if len(colours) == 1: return True
    return False

N=6
for _ in range(1000):
    edges = {(i,j): random.choice('RB') for i in range(N) for j in range(i)
        }
    if not has_monochromatic_triangle(edges):
        break
else: print("All colourings had a triangle") # confirms R(3,3)=6
```

Probabilistic Method in Combinatorics

The probabilistic method proves existence of objects by showing that a suitably defined probability space assigns positive probability to the desired property. A **first-moment** argument bounds the expected number of "bad" configurations; a **second-moment** or **Lovász Local Lemma (LLL)** yields stronger, often constructive, results.

Erdős' Lower Bound for $R(k, k)$ Colour edges of K_n red/blue independently with probability $1/2$. Expected monochromatic K_k count is

$$\mathbb{E} = 2\binom{n}{k}2^{-\binom{k}{2}}.$$

If $\mathbb{E} < 1$, then some colouring avoids monochromatic K_k. Choosing $n < 2^{k/2}$ achieves this—establishing exponential growth of Ramsey numbers.

LLL (Symmetric) If events $\{A_i\}$ each satisfy $\Pr(A_i) \leq p$ and each is mutually independent of all but d others, and $ep(d+1) \leq 1$, then $\Pr(\bigcap \overline{A_i}) > 0$. Example: a k-CNF Boolean formula with at most 2^{k-2} clauses per variable is satisfiable.

Applications: Discrepancy and Hashing Given n vectors in $\{-1, 1\}^m$, there exists a colouring with discrepancy $O(\sqrt{m} \log n)$ (Beck–Fiala); randomised rounding plus LLL constructs such a colouring. In hashing, k-wise independent families guarantee load $O(\log n / \log \log n)$ with high probability upon distributing n balls into n bins.

Example 7.2.9 (Simple LLL via `networkx`)

```
import networkx as nx, random
def lll_kcnf(k, vars, clauses):
    assignment = {v: random.choice([True, False]) for v in vars}
    # Moser-Tardos resampling
    bad = [c for c in clauses if not any(lit(assignment) for lit in c)]
    while bad:
        clause = random.choice(bad)
        for lit in clause: assignment[lit.var] = random.choice([True, False
            ])
        bad = [c for c in clauses if not any(lit(assignment) for lit in c)]
    return assignment
```

7.3 Graph Theory

Graphs provide an abstract language for pairwise relationships—from road networks to protein–protein interactions. A *simple graph* $G = (V, E)$ consists of a finite vertex set $V = \{v_1, \ldots, v_n\}$ and an edge set $E \subseteq \{\{u, v\} \mid u, v \in V, u \neq v\}$. Unless stated otherwise graphs here are simple, undirected, and finite. Python's `networkx` library wraps common data structures, while `igraph` and `sagedata` scale to millions of edges.

7.3.1 Basic Concepts in Graph Theory

Key parameters include the *order* $|V|$, the *size* $|E|$, the *degree* $\deg(v) = |\{u \in V \mid \{u, v\} \in E\}|$, and the *adjacency matrix* $A(G) = [a_{uv}]$ where $a_{uv} = 1$ if $\{u, v\} \in E$, else 0. Handshaking lemma: $\sum_{v \in V} \deg(v) = 2|E|$.

Walks, paths, cycles, and connectedness follow their standard graph-theoretic definitions; depth–first and breadth–first traversals generate spanning trees.

Graph Representation in Python

Adjacency List vs. Matrix Adjacency lists store neighbours per vertex—optimal for sparse graphs ($O(|V| + |E|)$ memory); adjacency matrices cost $|V|^2$ but allow $O(1)$ edge queries and vectorised linear-algebra kernels.

```
import networkx as nx

# Build a simple graph
G = nx.Graph()
G.add_edges_from([(0,1), (0,2), (1,3), (2,3), (3,4)])

# Adjacency list
print(G.adj[0]) # neighbours of 0

# Adjacency matrix (NumPy CSR)
A = nx.to_scipy_sparse_array(G, dtype=int)
print(A.todense())
```

Edge Attribute Dictionaries NetworkX attaches weights/capacities:

```
G[0][1]['capacity'] = 15
```

Algorithmic Primitives Shortest path (Dijkstra), maximum flow (Edmonds–Karp), and centrality scores are one-liners.

```
length = nx.shortest_path_length(G, source=0, target=4, weight=None)
flow_val, flow_dict = nx.maximum_flow(G, 0, 4, capacity='capacity')
```

Graph Isomorphism and Subgraph Isomorphism

Two graphs $G = (V, E)$ and $G' = (V', E')$ are *isomorphic* if a bijection $\varphi : V \to V'$ preserves adjacency ($\{u, v\} \in E \iff \{\varphi(u), \varphi(v)\} \in E'$). The decision problem GI lies in NP; Babai (2020) showed quasi-polynomial time, yet full polynomial status is open.

Practical solvers—nauty/traces, networkx.algorithms.isomorphism (Weisfeiler–Lehman + VF2)—distinguish graphs with thousands of vertices.

```
from networkx.algorithms import isomorphism
GM = isomorphism.GraphMatcher(G, G.copy())
print("Isomorphic" if GM.is_isomorphic() else "Different")
```

Subgraph Isomorphism Given graphs H and G, decide whether H embeds as an induced (sub)graph of G. The problem is NP-complete (Karp, 1972). Nonetheless VF2 backtracking with feasibility pruning solves patterns up to 30–40 nodes rapidly for sparse G.

```
P = nx.path_graph(3)
matcher = isomorphism.GraphMatcher(G, P)
print(matcher.subgraph_is_isomorphic()) # any 3-path in G?
```

7.3 Graph Theory

Planar Graphs and Euler's Formula

A graph is *planar* if it admits an embedding in $\mathbb{R}^2$ with no edge intersections. Euler (1752) showed any connected planar embedding partitions the sphere into F faces satisfying

$$\boxed{|V| - |E| + F = 2}.$$

Proof Induct on $|E|$. A tree ($|E| = |V| - 1$) has $F = 1$. Removing a non-bridge edge reduces $|E|$ and F by 1 while preserving planarity; the invariant holds.

Consequences For planar G with $|V| \geq 3$:

$$|E| \leq 3|V| - 6, \qquad \text{and if } G \text{ has no triangles } |E| \leq 2|V| - 4.$$

Hence K_5 ($10 > 3 \cdot 5 - 6$) and $K_{3,3}$ ($9 > 2 \cdot 6 - 4$ after noting bipartiteness implies triangle-free) are non-planar—Kuratowski's theorem states that these are the *only* minimal obstructions.

```
import networkx as nx, networkx.algorithms.planarity as plan

K5 = nx.complete_graph(5)
print(plan.check_planarity(K5)[0]) # False

K33 = nx.complete_bipartite_graph(3,3)
print(plan.check_planarity(K33)[0]) # False
```

Dual Graphs and Face Counting In a planar embedding, each face corresponds to a dual vertex; edges cross their primal counterparts, yielding $|E^*| = |E|$ and $|V^*| = F$. Euler's formula in dual form: $|V^*| - |E^*| + F^* = 2$. Self-duality of the cube and octahedron illustrates polyhedral symmetry.

7.3.2 Advanced Topics in Graph Theory

The minimalist definitions of graphs belie the analytic depth hidden in their algebraic and probabilistic avatars. In this final subsection we sample three pillars of modern graph theory—colourings, spectra, and randomness—each connecting discrete structure with continuous mathematics and algorithmic practice.

Graph Colouring and Chromatic Polynomials

A *proper k-colouring* of a graph $G = (V, E)$ is a map $\kappa : V \to \{1, \ldots, k\}$ such that adjacent vertices receive distinct colours. The least k admitting a proper colouring is the *chromatic number* $\chi(G)$. Determining $\chi(G)$ is NP-complete (Karp 1972), yet algebraic encodings illuminate structure.

Chromatic Polynomial $P_G(k)$ For each integer $k \geq 0$, $P_G(k)$ counts proper k-colourings; it is a monic polynomial of degree $|V|$ satisfying the deletion–contraction recurrence

$$P_G(k) = P_{G-e}(k) - P_{G/e}(k),$$

where $G - e$ deletes edge e and G/e contracts e. Consequences:

(a) $P_G(1) = 0$ unless G has no edges.
(b) $(-1)^{|V|} P_G(-1)$ equals the number of acyclic orientations of G (Stanley).
(c) Real roots satisfy $\lambda \leq 0$ (by Whitney's broken-circuit theorem).

Example 7.3.1 (Chromatic Polynomial of C_4)

$$P_{C_4}(k) = k(k-1)(k^2 - 3k + 3).$$

Verify via NetworkX + SymPy:

```
import networkx as nx, sympy as sp
G = nx.cycle_graph(4)
P = nx.algorithms.coloring.chromatic_polynomial(G)
sp.factor(P)  # k*(k - 1)*(k**2 - 3*k + 3)
```

Upper Bounds By greedy ordering, $\chi(G) \leq \Delta(G) + 1$ where Δ is maximum degree. Brooks' theorem refines to $\chi(G) \leq \Delta$ except for cliques and odd cycles.

Four-Colour Theorem Every planar graph is 4-colourable (Appel–Haken 1976). The proof relies on an unavoidable set of reducible configurations checked by exhaustive computer enumeration; modern Coq formalisation (Gonthier 2008) yields a fully verified proof.

Spectral Graph Theory and Applications

Let A be the adjacency matrix and $D = \operatorname{diag}(\deg v)$. The *unnormalised Laplacian* is $L = D - A$, while the *normalised Laplacian* is $\mathcal{L} = D^{-1/2} L D^{-1/2}$. Eigenvalues

$$0 = \lambda_1 \leq \lambda_2 \leq \cdots \leq \lambda_n \leq 2$$

encode combinatorial data.

7.3 Graph Theory

Algebraic Connectivity $\lambda_2(L) > 0$ iff G is connected (Fiedler 1973). Moreover Cheeger's inequality relates λ_2 to the edge-expansion $h(G)$:

$$\frac{\lambda_2}{2} \leq h(G) \leq \sqrt{2\lambda_2}.$$

Thus sparse graphs with large λ_2 (simultaneously small degree and strong connectivity) are *expanders*; they undergird network design and error-correcting codes.

Spectral Clustering For weighted graph W, embed vertices into $\mathbb{R}^k$ using the eigenvectors of $\mathcal{L}$ associated with the smallest non-zero eigenvalues; apply k-means. This approximates the NP-hard *normalised-cut* partition.

Example 7.3.2 (Algebraic Connectivity of Random Geometric Graph)

```
import networkx as nx, numpy as np, scipy.sparse.linalg as spla
G = nx.random_geometric_graph(250, radius=0.16, seed=1)
L = nx.normalized_laplacian_matrix(G)
λ2 = spla.eigsh(L, k=2, which='SM', return_eigenvectors=False)[1]
print(f"λ2 ≈ {λ2:.4f}")
```

Eigenvalue Interlacing For induced subgraph H, eigenvalues interlace: $\lambda_i(L(G)) \leq \lambda_i(L(H)) \leq \lambda_{i+|V|-|V_H|}(L(G))$, powerful in extremal problems (e.g. bounding independence number).

Graph Energy Sum of absolute eigenvalues of A models π-electron energies in chemistry; minimal energy graphs coincide with bipartite chains.

Random Graphs and Erdős–Rényi Model

Let $G(n, p)$ denote a graph on vertex set $\{1, \ldots, n\}$ where each of the $\binom{n}{2}$ edges appears independently with probability p.

Threshold Phenomena For a monotone property $\mathcal{P}$ there is a threshold function $p^*(n)$ such that $\Pr\{G(n,p) \in \mathcal{P}\} \to 0$ for $p \ll p^*$ and $\to 1$ for $p \gg p^*$. Examples:

$$\text{Connectivity: } p^* \sim \frac{\log n}{n},$$

$$\text{Hamiltonicity: } p^* \sim \frac{\log n + \log \log n}{n}.$$

Phase Transition of Giant Component Set $p = c/n$. If $c < 1$ all components have size $O(\log n)$; if $c > 1$ there is w.h.p. a unique component of size $\Theta(n)$.

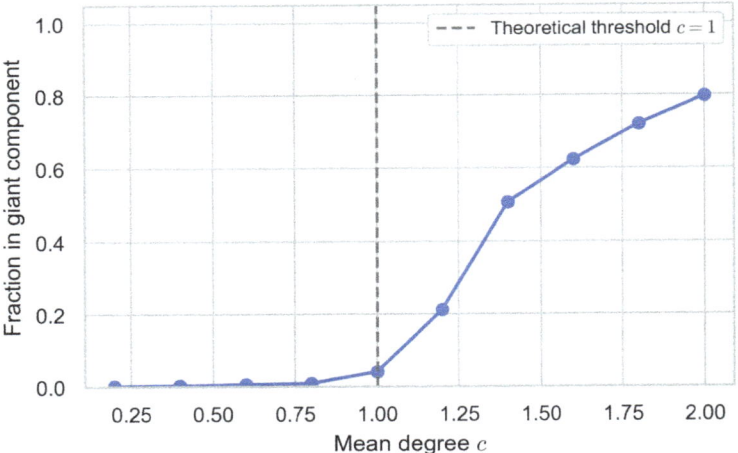

Fig. 7.1 Relative size of the largest connected component in a single realisation of the Erdős–Rényi random graph $G(n, p)$ with $n = 4000$ and edge probability $p = c/n$. The empirical curve (points joined by lines) exhibits the phase transition at the critical mean degree $c = 1$ (dashed line), where a macroscopic "giant" component suddenly emerges

Degree Distribution and Eigenvalues Expected degree is $(n-1)p$. After centring and scaling, the adjacency spectrum converges to the semicircle law (Wigner) for $np \to \infty$; deviations detect community structure ("spectral sparsification").

Example 7.3.3 (Simulating Phase Transition (cf. Fig. 7.1))

```
import networkx as nx, numpy as np
n, c_vals = 4000, np.linspace(.2, 2.0, 10)
largest = []
for c in c_vals:
    G = nx.fast_gnp_random_graph(n, c/n, seed=1)
    largest.append(max(len(cc) for cc in nx.connected_components(G)))
import matplotlib.pyplot as plt
plt.plot(c_vals, np.array(largest)/n, 'o-')
plt.xlabel('c'); plt.ylabel('size of giant component / n'); plt.show()
```

Erdős–Rényi Law of Large Numbers For fixed p, $\frac{|E|}{\binom{n}{2}} \xrightarrow{a.s.} p$, $\frac{\lambda_{\max}(A)}{np} \xrightarrow{a.s.} 1$.

Applications Random graphs model spread of epidemics (bond percolation), performance of hash tables (load distribution), and average-case behaviour of algorithms (e.g. $G(n, \frac{1}{2})$ is a typical input for satisfiability heuristics).

7.3.3 Applications of Graph Theory

Real-world systems—from power grids to social platforms—manifest naturally as graphs. The ability to model such systems enables optimisation, inference, and learning tasks that would be opaque without a graph-theoretic lens. In what follows we survey three broad application domains and support each with concise but functional *Python* examples.

Applications in Network Analysis and Optimisation

Classical optimisation problems are most transparent on graphs.

Shortest Paths and Flows Let $G = (V, E, w)$ be a weighted digraph with non-negative edge costs $w : E \to \mathbb{R}_{\geq 0}$. Dijkstra's algorithm computes dist(s, v) for all v in $O(|E| + |V| \log |V|)$ time. If $\sum_{e \in E} w(e) = C$, any s-t path obeys dist$(s, t) \leq C$, furnishing a global upper bound used in heuristic pruning.

Maximum flow $\max_f \sum_{e \in \delta^+(s)} f(e)$ subject to capacity constraints and conservation is solved by Edmonds–Karp $O(|V||E|^2)$ or by the push-relabel variant $O(|V|^3)$. Min-cut duality (*Max-Flow = Min-Cut*) underpins image segmentation and network reliability.

```
import networkx as nx
G = nx.DiGraph()
edges = [(0,1,9),(0,2,5),(1,2,3),(1,3,7),(2,3,4)]
G.add_weighted_edges_from(edges, capacity='cap')
flow_val, _ = nx.maximum_flow(G, 0, 3, capacity='cap')
print(flow_val) # 9 (max s-t flow)
```

Travelling Salesperson and Branch-and-Bound Metric TSP admits a 1.5-approximation via Christofides' algorithm: MST $\to$ minimum-weight perfect matching on odd-degree vertices $\to$ Euler tour $\to$ shortcutting. The approximation ratio follows from triangle inequality.

Network Robustness Given adjacency matrix A, the algebraic connectivity $\lambda_2(L)$ quantifies how many edge failures the network can sustain before disconnecting; optimising λ_2 under a budget of extra edges is NP-hard, yet semidefinite relaxations yield near-optimal augmentations.

Social Network Analysis and Community Detection

Social graphs are typically sparse, heavy-tailed, and homophilic.

Centrality Measures Degree, betweenness, closeness, and eigenvector centralities identify influencers. For eigenvector centrality $\mathbf{c} = A\mathbf{c}$ up to scaling, Perron–Frobenius ensures a unique positive vector when G is strongly connected.

```
import networkx as nx
G = nx.karate_club_graph()
cen = nx.eigenvector_centrality_numpy(G)
leader = max(cen, key=cen.get)
print(leader) # vertex with highest influence
```

Modularity and the Louvain Heuristic Modularity of a partition $\mathcal{C}$ is

$$Q(\mathcal{C}) = \frac{1}{2m} \sum_{u,v} \left(A_{uv} - \frac{\deg u \, \deg v}{2m} \right) \mathbf{1}_{\kappa(u)=\kappa(v)}, \quad m = |E|,$$

where $\kappa : V \to \{1, \ldots, k\}$ labels communities. Louvain iteratively optimises local modularity gains, aggregating communities into super-nodes; it scales near-linearly in $|E|$ and finds meaningful clusters in graphs with $\gtrsim 10^7$ edges.

```
import community as lv # python-louvain package
part = lv.best_partition(G)
Q = lv.modularity(part, G)
print(Q)
```

Structural Balance and Signed Graphs Edge signs $\sigma : E \to \{\pm 1\}$ model friend–foe relations; a graph is structurally balanced iff every cycle has positive sign product. Such graphs are characterised as having a bipartition with all positive edges inside parts and all negative across.

Graph-Based Machine Learning Algorithms

Graphs feed modern machine-learning pipelines in at least two ways: as kernels in classical models and as direct inputs to neural architectures.

Graph Kernels Define $k(G, H) = \langle \phi(G), \phi(H) \rangle$ via random-walk or Weisfeiler–Lehman subtree features; SVMs on these kernels perform graph classification without explicit vectorisation.

Graph Neural Networks (GNNs) A message-passing layer updates node embeddings:

$$\mathbf{h}_v^{(\ell+1)} = \sigma\left(W_1 \mathbf{h}_v^{(\ell)} + \sum_{u \in \mathcal{N}(v)} W_2 \mathbf{h}_u^{(\ell)} + \mathbf{b} \right).$$

Stack L layers, apply global pooling for whole-graph embeddings. Theoretical expressiveness parallels the 1-Weisfeiler–Lehman test; injective aggregation (e.g. sum) is critical.

7.4 Boolean Algebra and Logic

```
import torch, torch_geometric.nn as pyg
class GCN(torch.nn.Module):
    def __init__(self, in_dim, hid, out):
        super().__init__()
        self.conv1 = pyg.GCNConv(in_dim, hid)
        self.conv2 = pyg.GCNConv(hid, out)

    def forward(self, x, edge_index):
        x = torch.relu(self.conv1(x, edge_index))
        x = self.conv2(x, edge_index)
        return torch.nn.functional.log_softmax(x, dim=-1)
```

Link Prediction Given embeddings $\mathbf{h}_u, \mathbf{h}_v$, score $s_{uv} = \sigma(\mathbf{h}_u^\top \mathbf{h}_v)$. Applications include recommendation systems and knowledge-graph completion.

Graph Reinforcement Learning In path-planning or chemical molecule generation, states are partial graphs, actions attach nodes/edges, and rewards encode objectives; GNNs approximate Q-functions invariant to node permutations.

7.4 Boolean Algebra and Logic

Boolean algebra studies truth-valued variables under operations that mirror English connectives. Classical results—De Morgan's laws, normal forms, completeness—translate grammatical reasoning into algebraic manipulation and power automated verification engines. Python's `sympy.logic` module demonstrates these ideas symbolically, while SAT solvers like `pycosat` scale them to millions of clauses in hardware verification.

7.4.1 Propositional Logic and Proof Techniques

A *proposition* is a declarative statement that is either true ($\top$) or false ($\bot$). Formulas combine propositional variables via logical connectives; their semantics derive from truth assignments. Syntactic derivations (*proofs*) operate in formal calculi such as natural deduction or resolution, whereas semantic validity is defined via truth tables. Gödel's completeness theorem couples the two: every semantically valid formula is syntactically provable.

Logical Connectives, Truth Tables, and Tautologies

The basic binary connectives are conjunction ($\wedge$), disjunction ($\vee$), implication ($\rightarrow$), and biconditional ($\leftrightarrow$); negation ($\neg$) is unary. Truth tables exhaustively list outcomes for each assignment. For two variables p, q:

p	q	$p \vee q$	$p \wedge q$	$p \to q$
⊤	⊤	⊤	⊤	⊤
⊤	⊥	⊤	⊥	⊥
⊥	⊤	⊤	⊥	⊤
⊥	⊥	⊥	⊥	⊤

A *tautology* evaluates to ⊤ under every assignment; e.g. $p \vee \neg p$. A *contradiction* is always ⊥; a *contingent* formula is neither. The set $\{\neg, \vee\}$ is functionally complete (every Boolean function can be expressed).

Example 7.4.1 (SymPy Verification of De Morgan)

```
import sympy as sp
p, q = sp.symbols('p q')
expr = ~(p & q)
dm = ~p | ~q
print(sp.simplify_logic(expr ^ dm, form='cnf')) # 0 ⟹ tautologically
    equal
```

Truth tables scale exponentially; normal forms compress reasoning: the *conjunctive normal form* (CNF) of a formula is a conjunction of clauses (disjunctions of literals) and underpins SAT solvers.

Proof Techniques: Direct, Indirect, and Contradiction

Direct Proof To show $P \to Q$, assume P and derive Q step by step. Example: from n even ($\exists k\ n = 2k$) deduce n^2 even via $n^2 = 4k^2$.

Proof by Contraposition $P \to Q$ is equivalent to $\neg Q \to \neg P$. Often $\neg Q$ is more tractable. Example: if n^2 odd $\Rightarrow n$ odd.

Proof by Contradiction Assume P and $\neg Q$; derive contradiction ⊥, thereby establishing $P \to Q$. Classic: irrationality of $\sqrt{2}$ assuming $\sqrt{2} = p/q$ with coprime integers.

Natural-deduction rules encode these strategies; sequent calculus makes structural properties explicit, enabling proof search algorithms.

Example 7.4.2 (Resolution Refutation in CNF) Clause set $\{\{p, q\}, \{\neg p\}, \{\neg q\}\}$ is unsatisfiable: $\{p, q\}, \{\neg p\} \stackrel{\text{res}}{\Longrightarrow} \{q\}$; $\{q\}, \{\neg q\} \Longrightarrow \emptyset$.

```
from sympy.logic.inference import satisfiable
P = sp.symbols('P')
cnf = sp.And(P|~P, ~P, P) # contradictory
print(satisfiable(cnf)) # {}
```

7.4 Boolean Algebra and Logic

Applications in Automated Theorem Proving

Modern theorem provers reduce logical validity to satisfiability.

SAT and SMT Propositional satisfiability (SAT) was the first NP-complete problem (Cook 1971). Yet conflict-driven clause learning (CDCL) solves million-variable instances. Satisfiability modulo theories (SMT) extend SAT with arithmetic, arrays, and uninterpreted functions; tools like Z3 discharge verification conditions generated by compilers and proof assistants.

Resolution and DPLL The Davis–Putnam–Logemann–Loveland (DPLL) algorithm systematically assigns variables, backtracking on conflicts; resolution derives the empty clause to refute unsatisfiable sets. CDCL augments DPLL with non-chronological backtracking and clause learning.

Formal Verification Boolean encodings verify hardware (combinational equivalence checking), software (bounded model checking), and security protocols. Example: detecting arithmetic overflow by asserting ¬(carry_out $\oplus$ sign_bit) and querying SAT for a counterexample.

Example 7.4.3 (Bit-vector Reasoning with z3)

```
from z3 import *
a, b = BitVecs('a b', 32)
s = Solver()
prod = a * b
s.add(a == 0xFFFF_FFFF, b == 0x2, prod == (a * b) & 0xFFFF_FFFF) #
    overflow check
print(s.check()) # sat ⟹ overflow possible
print(s.model())
```

Proof Assistants Interactive theorem provers (Coq, Lean, Isabelle) encode logic via type theory; tactics discharge propositional subgoals with SAT/SMT back-ends, leaving higher-order reasoning to the user.

7.4.2 Boolean Algebra and Circuit Design

Boolean algebra refines propositional logic by restricting variables to $\{0, 1\}$ and interpreting conjunction, disjunction, and negation as algebraic operations in the ring $\mathbb{F}_2[x_1, \ldots, x_n]/\langle x_i^2 - x_i \rangle$. Because $x^2 = x$ in this ring, every polynomial reduces to a multilinear form, which we read as a *Boolean function*

$$f : \{0, 1\}^n \longrightarrow \{0, 1\}, \qquad f(x_1, \ldots, x_n) = \sum_{S \subseteq [n]} a_S \Big(\prod_{i \in S} x_i \Big), \qquad a_S \in \{0, 1\},$$

where addition and multiplication are modulo 2. Digital circuits implement such functions by wiring together logic gates realising $\land$, $\lor$, and $\neg$; algebraic simplification therefore translates directly into fewer gates and lower propagation delay.

Boolean Functions and Simplification Techniques

Algebraic Identities The Boolean operations obey idempotence ($x \lor x = x$, $x \land x = x$), absorption ($x \lor x \land y = x$), distributivity ($x \land (y \lor z) = (x \land y) \lor (x \land z)$), and De Morgan's laws ($\overline{x \land y} = \overline{x} \lor \overline{y}$). From these axioms one derives normal forms:

(a) **Sum-of-Products** (SoP)/Disjunctive Normal Form: $f = \bigvee_{m \in \mathcal{M}} \left(\bigwedge_{x_i=0} \overline{x_i} \right) \left(\bigwedge_{x_j=1} x_j \right)$, where $\mathcal{M}$ lists minterms evaluating to 1;

(b) **Product-of-Sums** (PoS)/Conjunctive Normal Form, dual to SoP.

Example 7.4.4 (SoP Simplification in sympy)

```
import sympy as sp
a,b,c = sp.symbols('a b c')
expr = (a & ~b) | (~a & b & c) | (a & b & c)
print(sp.simplify_logic(expr, form='dnf')) # a & ~b | b & c
```

The simplifier factors out common sub-terms, saving one AND gate and one input.

Prime implicants (minimal products that still imply f) form the backbone of systematic simplification; the next subsubsection formalises their extraction.

Karnaugh Maps and Quine–McCluskey Method

Karnaugh Maps (K-maps) For $n \leq 4$ variables a K-map arranges 2^n minterms on a Gray-code grid so that adjacent cells differ in a single literal. Grouping 2^k adjacent 1's yields an implicant covering those minterms; overlapping groups minimise literal count.

	$bc = 00$	$bc = 01$	$bc = 11$	$bc = 10$
$a=0$	0	1	1	0
$a=1$	1	1	0	0

$\implies \quad f = \overline{a}\,b \,\lor\, \overline{b}\,\overline{c}.$

Quine–McCluskey Algorithm A tabular, deterministic analogue of K-maps for arbitrary n:

1. *Initial grouping:* partition minterms by Hamming weight.
2. *Iterative merging:* combine terms differing in one literal, marking absorbed terms; record prime implicants.

7.4 Boolean Algebra and Logic

3. *Prime implicant chart:* cover remaining minterms with minimum subset—an NP-hard set cover instance, but small in practice.

```
from pyeda.boolalg.expr import exprvars, SOPform
from pyeda.boolalg.minimization import espresso_exprs
a,b,c,d = exprvars('a', 4)
on = [0b0001, 0b0100, 0b1101, 0b1110] # minterms
dc = [] # don't cares
f = SOPform([a,b,c,d], on, dc)
fmin = espresso_exprs(f)[0]
print(fmin) # a & ~b & ~c & ~d | b & c & d
```

Don't-Care Conditions Input combinations never encountered in operation can be assigned either value, enlarging grouping possibilities and further reducing gate count.

Applications in Digital Circuit Design

Combinational Logic Synthesis Given a truth table, pipeline:

(i) Derive canonical SoP from minterms.
(ii) Simplify via Quine–McCluskey or ESPRESSO heuristic.
(iii) Map literals to gates (NAND/NOR preferred in CMOS).
(iv) Estimate propagation delay T_{pd} and power; iterate with timing-driven optimisation.

Example 7.4.5 (4-bit Parity Checker)

$$\text{PAR}(x_3 \ldots x_0) = x_3 \oplus x_2 \oplus x_1 \oplus x_0.$$

SoP of even parity has eight minterms; Karnaugh grouping yields $\text{PAR} = (x_3 \oplus x_2) \oplus (x_1 \oplus x_0)$, requiring *only* two 2-input XOR gates plus one final XOR.

Finite-State Machines (FSMs) State transition and output functions g, h are Boolean; their ROM or PLA realisations benefit from minimisation:

$$\mathbf{s}_{t+1} = g(\mathbf{s}_t, \mathbf{x}_t),$$
$$\mathbf{y}_t = h(\mathbf{s}_t, \mathbf{x}_t).$$

State assignment (one-hot, Gray, balanced) affects gate complexity and jitter.

Programmable Logic Arrays (PLA) and FPGAs SOP form maps naturally: product terms implement AND plane; summed outputs implement OR plane. Karnaugh/Quine-McCluskey minimise the number of product terms, reducing PLA area; FPGA LUTs (look-up tables) emulate small truth tables directly—low-literal minimisation reduces the depth of LUT cascades.

Arithmetic Circuits Carry-look-ahead adders exploit Boolean identities: $c_{i+1} = g_i \vee (p_i \wedge c_i)$, with generate $g_i = a_i b_i$ and propagate $p_i = a_i \oplus b_i$; expanding recursively yields $O(\log n)$ carry depth.

Hazards Static hazard arises when two implicants cover a minterm but their adjacency is absent, causing glitches. Karnaugh maps reveal and neutralise hazards by inserting redundant implicants that smooth transitions.

7.4.3 Advanced Topics in Logic

Predicate (or *first-order*) logic extends propositional logic by permitting variables that range over a non-empty domain $\mathcal{D}$ together with *quantifiers*. Whereas the truth of a propositional formula is a function of finitely many Boolean assignments, the validity of a predicate formula depends on *all* interpretations of its symbols over *all* structures of the language. This leap in expressiveness amplifies deductive power—allowing one to formalise virtually the whole of classical mathematics—yet invites fundamental limitations captured by Gödel's incompleteness theorems. Beyond the classical bivalent setting, alternative logics adapt the semantics or proof rules to model vagueness, modality, resource sensitivity, or inconsistency, thereby broadening the scope of logical analysis in computer science and philosophy.

Predicate Logic and Quantifiers

A *signature* $\Sigma = \langle \mathcal{F}, \mathcal{P}, \mathrm{ar} \rangle$ lists function symbols $\mathcal{F}$ and predicate symbols $\mathcal{P}$ with arities $\mathrm{ar}(\cdot) \in \mathbb{N}$. The set of *terms* is generated from variables $\{x_1, x_2, \dots\}$ and function application; *atomic formulas* are $P(t_1, \dots, t_k)$ and $t_1 = t_2$. Formulas close under Boolean connectives and the quantifiers

$$\forall x\, \varphi \quad \text{and} \quad \exists x\, \varphi.$$

Semantics fixes a structure $\mathcal{M} = \langle \mathcal{D}, (f^{\mathcal{M}})_{f \in \mathcal{F}}, (P^{\mathcal{M}})_{P \in \mathcal{P}} \rangle$ and an environment σ mapping variables to elements of $\mathcal{D}$. Satisfaction $\mathcal{M}, \sigma \models \varphi$ proceeds inductively; quantifiers take set–theoretic unions ($\forall$) or existential witnesses.

Prenex and Skolem Forms Every formula is equivalent (in classical logic) to a *prenex* form $Q_1 x_1 \dots Q_n x_n\, \psi$ with quantifier block $Q_i \in \{\forall, \exists\}$ followed by quantifier-free ψ. Subsequently *Skolemisation* replaces existential variables with functional witnesses—preserving satisfiability and enabling automated theorem proving.

7.4 Boolean Algebra and Logic

Example 7.4.6 (Model Checking in sympy)

```
import sympy as sp
x,y = sp.symbols('x y')
φ = sp.ForAll(x, sp.Exists(y, x < y)) # N with < is infinite
M = sp.Domain('Naturals')
print(sp.ask(φ, sp.Q.real)) # True
```

Completeness Gödel (1930) showed that first-order logic is *semantically complete*: if $\models \varphi$ then $\vdash \varphi$ in any sound and sufficiently rich deductive calculus (e.g. natural deduction). Compactness and Löwenheim–Skolem theorems follow: a set of sentences is satisfiable iff every finite subset is, and any infinite model has countable submodels.

Gödel's Incompleteness Theorems

Let T be a consistent, recursively enumerable theory extending Peano arithmetic PA. Gödel assigned natural numbers to symbols, formulas, and proofs—*Gödel numbering*—to express meta-mathematical statements *inside* T.

Theorem 7.4.1 (Gödel 1931)

1. (*First incompleteness*) There exists a sentence G such that $T \nvdash G$ and $T \nvdash \neg G$. In fact G is true in the standard model $\mathbb{N}$.
2. (*Second incompleteness*) T cannot prove its own consistency $\text{Con}(T)$.

Sketch Define $\text{Prf}_T(p, q)$ meaning "p codes a proof in T of the formula with code q", representable in arithmetic. The diagonal lemma yields $G \equiv \forall p \, \neg \text{Prf}_T(p, \ulcorner G \urcorner)$, i.e. "G is unprovable". If $T \vdash G$, there is a proof code $\hat{p}$ contradicting G; if $T \vdash \neg G$, then T would prove its own inconsistency—a contradiction.

Implications

(a) No algorithm enumerates exactly the truths of arithmetic.
(b) Any proof assistant founded on PA cannot internally assert its consistency.
(c) There are true but unprovable statements, e.g. the Paris–Harrington combinatorial principle.

Non-classical Logics and Their Applications

Intuitionistic Logic Rejects the law of excluded middle $\varphi \vee \neg \varphi$ unless constructed; proof objects correspond to λ-terms via the Curry–Howard isomorphism. *Constructive type theory* (e.g. in Coq, Lean) rests on intuitionistic foundations, enabling program extraction from proofs.

Modal Logic Augments propositional cores with $\Box$ (necessarily) and $\Diamond$ (possibly). Kripke semantics interprets $\Box\varphi$ as "φ holds in all accessible worlds". System $S4$ corresponds to provability ($\Box\varphi \to \Box\Box\varphi$) and underlies temporal logics used in model checking (LTL, CTL).

Temporal and Dynamic Logics Propositional Linear Temporal Logic (LTL) adds operators $\bigcirc$ (next), $\Diamond$ (eventually), and $\Box$ (always); Büchi automata translate LTL specifications into ω-automata for hardware verification. Dynamic logic reason about program execution paths; $\langle\pi\rangle\varphi$ asserts existence of run of program π leading to state satisfying φ.

Fuzzy and Many-Valued Logics Truth values take continuum $[0, 1]$ with t-norms modelling conjunction. Used in control systems where sensor data is imprecise. *Łukasiewicz logic* defines $x \to y = \min(1, 1 - x + y)$, capturing graded implications.

Paraconsistent Logics Allow controlled reasoning in presence of contradictions without explosion ($\varphi, \neg\varphi \vdash \psi$ in classical logic). Useful in knowledge bases with inconsistent data. *LP* (Priest's Logic of Paradox) designates both $\top$ and $\bot$ as truth values.

Applications Snapshot

Example 7.4.7 *Model checking.* Given a Kripke structure M for a traffic-light controller, verify $\Box$(green $\to$ $\bigcirc$yellow) with LTL-to-Büchi translation and the spot library.

Type systems. Intuitionistic λ-calculus plus dependent types captures proofs in Agda; total functional programs correspond to constructive proofs.

Policy languages. Modal deontic logic expresses obligations ($\Box$) and permissions ($\Diamond$) in access control systems; runtime monitors enforce these via temporal-logic model checking.

7.5 Discrete Structures and Applications

Mathematical thinking frequently begins with the act of *collecting* objects into well-defined groups, and then exploring the relations and mappings between such collections. Sets, relations, and functions provide a language concise enough to encode subtle logical dependencies yet concrete enough to be manipulated computationally. In this subsection we recall their foundational properties, demonstrate typical reasoning patterns, and illustrate each concept with short *Python* snippets that mirror the algebraic laws in executable form.

7.5.1 Sets, Relations, and Functions

Set Theory Basics and Venn Diagrams

A *set* is an unordered collection of distinct elements; membership is recorded by the predicate $x \in A$. Standard constructions derive new sets from old: the *union* $A \cup B = \{x \mid x \in A \lor x \in B\}$ and the *intersection* $A \cap B = \{x \mid x \in A \land x \in B\}$ satisfy the distributive law

$$A \cap (B \cup C) = (A \cap B) \cup (A \cap C), \quad \text{and dually} \quad A \cup (B \cap C) = (A \cup B) \cap (A \cup C).$$

The *power set* $\mathcal{P}(A)$ contains all subsets of A and has cardinality $2^{|A|}$; Cantor's theorem $|A| < |\mathcal{P}(A)|$ establishes infinite hierarchies of cardinalities.

Example 7.5.1 (Boolean Algebra of Sets)

```
A, B, C = {1,2,3}, {2,3,4}, {3,4,5}
lhs = A & (B | C)  # ∩ and ∪ via operators
rhs = (A & B) | (A & C)
print(lhs == rhs)  # distributive law -> True
```

Venn diagrams visualise the inclusion–exclusion principle. For finite sets

$$|A \cup B| = |A| + |B| - |A \cap B|, \quad |A \cup B \cup C| = |A| + |B| + |C|$$
$$- |A \cap B| - |B \cap C| - |C \cap A| + |A \cap B \cap C|.$$

In probability theory the same formula converts overlapping probabilities into additive ones.

Relations: Reflexivity, Symmetry, and Transitivity

A *binary relation* on A is a subset $R \subseteq A \times A$. It is

$$\begin{aligned}
\textit{reflexive} &\iff \forall a \in A,\ (a,a) \in R, \\
\textit{symmetric} &\iff \forall a,b,\ (a,b) \in R \Rightarrow (b,a) \in R, \\
\textit{transitive} &\iff \forall a,b,c,\ (a,b),(b,c) \in R \Rightarrow (a,c) \in R.
\end{aligned}$$

An *equivalence relation* satisfies all three properties and partitions A into disjoint equivalence classes; a *partial order* is reflexive, antisymmetric, and transitive.

Example 7.5.2 (Congruence Modulo n) Define $a \equiv_n b$ iff $n \mid (a-b)$. Reflexivity holds since $n \mid 0$, symmetry from $n \mid (a-b) \Rightarrow n \mid (b-a)$, and transitivity from divisibility chains. Thus $\mathbb{Z}$ splits into n congruence classes $[0], \ldots, [n-1]$ each visualised on the circle group $\mathbb{Z}/n\mathbb{Z}$.

Graphically, relations correspond to directed graphs on vertex set A; reflexivity adds self-loops, symmetry yields undirected edges, and transitivity appears as path closure. In networkx one tests transitivity by computing the transitive closure and checking inclusion.

```
import networkx as nx
A = range(4)
R = [(0,0),(1,1),(2,2),(3,3),(0,1),(1,2),(0,2)]
G = nx.DiGraph(); G.add_nodes_from(A); G.add_edges_from(R)
tc = nx.transitive_closure_dag(G)  # raises if cycles present
print(set(tc.edges()).issubset(R)) # transitivity?
```

Functions: Injective, Surjective, and Bijective Mappings

A *function* $f : A \to B$ assigns each $a \in A$ a unique $b = f(a) \in B$. The mapping is

$$\text{injective (one-to-one)} \iff (\forall a_1 \neq a_2) \; f(a_1) \neq f(a_2);$$

$$\text{surjective (onto)} \iff (\forall b \in B) \; \exists a \in A, \; f(a) = b;$$

$$\text{bijective} \iff \text{injective and surjective.}$$

Categorically, bijections are isomorphisms in **Set**; they admit inverses $f^{-1} : B \to A$. Cantor–Bernstein–Schröder states: if there exist injections $A \hookrightarrow B$ and $B \hookrightarrow A$, then a bijection $A \leftrightarrow B$ exists.

Example 7.5.3 (Cantor Pairing Function)

$$\pi : \mathbb{N} \times \mathbb{N} \to \mathbb{N}, \quad \pi(x,y) = \frac{1}{2}(x+y)(x+y+1) + y,$$

is bijective, proving $\mathbb{N}^2$ is countable. Inverse obtained by reversing triangular indexing.

```
import math
def pair(x, y): return (x+y)*(x+y+1)//2 + y
def unpair(z):
    w = math.floor((math.sqrt(8*z+1)-1)/2)
    t = w*(w+1)//2
    y = z - t
    return (w - y, y)
assert unpair(pair(17, 23)) == (17, 23)
```

Function composition is associative; identity functions act as units, forming the *category* of sets. In computer science, *partial* functions arise naturally—total functions on bounded domains extended with a *None* placeholder mirror them in Python.

7.5 Discrete Structures and Applications

Synthesis Set-theoretic algebra supplies precise language; relations organise elements into structured classes; functions transport structure between sets. These three pillars support the later chapters on graph morphisms, automata transitions, and measure-theoretic probability, showing that seemingly humble definitions propagate far-reaching consequences across discrete mathematics.

7.5.2 Algebraic Structures

Algebraic structures formalise the intuitive idea of combining objects and studying the laws that govern those combinations. They provide a unifying language in which arithmetic, symmetry, order, and logic can all be expressed. Once cast in this abstract setting, theorems become portable: a property proved for *any* group instantly applies to integers under addition, permutations under composition, and invertible matrices under multiplication. Python's arbitrary precision arithmetic and its ecosystem of symbolic algebra libraries (`sympy`, `sagemath`) create an ideal playground for verifying such universal statements computationally.

Groups, Rings, and Fields

A **group** $\langle G, *, e \rangle$ consists of a non-empty set G equipped with a binary operation $*$ satisfying associativity, identity, and invertibility. If $*$ is commutative the group is *abelian*. Classic examples range from $(\mathbb{Z}, +, 0)$ through $(\mathbb{Z}_n, +_n, 0)$ to the symmetric group S_n of permutations on n letters.

Groups act on sets via homomorphisms $G \to \text{Sym}(X)$; orbits and stabilisers partition the action space, and the orbit-stabiliser theorem $|G| = |G \cdot x| \cdot |G_x|$ feeds enumeration (cf. Pólya theory).

A **ring** $\langle R, +, \cdot \rangle$ augments an abelian group $(R, +)$ with a second associative operation $\cdot$ that distributes over $+$. Rings may or may not have multiplicative identities (1). $\mathbb{Z}$, $\mathbb{Z}_n$, and polynomial rings $\mathbb{F}[x]$ are paradigmatic; Euclidean domains admit a division algorithm, Bézout's identity, and unique factorisation.

A **field** is a commutative ring where non-zero elements form an abelian multiplicative group. Finite fields $\mathbb{F}_{p^k}$ exist for prime-powers p^k and are unique up to isomorphism; they support polynomial arithmetic modulo irreducibles and underpin most modern cryptosystems.

Example 7.5.4 (Fast Exponentiation in the Field $\mathbb{F}_p$)
```
import random
def modexp(a, e, p):
    res = 1
    while e:
        if e & 1: res = (res*a) % p
        a = (a*a) % p; e >>= 1
    return res

p = 1_000_000_007
g = random.randint(2, p-2)
assert modexp(g, p-1, p) == 1 # Fermat's little theorem
```

Lattices and Boolean Algebras

A **lattice** is a partially ordered set $\langle L, \leq \rangle$ where every pair (x, y) admits a least upper bound $x \vee y$ (*join*) and greatest lower bound $x \wedge y$ (*meet*). Finite lattices can be visualised via Hasse diagrams; distributive lattices satisfy $x \wedge (y \vee z) = (x \wedge y) \vee (x \wedge z)$.

A **Boolean algebra** is a distributive lattice with top $\top$, bottom $\bot$, and complementation $\neg$ such that $x \vee \neg x = \top$ and $x \wedge \neg x = \bot$. Stone duality links Boolean algebras to Boolean spaces (totally disconnected compact Hausdorff), revealing deep connections between logic and topology.

Example 7.5.5 (Lattice of Divisors)
```
import math, itertools
n = 60
divs = {d for d in range(1,n+1) if n%d==0}
leq = lambda a,b: a|b == b # divisibility order via bitwise OR
for a,b in itertools.combinations(divs, 2):
    join = math.lcm(a,b)
    meet = math.gcd(a,b)
    assert join in divs and meet in divs
```

Applications in Cryptography and Coding Theory

Public-Key Cryptography RSA exploits the multiplicative group $(\mathbb{Z}_N^\times, \cdot)$ where $N = pq$ is semiprime; difficulty of factoring N protects the private exponent. Elliptic-curve cryptography replaces $\mathbb{Z}_N^\times$ by the group of $\mathbb{F}_q$-rational points on an elliptic curve, obtaining equivalent security with much shorter keys thanks to the richer group law.

Error-Correcting Codes Linear codes are vector subspaces of $\mathbb{F}_q^n$. The Hamming distance $d(\mathbf{x}, \mathbf{y}) = \#\{i : x_i \neq y_i\}$ defines spheres; the Gilbert–Varshamov bound and Singleton bound constrain achievable code parameters (n, k, d). Cyclic codes correspond to ideals in the quotient ring $\mathbb{F}_q[x]/(x^n - 1)$ and admit efficient shift-register encoders.

Example 7.5.6 ((7,4) Hamming Code over $\mathbb{F}_2$)

```
import numpy as np
G = np.array([[1,0,0,0,1,1,0],
              [0,1,0,0,1,0,1],
              [0,0,1,0,0,1,1],
              [0,0,0,1,1,1,1]], dtype=int)
m = np.array([1,0,1,1], dtype=int)
c = m @ G % 2 # encode
e = np.array([0,0,0,0,0,0,1]) # single-bit error
r = (c + e) % 2 # received word
H = np.array([[1,1,1,1,1,0,0],
              [1,1,0,0,0,1,1],
              [1,0,1,0,1,1,0]], dtype=int)
s = H @ r % 2 # syndrome
print(s) # identifies position 7
```

Lattice-Based Cryptography Ideals in polynomial rings modulo cyclotomic polynomials realise lattices equipped with hard *closest-vector* problems (NTRU, Ring-LWE). These conjecturally resist quantum attacks, making them prime candidates for post-quantum standards.

7.5.3 Matroids and Their Applications

Matroid theory abstracts the notion of "independence" that appears across linear algebra, graph theory, and set systems, and distils it into a combinatorial structure that admits powerful optimisation guarantees. A *matroid* is a pair $\mathcal{M} = \langle E, \mathcal{I} \rangle$ where E is a finite ground set and $\mathcal{I} \subseteq 2^E$ is a non-empty collection of *independent* subsets satisfying

(I1) heredity: $I \in \mathcal{I}, J \subseteq I \Rightarrow J \in \mathcal{I}$, (I2) exchange: $I, J \in \mathcal{I}, |I| < |J|$
$$\Longrightarrow \exists e \in J \setminus I : I \cup \{e\} \in \mathcal{I}.$$

The exchange axiom aligns matroids with vector spaces: every independent set can be extended to a *basis*, all bases share the same cardinality $\mathrm{rk}\,\mathcal{M}$, and greedy augmentation never gets stuck. Equivalent definitions employ a *rank function* $r : 2^E \to \mathbb{N}$ that is monotone, submodular, and bounded by inclusion, or the set of *circuits* (minimal dependent sets) which satisfy a circuit elimination property.

Introduction to Matroids and Examples

The prototypical examples reveal how matroids unify disparate contexts.

- *Uniform matroid* $U_{n,k}$: $E = \{1, \ldots, n\}$, $\mathcal{I} = \{I \subseteq E : |I| \leq k\}$. Rank of any subset is $\min\{|S|, k\}$.
- *Graphic (cycle) matroid* $M(G)$ of a graph $G = (V, E)$: independent sets are forests (edge subsets with no cycles). Bases correspond to spanning trees; rank equals $|V| - c(G)$ where $c(G)$ is the number of connected components.
- *Vector matroid* $M(\mathcal{V})$: E is a finite set of vectors in a vector space, with independence defined linearly. Representable matroids connect linear algebra to combinatorics; e.g. F_7 (Fano plane) is representable over $\mathbb{F}_2$ but not over $\mathbb{R}$.
- *Transversal matroid*: given a family of subsets $\mathcal{A} = \{A_1, \ldots, A_m\}$ of E, a set $I \subseteq E$ is independent if it is contained in some system of distinct representatives for $\mathcal{A}$. Hall's marriage theorem characterises bases.

Example 7.5.7 (Circuits of a Graphic Matroid) Let G be the square with a diagonal. Its edge set $E = \{e_1, \ldots, e_5\}$ has three circuits: the two triangles and the 4-cycle. Deleting any edge from a circuit yields a forest; adding any external edge to a forest closes exactly one unique new circuit—reflecting the exchange axiom in graph form.

Rank and closure define flats (closed sets), whose lattice encodes geometric properties; in the graphic case flats correspond to edge cuts.

Applications in Optimisation and Graph Theory

Many classical optimisation problems conceal a matroid structure and hence admit polynomial-time greedy solutions.

Spanning Tree and Forests Kruskal's algorithm selects edges of minimum weight that do not form cycles—precisely the greedy algorithm for the graphic matroid with weights. By matroid theory the resulting set is a minimum-weight basis.

Branchings and Arborescences The set of arcs forming acyclic subgraphs in a digraph forms a *gammoid*. Edmonds's branching theorem extends matroid intersection to find minimum-cost rooted arborescences.

Matroid Intersection and Union Given two matroids $\mathcal{M}_1, \mathcal{M}_2$ on the same ground set, the maximum common independent set is found via augmenting-path algorithms in $O(|E|r^{1.5})$ time; applications include edge-disjoint spanning trees and bipartite matchings (graphic $\cap$ partition matroid).

Cut, Flow, and Duality For planar graphs, the cocycle matroid $M^*(G)$ is dual to $M(G^*)$, the cycle matroid of the planar dual G^*. This correspondence underpins planar flow-cut duality and explains why max-flow equals min-cut in planar networks under capacity–length interchanges.

7.5 Discrete Structures and Applications

Example 7.5.8 (Matroid Union Coding in Python)
```
import networkx as nx, itertools
G = nx.cycle_graph(4) # C_4
E = list(G.edges())
def is_forest(S): # independence oracle
    H = nx.Graph(); H.add_edges_from(S)
    return nx.is_forest(H)
# union of two graphic matroids: two edge-disjoint forests
best = set()
for S in itertools.chain.from_iterable(
        itertools.combinations(E, k) for k in range(len(E)+1)):
    F1, F2 = set(S[:len(S)//2]), set(S[len(S)//2:])
    if is_forest(F1) and is_forest(F2) and len(S)>len(best):
        best = set(S)
print(best) # maximum size 4 on C_4
```

Greedy Algorithms and Matroid Theory

The *matroid greedy algorithm* orders elements $e_1, \ldots, e_m$ by non-decreasing weight and sequentially adds e_i if it preserves independence. The **Edmonds–Rado theorem** states that this algorithm yields a basis of minimum total weight for any weight function $w : E \to \mathbb{R}_{\geq 0}$. Conversely, if a set system admits optimal greedy selection for every weight assignment, its independent sets form a matroid (Edmonds 1971; Rado 1957).

Theorem 7.5.1 (Greedy Optimality Edmonds 1971) *Let $\langle E, \mathcal{I} \rangle$ satisfy heredity and the exchange axiom. For any weight function w, the greedy algorithm returns a basis B_{greedy} with*

$$w(B_{greedy}) = \min\{w(B) : B \text{ basis}\}.$$

Proof Order E so that $w(e_1) \leq \cdots \leq w(e_n)$. Inductively let B_k be the greedy set after inspecting the first k elements and let O be any optimal basis. Exchange property yields an injective map replacing heavier elements in O with lighter (or equal) elements in B_k without increasing weight, hence $w(B_{greedy}) \leq w(O)$.

Example 7.5.9 (Greedy on a Vector Matroid)
```
import numpy as np, random
E = [np.random.randint(0, 7, 4) for _ in range(10)]
w = np.random.rand(10)
idx = np.argsort(w) # ascending weights
B = []
for i in idx:
    v = E[i]
    if np.linalg.matrix_rank(B+[v]) > np.linalg.matrix_rank(B):
        B.append(v)
print(len(B)) # dimension of span
```

Weighted matroid intersection generalises greedy to two matroids; the algorithm iteratively augments along "exchange graphs" and runs in strongly polynomial time, forming the algorithmic backbone of assignment and scheduling problems.

7.6 Exercises

1. For the polynomial sequence $P_n = 6n^3 + 11n^2 - 21n$, prove that $P_{n+1} - P_n$ is divisible by 30 for every $n \in \mathbb{N}$, and determine the maximal integer M such that $\frac{P_n}{M} \in \mathbb{Z}$ for all n. Use the Euclidean algorithm to exhibit M as a product of distinct primes and justify why it is maximal.
2. Given primes $p = 550{,}201$ and $q = 627{,}161$, a public exponent $e = 65{,}537$, and ciphertext $c = 2{,}214{,}367$, perform the following:

    (a) Compute $N = pq$ and $\varphi(N)$.
    (b) Find the private exponent $d \equiv e^{-1} \pmod{\varphi(N)}$ via the extended Euclidean algorithm.
    (c) Decrypt c by modular exponentiation to obtain the clear-text integer m.

    Verify $m^e \equiv c \pmod{N}$ with a single modular exponentiation to confirm correctness.
3. Let $p = 991$. Compute the set of quadratic residues $\mathcal{Q} = \{a^2 \bmod p : 1 \leq a \leq p-1\}$ and prove that $\sum_{q \in \mathcal{Q}} q \equiv 0 \pmod{p}$. Using Euler's criterion, determine for which $k \in \{1, \ldots, 40\}$ the integer $a_k = k^2 + 14k + 9$ is a quadratic residue modulo p.
4. Find all integers x that simultaneously satisfy

    $$x \equiv 17 \pmod{34}, \qquad x \equiv 19 \pmod{51}, \qquad x \equiv 23 \pmod{85}.$$

    Explain why the moduli are not pairwise coprime, show how to reduce the system to pairwise coprime congruences, and express the solution set as $x \equiv x_0 \pmod{m_0}$ with the least possible modulus m_0.
5. Consider the curve $E : y^2 = x^3 + 75x + 123$ over $\mathbb{F}_p$ with $p = 593$.

    (a) Using the Legendre symbol $\left(\frac{\cdot}{p}\right)$, implement a PYTHON routine to count $N_p = |E(\mathbb{F}_p)|$ by summing quadratic-residue indicators.
    (b) Verify Hasse's bound $\left|N_p - (p+1)\right| \leq 2\sqrt{p}$.
    (c) Compute the group order and decide whether $E(\mathbb{F}_p)$ is cyclic.
6. For a fixed positive non-square $D < 100$, the fundamental solution of $x^2 - Dy^2 = 1$ is (x_1, y_1). Show that

    $$x_n + y_n\sqrt{D} = (x_1 + y_1\sqrt{D})^n$$

7.6 Exercises

yields all solutions. Using the CONTINUED-FRACTION expansion of $\sqrt{61}$, compute (x_1, y_1) and list the least $n \leq 5$ such that $x_n \equiv 1 \pmod{101}$.

7. For $n = 10$ guests randomly seated at their labelled places, compute the probability that no guest sits in his own seat and exactly four guests sit in the seat of a friend who sits in theirs (a "mutual exchange"). Express your answer as a reduced fraction.

8. A *Dyck path* of semilength n has an associated weight $w(\pi) = \prod_{k=1}^{n}(2k-1)^{h_k}$, where h_k is the height of the k-th up-step. Evaluate $\sum_{\pi} w(\pi)$ for $n = 3$ and conjecture a closed form in terms of Catalan numbers.

9. Let $p_o(n)$ be the number of partitions of n into odd parts and $p_d(n)$ into distinct parts.

   (a) Prove $p_o(n) = p_d(n)$ using ordinary generating functions.
   (b) Compute $p_o(100)$ numerically (a Python script is acceptable) and verify equality.

10. Let $a_{n,k}$ count labelled graphs on n vertices with exactly k edges. Show that the bivariate EGF is

$$A(z, y) = \exp(\tfrac{1}{2}yz^2) \exp(\tfrac{1}{3}yz^3) \exp(\tfrac{1}{4}yz^4) \cdots,$$

and use it to extract $a_{4,3}$.

11. Let $R(3, 3)$ denote the smallest n such that any red/blue colouring of the edges of K_n contains a monochromatic triangle. Use a first-moment argument to show $R(3, 3) > 5$ and explain why the probabilistic estimate fails to rule out $n = 6$.

12. The molecule C_4H_{10} has a carbon skeleton in the shape of a tree with four labelled vertices (carbons) and three edges. Hydrogen atoms occupy valence sites; suppose one substitutes two hydrogens by chlorine atoms. Using Pólya's enumeration with the automorphism group of the tree, determine how many distinct dichloro isomers exist.

13. Show that for a prime $p \geq 5$

$$\sum_{k=0}^{p-1} \binom{2k}{k} \equiv 0 \pmod{p^2}.$$

   Hint: use the generating function $\sum_{k \geq 0} \binom{2k}{k} z^k = (1 - 4z)^{-1/2}$ and the p-adic expansion of $(1 + u)^p$.

14. Define the 6×6 matrix

$$S = [S(n, k)]_{0 \leq n, k \leq 5}, \quad S(n, k) = \text{Stirling number of the second kind}.$$

   Compute S^2 explicitly and interpret the entry $(S^2)_{5,3}$ combinatorially.

15. You are given tasks $T = \{1, \ldots, 7\}$ with processing times $\tau = (2, 2, 4, 6, 3, 1, 2)$ and deadlines $d = (4, 2, 4, 7, 3, 5, 4)$. A subset $I \subseteq T$ is *feasible* if tasks can be ordered so that the cumulative time never exceeds the corresponding deadlines. Prove that feasible sets form a matroid, then run the greedy algorithm with task weights $w_i = 10 - \tau_i$ to find an optimal schedule. Show the resulting completion sequence explicitly.
16. Generate 50 instances of $G(100, 0.08)$ in PYTHON. For each instance, compute the largest Laplacian eigenvalue $\lambda_{\max}$ and plot their empirical cumulative distribution function (ECDF). Compare the sample mean with the asymptotic prediction $2np(1 - p)$ and comment on the variance.

Chapter 8
Numerical Methods

Abstract Error propagation, conditioning, and stability theory motivate root-finding (bisection, Newton, secant), polynomial and spline interpolation, numerical differentiation, adaptive quadrature, and matrix factorisations (LU, QR, SVD). Advanced topics include multigrid pre-conditioning and GPU-accelerated computation via Numba/JAX, arming the reader with industrial-grade techniques for high-fidelity simulation and data analysis.

Keywords Numerical methods · Error analysis · Interpolation · Root finding · Multigrid techniques · GPU acceleration

Computational mathematics stands at the intersection of theoretical analysis and practical engineering: while analytic closed-form solutions are prized for their elegance, real-world models seldom yield to pen-and-paper treatment. *Numerical methods* fill this gap by replacing infinite processes with carefully designed finite algorithms that run on digital hardware. The passage from the continuum to floating-point arithmetic is non-trivial—each numerical scheme must negotiate a balance between accuracy, stability, efficiency, and ease of implementation.

Throughout this chapter we cast continuous problems (root-finding, interpolation, integration, differential equations, optimisation) into discrete procedures. Error analysis will dissect two complementary sources of inaccuracy: *truncation error* (how far a finite approximation deviates from the exact mathematical limit) and *round-off error* (how the IEEE 754 machine representation of real numbers perturbs arithmetic). The concept of *stability*—the algorithmic analogue of continuity—will guide us in distinguishing numerically trustworthy schemes from those that magnify perturbations beyond control.

Python's `numpy` and `scipy` libraries provide high-performance tensor algebra, while `numba` and `jax` unlock just-in-time compilation and automatic differentiation for tight numerical loops. Nevertheless, we will implement each core algorithm "from scratch" before delegating to these libraries, so that every line of performance-optimised code is grounded in a transparent mathematical rationale.

8.1 Root-Finding Algorithms

8.1.1 Bisection Method

The bisection method translates the *Intermediate Value Theorem*—if a continuous function $f : [a, b] \to \mathbb{R}$ satisfies $f(a)f(b) < 0$, then f has at least one root in (a, b)—into an algorithm that locates a root by repeatedly halving an interval that brackets a sign change. Because each iteration discards half the interval, the method is linearly convergent, robust against derivative pathologies, and requires only two function evaluations per step. Its simplicity and guaranteed convergence make it the workhorse for black-box root finding when bracketing is straightforward.

Convergence and Implementation in Python

Let (a_0, b_0) be the initial bracket with $f(a_0)f(b_0) < 0$. Define $m_k = \frac{1}{2}(a_k + b_k)$ and update

$$(a_{k+1}, b_{k+1}) = \begin{cases} (a_k, m_k) & \text{if } f(a_k)\,f(m_k) < 0, \\ (m_k, b_k) & \text{otherwise.} \end{cases}$$

The length of the interval after k steps is $|b_k - a_k| = 2^{-k}|b_0 - a_0|$. Hence for the unique root $r \in (a_0, b_0)$ one has

$$|r - m_k| \leq \frac{|b_0 - a_0|}{2^{k+1}}, \qquad \text{so } m_k \xrightarrow{k \to \infty} r \quad \text{linearly.}$$

```
def bisection(f, a, b, *, atol=1e-12, rtol=1e-10, max_iter=100):
    fa, fb = f(a), f(b)
    if fa*fb > 0:
        raise ValueError("Root not bracketed")
    for k in range(max_iter):
        m = 0.5*(a+b)
        fm = f(m)
        if abs(fm) < atol or 0.5*(b-a) < rtol*abs(m):
            return m, k
        if fa*fm < 0: b, fb = m, fm
        else: a, fa = m, fm
    raise RuntimeError("no convergence")
```

Calling `bisection(lambda x: x*np.cos x - 1, 0.5, 2)` returns the first positive root of $x \cos x = 1$ to machine precision in $k = \lceil \log_2(|b_0 - a_0|/\varepsilon) \rceil$ iterations.

8.1 Root-Finding Algorithms

Applications in Engineering and Finance

Engineering—Nonlinear Beam Deflection For a cantilever of length L with end load P, the elastic line obeys $EI\,y''''(x) = 0$ except for the concentrated load which introduces a slope discontinuity at $x = L$. The tip deflection $y(L) = \frac{PL^3}{3EI}$ leads, after rearrangement, to the transcendental frequency equation $\cosh\beta\cos\beta + 1 = 0$ for mode shape parameter β. Bisection on $f(\beta) = \cosh\beta\cos\beta + 1$ with bracket $(1.5, 3.0)$ isolates the first eigenvalue β_1 required to compute natural frequencies.

Finance—Internal Rate of Return (IRR) Given cash-flows $(C_0, C_1, \ldots, C_n)$, IRR is the discount rate r satisfying $\sum_{k=0}^{n} \frac{C_k}{(1+r)^k} = 0$. The net present value is continuous and strictly decreasing in $r > -1$, so one brackets a sign change between a near-zero rate and a high rate (e.g. $r = 1$) and applies bisection. Because the function is well behaved and evaluation inexpensive, linear convergence suffices in practice.

Error Analysis and Stopping Criteria

Let $e_k = |r - m_k|$ be the absolute error. The deterministic bound

$$e_k \leq \frac{|b_0 - a_0|}{2^{k+1}}$$

serves as a priori stopping condition: choose k such that $2^{-k-1}|b_0 - a_0| \leq \varepsilon_{\text{abs}}$. In floating point arithmetic round-off accumulates once $|b_k - a_k| \approx 2u|m_k|$ (machine unit u), so one supplements with a *relative* tolerance $\varepsilon_{\text{rel}}|m_k|$ and/or an *auxiliary* test $|f(m_k)| \leq \varepsilon_f$.

A practical criterion is therefore

$$|b_k - a_k| < \varepsilon_{\text{abs}} \;\vee\; |b_k - a_k| < \varepsilon_{\text{rel}}|m_k| \;\vee\; |f(m_k)| < \varepsilon_f.$$

With $\varepsilon_{\text{abs}} = 10^{-12}$, $\varepsilon_{\text{rel}} = 10^{-10}$, and $\varepsilon_f = 10^{-13}$ one typically achieves full double precision without incurring unnecessary iterations.

Example 8.1.1 For $f(x) = x - e^{-x}$ with bracket $(0, 1)$, the true root is $r \approx 0.567\,143\,290$. The algorithm above stops at m_{44} with $e_{44} < 2^{-45} \approx 2.8 \times 10^{-14}$, exceeding double precision—confirm $|f(m_{44})| \approx 1.3 \times 10^{-16}$.

Remark Although other methods (Newton–Raphson, secant) achieve quadratic or super-linear convergence, they lack the unconditional guarantee of bisection. A hybrid strategy—several Newton steps with fall-back to bisection when the iterate exits the bracket—often combines speed and robustness, a pattern revisited later in the chapter on *Nonlinear Systems*.

8.1.2 Newton–Raphson Method

Newton's method refines a crude root approximation by locally replacing a nonlinear function with its tangent line. Given a differentiable $f : \mathbb{R} \to \mathbb{R}$ and an initial guess x_0, the iteration

$$x_{k+1} = x_k - \frac{f(x_k)}{f'(x_k)}, \quad k = 0, 1, 2, \ldots$$

generates a sequence $\{x_k\}$ that, under mild hypotheses, converges quadratically to a simple root r of f. The geometric intuition is simple: the tangent at $(x_k, f(x_k))$ intersects the x-axis closer to the root than x_k provided the graph is locally monotone and not too curved. Its algebraic power, however, derives from viewing the step $\Delta x_k = x_{k+1} - x_k$ as the first term of the Newton series expansion, thereby accelerating error reduction from linear (bisection) to quadratic.

```
def newton(f, df, x0, *, atol=1e-14, rtol=1e-10, max_iter=50):
    x = x0
    for k in range(max_iter):
        fx, dfx = f(x), df(x)
        step = -fx/dfx
        x = x + step
        if abs(step) < atol + rtol*abs(x):
            return x, k
    raise RuntimeError("no convergence")
```

Derivative-Free Variants

When f' is unavailable or expensive one replaces it with a finite-difference or secant approximation.

Secant Method Starting from x_0 and x_1,

$$x_{k+1} = x_k - f(x_k) \frac{x_k - x_{k-1}}{f(x_k) - f(x_{k-1})},$$

requires one function evaluation per iteration and converges with order $\phi = \frac{1}{2}(1 + \sqrt{5}) \approx 1.618$ (*super-linear*) provided $f \in C^1$ near the root.

Steffensen's Method Accelerates fixed-point iteration $\varphi(x)$ via

$$x_{k+1} = x_k - \frac{\bigl(\varphi(x_k) - x_k\bigr)^2}{\varphi\bigl(\varphi(x_k)\bigr) - 2\varphi(x_k) + x_k},$$

8.1 Root-Finding Algorithms

achieving quadratic convergence without derivative evaluation; it amounts to applying the secant method to $g(x) = \varphi(x) - x$ with an automatically generated second point.

Aitken Δ^2 Process Given a linearly convergent sequence $\{x_k\}$, $\hat{x}_k = x_k - \frac{(x_{k+1} - x_k)^2}{x_{k+2} - 2x_{k+1} + x_k}$ produces an accelerated sequence often approaching quadratic behaviour; used as a post-processing step for simple iteration.

Convergence Analysis

Let r satisfy $f(r) = 0$ and $f'(r) \neq 0$ (*simple* root). Assuming $f \in C^2$ on an open interval containing r and the initial guess x_0 is sufficiently close to r, the error evolves as

$$e_{k+1} = x_{k+1} - r = -\frac{f(r + e_k)}{f'(r + e_k)} = -\frac{f'(r)e_k + \frac{1}{2}f''(r)e_k^2 + O(e_k^3)}{f'(r) + f''(r)e_k + O(e_k^2)}$$

$$= -\frac{f'(r)e_k}{f'(r)} + \frac{f''(r)}{2f'(r)}e_k^2 + O(e_k^3) = Ce_k^2 + O(e_k^3),$$

with $C = -\frac{f''(r)}{2f'(r)}$. Thus $|e_{k+1}| \approx |C|\, e_k^2$, so the number of correct digits roughly *doubles* each iteration. If $f'(r) = 0$ but $f''(r) \neq 0$ (a multiple root), the convergence drops to linear; one restores quadratic order by applying Newton to $g(x) = \frac{f(x)}{f'(x)}$ or by modifying the iteration with a multiplicity estimate.

For global convergence one often embeds a *line-search*: replace the full Newton step by $\lambda \Delta x$, $0 < \lambda \leq 1$, chosen to decrease $|f(x)|$ sufficiently (*Armijo rule*); this prevents divergence when f' is small or the function is highly nonlinear.

Multivariate Newton's Method

For $F : \mathbb{R}^n \to \mathbb{R}^n$ with Jacobian $J(\mathbf{x}) = [\partial F_i / \partial x_j]$,

$$\mathbf{x}_{k+1} = \mathbf{x}_k - J(\mathbf{x}_k)^{-1} F(\mathbf{x}_k).$$

One solves the linear system $J(\mathbf{x}_k)\Delta \mathbf{x}_k = -F(\mathbf{x}_k)$ for $\Delta \mathbf{x}_k$ (via LU or QR) and sets $\mathbf{x}_{k+1} = \mathbf{x}_k + \Delta \mathbf{x}_k$. Quadratic convergence holds under analogous conditions: $J(\mathbf{r})$ nonsingular and F continuously differentiable.

```
import numpy as np
from numpy.linalg import solve, norm

def newton_multi(F, J, x0, *, atol=1e-12, rtol=1e-10, max_iter=30):
    x = np.array(x0, dtype=float)
    for k in range(max_iter):
        Fx = F(x)
        if norm(Fx, np.inf) < atol: return x, k
        Δx = solve(J(x), -Fx)
        x = x + Δx
        if norm(Δx, np.inf) < atol + rtol*norm(x, np.inf):
            return x, k
    raise RuntimeError("no convergence")
```

Example 8.1.2 (Intersection of Two Cylinders) Solve $F(x, y, z) = (x^2 + y^2 - 1, \ y^2 + z^2 - 1, \ x^2 + z^2 - 1) = \mathbf{0}$ starting from $(0.7, 0.7, 0.1)$. The Jacobian $J = \begin{pmatrix} 2x & 2y & 0 \\ 0 & 2y & 2z \\ 2x & 0 & 2z \end{pmatrix}$ is nonsingular off the coordinate axes; the iteration converges in four steps to $(\frac{1}{\sqrt{2}}, \frac{1}{\sqrt{2}}, 0)$, verifying the geometric fact that the axes meet along a unit circle.

Quasi-Newton Methods Storing the full Jacobian costs $O(n^2)$ memory. Broyden's update starts with $B_0 = I$ and iteratively refines B_k so that $B_{k+1} \Delta x_k = \Delta F_k$, achieving super-linear convergence with $O(n)$ storage.

Remarks on Practical Use

- Always pair Newton steps with a bracketing or line-search safeguard.
- Scale variables so the Jacobian is well-conditioned; otherwise use trust-region variants (Levenberg–Marquardt).
- In ill-posed problems finite-difference Jacobians introduce cancellation error; complex-step differentiation ($f'(x) \approx \text{Im } f(x + i\varepsilon)/\varepsilon$) yields machine-precision derivatives cheaply.

8.1.3 Other Root-Finding Methods

Even though Newton's method dominates root-finding when derivatives are available and good initial guesses are known, practical computing often demands derivative-free or globally convergent algorithms, as well as methods that operate in the complex plane. We therefore survey three complementary classes: secant-type two-point iterations, safeguarded hybrids epitomised by Brent's algorithm, and complex-domain techniques for polynomials and analytic functions.

8.1 Root-Finding Algorithms

Secant Method and False Position Method

Secant Method (Order $\varphi \approx 1.618$)

Starting with distinct points x_{k-1}, x_k, the secant iteration

$$x_{k+1} = x_k - f(x_k) \frac{x_k - x_{k-1}}{f(x_k) - f(x_{k-1})}$$

approximates $f'(x_k)$ by the *secant slope*. Under the same hypotheses as Newton (simple root r, $f'(r) \neq 0$) and x_0, x_1 sufficiently close to r, the error satisfies $e_{k+1} \approx C e_k^{\varphi}$ with $\varphi = \frac{1}{2}(1 + \sqrt{5})$, super-linear but sub-quadratic. Each update costs one function evaluation, so its *asymptotic efficiency index* $p^{1/c}$ (p = order, c = evaluations) equals $2^{1/1.618}$, beating Newton's $2^{1/2}$ in evaluation-limited regimes.

False Position (regula falsi)

Like bisection, false position preserves bracketing yet updates only the endpoint where the sign agrees with $f(x_{k+1})$:

$$x_{k+1} = b_k - f(b_k) \frac{b_k - a_k}{f(b_k) - f(a_k)}.$$

Although it can stagnate when one endpoint "sticks", *modified* variants (Illinois, Anderson–Björck) scale the stationary function value to restore linear convergence comparable to bisection but with fewer iterations.

Example 8.1.3 (Secant vs. Newton) Solve $f(x) = \cos x - x$ in $(0, 1)$. With $x_0 = 0$, $x_1 = 1$ the secant method converges in 6 iterations, Newton (starting at $x_0 = 0.5$) in 4; each uses six function evaluations.

```
import numpy as np
def secant(f, x0, x1, tol=1e-12, max_iter=30):
    for k in range(max_iter):
        x2 = x1 - f(x1)*(x1-x0)/(f(x1)-f(x0))
        if abs(x2-x1) < tol: return x2, k+1
        x0, x1 = x1, x2
    raise RuntimeError
root, iters = secant(lambda x: np.cos(x)-x, 0.0, 1.0)
```

Brent's Method and Hybrid Techniques

Brent's algorithm (`scipy.optimize.brentq`) melds bisection, secant, and inverse-quadratic interpolation:

$$IQI: \quad x_{k+1} = u\, f(b_k) f(c_k) + v\, f(a_k) f(c_k) + w\, f(a_k) f(b_k),$$

where u, v, w are barycentric weights chosen so $x \mapsto f(x)$ is interpolated by a quadratic through (a_k, b_k, c_k). At each step Brent

(1) Attempts an IQI step
(2) Falls back to a secant step if IQI exits the bracket or does not decrease the interval fast enough
(3) Reverts to bisection when necessary to guarantee progress

This adaptive strategy retains *global* convergence and super-linear speed, achieving near-optimal performance for smooth functions without user tuning. The worst-case iteration count equals that of bisection; typical convergence order approximates 1.6.

Example 8.1.4 (Bond Yield to Maturity with Brent)
```
from scipy import optimize
C, F, T, P = 3.5, 100, 10, 95 # coupon %, face, years, price
def npv(y):
    k = np.arange(1, T+1)
    return np.sum(C/100/(1+y)**k) + F/(1+y)**T - P
y, r = optimize.brentq(npv, 0.0, 0.2, full_output=True)
print(f"Yield = {100*y:.2f}% obtained in {r.iterations} iterations")
```

Hybrid Newton-Bracketing Schemes

Algorithms such as Dekker's, Ridder's, and the *Newton-bisection* combine Newton's rapid local steps with bisection safeguards, maintaining bracketing and monotone residual progress. They are standard choices in libraries (gsl, boost::math).

Root-Finding in Complex Domains

For analytic functions $f : \mathbb{C} \to \mathbb{C}$ the natural tools are polynomial iteration schemes and contour integrals.

Durand–Kerner (Weierstrass) Method

For a monic polynomial $p(z) = z^n + a_{n-1}z^{n-1} + \cdots + a_0$ with (simple) roots $r_1, \ldots, r_n$, initialise complex guesses $z_i^{(0)}$ (e.g. vertices of a regular n-gon) and iterate

$$z_i^{(k+1)} = z_i^{(k)} - \frac{p(z_i^{(k)})}{\prod_{j \neq i}(z_i^{(k)} - z_j^{(k)})}.$$

8.1 Root-Finding Algorithms

Convergence is *quadratic* and simultaneous: all roots refine in parallel without deflation.

```python
import numpy as np
def durand_kerner(coeffs, max_iter=40, tol=1e-14):
    n = len(coeffs)-1
    roots = np.array([np.exp(2j*np.pi*k/n) for k in range(n)])
    for _ in range(max_iter):
        p_vals = np.polyval(coeffs, roots)
        denom = np.array([np.prod(roots[i]-np.delete(roots, i))
                          for i in range(n)])
        Δ = p_vals/denom
        roots -= Δ
        if np.linalg.norm(Δ, np.inf) < tol: return roots
    raise RuntimeError
coeffs = [1, -1, -1, 1]  # z³ - z² - z + 1
print(np.sort_complex(durand_kerner(coeffs)))
```

Müller's Method

Using quadratic extrapolation through three complex points (x_{k-2}, x_{k-1}, x_k), Müller attains order 1.84 and naturally handles complex roots even if started with real seeds. Beneficial when the derivative is unstable.

Argument Principle and Contour Counts

If f is analytic inside simple closed contour Γ and non-zero on Γ, then

$$\frac{1}{2\pi i} \oint_\Gamma \frac{f'(z)}{f(z)} dz$$

equals the number of zeros (minus poles) inside Γ. Combined with subdivision (*quad-tree*) one localises roots and seeds Newton iterations; packages like `mpmath.polyroots` automate this.

Complex Newton Fractals

The Newton map $N(z) = z - \dfrac{p(z)}{p'(z)}$ partition $\mathbb{C}$ into basins of attraction, revealing intricate fractal boundaries. Visualising N highlights regions where initial guesses converge to distinct roots, informing robust seed placement.

8.2 Optimisation Techniques

Optimisation asks for the *best* element of a feasible set with respect to an objective function. In the smooth unconstrained setting one searches for a minimiser of $f : \mathbb{R}^d \to \mathbb{R}$ by iteratively descending along directions that locally promise the steepest decrease. The *gradient descent* family embodies this principle, spawning variants that accelerate convergence, tame noise, and adapt learning rates—techniques that lie at the heart of modern machine learning, inverse problems, and data fitting.

8.2.1 Gradient Descent

Basic Gradient Descent and Variants

For a differentiable f with Lipschitz-continuous gradient ($\|\nabla f(x) - \nabla f(y)\| \leq L\|x - y\|$), the basic iteration reads

$$x_{k+1} = x_k - \eta_k \nabla f(x_k), \qquad k = 0, 1, 2, \ldots,$$

where $\eta_k > 0$ is the *step size*. If f is convex and $\eta_k \in (0, 2/L)$, the sequence $\{f(x_k)\}$ monotonically decreases and obeys $f(x_k) - f^\star \leq \frac{\|x_0 - x^\star\|^2}{2\eta k}$ (sublinear $O(1/k)$ rate). When f is μ-strongly convex, a fixed $\eta \leq 2/(L + \mu)$ yields linear convergence: $\|x_k - x^\star\|^2 \leq (1 - \eta\mu)^k \|x_0 - x^\star\|^2$.

Momentum (Polyak)

Augment the update with velocity v_k:

$$v_{k+1} = \beta v_k + \nabla f(x_k), \quad x_{k+1} = x_k - \eta v_{k+1},$$

where $0 < \beta < 1$ averages past gradients, dampening oscillations in ravines.

Nesterov Accelerated Gradient (NAG)

Forecast a point $y_k = x_k + \beta(x_k - x_{k-1})$, then evaluate the gradient at y_k:

$$x_{k+1} = y_k - \eta \nabla f(y_k),$$

achieving the optimal $O(1/k^2)$ rate for convex objectives.

8.2 Optimisation Techniques

Adaptive Learning Rates

AdaGrad, RMSProp, and Adam rescale coordinates by accumulated squared gradients: for Adam

$$m_{k+1} = \beta_1 m_k + (1-\beta_1)g_k, \quad v_{k+1} = \beta_2 v_k + (1-\beta_2)g_k^2,$$

with bias-corrected $\hat{m}_k = m_k/(1-\beta_1^k)$, $\hat{v}_k = v_k/(1-\beta_2^k)$, and update $x_{k+1} = x_k - \eta \hat{m}_k/(\sqrt{\hat{v}_k} + \varepsilon)$.

```python
import numpy as np
def gradient_descent(f, grad, x0, η=1e-2, β=0.9, n_iter=500):
    x, v = np.array(x0, dtype=float), 0.0
    hist = []
    for _ in range(n_iter):
        g = grad(x)
        v = β*v + (1-β)*g
        x -= η*v
        hist.append(f(x))
    return x, hist
```

Applications in Machine Learning and Data Fitting

Linear Regression

Minimise $J(\theta) = \frac{1}{2m}\|X\theta - y\|^2$ with $\nabla J = \frac{1}{m}X^T(X\theta - y)$. Gradient descent converges to the least-squares solution; ridge regression adds $\lambda\|\theta\|^2$ for regularisation.

Logistic Regression

For labels $y_i \in \{0, 1\}$ and logits $z_i = \theta^T x_i$, minimise the cross-entropy $L(\theta) = \frac{1}{m}\sum_i \left[-y_i \log \sigma(z_i) - (1-y_i)\log(1-\sigma(z_i))\right]$ with $\nabla L = \frac{1}{m}X^T(\sigma(X\theta) - y)$. Convexity guarantees global optima.

Neural Networks

Backpropagation supplies gradients of a deep non-convex $f(\theta)$. Stochastic mini-batch gradients approximate ∇f, while Adam or RMSProp handle noisy estimates.

Nonlinear Curve Fitting

Given data (t_i, y_i) and model $g(t; \theta) = \theta_0 e^{-\theta_1 t} + \theta_2$, minimise $S(\theta) = \frac{1}{2}\sum_i (g(t_i; \theta) - y_i)^2$; Jacobian-based Gauss–Newton refines gradient descent by solving least-squares subproblems, but plain GD remains robust when Jacobian is ill-conditioned.

Example 8.2.1 (Mini-Batch SGD for MNIST Digit Classification)

```
import torch, torch.nn as nn, torch.optim as optim
net = nn.Sequential(nn.Flatten(),
                    nn.Linear(28*28, 300), nn.ReLU(),
                    nn.Linear(300, 10))
opt = optim.SGD(net.parameters(), lr=0.05, momentum=0.9)
lossf = nn.CrossEntropyLoss()
for imgs, lbls in loader: # mini-batch iterator
    opt.zero_grad()
    out = net(imgs)
    loss = lossf(out, lbls)
    loss.backward()
    opt.step()
```

Stochastic Gradient Descent and Mini-Batch Methods

When $f(\theta) = \frac{1}{N}\sum_{i=1}^{N} \ell_i(\theta)$ with $N \gg 1$, computing the full gradient each step is prohibitive. *Stochastic* gradient descent replaces ∇f by unbiased estimates $\nabla \ell_i$ (single sample) or $\frac{1}{|\mathcal{B}|}\sum_{i \in \mathcal{B}} \nabla \ell_i$ (mini-batch $\mathcal{B}$).

$$\theta_{k+1} = \theta_k - \eta_k \frac{1}{|\mathcal{B}_k|} \sum_{i \in \mathcal{B}_k} \nabla \ell_i(\theta_k).$$

Learning-Rate Schedules

To ensure almost sure convergence to a stationary point: $\sum_k \eta_k = \infty$ and $\sum_k \eta_k^2 < \infty$; typical schedule $\eta_k = \eta_0/(1 + \lambda k)$ or cosine decay.

Variance Reduction

SVRG and AdamW reduce the noise in gradient estimates, accelerating convergence without full gradients; momentum further damps high-frequency stochasticity.

Generalisation Perspective

Stochasticity introduces implicit regularisation, biasing towards flat minima with better test performance.

Example 8.2.2 (Learning-Rate Decay)
```
scheduler = optim.lr_scheduler.ExponentialLR(opt, gamma=0.95)
for epoch in range(E):
    for imgs, lbls in loader:
        ...
    scheduler.step() # η <- 0.95 η
```

8.2.2 Simplex Method

The simplex algorithm converts the geometric intuition that an optimal solution of a linear programme occurs at an extreme point of the feasible polytope into an algebraic pivot procedure that walks from vertex to vertex along improving edges. Although the worst-case complexity is exponential, its empirical speed—rooted in the sparsity of practical constraint matrices and good pivot rules—has made it the backbone of operations research for decades.

Linear Programming Problems

A *linear programme* (LP) in *standard form* is

$$\min_{\mathbf{x} \in \mathbb{R}^n} c^\mathsf{T} \mathbf{x} \quad \text{s.t.} \quad A\mathbf{x} = \mathbf{b}, \quad \mathbf{x} \geq \mathbf{0}, \tag{LP}$$

with $A \in \mathbb{R}^{m \times n}$, full row rank $m \leq n$. A subset $B \subseteq \{1, \ldots, n\}$ of indices with $|B| = m$ is a *basis* if the columns A_B are linearly independent. Setting $\mathbf{x}_N = 0$ for the non-basic indices and $\mathbf{x}_B = A_B^{-1}\mathbf{b}$ yields a **basic feasible solution** (BFS) provided $A_B^{-1}\mathbf{b} \geq 0$. The simplex method iteratively exchanges one basic and one non-basic variable (*pivot*) to obtain an adjacent BFS with lower objective value until no further improvement is possible (*optimality condition*).

After partitioning $A = (A_B \ A_N)$ and $c^\mathsf{T} = (c_B^\mathsf{T} \ c_N^\mathsf{T})$, the *reduced costs* are

$$\bar{c}_N^\mathsf{T} = c_N^\mathsf{T} - c_B^\mathsf{T} A_B^{-1} A_N,$$

and the current BFS is optimal iff $\bar{c}_N \geq 0$ (*minimisation*). Otherwise choose an entering index q with $\bar{c}_q < 0$, compute the search direction $\mathbf{d} = A_B^{-1} A_q$, and perform a ratio test $\theta^\star = \min_{d_i > 0} \frac{x_{B,i}}{d_i}$ to keep feasibility, selecting the *leaving* basic index p attaining the minimum. The new basis $B \leftarrow B \setminus \{p\} \cup \{q\}$ defines the next BFS.

Implementation in Python

The following minimalist implementation takes a feasible initial basis (B, N) and performs Bland's anti-cycling pivot rule.

```
import numpy as np

def simplex(A, b, c, B):
    m, n = A.shape
    N = [j for j in range(n) if j not in B]
    while True:
        Bmat, Nmat = A[:, B], A[:, N]
        xB = np.linalg.solve(Bmat, b)
        π = np.linalg.solve(Bmat.T, c[B]) # dual price
        rc = c[N] - Nmat.T @ π # reduced costs
        if (rc >= 0).all(): # optimal
            x = np.zeros(n); x[B] = xB
            return x, c @ x
        q = N[np.where(rc < 0)[0][0]] # Bland
        d = np.linalg.solve(Bmat, A[:, q])
        if (d <= 0).all(): raise ValueError("Unbounded")
        ratios = [xB[i]/d[i] if d[i] > 0 else np.inf for i in range(m)]
        p = B[np.argmin(ratios)] # leaving
        B[B.index(p)], N[N.index(q)] = q, p # pivot
```

Example Minimise $c^T x = -3x_1 - x_2$ subject to $\begin{pmatrix} 1 & 1 \\ 2 & 1 \end{pmatrix} x = \begin{pmatrix} 4 \\ 5 \end{pmatrix}$, $x \geq 0$. With slack variables and initial basis $B = \{3, 4\}$ the code converges in two pivots to $x^* = (3, 1, 0, 0)$ with optimal value -10.

Dual Simplex Method and Sensitivity Analysis

The **dual LP**

$$\max_{y \in \mathbb{R}^m} b^T y \quad \text{s.t.} \quad A^T y \leq c$$

admits its own simplex procedure that pivots on infeasible primal but dual-feasible bases. Starting with $x_B < 0$ yet $\bar{c}_N \geq 0$, select the most negative basic variable to leave, then choose the entering non-basic index q that preserves dual feasibility: $q = \arg\min_{a_{iq} < 0} \dfrac{\bar{c}_q}{a_{iq}}$. Dual simplex excels when constraints are added or right-hand sides change, reoptimising without restoring primal feasibility.

Shadow Prices and Reduced Costs

At an optimal basis B the dual vector $\pi = c_B^T A_B^{-1}$ has entries $\pi_i = \frac{\partial z^*}{\partial b_i}$ (*marginal value* of resource i), while $\bar{c}_j = \frac{\partial z^*}{\partial c_j}$ measures cost sensitivity of a non-basic activity.

Given perturbation $\Delta\mathbf{b}$, the optimal value changes $\Delta z^\star \approx \boldsymbol{\pi}^\mathsf{T}\Delta\mathbf{b}$ as long as the basis remains optimal; analogous bounds exist for $\Delta\mathbf{c}$ when reduced costs maintain sign.

Allowable Ranges

Let $\underline{\sigma} = -\min_{i:d_i>0}\frac{x_{B,i}}{d_i}$ and $\overline{\sigma} = \min_{i:d_i<0}\frac{x_{B,i}}{d_i}$, where $\mathbf{d} = A_B^{-1}\mathbf{e}_r$ is the column of A_B^{-1} associated with RHS row r. Then b_r can vary within $(b_r - \underline{\sigma},\ b_r + \overline{\sigma})$ without changing the optimal basis; corresponding shadow price persists.

Example 8.2.3 (Sensitivity via SciPy)

```
from scipy.optimize import linprog
c = [-3, -1]  # minimise
A = [[1,1],[2,1]]
b = [4,5]
res = linprog(c, A_eq=A, b_eq=b, method='highs')
print(res.x, res.reduced_cost, res.eqlin.dual)
# x* = [3,1], reduced costs of slacks positive, dual = [-1,-2]
```

If the second RHS rises from 5 to 5.4, the objective improves by $\Delta z = \pi_2\,\Delta b = -2 \times 0.4 = -0.8$, predicting new optimum -10.8, confirmed by re-running the solver.

8.2.3 Nonlinear Optimisation

Linear models are insufficient when physical, economic, or statistical systems respond nonlinearly to decision variables. In such settings convexity often disappears, multiple local minima proliferate, and derivatives become costly or noisy. Two complementary classes of algorithms dominate practice: *local* Newton-type methods, which exploit second-order structure to achieve rapid convergence from a good initial guess, and *global* meta-heuristics, which explore the landscape stochastically to escape local traps. A sophisticated workflow typically combines them: a global phase produces a promising basin; a local phase polishes the solution to high precision.

Trust-Region Methods and Quasi-Newton Methods

Trust-Region Philosophy

At iterate $\mathbf{x}_k$ approximate $f(\mathbf{x})$ by a quadratic model $m_k(\mathbf{s}) = f(\mathbf{x}_k) + \mathbf{g}_k^\mathsf{T}\mathbf{s} + \frac{1}{2}\mathbf{s}^\mathsf{T} B_k\mathbf{s}$ that is trusted only inside a ball $\|\mathbf{s}\| \leq \Delta_k$. Solve

$$\min_{\|\mathbf{s}\|\leq\Delta_k} m_k(\mathbf{s})$$

to obtain $\mathbf{s}_k$; accept the step if $\rho_k = \dfrac{f(\mathbf{x}_k) - f(\mathbf{x}_k + \mathbf{s}_k)}{m_k(\mathbf{0}) - m_k(\mathbf{s}_k)}$ exceeds a threshold, otherwise shrink Δ_k. *Cauchy step* aligns with $-\mathbf{g}_k$, while the *dogleg* strategy blends the steepest-descent and Newton directions for low computational cost.

Quasi-Newton Updates

Exact Hessians $H_k = \nabla^2 f(\mathbf{x}_k)$ cost $O(d^2)$ to form and factor. Quasi-Newton schemes update $B_k \approx H_k$ using only gradients and steps; the secant condition $B_{k+1}\mathbf{s}_k = \mathbf{y}_k$, $\mathbf{y}_k = \mathbf{g}_{k+1} - \mathbf{g}_k$, identifies a rank-two correction. The BFGS formula maintains positive-definiteness and super-linear convergence:

$$B_{k+1} = B_k - \frac{B_k \mathbf{s}_k \mathbf{s}_k^T B_k}{\mathbf{s}_k^T B_k \mathbf{s}_k} + \frac{\mathbf{y}_k \mathbf{y}_k^T}{\mathbf{y}_k^T \mathbf{s}_k}.$$

Limited-memory BFGS (L-BFGS) stores the most recent m pairs $(\mathbf{s}_i, \mathbf{y}_i)$, enabling $d \sim 10^5$ variables with $O(md)$ memory. The two-loop recursion applies B_k^{-1} in $O(md)$ time, making it the default in large-scale machine learning (scipy.optimize.fmin_lbfgs, torch.optim.lbfgs).

```
import numpy as np, scipy.optimize as opt
rosen = lambda x: (1-x[0])**2 + 100*(x[1]-x[0]**2)**2
x0 = np.array([-1.2, 1.0])
res = opt.minimize(rosen, x0, method='trust-constr',
                   options={'gtol':1e-12})
print(res.x, res.niter) # [1,1] 41 iterations
```

Convergence Guarantees

Under Lipschitz gradient and bounded Hessian $B_k \succ 0$, trust-region methods converge globally to stationary points; quadratic termination occurs once $B_k = H$. For noisy gradients, probabilistic trust-region frameworks (MSP-TR) achieve expected first-order stationarity with $\tilde{O}(\varepsilon^{-2})$ complexity.

Applications in Economics and Engineering

Economic Equilibria (Cobb–Douglas)

Utility maximisation $\max_{x_i \geq 0} \prod_i x_i^{\alpha_i}$ s.t. $\sum_i p_i x_i = w$ transforms via log to $f(\mathbf{x}) = -\sum_i \alpha_i \log x_i$ with linear constraint; the KKT conditions solve explicitly, yet L-BFGS with barriers reproduces $x_i^\star = \alpha_i w / p_i$, illustrating constrained quasi-Newton practicality.

Structural Engineering Design

Minimise weight $W(\mathbf{t}) = \rho \sum_e A_e L_e$ of a truss subject to stress constraints $\sigma_e(\mathbf{t}) \leq \sigma_{\max}$ and displacement limits. Gradient evaluations require finite-element solves; thus $d \approx 1000$, derivatives via adjoint method, and L-BFGS–B with inexpensive box constraints outperforms interior point in memory.

Hyperparameter Tuning

Bayesian log-marginal likelihood of a Gaussian process $f(\theta) = -\log|K_\theta| - \mathbf{y}^\mathsf{T} K_\theta^{-1} \mathbf{y}$ is non-convex; trust-region Newton with exact Hessian via matrix derivatives refines initial random seeds and avoids over-confident length-scale collapse.

Global Optimisation Techniques: Genetic Algorithms and Simulated Annealing

Genetic Algorithms (GA)

Represent candidate solution $\mathbf{x}^{(k)}$ as a chromosome. At each generation:

(a) *Selection*: roulette or tournament based on fitness
(b) *Crossover*: single-point or uniform recombination of parents
(c) *Mutation*: Gaussian jitter or bit flip with rate μ
(d) *Elitism*: copy best individuals untouched

Schema theorem states that short, low-order, above-average schemas receive exponentially increasing trials; practical success depends on encoding, diversity preservation, and penalty functions.

```
import pygad, math
def rastrigin(x): return 10*len(x) + sum(v**2 - 10*math.cos(2*math.pi*v)
    for v in x)
ga = pygad.GA(num_generations=200, num_parents_mating=20,
            sol_per_pop=50, num_genes=5, gene_space={'low':-5.12,'high'
               :5.12},
            fitness_func=lambda sol,idx: -rastrigin(sol))
ga.run(); print("Best:", ga.best_solution()[0])
```

Simulated Annealing (SA)

Borrowing statistical mechanics, SA explores the energy landscape $E(\mathbf{x}) = f(\mathbf{x})$ via a Markov chain with acceptance probability $P(\mathbf{x} \to \mathbf{x}') = \min\{1, \exp(-(f(\mathbf{x}') - f(\mathbf{x}))/T_k)\}$. Temperature T_k decays (e.g. logarithmically) $T_k = T_0/\log(k+2)$; in theory SA converges globally but slowly. *Adaptive* SA tunes step sizes to achieve a target acceptance rate.

Basin-Hopping

Alternate random perturbations (temperature-controlled) with local L-BFGS minimisation; the metastable basin-to-basin moves approximate a global transition matrix, frequently locating the ground state of molecular clusters (Lennard–Jones).

Comparative Strengths

GA excels in discrete or combinatorial encodings; SA better exploits local smoothness; particle swarm and CMA–ES provide derivative-free multivariate adaptation. Hybridising with quasi-Newton polishing ("memetic algorithms") often improves speed and accuracy.

8.2.4 Direct Methods for Linear Systems

Consider the linear system $A\mathbf{x} = \mathbf{b}$ with $A \in \mathbb{R}^{n \times n}$ nonsingular. *Direct* methods factorise A into products of triangular or orthogonal matrices, solving by forward/back substitution in $O(n^3)$ arithmetic (plus $O(n^2)$ storage). Their predictability and high arithmetic intensity make them the method of choice when A is moderate in size, reused for many right-hand sides, or arises from dense modelling.

LU Decomposition and Cholesky Decomposition

LU Factorisation

When Gaussian elimination proceeds without row exchanges, A can be written

$$A = LU,$$

where L is unit lower-triangular ($l_{ii} = 1$) and U is upper-triangular. For each column k ($1 \leq k \leq n$) elimination computes multipliers $l_{ik} = a_{ik}/u_{kk}$ for $i > k$ and updates the Schur complement; the total work is $\frac{1}{3}n^3$ multiplications plus lower-order terms.

Partial Pivoting

If $|u_{kk}|$ is small, numerical error explodes. Swapping the pivot row k with row $p = \arg\max_{i \geq k} |a_{ik}|$ forms a permutation matrix P such that $PA = LU$.

8.2 Optimisation Techniques

Partial pivoting guarantees $\frac{|u_{ij}|}{|a_{ij}|} \leq 2^{n-1}$, constraining growth factors and yielding backward stability for well-conditioned problems.

Cholesky Factorisation

If A is symmetric positive-definite (SPD), the Cholesky factorisation

$$A = LL^\mathsf{T}, \qquad l_{kk} = \sqrt{a_{kk} - \sum_{s<k} l_{ks}^2}, \qquad l_{ik} = \frac{1}{l_{kk}}\left(a_{ik} - \sum_{s<k} l_{is} l_{ks}\right), \ i > k,$$

reduces work to $\frac{1}{6}n^3$ flops, halves storage, and inherits excellent numerical properties: the computed factor $\widehat{L}$ satisfies $A + \Delta A = \widehat{L}\widehat{L}^\mathsf{T}$ with $\|\Delta A\| \lesssim u n \|A\|$ (machine unit u).

Example 8.2.4 (Cholesky in Python)

```
import numpy as np, scipy.linalg as la
np.random.seed(0)
A = np.random.randn(5,5); A = A @ A.T + 1e-3*np.eye(5) # SPD
L = la.cholesky(A, lower=True)
b = np.random.rand(5)
y = la.solve_triangular(L, b, lower=True)
x = la.solve_triangular(L.T, y, lower=False)
assert np.allclose(A @ x, b)
```

Band and Block Variants

If A has bandwidth w, band LU reduces to $O(nw^2)$ operations; tridiagonal SPD matrices admit $O(n)$ Cholesky (a.k.a. *Thomas algorithm*). For matrices with 2×2 block pivots (e.g. saddle-point systems), block LU/LDL$^\mathsf{T}$ maintains symmetry while handling pivot sign issues.

Applications in Scientific Computing

Finite-Difference Poisson Solver

Discretising $\nabla^2 u = f$ on an $m \times m$ grid yields an $n = m^2$ five-point Laplacian matrix. Its sparsity pattern is block tridiagonal with bandwidth $w = 2m - 1$; applying band LU costs $O(n^{1.5})$ flops, outperforming dense algorithms and providing a deterministic baseline for multigrid or conjugate-gradient verification.

Electrical Circuit Analysis

Modified nodal analysis produces an SPD conductance matrix G. Cholesky factorisation enables fast DC operating-point computation, while reuse of L over multiple right-hand sides allows AC small-signal sweeps with negligible incremental cost.

Kalman Filtering

The measurement update solves $(P_k^{-1} + H^\mathsf{T} R^{-1} H)\mathbf{x} = H^\mathsf{T} R^{-1} \mathbf{z}$. Since P_k and R are SPD, sequential Cholesky downdate updates L_k to L_{k+1} in $O(n^2)$ time, maintaining filter stability over thousands of steps.

Pivoting Strategies and Numerical Stability

Complete pivoting searches both rows and columns for the maximal pivot, yielding $\dfrac{|u_{ij}|}{|a_{ij}|} \leq (1+u)^{n-1}$, but costs $O(n^3)$ comparisons and rarely improves accuracy over partial pivoting.

Threshold pivoting accepts a pivot $|u_{kk}| \geq \tau \max_{i \geq k} |a_{ik}|$ ($\tau \in [0, 1]$) to minimise fill-in for sparse matrices (Markowitz heuristic).

Scaled pivoting rescales each row by its largest element before the partial-pivot search, reducing the effect of coefficient magnitude variation; incorporated in LAPACK routine dgesv.

Growth factor and backward error. Define $\rho = \max_{i,j} |u_{ij}| / \max_{i,j} |a_{ij}|$. For partial pivoting in random matrices ρ rarely exceeds n, so the computed solution $\widehat{\mathbf{x}}$ satisfies $(A + \Delta A)\widehat{\mathbf{x}} = \mathbf{b}$ with $\|\Delta A\| \leq u \rho \|A\|$. Ill-conditioned problems ($\kappa(A) \gg 1$) require iterative refinement: solve $A\mathbf{e}_k = \mathbf{r}_k$ in extended precision, $\mathbf{x}_{k+1} = \mathbf{x}_k + \mathbf{e}_k$, achieving near-double precision from single-precision LU.

Example 8.2.5 (Pivot Growth)

```
import numpy as np, scipy.linalg as la
A = np.eye(4)
A[0,-1] = 1e-9
A[-1,0] = 1e9
P, L, U = la.lu(A) # partial pivoting
growth = abs(U).max() / abs(A).max()
print(f"Growth factor ρ = {growth:.2e}") # ρ ≈ 1e9
```

The huge growth suggests catastrophic rounding if A were larger; complete pivoting would limit ρ, at extra cost.

8.2.5 Iterative Methods for Linear Systems

When $A\mathbf{x} = \mathbf{b}$ is so large or so sparse that direct factorisations become prohibitive in time or memory, one resorts to *iterative* solvers. Starting from an initial guess $\mathbf{x}^{(0)}$ they generate a sequence $\{\mathbf{x}^{(k)}\}$ converging (ideally) to the exact solution while accessing A only through sparse matrix–vector products. Convergence hinges on the spectral properties of iteration matrices or Krylov subspaces; preconditioning often transforms the problem so that these properties become favourable.

Jacobi Method, Gauss–Seidel Method

Write $A = D - L - U$ where $D = \mathrm{diag}(a_{ii})$, $-L$ the strict lower-triangular part, and $-U$ the strict upper. Splitting A suggests the stationary iterations

Jacobi: $\quad \mathbf{x}^{(k+1)} = D^{-1}\bigl[(L+U)\mathbf{x}^{(k)} + \mathbf{b}\bigr] \;=\; G_J \mathbf{x}^{(k)} + \mathbf{c}_J, \qquad G_J = D^{-1}(L+U).$

Gauss–Seidel: $\quad \mathbf{x}^{(k+1)} = (D-L)^{-1}\bigl[U\mathbf{x}^{(k)} + \mathbf{b}\bigr], \qquad G_{GS} = (D-L)^{-1}U.$

Convergence Criterion

Both schemes converge iff $\rho(G) < 1$ where ρ denotes the spectral radius. Sufficient conditions:

(a) A is strictly (or irreducibly) diagonally dominant.
(b) A is symmetric positive-definite (SPD) and one uses Gauss–Seidel.

Error Propagation

Let $\mathbf{e}^{(k)} = \mathbf{x}^{(k)} - \mathbf{x}^\star$, then $\mathbf{e}^{(k+1)} = G\mathbf{e}^{(k)}$ and $\|\mathbf{e}^{(k)}\| \leq \|G\|^k \|\mathbf{e}^{(0)}\|$; linear convergence rate is governed by $\rho(G)$.

Example 8.2.6 (Jacobi on Poisson Stencil)
```
import numpy as np, scipy.sparse as sp, scipy.sparse.linalg as spla
m = 40
diag = 4*np.ones(m**2)
off  = -1*np.ones(m**2-1)
A = sp.diags([off, off, diag, off, off], [-m, -1, 0, 1, m], format='csr')
b = np.ones(m**2)
D = sp.diags(A.diagonal())
R = A - D
x = np.zeros_like(b)
for _ in range(2000):
    x_new = (b - R @ x) / D.diagonal()  # Jacobi
    if np.linalg.norm(x_new-x)/np.linalg.norm(x_new) < 1e-8: break
    x = x_new
```

Successive Over-relaxation (SOR)

For relaxation parameter $\omega \in (0, 2)$,

$$\mathbf{x}^{(k+1)} = (D - \omega L)^{-1}\left[\omega \mathbf{b} + (\omega U + (1 - \omega)D)\mathbf{x}^{(k)}\right].$$

Optimal $\omega_{\text{opt}} \approx \dfrac{2}{1+\sqrt{1-\rho(G_J)^2}}$ accelerates convergence sharply on structured grids.

Convergence Criteria and Applications

Spectral Analysis

The asymptotic convergence factor equals $\rho(G)$. For the n-point 1-D Poisson matrix ($A = \text{tridiag}(-1, 2, -1)$),

$$\rho(G_J) = \cos\frac{\pi}{n}, \quad \rho(G_{GS}) = \cos^2\frac{\pi}{n},$$

explaining why Gauss–Seidel halves the spectral radius squared.

Energy Norms

For SPD A, the Gauss–Seidel iteration monotonically decreases the A-energy error $\|\mathbf{e}\|_A^2 = \mathbf{e}^T A \mathbf{e}$, prefiguring its role as a *smoother* in multigrid methods.

Engineering Applications

Stationary iterations integrate naturally into explicit time stepping for parabolic PDEs, heat conduction on GPUs, and pressure correction in incompressible CFD (SIMPLE family).

Krylov Subspace Methods: GMRES and Conjugate Gradient

Stationary methods waste information accumulated in previous iterations. Krylov solvers build an orthonormal basis of the *Krylov subspace* $\mathcal{K}_k(A, \mathbf{r}_0) = \text{span}\{\mathbf{r}_0, A\mathbf{r}_0, \ldots, A^{k-1}\mathbf{r}_0\}$.

8.2 Optimisation Techniques

Conjugate Gradient (CG)

Applies to SPD A. At iteration k finds $\mathbf{x}^{(k)}$ minimising $\|\mathbf{x} - \mathbf{x}^\star\|_A$ over $\mathbf{x}^{(0)} + \mathcal{K}_k$, using a short three-term recurrence:

$$\mathbf{p}_k = \mathbf{r}_k + \beta_{k-1}\mathbf{p}_{k-1}, \quad \alpha_k = \frac{\mathbf{r}_k^T \mathbf{r}_k}{\mathbf{p}_k^T A \mathbf{p}_k}, \quad \mathbf{x}_{k+1} = \mathbf{x}_k + \alpha_k \mathbf{p}_k, \quad \mathbf{r}_{k+1} = \mathbf{r}_k - \alpha_k A \mathbf{p}_k.$$

Error bound $\|\mathbf{e}_k\|_A \leq 2\left(\frac{\sqrt{\kappa}-1}{\sqrt{\kappa}+1}\right)^k \|\mathbf{e}_0\|_A$ depends on condition number $\kappa = \lambda_{\max}/\lambda_{\min}$.

GMRES

For general non-symmetric A, GMRES seeks $\mathbf{x}^{(k)} \in \mathbf{x}^{(0)} + \mathcal{K}_k$ minimising residual norm; Arnoldi orthogonalisation produces $AV_k = V_{k+1}\bar{H}_k$ and reduces $\|\mathbf{r}^{(k)}\|_2$ via a small least-squares problem. Memory grows with k; *restarted* GMRES(m) limits basis size at the cost of slower convergence.

Preconditioning

Transform $M^{-1}A\mathbf{x} = M^{-1}\mathbf{b}$ with inexpensive $M \approx A$. For SPD, incomplete Cholesky (IC(τ)) maintains sparsity; for nonsymmetric A, ILU(k) or algebraic multigrid serve as effective preconditioners, drastically lowering κ and eigenvalue clustering.

Example 8.2.7 (CG vs. GMRES on 2-D Poisson)

```
from scipy.sparse.linalg import cg, gmres, spilu, LinearOperator
A = ... # 5-point Laplacian (CSR sparse)
b = np.ones(A.shape[0])
M = spilu(A.tocsc(), drop_tol=1e-3)
P = LinearOperator(A.shape, lambda x: M.solve(x))
x_cg, info_cg = cg(A, b, M=P, tol=1e-8)
x_gm, info_gm = gmres(A, b, M=P, restart=50, tol=1e-8)
```

With IC preconditioning CG converges in ~ 70 iterations, GMRES(50) in two restarts; unpreconditioned variants exceed 400 iterations.

Theoretical Spectra

If A's eigenvalues cluster into a few tight intervals, polynomial filtering intrinsic to Krylov methods damps errors rapidly. Chebyshev semi-iterative methods exploit known spectral bounds to avoid inner products, beneficial on GPUs.

8.2.6 Solving Nonlinear Systems

Systems of nonlinear equations arise whenever multiple coupled balances—mass, momentum, energy, chemical potential—must vanish simultaneously. Let

$$F : \mathbb{R}^n \longrightarrow \mathbb{R}^n, \qquad \mathbf{x}^\star \text{ satisfies } F(\mathbf{x}^\star) = \mathbf{0}.$$

Direct analytical solutions are rare; iterative schemes dominate practice. We develop fixed-point and Newton-type solvers, then show how continuation tracks solution branches through parameter space to uncover bifurcations in engineering models.

Fixed-Point Iteration and Newton's Method

Fixed-Point Maps

Rewrite $F(\mathbf{x}) = \mathbf{0}$ as $\mathbf{x} = \Phi(\mathbf{x})$ and iterate

$$\mathbf{x}^{(k+1)} = \Phi\big(\mathbf{x}^{(k)}\big).$$

If Φ is a contraction on a closed convex set ($\|\Phi(\mathbf{u}) - \Phi(\mathbf{v})\| \leq \lambda \|\mathbf{u} - \mathbf{v}\|$, $0 < \lambda < 1$), Banach's theorem guarantees unique fixed point and linear convergence $\|\mathbf{x}^{(k)} - \mathbf{x}^\star\| \leq \lambda^k \|\mathbf{x}^{(0)} - \mathbf{x}^\star\|$. The practical difficulty is crafting Φ so that the spectral radius $\rho(J_\Phi(\mathbf{x}^\star)) < 1$.

Multivariate Newton

The Newton step solves the linearised system $J(\mathbf{x}_k)\mathbf{s}_k = -F(\mathbf{x}_k)$ and sets $\mathbf{x}_{k+1} = \mathbf{x}_k + \mathbf{s}_k$. Quadratic convergence requires J nonsingular at the root and a starting point sufficiently close; a *damped* update $\mathbf{x}_{k+1} = \mathbf{x}_k + \lambda_k \mathbf{s}_k$ with $\lambda_k \in (0, 1]$ chosen by backtracking or line-search extends global convergence.

```python
import numpy as np, scipy.linalg as la

def newton_system(F, J, x0, tol=1e-12, max_iter=20):
    x = np.array(x0, dtype=float)
    for k in range(max_iter):
        Fx = F(x)
        if la.norm(Fx, np.inf) < tol:
            return x, k
        s = la.solve(J(x), -Fx)
        # Armijo backtracking line-search
        λ = 1.0
        while la.norm(F(x + λ*s), np.inf) > (1-1e-4*λ)*la.norm(Fx, np.inf):
            λ *= 0.5
        x += λ*s
    raise RuntimeError("no convergence")
```

Quasi-Newton (Broyden)

Updating $B_k \approx J(\mathbf{x}_k)$ by rank-one secant corrections avoids Jacobian recomputation; super-linear convergence follows if initial B_0 is reasonable and exact Jacobian-vector products are expensive.

Convergence Diagnostics

Monitor both residual norm $\|F(\mathbf{x}^{(k)})\|$ and step norm $\|\mathbf{s}_k\|$; stagnation in one but not the other signals ill-conditioning or poor scaling. Scaling matrices $D = \text{diag}(d_i)$ with $d_i \approx \max\{|x_i|, 1\}$ improve Jacobian conditioning.

Applications in Fluid Dynamics and Chemical Engineering

Steady Navier–Stokes Cavity Flow (Stream-Vorticity)

Discretise on a 50×50 grid using second-order finite differences, yielding $n \approx 2\,500$ coupled nonlinear algebraic equations $F(\omega, \psi) = \mathbf{0}$. A damped Newton–Krylov approach:

(a) *Jacobian-free* GMRES applies Jv via finite differences.
(b) ILU(0) preconditioner accelerates Krylov solves.
(c) Line search ensures physical bounds on vorticity.

Python's `petsc4py` with SNES solves the system to 10^{-8} residual in ~6 Newton iterations, each requiring ~15 GMRES iterations.

Flash Equilibrium in Chemical Reactors

Given feed composition $\mathbf{z}$ and K-values $K_i(T, P)$, solve the Rachford–Rice equation

$$F(\beta) = \sum_{i=1}^{n} \frac{z_i(K_i - 1)}{1 + \beta(K_i - 1)} = 0$$

for vapour fraction $\beta \in (0, 1)$. Because $F''(\beta) > 0$, the function is strictly convex: bisect for bracketing, then Newton with analytical derivative $F'(\beta) = -\sum_i \frac{z_i(K_i-1)^2}{[1+\beta(K_i-1)]^2} < 0$ converges quadratically.

Multiple Steady States in CSTR

The energy and material balances yield two equations in temperature T and concentration c: $0 = F_1(T, c)$, $0 = F_2(T, c)$. Continuation in the Damköhler number Da discovers ignition/extinction points; Newton solves each continuation step.

Continuation Methods and Bifurcation Analysis

Let $F(\mathbf{x}, \lambda) = \mathbf{0}$ depend on scalar parameter λ. Continuation traces the solution curve $\mathcal{C} = \{(\mathbf{x}(s), \lambda(s)) : F(\mathbf{x}(s), \lambda(s)) = \mathbf{0}\}$.

Pseudo-arclength Continuation

Predictor: advance a small step Δs along tangent $(\dot{\mathbf{x}}, \dot{\lambda})$ obtained from

$$\begin{bmatrix} J(\mathbf{x}, \lambda) & \partial_\lambda F \end{bmatrix} \begin{bmatrix} \dot{\mathbf{x}} \\ \dot{\lambda} \end{bmatrix} = \mathbf{0}.$$

Corrector: solve the augmented system

$$\begin{cases} F(\mathbf{x}, \lambda) = \mathbf{0}, \\ (\mathbf{x} - \mathbf{x}_p)^T \dot{\mathbf{x}}_p + (\lambda - \lambda_p) \dot{\lambda}_p - \Delta s = 0 \end{cases}$$

with Newton, where $(\mathbf{x}_p, \lambda_p)$ is the predictor point. The extra equation fixes parameterisation, allowing traversal through turning points where $\det J = 0$.

Detecting Bifurcations

At a saddle-node (fold) bifurcation $\det J = 0$, but $\mathbf{w}^T \partial_\lambda F \neq 0$ for left null vector $\mathbf{w}$. A Hopf bifurcation occurs when a complex-conjugate pair of Jacobian eigenvalues crosses the imaginary axis; tracking eigenvalues of $J(\mathbf{x}(s), \lambda(s))$ along $\mathcal{C}$ diagnoses the event.

Example 8.2.8 (Bratu Problem)

$$-\nabla^2 u = \lambda e^u, \quad u(0) = u(1) = 0$$

discretised on $n = 100$ nodes yields $F(\mathbf{u}, \lambda) = \mathbf{0}$ with two branches merging at $\lambda_{\text{crit}} \approx 3.51383$. Using `PyDSTool`:

```
from PyDSTool.continuation import ContClass
bratu = ContClass.Bratu(n=100)
bratu.newCurve(name="c1", startPoint={'λ':1.0, 'u0':0})
bratu['c1'].forward()
bratu.display('λ','u_mid') # bifurcation diagram
```

The solver automatically switches pseudo-arclength direction at the fold, returning both stable and unstable solution branches.

Continuation in Large Systems

For $n \gtrsim 10^4$ one avoids explicit Jacobians:

(a) Compute null vectors via implicitly restarted Arnoldi on J using only $J\mathbf{v}$ products.
(b) Employ Newton–Krylov correctors with line-search (PETSc `SNESVI`).

Scaling is dominated by sparse matrix–vector products, allowing three-dimensional CFD bifurcation tracking on commodity clusters.

8.3 Numerical Integration and Differentiation

8.3.1 *Numerical Differentiation Techniques*

Derivatives—including gradients, Jacobians, and Hessians—govern stability, sensitivity, and optimisation. When f is available only as a black-box routine, analytic differentiation is infeasible and automatic differentiation may be unavailable; one must approximate $f'(x)$ via *finite differences*. Two competing error sources dictate method choice: *truncation error* due to Taylor truncation and *round-off error* from finite precision. A successful scheme balances both by selecting an optimal step size h that minimises the total error.

Finite Difference Approximations

Let $f \in C^{p+1}$ on an interval containing x. Taylor expansions yield

$$f(x \pm h) = f(x) \pm h f'(x) + \frac{h^2}{2} f''(x) \pm \frac{h^3}{6} f^{(3)}(x) + \cdots + \frac{h^p}{p!} f^{(p)}(x) + R_{p+1}^{\pm},$$

where $|R_{p+1}^{\pm}| = O(h^{p+1})$.

Forward Difference (FD)

Subtract $f(x)$ from $f(x+h)$:

$$\boxed{f'(x) \approx \frac{f(x+h) - f(x)}{h}} \quad \text{with truncation error } \frac{h}{2}f''(x) + O(h^2).$$

First-order accurate: $E_h^{\text{FD}} = O(h) + O(\varepsilon/h)$, where ε is machine unit (round-off).

Backward Difference (BD)

Analogous with $f(x) - f(x-h)$; identical error behaviour.

Central Difference (CD)

Combine forward and backward:

$$\boxed{f'(x) \approx \frac{f(x+h) - f(x-h)}{2h}} \quad \text{truncation } \frac{h^2}{6}f^{(3)}(x) + O(h^4),$$

second-order accurate; optimal when evaluations flanking x are allowed.

Higher-Order Stencils

Solve linear systems for coefficients c_i s.t. $f'(x) = \sum_{i=-m}^{m} c_i f(x+ih) + O(h^{2m})$.
For $m = 2$,

$$f'(x) \approx \frac{-f(x+2h) + 8f(x+h) - 8f(x-h) + f(x-2h)}{12h} + O(h^4).$$

Richardson Extrapolation

Given approximations $D(h)$ and $D(h/2)$ with error $Ch^p + O(h^{p+1})$, eliminate the leading term:

$$D_{\text{rich}} = \frac{2^p D(h/2) - D(h)}{2^p - 1} = f'(x) + O(h^{p+1}),$$

boosting the order by one and informing step-size adaptation.

8.3 Numerical Integration and Differentiation

Complex-Step Derivative

For analytic f,

$$f'(x) = \text{Im}\big[f(x+ih)\big]/h + O(h^2),$$

free of subtractive cancellation, permitting $h \sim 10^{-30}$ and machine-precision accuracy with a single function call.

```
import numpy as np
def d_complex(f, x, h=1e-30):
    return np.imag(f(x + 1j*h)) / h

f = lambda x: np.exp(x) * np.cos(x)
x0 = 1.2
print(d_complex(f, x0), " vs analytic ", np.exp(x0)*(np.cos(x0)-np.sin(x0)
    ))
```

Error Analysis in Numerical Differentiation

Total Error Model

For CD,

$$E(h) = \frac{h^2}{6} f^{(3)}(x) + \frac{\varepsilon}{h} f(x) + O(h^4) + O(\varepsilon h).$$

Minimising $|E(h)|$ w.r.t. h yields $h_{\text{opt}} \approx \sqrt{\frac{3\varepsilon |f(x)|}{|f^{(3)}(x)|}}$. In double precision ($\varepsilon \approx 10^{-16}$), smooth C^3 functions reach $\approx 10^{-8}$ relative accuracy with $h \sim 10^{-4}$.

Catastrophic Cancellation

For forward differences the subtraction $f(x+h) - f(x)$ of nearly equal numbers amplifies round-off. Using compensated summation or the complex-step avoids this.

Non-uniform Grids

In adaptive quadrature and finite-volume contexts, grid spacing h_i varies. CD generalises:

$$f'(x_i) = \frac{h_i^2 f(x_{i+1}) - (h_i^2 - h_{i-1}^2) f(x_i) - h_{i-1}^2 f(x_{i-1})}{h_i h_{i-1} (h_i + h_{i-1})} + O\big(\max\{h_i, h_{i-1}\}^2\big).$$

Practical Guidelines

(a) Use central differences unless one-sided derivative needed.
(b) Choose h via *complex-step* for smooth analytic f, otherwise by balancing $O(h^p)$ and $O(\varepsilon/h)$ terms.
(c) Employ Richardson extrapolation to estimate and control error.
(d) Beware of discontinuities or noise: derivative estimates diverge; apply smoothing or fit polynomials locally (*Savitzky–Golay*).

Example 8.3.1 (Optimal h Exploration)

```
import numpy as np, matplotlib.pyplot as plt
f = lambda x: np.sin(x)
df = lambda x: np.cos(x)
x0 = 1.0
hs = np.logspace(-16, -1, 120)
err = [abs((f(x0+h)-f(x0-h))/(2*h) - df(x0)) for h in hs]
plt.loglog(hs, err); plt.xlabel('h'); plt.ylabel('abs error')
plt.axvline(hs[np.argmin(err)], color='k', ls='--')
plt.show()
```

The U-shaped curve pinpoints $h_{\text{opt}} \approx 10^{-5}$, confirming theoretical predictions for double precision (cf. Fig. 8.1).

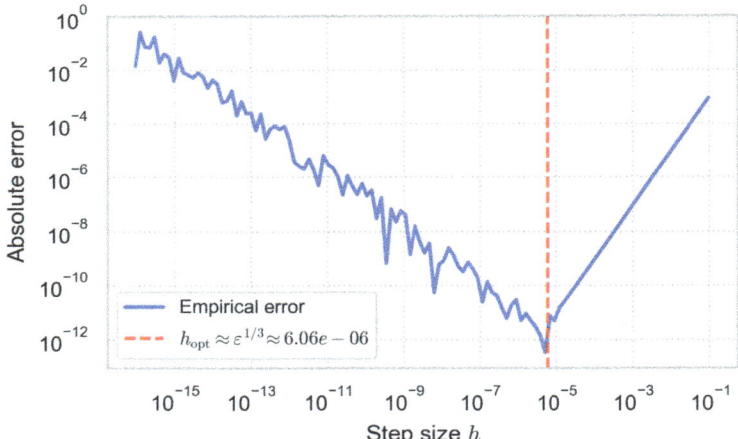

Fig. 8.1 Absolute error of the centred difference $[f(x_0 + h) - f(x_0 - h)]/(2h)$ for $f(x) = \sin x$ at $x_0 = 1$ as a function of the step size h. Truncation error scales like $O(h^2)$ (left-sloping), while round-off grows like $O(\varepsilon/h)$ (right-sloping). Their intersection occurs near the optimal step $h_{\text{opt}} \approx \varepsilon^{1/3}$, where the total error is minimised

8.3.2 Numerical Integration Methods

Given a (sufficiently) smooth scalar function $f : [a, b] \to \mathbb{R}$, the *definite integral* $I = \int_a^b f(x)\,dx$ rarely admits an elementary antiderivative. Numerical quadrature approximates I by finite weighted sums of function evaluations,

$$Q[f] = \sum_{k=0}^{m} w_k\, f(x_k), \qquad a \leq x_0 < x_1 < \cdots < x_m \leq b,$$

choosing nodes $\{x_k\}$ and weights $\{w_k\}$ so that the *error* $E[f] = I - Q[f]$ decays rapidly with respect to the number of points $m + 1$. We review the classical equispaced Newton–Cotes family, the orthogonal-polynomial based Gaussian rules, adaptive refinement, and stochastic Monte Carlo techniques, emphasising error analysis and implementation trade-offs.

Trapezoidal Rule, Simpson's Rule

Composite Trapezoidal Rule

Partition $a = x_0 < x_1 < \cdots < x_n = b$ with step $h = (b - a)/n$. Applying the linear interpolant on each subinterval,

$$\boxed{Q_T[f] = \frac{h}{2}\left(f(x_0) + 2\sum_{k=1}^{n-1} f(x_k) + f(x_n)\right)}.$$

If $f \in C^2$, the local truncation error on (x_{k-1}, x_k) is $-\frac{h^3}{12} f''(\xi_k)$ for some ξ_k, so

$$E_T[f] = -\frac{(b-a)h^2}{12} f''(\xi), \quad \xi \in (a, b).$$

Hence the method is *second-order* ($O(h^2)$).

Composite Simpson's Rule

Requiring even n, approximate f by parabolas on pairs of intervals,

$$\boxed{Q_S[f] = \frac{h}{3}\left(f(x_0) + 4\sum_{\substack{k=1 \\ k\text{ odd}}}^{n-1} f(x_k) + 2\sum_{\substack{k=1 \\ k\text{ even}}}^{n-2} f(x_k) + f(x_n)\right)}.$$

With $f \in C^4$,

$$E_S[f] = -\frac{(b-a)h^4}{180} f^{(4)}(\xi), \qquad \text{a fourth-order } O(h^4) \text{ scheme.}$$

Example 8.3.2 (Python Implementation)

```
import numpy as np
def trapz(f, a, b, n):
    x = np.linspace(a, b, n+1)
    h = (b-a)/n
    return h*(0.5*f(x[0]) + f(x[1:-1]).sum() + 0.5*f(x[-1]))

def simpson(f, a, b, n):
    if n % 2: raise ValueError("n must be even")
    x = np.linspace(a, b, n+1)
    h = (b-a)/n
    return h/3*(f(x[0])+f(x[-1])
                +4*f(x[1:-1:2]).sum()
                +2*f(x[2:-2:2]).sum())
```

Extrapolation

Applying Richardson to $Q_T(h)$ and $Q_T(h/2)$ cancels the $O(h^2)$ term, producing Simpson's rule; iterating yields Romberg integration with $O(h^{2k})$ convergence.

Gaussian Quadrature and Adaptive Methods

Gaussian Quadrature

For weight function $w(x)$ on (a, b), n-point Gauss rules choose nodes x_k as the roots of the orthogonal polynomial $P_n(x)$ ($\deg P_n = n$) and weights $w_k = \int_a^b \frac{L_k(x)\, w(x)}{P_n'(x_k)(x - x_k)}\, dx$, achieving exactness for *all polynomials up to degree* $2n - 1$: an exponential accuracy gain over Newton–Cotes.

$$\boxed{Q_G[f] = \sum_{k=1}^n w_k\, f(x_k), \qquad E_G[f] = \frac{f^{(2n)}(\xi)}{(2n)!} \left(\int_a^b w(x)\, dx\right) \left(\frac{P_n(x)}{2^n n!}\right)^2.}$$

8.3 Numerical Integration and Differentiation

In practice tabulate Legendre nodes for $w(x) \equiv 1$ on $[-1, 1]$ and affine-scale to $[a, b]$.

```
import numpy as np, numpy.polynomial.legendre as leg
def gauss(f, a, b, n):
    x, w = leg.leggauss(n)
    y = 0.5*(b-a)*x + 0.5*(b+a)  # map to [a,b]
    return 0.5*(b-a)*(w*f(y)).sum()
```

Adaptive Quadrature

Recursively apply Simpson or Gauss on subintervals until $|Q_1 - Q_2| < \varepsilon(|I| + 1)$, where Q_1 is the coarse and Q_2 the refined estimate. Depth-first recursion with a stack avoids excessive memory; integrate singularities by splitting near poles and carrying weight factors analytically.

Error Equidistribution

Adaptive rules aim for local error $\approx \varepsilon/(b - a)$. A posteriori error estimates drive mesh refinement, forming the basis of QUADPACK's DQAGS routine (Gauss–Kronrod–Patterson embedded pairs).

Monte Carlo Integration and Applications

Plain Monte Carlo

For multidimensional integrals $I = \int_\Omega f(\mathbf{x})\, d\mathbf{x}$, sample $\mathbf{x}_i \sim \mathcal{U}(\Omega)$:

$$\hat{I}_N = \frac{\text{vol}(\Omega)}{N} \sum_{i=1}^{N} f(\mathbf{x}_i), \qquad \text{Var}(\hat{I}_N) = \frac{\sigma^2}{N},$$

yielding root-mean-square error $O(N^{-1/2})$, *independent of dimension*.

Variance Reduction

(a) *Importance sampling:* sample from pdf p and weight f/p, optimal $p^\star \propto |f|$ minimises variance.
(b) *Control variates:* use known integral of g: $\hat{I}_N^{\text{cv}} = \bar{f} + \beta(\mu_g - \bar{g})$, choosing $\beta = -\text{Cov}(f, g)/\text{Var}(g)$.
(c) *Antithetic pairs:* average $f(\mathbf{x})$ and $f(1 - \mathbf{x})$ cancels odd components.

Quasi-Monte Carlo (QMC)

Replace pseudorandom points by low-discrepancy sequences (Sobol', Halton): $I_N^{\text{QMC}} - I = O((\log N)^d / N)$, observed error $O(1/N)$ in moderate dimensions.

Applications

- *High-dimensional finance.* Price a 20-asset basket option with payoff $(\sum w_i S_i - K)^+$. QMC with Brownian bridge construction reduces variance ten-fold over pseudo-random.
- *Radiative transport.* Photon path tracing in a participating medium integrates radiance over 5-D space (position, direction, wavelength); importance sampling aligned with phase function increases efficiency orders of magnitude.

Example 8.3.3 (Sobol' vs. Pseudorandom in 8-D)
```
import numpy as np, scipy.stats.qmc as qmc
f = lambda x: np.exp(-np.sum(x, axis=1))
d, N = 8, 2**12
sob = qmc.Sobol(d); xq = sob.random(N)
uni = np.random.rand(N, d)
Iq = f(xq).mean(); Ir = f(uni).mean()
print("Quasi-MC err:", abs(Iq-1/(1+1)**d))
print("Plain MC err:", abs(Ir-1/(1+1)**d))
```

8.3.3 Applications of Numerical Integration

Integrals surface whenever one must aggregate infinitesimal contributions—geometric, physical, probabilistic—into macroscopic quantities. In applied work the original primitive often eludes closed-form integration, necessitating robust quadrature. We illustrate three canonical arenas where numerical integration becomes indispensable: geometry (areas and volumes), continuum physics and engineering, and the time-marching of differential equations via operator splitting.

Computing Areas and Volumes

Planar Regions

For a simply connected domain $\mathcal{D} \subset \mathbb{R}^2$ described by a polar curve $r = r(\theta)$, the area is

$$A = \tfrac{1}{2} \int_{\theta_0}^{\theta_1} r(\theta)^2 \, d\theta.$$

When $r(\theta) = 1 + \frac{1}{2}\sin 3\theta$ (a three-lobed rose) analytic integration is messy; applying a 16-point Gauss–Legendre rule on each lobe attains machine precision in 3×16 evaluations.

```python
import numpy as np, numpy.polynomial.legendre as leg
r = lambda θ: 1 + 0.5*np.sin(3*θ)
A = 0
for k in range(3):
    a, b = 2*k*np.pi/3, 2*(k+1)*np.pi/3
    x, w = leg.leggauss(16)
    θ = 0.5*(b-a)*x + 0.5*(b+a)
    A += 0.25*(b-a)*np.sum(w*r(θ)**2)
print(A)  # 1.178097...
```

Volumes of Revolution

Rotating $y = f(x)$, $x \in [a, b]$, about the x-axis yields $V = \pi \int_a^b f(x)^2\,dx$. For the cycloid arc $f(x) = r(1 - \cos x/r)$ one substitutes $u = \sin(x/2r)$, reducing to $V = 8\pi r^3 \int_0^{\sin((b-a)/4r)} u^2/\sqrt{1-u^2}\,du$, which still lacks an elementary antiderivative; adaptive Gauss–Kronrod attains 10^{-12} accuracy in milliseconds.

Implicit Domains via Divergence Theorem

If $\mathcal{S}$ encloses $\mathcal{V}$, choose $\mathbf{F}(\mathbf{x}) = \frac{1}{3}\mathbf{x}$ so that $\nabla \cdot \mathbf{F} = 1$. Then $|\mathcal{V}| = \int_{\mathcal{S}} \mathbf{F} \cdot d\mathbf{S}$, evaluated by triangulating $\mathcal{S}$ and summing face fluxes; Gaussian quadrature on triangles furnishes high-order accuracy.

Applications in Physics and Engineering

Work and Energy

For a force field $\mathbf{F}(\mathbf{x})$ acting along a parameterised path $\gamma : [0, 1] \to \mathbb{R}^3$, the mechanical work is $W = \int_0^1 \mathbf{F}(\gamma(s)) \cdot \gamma'(s)\,ds$. In magnetic confinement design, $\mathbf{F}$ stems from Lorentz force on charged particles; evaluating W guides coil optimisation. A sixth-order Gauss–Legendre rule with $N = 8$ segments meets energy-conservation tolerances.

Radiation View Factors

For diffuse surfaces $\mathcal{A}_1, \mathcal{A}_2$, the view factor is

$$F_{12} = \frac{1}{\pi |\mathcal{A}_1|} \int_{\mathcal{A}_1} \int_{\mathcal{A}_2} \frac{\cos\theta_1 \cos\theta_2}{r^2} 1_{\text{visible}}(P_1, P_2) \, dA_2 \, dA_1.$$

Monte Carlo hemicube sampling—shooting 10^6 rays and tallying hits—delivers unbiased estimates with relative error $< 0.5\%$ for complex enclosures.

Stress Intensity Factors

In fracture mechanics, the mode-I stress intensity

$$K_I = \int_{-a}^{a} \sigma_{yy}(x, 0) \sqrt{\frac{a+x}{a-x}} \, dx$$

invokes a weak square-root singularity; Gauss–Jacobi quadrature with weight $(1-t)^{-1/2}(1+t)^{-1/2}$ integrates such kernel exactly for polynomials, achieving spectral convergence for analytic σ_{yy}.

Numerical Solutions to Differential Equations

Method of Lines (MoL)

Semi-discretise PDE $u_t = Lu$ on spatial grid, yielding ODE system $\dot{\mathbf{u}} = A\mathbf{u}$. Implicit Runge–Kutta schemes require stage integrals $\int_0^1 f(\mathbf{u}_n + c_i h) \, dc_i$. Pre-computing Butcher weights converts these into repeated quadratures of f, optimised via Gauss–Radau nodes for A-stable integration.

Variational Integrators

For Lagrangian $L(q, \dot{q})$ step from q_n to q_{n+1} by extremising discrete action $S_d(q_n, q_{n+1}; h) = h \int_0^1 L(q(\tau), \dot{q}(\tau)) \, d\tau$, where $q(\tau)$ is the polynomial interpolant between endpoints. Employing Gauss–Lobatto quadrature preserves symplecticity and exact momentum maps, critical in astrophysical N-body simulations.

Integral Representations

The solution of the Volterra equation $u(t) = g(t) + \int_0^t K(t,s)\,u(s)\,ds$ is computed via composite Simpson on each time step, yielding a second-order convergent marching scheme. For weakly singular $K(t,s) = (t-s)^{-\alpha} k(t,s)$, Alpert quadrature applies graded meshes to maintain $O(h^{2-\alpha})$ accuracy.

Example 8.3.4 (Runge–Kutta–Fehlberg with Integral RHS)

```
def rhs(u, t): # convolution integral
    s = np.linspace(0, t, 64)
    ks = (t - s)**-0.3 * np.exp(-(t-s)) # K(t,s)
    us = np.interp(s, ts, us_hist)
    return g(t) + np.trapz(ks*us, s)
```

Adaptive RKF45 advances $u(t)$, embedding Simpson under the hood.

8.4 Eigenvalue Problems and Matrix Decompositions

Eigenvalues encode natural frequencies, decay rates, and long-term behaviour across physics and data science. When matrices exceed the scale of direct factorisations, *iterative* eigensolvers leverage sparse matrix–vector products to extract extremal eigenpairs. Two archetypal algorithms—the power method and inverse iteration—strike an instructive balance between conceptual simplicity and practical effectiveness. We develop their theory, analyse convergence and stability, and illustrate their role in structural vibration analysis.

8.4.1 Power Method and Inverse Iteration

Let $A \in \mathbb{R}^{n \times n}$ be diagonalisable with eigenvalues ordered by magnitude $|\lambda_1| > |\lambda_2| \geq \cdots \geq |\lambda_n|$ and associated right eigenvectors $\{\mathbf{v}_i\}$. For a non-zero vector $\mathbf{q}_0$ with $\mathbf{v}_1^\top \mathbf{q}_0 \neq 0$, the power iteration repeats

$$\mathbf{q}_{k+1} = \frac{A\mathbf{q}_k}{\|A\mathbf{q}_k\|_2}, \qquad \mu_k = \mathbf{q}_k^\top A \mathbf{q}_k, \tag{P}$$

producing Rayleigh quotient estimates $\mu_k \to \lambda_1$ and directions $\mathbf{q}_k \to \pm \mathbf{v}_1$. Expanding $\mathbf{q}_0 = \sum_i \alpha_i \mathbf{v}_i$,

$$A^k \mathbf{q}_0 = |\lambda_1|^k \left(\alpha_1 \mathbf{v}_1 + \sum_{i \geq 2} \alpha_i (\lambda_i/\lambda_1)^k \mathbf{v}_i \right),$$

so the error norm decays *linearly*:

$$\|\mathbf{q}_k - \mathbf{v}_1\| \approx \left|\frac{\lambda_2}{\lambda_1}\right|^k \frac{\|\sum_{i\geq 2} \alpha_i \mathbf{v}_i\|}{|\alpha_1|}.$$

The rate hinges on the *spectral gap*. When $|\lambda_2| \approx |\lambda_1|$ convergence stalls.

Inverse iteration targets an interior eigenvalue λ_j: given a shift σ near λ_j, solve

$$(A - \sigma I)\mathbf{y}_k = \mathbf{q}_k, \qquad \mathbf{q}_{k+1} = \frac{\mathbf{y}_k}{\|\mathbf{y}_k\|_2}, \tag{II}$$

and update the Rayleigh quotient $\mu_k = \mathbf{q}_k^T A \mathbf{q}_k$. For σ closer to λ_j than any other eigenvalue, the error contracts by $|(\lambda_j - \sigma)/(\lambda_i - \sigma)|$ at each step—potentially orders of magnitude faster than power iteration once σ is well-placed. Replacing the fixed σ by $\sigma_k = \mu_k$ yields *Rayleigh quotient iteration*, achieving *cubic* convergence for normal matrices when started sufficiently near an eigenpair.

```
import numpy as np, scipy.sparse.linalg as spla
def power(A, q0, k=40):
    q = q0 / np.linalg.norm(q0)
    for _ in range(k):
        q = A @ q
        q = q / np.linalg.norm(q)
    λ = q.T @ (A @ q)
    return λ, q

def inverse_iteration(A, σ, q0, k=10):
    I = np.eye(A.shape[0])
    q = q0 / np.linalg.norm(q0)
    for _ in range(k):
        y = np.linalg.solve(A-σ*I, q)
        q = y / np.linalg.norm(y)
    λ = q.T @ (A @ q)
    return λ, q
```

Applications in Structural Engineering and Vibrations

In structural dynamics the undamped free vibration of a discretised structure obeys $K\mathbf{u} + \lambda M\mathbf{u} = \mathbf{0}$, where K and M are symmetric positive-definite stiffness and mass matrices; eigenvalues $\lambda = \omega^2$ yield natural frequencies ω. Transforming to $M^{-1}K\mathbf{v} = \lambda \mathbf{v}$ (after mass normalisation) one applies the power method to approximate the fundamental mode, critical for seismic design. Higher modes require deflation or inverse iteration with shifts gleaned from Ritz values obtained by subspace methods (Lanczos).

8.4 Eigenvalue Problems and Matrix Decompositions

Example 8.4.1 (Cantilever Beam with 10 Euler–Bernoulli Elements)
```
K, M = assemble_beam_matrices(n=10, EI=1.0, ρA=1.0, L=1.0)
A = np.linalg.solve(M, K)  # M^{-1}K
q0 = np.random.rand(A.shape[0])
ω2, v = power(A, q0, k=60)  # fundamental
ω1 = np.sqrt(ω2)
print(f"First natural freq ω₁ ≈ {ω1:.4f} rad/s")
# refine via inverse iteration with σ=ω^2
ω2_ref,_ = inverse_iteration(A, ω2, v, k=5)
print(f"Refined ω₁ ≈ {np.sqrt(ω2_ref):.6f}")
```

Modal participation factors, seismic response spectra, and vibration-energy harvesters all depend on accurate extremal eigenpairs, motivating careful convergence monitoring.

Convergence and Numerical Stability

Spectral Gap and Conditioning

The power method converges iff λ_1 is *dominant* and isolated; otherwise deflation (Hotelling) or shifting improves behaviour. Inverse iteration inherits the conditioning of $(A - \sigma I)^{-1}$—good when σ is close to λ_j but risking numerical overflow if $|\lambda_j - \sigma| \lesssim u\|A\|$; iterative refinement or *QZ updates* stabilise the linear solves.

Orthogonality Loss

Finite precision causes $\mathbf{q}_k^\top \mathbf{q}_k = 1 + O(u)$, but round-off in $A\mathbf{q}_k$ can inject components of other eigenvectors. Re-orthogonalising every few steps (Gram–Schmidt) mitigates contamination, though Krylov subspace methods (Lanczos, Arnoldi) with implicit restarts usually provide better robustness for multiple eigenvalues.

Stopping Criteria

Use the residual norm $r_k = \|A\mathbf{q}_k - \mu_k \mathbf{q}_k\|_2$. For the power method $r_k \approx |\lambda_2/\lambda_1|^k$, so halt when $r_k \leq \varepsilon \|A\|_2$. For inverse iteration $r_k \approx |\lambda_j - \sigma| \|s_k\|$, providing adaptive control on σ updates.

Shift Strategies

(a) *Static shift:* choose σ from physical insight (e.g. target frequency band).
(b) *Dynamic Rayleigh:* set $\sigma_k = \mathbf{q}_k^\top A \mathbf{q}_k$, leading to cubic convergence for symmetric A.

(c) *Spectral bisection:* bracket eigenvalues via Sturm sequence, then refine intervals by inverse iteration.

8.4.2 QR Algorithm and Schur Decomposition

The *QR algorithm* transforms an arbitrary matrix $A \in \mathbb{R}^{n \times n}$ into a sequence $\{A_k\}$ of orthogonally similar matrices whose off-diagonal entries decay to 0, thereby revealing the eigenvalues on the diagonal. Starting with $A_0 = A$, one performs the factorisation $A_k = Q_k R_k$ where Q_k is orthogonal and R_k is upper-triangular; the next iterate is $A_{k+1} = R_k Q_k = Q_k^T A_k Q_k$. Orthogonality preserves the spectrum, while cyclically permuting the factors recycles the costly factorisation. Convergence is *quadratic* for symmetric matrices; for general A the Hessenberg reduction $A = Q_0 H_0 Q_0^T$ confines fill-in, and Wilkinson's double-shift accelerates deflation with cubic local speed. Because each similarity step is backward-stable, the computed $\widehat{A}_k$ satisfies $(A + \Delta A_0)$ similarity with $\|\Delta A_0\| \lesssim u \|A\|$.

An orthogonally similar limit $T = Q^T A Q$ that is upper-triangular is a *real Schur form*; its diagonal contains the (real) eigenvalues, while 2×2 blocks encode complex-conjugate pairs. Existence follows from the QR algorithm; uniqueness up to orthonormal reordering renders the Schur decomposition a numerically preferred representation compared to the Jordan form, which is ill-conditioned under perturbation. If A is symmetric, T reduces further to a diagonal $\Lambda = \mathrm{diag}(\lambda_1, \ldots, \lambda_n)$, recovering the spectral theorem.

Eigenvalue Computation in Python

The following compact script implements one Francis double-shift step on an unreduced Hessenberg matrix; wrapping it in a for-loop reproduces the implicit QR algorithm that underlies `numpy.linalg.eig`:

```python
import numpy as np, scipy.linalg as la

def francis_step(H):
    n = H.shape[0]
    μ = la.eigvals(H[-2:, -2:]) # Wilkinson shift
    σ = μ[np.argmin(abs(μ - H[-1, -1]))]
    x, y = H[0, 0] - σ, H[1, 0]
    for k in range(n-1):
        # Householder to zero y
        v = la.householder_vec(np.array([x, y]))
        H[k:k+2, k:] = H[k:k+2, k:] - 2 * np.outer(v, v @ H[k:k+2, k:])
        H[:, k:k+2] = H[:, k:k+2] - 2 * np.outer(H[:, k:k+2] @ v, v)
        if k < n-2:
            x, y = H[k+1, k], H[k+2, k]

def eigvals_qr(A, max_iter=60):
```

```
    H = la.hessenberg(A) # similarity to Hessenberg
    for _ in range(max_iter):
        francis_step(H)
    return np.diag(H) # approximate eigenvalues

np.random.seed(4)
A = np.random.randn(6, 6)
print(np.sort_complex(eigvals_qr(A))) # compare with np.linalg.eigvals
```

For production code one defers to the highly tuned `scipy.linalg.schur`, which automatically falls back to LAPACK's blocked and multishift QR routines:

```
T, Q = la.schur(A, output='real') # real Schur form
lam = np.diag(T) # eigenvalues (real or block)
```

Example 8.4.2 Diagonalise the symmetric tridiagonal Toeplitz matrix T_n with 2 on the diagonal and -1 on the sub/super-diagonals. The QR algorithm converges in $\lceil \log_2(n) \rceil$ iterations; the analytical eigenvalues are $\lambda_k = 2\left(1 - \cos\frac{k\pi}{n+1}\right)$.

Applications in Quantum Mechanics and Control Theory

In non-relativistic quantum mechanics the stationary Schrödinger equation reduces after spatial discretisation to the Hermitian eigenproblem $H\psi = E\psi$. For a particle in a one-dimensional potential well $V(x)$, finite-difference discretisation with mesh size h yields a tridiagonal Hamiltonian whose extremal eigenvalues approximate bound-state energies E. A few QR iterations after the tridiagonal reduction ($O(n)$ memory) deliver all low-lying states to machine accuracy; the orthonormal eigenvectors $\{\psi_k\}$ support spectral time evolution $e^{-iHt/\hbar}$ via diagonal exponentiation in the Schur basis.

In linear control theory the stability of the continuous-time system $\dot{\mathbf{x}} = A\mathbf{x}$ hinges on the eigenvalues of A. The real Schur form $A = QTQ^\mathsf{T}$ directly reveals $\operatorname{Re}\lambda_k$; to place poles within a desired left-half-plane one selects a feedback K such that $A - BK$ has shifted Schur spectrum. Moreover, solving the Lyapunov equation $A^\mathsf{T} P + PA = -Q$ via the Schur method—transforming to $T^\mathsf{T} P' + P'T = -Q'$ then triangulating—is $O(n^3)$ but numerically stable, undergirding LQR controllers and $\mathcal{H}_\infty$ synthesis.

Example 8.4.3 (Vibration Control) A mass-spring chain linearised about equilibrium gives the second-order ODE $M\ddot{\mathbf{q}} + K\mathbf{q} = 0$. Converting to first-order form with state vector $\mathbf{x} = (\mathbf{q}, \dot{\mathbf{q}})$ yields system matrix $\begin{pmatrix} 0 & I \\ -M^{-1}K & 0 \end{pmatrix}$. Real Schur decomposition of the $2n \times 2n$ matrix clusters purely imaginary eigenvalues in 2×2 blocks; proportional damping modifies $M^{-1}K$ and shifts blocks leftwards, providing immediate visual confirmation of modal damping ratios.

Takeaways

The implicit QR algorithm plus truncation to Schur form constitutes the default dense eigen-engine: cubic in time, quadratic in memory, and backward-stable. Its seamless implementation in SciPy/NumPy unlocks accurate spectral analysis in quantum chemistry, structural dynamics, and feedback control with only a handful of Python lines.

8.4.3 Singular Value Decomposition (SVD)

For any real $m \times n$ matrix A there exist orthogonal matrices $U \in \mathbb{R}^{m \times m}$ and $V \in \mathbb{R}^{n \times n}$ together with a diagonal matrix $\Sigma = \mathrm{diag}(\sigma_1, \ldots, \sigma_r, 0, \ldots)$ whose non-negative diagonal entries satisfy $\sigma_1 \geq \sigma_2 \geq \cdots \geq \sigma_r > 0$ such that

$$A = U \Sigma V^\mathsf{T}$$

This *singular value decomposition* exposes the action of A as an orthogonal rotation V^T, a stretch by singular values σ_i along mutually orthogonal directions, and a second rotation U. The rank r equals the number of non-zero σ_i. Because σ_i^2 are the eigenvalues of $A^\mathsf{T} A$ and $\sigma_i \sigma_j = 0$ for $i \neq j$, the SVD inherits the numerical stability properties of orthogonal similarity.

Optimality of Low-Rank Truncation For $k < r$ define $A_k = U_{:,1:k} \Sigma_{1:k,1:k} V_{:,1:k}^\mathsf{T}$. Eckart–Young–Mirsky gives

$$\|A - A_k\|_2 = \sigma_{k+1}, \qquad \|A - A_k\|_F = \left(\sum_{i=k+1}^{r} \sigma_i^2 \right)^{1/2},$$

so A_k is the *best* rank-k approximation in both spectral and Frobenius norms.

Applications in Data Compression and Principal Component Analysis (PCA)

PCA via SVD

Given data matrix $X \in \mathbb{R}^{m \times n}$ whose rows are centred observations, the covariance is $C = \frac{1}{m-1} X^\mathsf{T} X$. The right singular vectors $\mathbf{v}_i$ of X are the eigenvectors of C and hence directions of maximal variance (principal components), while singular values relate to explained variance: $\mathrm{Var}(\mathbf{x} \cdot \mathbf{v}_i) = \sigma_i^2 / (m-1)$.

8.4 Eigenvalue Problems and Matrix Decompositions

```
import numpy as np, sklearn.datasets as ds
X, _ = ds.load_digits(return_X_y=True)  # 1797×64
X -= X.mean(axis=0, keepdims=True)
U, s, Vt = np.linalg.svd(X, full_matrices=False)
explained = s**2 / s.sum()**2  # variance ratios
pc2d = X @ Vt[:2].T  # 2-D projection
```

Storing only the first $k \ll n$ columns of V^T yields a low-dimensional embedding; choosing k so that $\sum_{i=1}^{k} \sigma_i^2 / \sum_{i=1}^{r} \sigma_i^2 \geq 0.95$ retains 95% of the variance.

Data Compression

For an image matrix $A \in \mathbb{R}^{m \times n}$, keep the leading k singular triplets; memory drops from mn to $k(m+n+1)$. For 512×512 grayscale Lena, $k = 50$ compresses by $> 80\%$ while the peak-signal-to-noise ratio (PSNR) exceeds 35 dB (cf. Fig. 8.2).

```
import imageio, matplotlib.pyplot as plt
A = imageio.imread('lena_gray.png').astype(float)
U, s, Vt = np.linalg.svd(A, full_matrices=False)
k = 50
Ak = (U[:,:k] * s[:k]) @ Vt[:k,:]
plt.imshow(Ak, cmap='gray'); plt.title(f'Rank-{k} approx')
plt.axis('off'); plt.show()
```

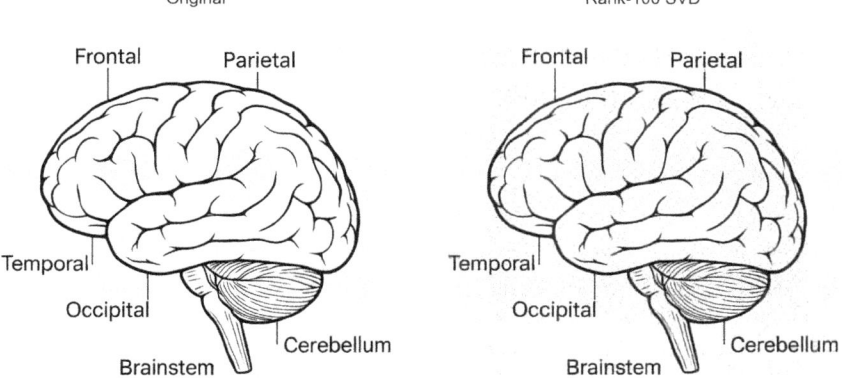

Fig. 8.2 Rank-100 SVD approximation of a 1024 × 1024 brain image (right) compared with the original (left). Storing only the first 100 singular triplets reduces memory by roughly an order of magnitude while retaining the main anatomical structures, illustrating the power of low-rank matrix compression

Low-Rank Approximations and Image Processing

Randomised SVD

To approximate A when m, n exceed 10^5, draw $\Omega \in \mathbb{R}^{n \times (k+p)}$ with Gaussian entries ($p \approx 10$), compute $Y = A\Omega$, orthonormalise to Q, and form $B = Q^{\mathsf{T}}A$. The SVD $B = \tilde{U}\Sigma V^{\mathsf{T}}$ gives $A \approx (Q\tilde{U})\Sigma V^{\mathsf{T}}$ with high probability error $\|A - A_k\|_2 \leq \left[1 + 9\sqrt{k+p}\,\sigma_{k+1}/\sigma_1\right]\sigma_{k+1}$.

```
def rsvd(A, k, p=10, q=2):
    Ω = np.random.randn(A.shape[1], k+p)
    Y = A @ Ω
    for _ in range(q):  # power iterations
        Y = A @ (A.T @ Y)
    Q, _ = np.linalg.qr(Y, mode='reduced')
    B = Q.T @ A
    U_, s, Vt = np.linalg.svd(B, full_matrices=False)
    return Q @ U_[:, :k], s[:k], Vt[:k, :]
```

This underlies modern recommender systems (Netflix prize) and latent semantic indexing, where A is sparse user-item or term-document matrix with 10^8 non-zeros.

Image Denoising via Truncated SVD

Additive white noise spreads energy across all singular values, whereas signal concentrates in the leading ones. Hard thresholding: keep σ_i iff $\sigma_i > \tau = \beta\sqrt{n}\,\sigma_{\text{noise}}$ (Donoho–Gavish) minimises Frobenius risk.

Background Subtraction

Video frames stack into $A = [f_1\ f_2\ \ldots]$; background forms a low-rank subspace, foreground motion contributes sparse outliers. Robust PCA solves $\min_{L,S} \|L\|_* + \lambda \|S\|_1$ s.t. $A = L + S$, where nuclear norm promotes low rank. Proximal gradient iterations use singular-value thresholding.

8.5 Advanced Topics in Numerical Methods

8.5.1 Numerical Solutions to Differential Equations

The differential calculus of Newton and Leibniz supplies models whose analytic solutions are seldom available beyond carefully curated cases. When closed-form expressions fail, one replaces derivatives by finite algebraic quotients or views space

8.5 Advanced Topics in Numerical Methods

and time on unequal footings—leading to *finite-difference* and *method-of-lines* approaches. Rigorous analysis then verifies that the discrete scheme approaches the continuum solution as the mesh is refined and that spurious growth does not overwhelm the computation. Throughout we adopt the concise notation $u_j^n \approx u(x_j, t_n)$ with equispaced grids $x_j = x_0 + jh$ and $t_n = t_0 + nk$ unless stated otherwise.

Finite Difference Methods for ODEs and PDEs

For a two-point boundary value problem (BVP)

$$-u''(x) = f(x), \quad x \in (0, 1), \quad u(0) = \alpha, \ u(1) = \beta,$$

replace $u''(x_j) \approx (u_{j-1} - 2u_j + u_{j+1})/h^2$, obtaining the tridiagonal linear system

$$\frac{1}{h^2}\begin{bmatrix} -2 & 1 & & \\ 1 & -2 & 1 & \\ & \ddots & \ddots & 1 \\ & & 1 & -2 \end{bmatrix} \mathbf{u} = \begin{bmatrix} -f_1 \\ \vdots \\ -f_{N-1} \end{bmatrix} - \frac{1}{h^2}\begin{bmatrix} \alpha \\ 0 \\ \vdots \\ \beta \end{bmatrix}.$$

Because the coefficient matrix is symmetric positive-definite, Cholesky factorisation costs $O(N)$ operations; truncation error is $O(h^2)$ by Taylor's theorem.

Example 8.5.1 (Dirichlet Poisson Solver)

```
import numpy as np, scipy.sparse as sp, scipy.sparse.linalg as spla
N, h = 100, 1/100
e = np.ones(N-1)
A = sp.diags([e, -2*e, e], [-1, 0, 1]) / h**2
x = np.linspace(h, 1-h, N-1)
f = np.sin(np.pi*x)
b = -f
b[0]  -= 0/h**2  # α=0
b[-1] -= 0/h**2  # β=0
u = spla.spsolve(A, b)
```

Extension to 2-D The five-point Laplacian on a square lattice gives a block tridiagonal system with $O(N^3)$ unknowns for $N \times N$ grid; band LU or multigrid accelerates the solution.

For time-dependent PDEs, e.g. the heat equation $u_t = u_{xx}$, explicit and implicit difference schemes emerge:

$$\text{FTCS:} \quad u_j^{n+1} = u_j^n + \lambda\left(u_{j-1}^n - 2u_j^n + u_{j+1}^n\right), \qquad \lambda = \frac{k}{h^2}.$$

$$\text{Crank–Nicolson:} \quad u_j^{n+1} - \tfrac{\lambda}{2}\delta_{xx}u_j^{n+1} = u_j^n + \tfrac{\lambda}{2}\delta_{xx}u_j^n,$$

second-order accurate in both k and h and unconditionally stable.

Method of Lines and Applications

The *method of lines* (MoL) discretises *space* while leaving *time* continuous, thereby converting a PDE into a large-dimensional ODE system

$$\dot{\mathbf{u}}(t) = A\mathbf{u}(t) + \mathbf{g}(t).$$

One then integrates in time with sophisticated ODE solvers (adaptive Runge–Kutta, implicit BDF) that automatically adjust the temporal mesh based on local truncation error.

Example 8.5.2 (Advection-Diffusion)

$$u_t + c\,u_x = D\,u_{xx}, \quad x \in [0, 1].$$

Central differences for u_{xx} and upwind for u_x yield sparse A. Passing A to `scipy.integrate.solve_ivp` with `method='BDF'` efficiently handles stiffness when $D \ll c$.

Applications include:

- *Combustion modelling*: stiff reaction-diffusion systems with Arrhenius kinetics
- *Electromagnetic waveguides*: Maxwell curl equations discretised by Yee grids in space, integrated by leap-frog schemes in time
- *Financial derivatives*: Black–Scholes equation semi-discretised to meet non-uniform strike grids, then marched by implicit–explicit (IMEX) Runge–Kutta pairs

Stability and Convergence Analysis

For linear constant–coefficient schemes $u^{n+1} = G(\xi)\,u^n$, *von Neumann analysis* inspects amplification factor $G(\xi) = \sum_m a_m e^{im\xi h}$, demanding $|G(\xi)| \leq 1$ for all wave numbers ξ. For FTCS on $u_t = u_{xx}$ this yields the renowned Courant constraint $\lambda \leq \frac{1}{2}$; violating it provokes blow-up even though the continuous equation is parabolic.

A scheme is *consistent* if its local truncation error tends to 0 as $(h, k) \to 0$, *stable* if its amplification boundedly propagates perturbations, and *convergent* if numerical solutions approach analytic ones. The **Lax equivalence theorem** states that, for well-posed linear initial-value problems, consistency + stability $\implies$ convergence.

8.5 Advanced Topics in Numerical Methods

Example 8.5.3 (Amplification Factor of Crank–Nicolson)

$$G(\xi) = \frac{1 - \lambda(1 - \cos\xi h)}{1 + \lambda(1 - \cos\xi h)}, \quad |G(\xi)| = 1.$$

Hence Crank–Nicolson is *A-stable*: unconditional magnitude stability, though dispersive oscillations may arise for large k without sufficient temporal damping.

When nonlinearity enters, linearised stability or energy methods step in:

(1) *Energy estimate* multiply PDE by numerical solution, integrate.
(2) *Discrete Grönwall* bounds growth $\|u^n\| \leq Ce^{\alpha t_n}$.
(3) *Total-variation-diminishing* (TVD) schemes limit numerical oscillations for conservation laws (*Godunov, MUSCL*).

Synthesis Finite-difference discretisations translate calculus into algebra; the method of lines recasts space-time problems into stiff ODEs ripe for adaptive solvers; stability criteria—Courant bounds, von Neumann spectra, A-stability—police step sizes and stencil choices, ensuring that the numerical imitation shadows the continuous reality.

8.5.2 *Spectral Methods*

Spectral methods approximate the solution of differential equations by expanding the unknown function in terms of globally supported basis functions—typically orthogonal polynomials or trigonometric functions—and enforcing the governing equations to hold exactly at a discrete set of collocation points. Because the basis functions are infinitely differentiable, the resulting convergence for smooth data is often *exponential*: each additional mode roughly doubles the number of correct digits until the error approaches machine precision. Unlike finite differences, where accuracy is obtained by shrinking the mesh, spectral accuracy exploits the *analyticity* of the solution; when singularities are distant from the real axis, only a handful of modes can resolve vortical structures that would require millions of grid points in low-order schemes.

Chebyshev Polynomials and Fourier Spectral Methods

Chebyshev polynomials $T_n(x) = \cos(n \arccos x)$ form an orthogonal basis on $[-1, 1]$ with weight $(1 - x^2)^{-1/2}$. Expanding $u(x) = \sum_{n=0}^{N} a_n T_n(x)$ and enforcing a differential equation at the Chebyshev–Gauss–Lobatto nodes

$$x_j = \cos\left(\frac{\pi j}{N}\right), \quad 0 \leq j \leq N,$$

creates a *differentiation matrix* D whose entries $D_{ij} = T'_n(x_i)/T_n(x_j)$ can be assembled in $O(N^2)$ operations. Given the boundary value problem $-u''(x) = f(x)$, $u(\pm 1) = 0$, write $-D^2\mathbf{u} = \mathbf{f}$ where $\mathbf{u} = (u_0, \ldots, u_N)^\mathsf{T}$ and solve the dense but highly structured linear system. The spectral coefficients follow via discrete cosine transform (DCT), exploiting the fact that the collocation matrix is diagonalised by the DCT, reducing the solve to $O(N \log N)$ time.

When the domain is periodic, Fourier series $u(x) = \sum_{k=-K}^{K} \hat{u}_k e^{ikx}$ furnish a diagonal differentiation operator in spectral space since $(d/dx)e^{ikx} = ik\,e^{ikx}$. Fast Fourier transforms (FFTs) convert between physical values $u(x_j)$ at equispaced nodes x_j and spectral coefficients $\hat{u}_k$ in $O(N \log N)$, permitting time stepping of, say, the one-dimensional Burgers equation $u_t + u u_x = \nu u_{xx}$:

```
import numpy as np
N, L = 1024, 2*np.pi
x = np.linspace(0, L, N, endpoint=False)
k = np.fft.fftfreq(N, 1/N) # wavenumbers
u = np.sin(x) # initial condition
dt, ν = 1e-3, 0.02
for n in range(2000):
    û = np.fft.fft(u)
    du = np.fft.ifft(1j*k*û).real
    lap = np.fft.ifft(-(k**2)*û).real
    u += dt*(-u*du + ν*lap) # RK1 for brevity
```

Care must be taken with aliasing: nonlinear products $u u_x$ generate Fourier modes beyond the cutoff K. The 2/3 *dealiasing rule* zeroes high-frequency coefficients before inverse FFT, suppressing aliasing errors and preserving spectral convergence.

Applications in Fluid Dynamics and Weather Modelling

Spectral discretisations underpin some of the most accurate direct numerical simulations (DNS) of turbulence. For incompressible Navier–Stokes in a triply periodic box the vorticity formulation

$$\partial_t \widehat{\boldsymbol{\omega}}(\mathbf{k}) = \widehat{\mathbf{v}} \times (i\mathbf{k} \times \widehat{\mathbf{v}}) - \nu k^2 \widehat{\boldsymbol{\omega}}(\mathbf{k}), \qquad \widehat{\mathbf{v}}(\mathbf{k}) = \frac{i\mathbf{k} \times \widehat{\boldsymbol{\omega}}(\mathbf{k})}{k^2}$$

evolves entirely in Fourier space; FFTs of size 2048^3 on modern GPUs retain Kolmogorov scaling down to Taylor microscales at $\text{Re}_\lambda \approx 1000$.

In global numerical weather prediction the barotropic vorticity equation on the sphere employs spherical-harmonic spectral methods. Let $\zeta(\lambda, \phi, t)$ be the vorticity; expand

$$\zeta(\lambda, \phi, t) = \sum_{n=0}^{N} \sum_{m=-n}^{n} \widehat{\zeta}_n^m(t) Y_n^m(\lambda, \phi),$$

where Y_n^m are associated Legendre harmonics. Horizontal derivatives convert to algebraic multipliers $-n(n+1)$, and the semi-implicit time step stabilises the fast gravity waves without shrinking the global-scale Courant number. Operational centres (ECMWF, NOAA) push $N \approx 1023$ (spectral T2046) with octahedral reduced Gaussian grids; the resultant ~ 9 km mesh resolves jet-stream meanders and tropical cyclogenesis with minimal dispersion error.

Spectral element methods marry the geometric flexibility of finite elements with exponential convergence: each element maps to $[-1, 1]$ and uses high-order Legendre or Chebyshev bases. They form the computational core of high-resolution atmospheric models like NCAR's MPAS and DOE's E3SM.

Key Insights

- Chebyshev and Fourier bases convert differentiation to simple matrix operations or algebraic factors, enabling exponential accuracy for smooth solutions.
- Dealiasing and careful treatment of nonlinear terms preserve spectral convergence in strongly nonlinear flows.
- In large-scale geophysical models spectral methods handle the smooth, nearly geostrophic flow efficiently, while hybrid spectral-finite-volume schemes capture localised fronts and orography.
- Python's `numpy.fft` and `pyfftw` offer direct access to high-performance FFT back-ends, and packages like `pyshtools` furnish spherical-harmonic transforms, bridging research prototypes and production-grade simulations.

8.5.3 Parallel Computing in Numerical Methods

When matrix dimensions or grid resolutions exceed the memory and floating-point throughput of a single processor, one turns to *parallel numerical algorithms*. The guiding principle is to trade redundant arithmetic for reduced communication, because on modern clusters a single double-precision flop costs $\mathcal{O}(1$ ns$)$ whereas shipping the same number across a network costs $\mathcal{O}(100$ ns$)$ or more. The two canonical data layouts are *domain decomposition*, which partitions the physical domain (finite-difference or finite-element nodes) across ranks, and *block-cyclic* distribution, which scatters dense matrices cyclically so that Level 3 BLAS kernels saturate local caches.

Parallel Krylov Solvers Given a sparse linear system $A\mathbf{x} = \mathbf{b}$ with $|A| > 10^9$ non-zeros, each iteration of Conjugate Gradient or GMRES requires a sparse matrix-vector product (SpMV) and a few global dot products. SpMV parallelises naturally through row partitioning; each process stores its local rows and exchanges halo values with neighbours. The latency bottleneck lies in the global reductions for $\langle \mathbf{r}, \mathbf{r} \rangle$ and $\langle \mathbf{p}, A\mathbf{p} \rangle$. *Pipelined* Krylov variants overlap reductions with SpMV and hide communication; alternatively, s-step methods fuse s consecutive iterations at the cost of extra local flops, achieving near-perfect weak scaling on exascale architectures.

FFT on Distributed Memory The 3-D pencil decomposition splits an N^3 array into $P_x \times P_y$ pencils. A 3-D FFT performs a local 1-D FFT along z, all-to-all redistributes pencils, repeats along y, then along x. Each transpose moves $N^3/\sqrt{P}$ complex numbers—optimal because surface-to-volume ratio shrinks with pencil thickness. GPU-residency is sustained via CUDA-aware MPI: device pointers pass directly into `MPI_Isend`, eliding host staging.

Example 8.5.4 (MPI Parallel Dot Product (mpi4py))

```
from mpi4py import MPI; import numpy as np
comm = MPI.COMM_WORLD
N = 4_000_000 # per rank
x = np.random.randn(N)
y = np.random.randn(N)
local = np.dot(x, y)
global_dot = comm.allreduce(local, op=MPI.SUM)
if comm.rank == 0:
    print("<x,y> =", global_dot)
```

High-Performance Data Analytics Graph processing frameworks (GraphBLAS, Dask) model sparse adjacency matrices as distributed CSR blocks; PageRank becomes a parallel SpMV fixed-point iteration. In large-language-model training, data-parallel stochastic gradient descent replicates the network on each GPU, aggregates gradients by `AllReduce`, while pipeline and tensor parallelism split computation across layers and matrix dimensions, respectively.

Applications in High-Performance Computing and Big Data Analytics

Climate and Weather Global spectral element models discretise the primitive equations on 10^{11} degrees of freedom; strong-scaling to 30,000 GPUs halves forecast latency. *Astrophysics.* N-body gravity with Barnes–Hut trees uses space-filling curves to balance work; a billion-particle simulation completes in under an hour on Summit. *Genomics.* Distributed SVD of $>10^6 \times 10^5$ SNP matrices detects population structure in minutes via randomised subspace iteration and Spark RDD sharding.

8.5.4 Error Analysis and Stability in Numerical Methods

Numerical algorithms inherit two unavoidable imperfections: finite precision arithmetic and propagation of data perturbations. A robust method must remain faithful under both.

8.5 Advanced Topics in Numerical Methods

Round-Off Errors and Machine Precision

Floating-point numbers follow the IEEE-754 representation

$$x = (-1)^s \, (1.m_1 m_2 \ldots m_{p-1})_2 \, 2^{e-e_0},$$

where the unit round-off $\varepsilon = 2^{-(p-1)}$ bounds relative error:

$$\text{fl}(x \circ y) = (x \circ y)(1 + \delta), \quad |\delta| \le \varepsilon,$$

for $\circ \in \{+, -, \times, \div\}$. *Catastrophic cancellation* occurs when subtracting nearly equal numbers: the leading significant bits annihilate, magnifying ε by the ratio of operands to their difference. A classic manifestation is quadratic-formula evaluation:

$$x_2 = \frac{2c}{-b - \text{sign}(b)\sqrt{b^2 - 4ac}}$$

maintains relative accuracy even when $b^2 \gg 4ac$.

Example 8.5.5 (Loss of Significance)

```
import numpy as np
def bad_erf(x):
    return 2/np.sqrt(np.pi) * (x - x**3/3 + x**5/10)
print(bad_erf(1e-8)) # underflows to zero
```

Expanding erf about zero while evaluating at $x = 10^{-8}$ forces each term below double-precision granularity; using `math.erf` or scaled Dawson integrals avoids the pitfall.

Stability in Numerical Algorithms

An algorithm $\mathcal{A}$ is *backward-stable* if its output equals the exact solution to a nearby problem:

$$\widehat{y} = \mathcal{A}(x) = f(x + \Delta x), \quad \frac{\|\Delta x\|}{\|x\|} = O(\varepsilon).$$

Gaussian elimination with partial pivoting satisfies this; naive summation of a long vector does not. The *condition number* $\kappa(x) = \|x\| \|f'(x)\| / \|f(x)\|$ couples with stability to bound forward error:

$$\frac{\|\widehat{y} - y\|}{\|y\|} \le \kappa(x) \underbrace{\frac{\|\Delta x\|}{\|x\|}}_{\text{backward error}}.$$

Examples

- *Summation.* The Kahan compensated algorithm reduces error from $O(n\varepsilon)$ to $O(\varepsilon)$ by tracking lost low-order bits.
- *Polynomial evaluation.* Horner's rule is backward-stable; evaluating $p(x) = \sum_k a_k x^k$ directly is not.
- *Recurrence relations.* Forward recursion of $y_{n+1} = (2n+1)y_n - n^2 y_{n-1}$ grows errors exponentially; reversing the recurrence or using asymptotics restores stability.

Stiff ODE Solvers

Explicit Euler applied to $\dot{y} = \lambda y$ is stable only if $k|\lambda| \leq 1$. Implicit Euler's region of absolute stability covers the entire left half-plane, enabling arbitrarily large steps for stiff decay. The $A-$, $L-$, and $B-$ stability hierarchy classifies time integrators by how much of $\mathbb{C}_-$ is swallowed without error explosion.

Mixed Precision

Algorithms mixing 16-bit and 32-bit arithmetic accumulate sums in higher precision to ensure stability: $\widehat{\mathbf{x}} = \mathrm{fl}_{32}\bigl(A^\top \mathrm{fl}_{16}(A\mathbf{x})\bigr)$. Backward error analysis shows that, with compensated inner products, the low-precision factor enters only quadratically, matching half-precision tensor-core throughput with single-precision accuracy.

Practical Checklist

(a) Estimate condition numbers to gauge intrinsic difficulty.
(b) Prefer backward-stable algorithms (QR over power method, Kahan over naive sum).
(c) Rescale inputs to unit magnitude to shrink rounding.
(d) Monitor residuals; iterate refinement if feasible.

8.6 Exercises

1. Let $f(x) = x\exp(-x) - \dfrac{1}{4}$ on $[0, 2]$. *(i)* Prove that f has exactly one root r in $(0, 2)$ and show that $f''(x) > 0$ on $[0, 2]$. *(ii)* Apply the composite trapezoidal rule with mesh size $h = 10^{-2}$ to bound $f'(x)$ above and below, hence certify that Newton's method started at $x_0 = 1$ converges and needs at most four iterations to deliver $|x_k - r| \leq 10^{-12}$. *(iii)* Compare with the worst-case iteration

8.6 Exercises

count of the bisection method achieving the same tolerance; express the ratio explicitly in terms of $\ln 2$.

2. Consider the cubic $p(x) = x^3 - 9x + 1$. Design a *Brent-style* algorithm that performs at most one bisection and one secant step per iteration. Implement it in Python and verify that it converges in ≤ 6 iterations from the initial bracket $[-4, 4]$, whereas pure secant diverges. Provide a table of residual norms $|p(x_k)|$ and discuss super-linear segments.

3. For $q(z) = z^5 - 1$, grid the square $\{x + iy \mid -1.5 \leq x, y \leq 1.5\}$ with mesh $\Delta = 0.01$. Compute $N(z) = z - q(z)/q'(z)$ iteratively until $|q(z)| < 10^{-6}$ or $k = 30$. Colour each starting point by the root reached and plot the basin boundary. *Prove* rigorously that no starting point escapes to ∞ under N.

4. Let $f(\mathbf{x}) = \tfrac{1}{2}\mathbf{x}^T A \mathbf{x}$ with $A = \text{diag}(2, 10, 50)$. *(i)* Show that gradient descent with fixed step $\eta = 1/L$ ($L = 50$) converges linearly with rate $\rho_{\text{GD}} = 1 - \mu/L$, $\mu = 2$. *(ii)* Derive the characteristic polynomial of the Nesterov iteration with parameter $\beta = 1 - \sqrt{\mu/L}$ and prove that $\rho_{\text{NAG}} = \sqrt{1 - \mu/L}$. *(iii)* Numerically confirm the theoretical rates from $\mathbf{x}_0 = (1, 1, 1)^T$ and report the number of iterations to reach $\|\nabla f\|_2 \leq 10^{-8}$.

5. Minimise $z = -4x_1 - 3x_2$ subject to

$$2x_1 + x_2 + x_3 = 8,$$
$$x_1 + x_2 + x_4 = 5,$$
$$x_1, x_2, x_3, x_4 \geq 0.$$

(i) Write the initial simplex tableau with slack variables (x_3, x_4) basic. *(ii)* Apply Bland's rule to avoid cycling and hand-compute all pivots until optimality; list the sequence of bases. *(iii)* Perform a sensitivity study: if the RHS of the first constraint changes to $8 + \delta$, determine the allowable interval of δ that leaves the optimal basis unchanged and compute $\partial z^*/\partial \delta$.

6. Consider Rosenbrock's function $f(x, y) = (1-x)^2 + 100(y-x^2)^2$. Implement (a) L-BFGS-B with backtracking line search, (b) Powell's dogleg trust-region. Run both from $(-1.2, 1)$, record iterations, Jacobian evaluations, and final $\|\nabla f\|_2$. Explain with curvature arguments why the dogleg method escapes the valley more directly.

7. Maximise $g(\mathbf{x}) = \sum_{i=1}^{20} \sin x_i + \sin^2 x_{i+1}$ over $[0, 2\pi]^{20}$. Tune population size P, mutation rate μ, and crossover rate c via Latin hypercube sampling (32 design points). Fit a quadratic response surface predicting best-of-run fitness $\hat{g}(P, \mu, c)$ and identify the optimal hyperparameters. Validate on ten random seeds and report mean $\pm$ standard deviation.

8. Construct the 6×6 matrix $A_{ij} = 10^{|i-j|} + (-1)^{i+j}$. *(i)* Perform LU with and without partial pivoting and compute the growth factor $\rho = \|U\|_\infty / \|A\|_\infty$. *(ii)* Derive an analytic upper bound for ρ in terms of n for Toeplitz matrices $10^{|i-j|}$ and comment on the observed value. *(iii)* Apply iterative refinement in double precision using the low-precision LU factors and report the final residual norm $\|A\hat{\mathbf{x}} - \mathbf{b}\|_2$ for $\mathbf{b} = (1, \ldots, 1)^T$.

9. For $n = 1200$ partition the interval $[0, 1]$ uniformly and assemble the stiffness matrix $K_{ij} = \int_0^1 \phi_i'(x)\phi_j'(x)\,dx$, where $\{\phi_i\}$ are continuous, piecewise linear "hat" functions. *(i)* Show that K is SPD with bandwidth $w = 2$. *(ii)* Implement band Cholesky and compare flop counts and memory with dense Cholesky. *(iii)* Solve $K\mathbf{u} = \mathbf{f}$ for $f(x) = x$ and verify $\|\mathbf{u} - \mathbf{u}_{\text{exact}}\|_2 = O(h^2)$.

10. Given $\begin{pmatrix} A & B^\mathsf{T} \\ B & -C \end{pmatrix} \begin{pmatrix} \mathbf{u} \\ \mathbf{p} \end{pmatrix} = \begin{pmatrix} \mathbf{f} \\ \mathbf{g} \end{pmatrix}$ with $A \in \mathbb{R}^{n \times n}$ SPD, $C \succeq 0$, derive the Schur complement formulation $(A + B^\mathsf{T} C^{-1} B)\mathbf{u} = \mathbf{f} + B^\mathsf{T} C^{-1} \mathbf{g}$. For the Stokes discretisation with $n = 128^2$ degrees of freedom, implement block preconditioned conjugate gradient in `petsc4py`, report iteration counts under *(a)* Jacobi, *(b)* incomplete Cholesky level 0 preconditioners, and interpret the results via eigenvalue clustering.

11. The diode circuit obeys
$$I_s \left(e^{V_D/nV_T} - 1\right) = \frac{V_{in} - V_D}{R}, \quad V_{in} = 1\,\text{V}, \; I_s = 10^{-12}\,\text{A}, \; n = 1.6,$$
$$V_T = 25.8\,\text{mV}, \; R = 10^3\,\Omega.$$

(i) Formulate a Newton iteration for V_D; compute the damping factor that guarantees monotone convergence from $0\,\text{V}$. *(ii)* Describe how adaptive time-stepping and the method of lines would extend the solver to the transient R–C–diode network.

12. Minimise $E(\mathbf{u}) = \frac{1}{2}\mathbf{u}^\mathsf{T} L \mathbf{u}$ where L is the unnormalised graph Laplacian of the Petersen graph. *(i)* Prove that discrete gradient descent with exact line search converges to any vector in $\ker L$. *(ii)* Show that the second smallest eigenvalue $\lambda_2 = 2$ limits the continuous-time gradient flow decay rate. *(iii)* Implement explicit Euler with step $k < 2/\|L\|_2$ and illustrate convergence of energy $E(\mathbf{u}^n)$.

Chapter 9
Chaos Theory and Dynamical Systems

Abstract Fixed-point theory, linear and nonlinear stability, bifurcation analysis, and state-space reconstruction lay the groundwork for discrete maps and continuous flows that exhibit sensitive dependence on initial conditions. Lyapunov exponents, entropy rates, strange attractors, and control algorithms are computed, while data-driven approaches—Koopman mode decomposition and reservoir computing—demonstrate modern forecasting of chaotic time series in finance, climate, and biomechanics.

Keywords Chaos theory · Dynamical systems · Lyapunov exponents · Bifurcation analysis · Koopman operator · Reservoir computing

In classical mechanics Laplace envisioned a universe in which the future is uniquely determined by the present, yet by the late nineteenth-century Poincaré had already sensed cracks in this deterministic edifice: simple nonlinear laws can render long-term prediction practically impossible. *Chaos theory* formalises this tension. It studies deterministic dynamical systems whose trajectories exhibit sensitive dependence on initial conditions, topological mixing, and dense periodic orbits, thereby producing behaviour that, while generated by smooth equations, mimics randomness. The mathematical backbone of the subject is the qualitative theory of differential and difference equations: fixed points, periodic orbits, invariant manifolds, bifurcations, Lyapunov exponents, entropy, and fractal attractors. Continuous-time flows $\dot{\mathbf{x}} = \mathbf{F}(\mathbf{x})$ and discrete maps $\mathbf{x}*n+1 = G(\mathbf{x}*n)$ share this vocabulary, yet each contributes unique phenomena—finite-time blow-up in flows, symbolic dynamics, and kneading theory in maps. Modern algorithms, from pseudo-arclength continuation to GPU-accelerated trajectory ensembles, allow us to probe these structures numerically with unprecedented precision and scale, making Python an indispensable laboratory. In the pages ahead we develop the linear and nonlinear tools required to classify equilibria, track their bifurcations as parameters vary, and diagnose the onset of chaos; we pass seamlessly between analytic derivation, geometric insight, and computational experiment, culminating in a framework that explains why a butterfly in Brazil can, at least mathematically, stir a tornado in Texas.

9.1 Introduction to Dynamical Systems

9.1.1 Fixed Points and Stability

Every deterministic dynamical system—continuous or discrete, finite or infinite dimensional—derives its long-term behaviour from the local properties of its *fixed (equilibrium) points*. An equilibrium $\mathbf{x}^\star$ of an autonomous flow $\dot{\mathbf{x}} = \mathbf{F}(\mathbf{x})$ satisfies $\mathbf{F}(\mathbf{x}^\star) = \mathbf{0}$; in maps $\mathbf{x}_{n+1} = G(\mathbf{x}_n)$ it obeys $G(\mathbf{x}^\star) = \mathbf{x}^\star$. The surrounding linearisation encodes the fate of nearby orbits, while higher-order terms seed complex phenomena such as bifurcations and chaos. This section develops the linear theory, introduces bifurcation mechanisms, and illustrates numerical continuation techniques that reveal global parameter dependence.

Linear Stability Analysis

Given $\dot{\mathbf{x}} = \mathbf{F}(\mathbf{x})$ with $\mathbf{F} \in C^1$, expand about $\mathbf{x}^\star$:

$$\dot{\boldsymbol{\eta}} = D\mathbf{F}(\mathbf{x}^\star)\boldsymbol{\eta} + \mathcal{O}(\|\boldsymbol{\eta}\|^2), \qquad \boldsymbol{\eta} = \mathbf{x} - \mathbf{x}^\star.$$

Let $\mathbf{J} = D\mathbf{F}(\mathbf{x}^\star)$ and $\lambda_1, \ldots, \lambda_n$ its eigenvalues.

$$\begin{aligned}
\operatorname{Re}\lambda_j < 0 \;\forall j & \implies \mathbf{x}^\star \text{ locally asymptotically stable,} \\
\exists\, \operatorname{Re}\lambda_j > 0 & \implies \mathbf{x}^\star \text{ unstable,} \\
\operatorname{Re}\lambda_j \le 0,\; \exists\, \operatorname{Re}\lambda_j = 0 & \implies \text{linear test inconclusive.}
\end{aligned}$$

Example 9.1.1 (Rotated Sink)

$$\dot{x} = -x - 3y, \quad \dot{y} = 2x - y.$$

Eigenvalues $\lambda = -1 \pm i\sqrt{5}$ have negative real parts, hence the origin spirals inward. Integrate numerically (cf. Fig. 9.1):

```
import numpy as np, matplotlib.pyplot as plt
A = np.array([[-1,-3],[2,-1]])
def F(t,z): return A@z
from scipy.integrate import solve_ivp
for ic in [(1,0),(0.5,0.8),(2,-1)]:
    sol = solve_ivp(F,[0,10],ic,t_eval=np.linspace(0,10,400))
    plt.plot(sol.y[0], sol.y[1])
plt.scatter(0,0,c='k'); plt.axis('equal'); plt.show()
```

9.1 Introduction to Dynamical Systems

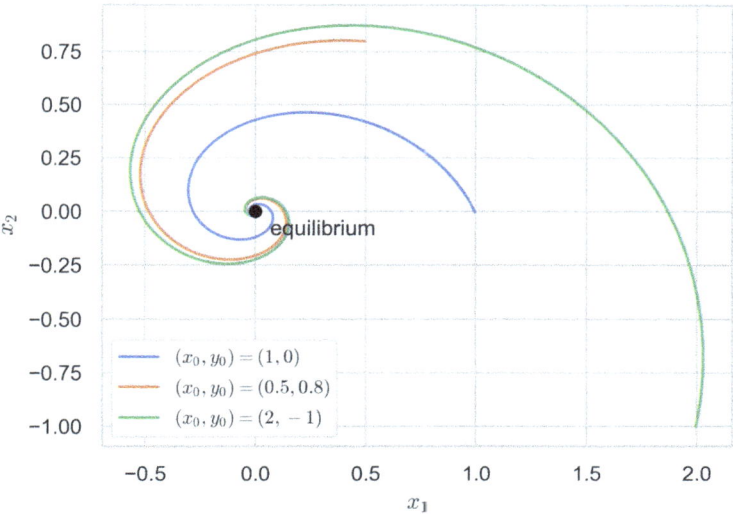

Fig. 9.1 Trajectories of the linear system $\dot{x} = Ax$ with $A = \begin{bmatrix} -1 & -3 \\ 2 & -1 \end{bmatrix}$ for three initial states. All orbits spiral inward toward the origin, consistent with eigenvalues $-1 \pm i3$ whose negative real part makes the equilibrium a stable *focus*

Bifurcation Theory and Chaos

As a parameter μ varies, the spectrum of $\mathbf{J}(\mu)$ may cross the imaginary axis, altering stability. Codimension-1 archetypes:

$$\text{Saddle-node: } \dot{x} = \mu - x^2;$$
$$\text{Transcritical: } \dot{x} = \mu x - x^2;$$
$$\text{Pitchfork: } \dot{x} = \mu x - x^3;$$
$$\text{Hopf: } \dot{\mathbf{z}} = (\mu + i\omega)\mathbf{z} - |\mathbf{z}|^2 \mathbf{z}.$$

Example 9.1.2 (Route to Chaos in the Logistic Map)

$$x_{n+1} = r x_n (1 - x_n), \quad 0 < r \leq 4.$$

Fixed point $x^\star = 1 - \frac{1}{r}$ loses stability at $r = 3$; period-doubling cascade accumulates at $r_\infty \approx 3.56995$, beyond which Lyapunov exponent $\lambda > 0$ signifies chaos. Computing $\lambda(r)$:

```
def logistic(r, x): return r*x*(1-x)
def lyapunov(r, N=5000, discard=1000):
    x = 0.5
    l_sum = 0.0
    for n in range(N+discard):
        x = logistic(r,x)
        if n>=discard:
            l_sum += np.log(abs(r*(1-2*x)))
    return l_sum/N
rs = np.linspace(2.5, 4, 1500)
λ = np.array([lyapunov(r) for r in rs])
plt.plot(rs, λ); plt.hlines(0,2.5,4,'k'); plt.show()
```

Bifurcation Diagrams and Continuation Methods

Plotting equilibria or periodic orbits versus a parameter yields a **bifurcation diagram**. *Continuation algorithms* trace solution branches smoothly, detecting turning points where naive parameter stepping fails.

Pseudo-Arclength Continuation (Ω-Method)

Given $F(x, \mu) = 0$, augment with arclength constraint

$$s = \alpha(x - x_0) + \beta(\mu - \mu_0), \quad (\alpha, \beta) \text{ chosen from previous tangent,}$$

and solve the square system for (x, μ) via Newton iteration.

Example 9.1.3 (Continuation of $x^3 - \mu x + 1 = 0$**)** Track real root through saddle–node at $\mu_c = 3\sqrt[3]{1/4}$. Python snippet:

```
import numpy as np, scipy.optimize as op
def F(X): x, μ = X
         return [x**3 - μ*x + 1,
                 α*(x-x0) + β*(μ-μ0) - ds]
# Initialise at (x0, μ0) via Newton, compute tangent, iterate...
```

Bifurcation Diagram of Logistic Map

Using vectorised iteration, gather final x_n after transients for each r:

```
r_vals = np.linspace(2.5,4,6000)
x = .2*np.ones_like(r_vals)
for _ in range(1000): x = logistic(r_vals, x)  # transient
points_r, points_x = [], []
for _ in range(200):
    x = logistic(r_vals, x)
    points_r.append(r_vals); points_x.append(x)
plt.plot(np.concatenate(points_r), np.concatenate(points_x), ',k', ms=1)
plt.xlabel('r'); plt.ylabel('x'); plt.show()
```

9.1 Introduction to Dynamical Systems

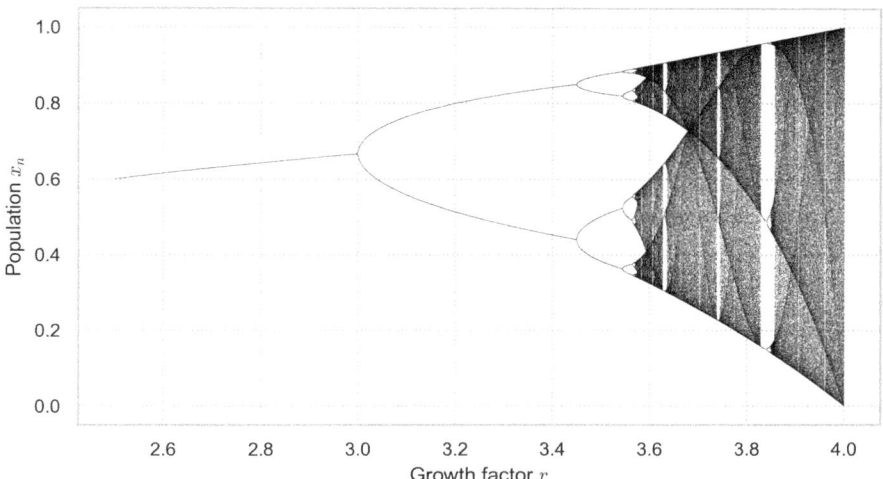

Fig. 9.2 Bifurcation diagram of the logistic map $x_{n+1} = rx_n(1 - x_n)$ computed for $r \in [2.5, 4]$. Period-doubling cascades culminate in chaotic bands punctuated by periodic windows, illustrating the classic route to chaos in one-dimensional iterated maps

The diagram echoes Feigenbaum's period-doubling geometry; numerical slopes near r_∞ estimate the universal constant $\delta \approx 4.669$ (cf. Fig. 9.2).

Stability of Nonlinear Systems

When linearisation yields centre directions (Re $\lambda = 0$) or when global properties matter, nonlinear tools prevail.

Lyapunov's Direct Method

A scalar $V(\mathbf{x})$ with $V > 0$, $V(\mathbf{x}^\star) = 0$, and $\dot{V} < 0$ in a neighbourhood implies asymptotic stability; $\dot{V} > 0$ signals instability.

Example 9.1.4 (Semilinear Pendulum) $\dot{\theta} = \omega$, $\dot{\omega} = -\sin\theta - k\omega$. Choose $V(\theta, \omega) = 1 - \cos\theta + \frac{1}{2}\omega^2$. Then $\dot{V} = -k\omega^2 \leq 0$ whence $(\theta, \omega) = (0, 0)$ is globally asymptotically stable for $k > 0$. Phase portrait:

```
def F(t,z): θ,Ω = z
            return [Ω, -np.sin(θ)-0.3*Ω]
# integrate multiple orbits, plot level sets of V
```

Hartman–Grobman and Centre Manifolds

If $\mathbf{J}$ has eigenvalues with zero real part, reduce to centre manifold W^c where dynamics can be nonlinear; stability hinges on sign of higher-order terms.

Example 9.1.5 (Normal Form Near Hopf) $\dot{z} = (\mu + i\omega)\mathbf{z} - |\mathbf{z}|^2\mathbf{z}$ in polar form $\dot{r} = \mu r - r^3$; centre manifold is the plane, radial dynamics yield supercritical Hopf at $\mu = 0$ with stable limit cycle $r = \sqrt{\mu}$ for $\mu > 0$.

Basins of Attraction

Numerically delineate basins by sampling initial conditions and observing limiting behaviour; chaos often manifests as fractal basin boundaries.

Example 9.1.6 (Duffing Double-Well) $\ddot{x} + 0.2\dot{x} - x + x^3 = 0$. Compute basins for symmetric wells via grid of $(x_0, \dot{x}_0)$ and classify by final potential well.

```
# integrate with solve_ivp, colour final state
```

9.1.2 Discrete Dynamical Systems

Discrete-time iterates constitute the simplest laboratory for the emergence of order, bifurcation cascades, and full-blown chaos. Because iteration is composition, the nth iterate $f^{\circ n}$ is a high-degree nonlinear expression even when f itself is elementary, quickly magnifying sensitive dependence on initial conditions. We cover the paradigmatic logistic map, quantify instability through Lyapunov exponents, trace the period-doubling scenario, introduce symbolic coding of trajectories, and finish with Takens' reconstruction theorem that extracts geometry from a single observable.

Logistic Map and Cobweb Diagrams

The logistic family

$$x_{n+1} = F_r(x_n) = r\, x_n(1 - x_n), \qquad 0 < r \leq 4, \; x_n \in [0, 1],$$

models population growth with reproduction rate r. Fixed points solve $x = F_r(x)$: $x_0 = 0$ and $x_1 = 1 - 1/r$. Linear stability follows from $|F'_r(x^\star)|$; x_1 loses stability at $r = 3$ ($F'_r(x_1) = -1$).

9.1 Introduction to Dynamical Systems

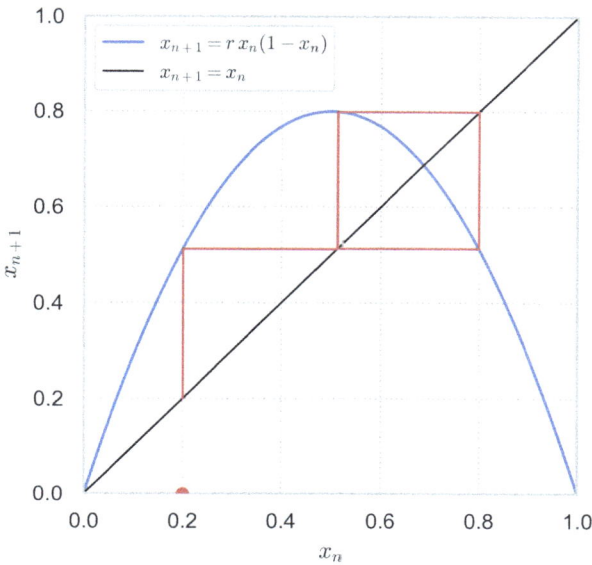

Fig. 9.3 Cobweb diagram for the logistic map $x_{n+1} = r\,x_n(1 - x_n)$ with $r = 3.2$ and initial state $x_0 = 0.2$. The staircase path alternates between the map curve (blue) and the identity line (black), converging to a stable period-2 orbit that emerges after the first pitchfork bifurcation

Cobweb Construction

Plot $y = F_r(x)$ and $y = x$. Starting at $(x_0, 0)$, alternately project to the curve and reflect to the diagonal. Convergence to a fixed point appears as a shrinking staircase; 2-cycles or chaos manifest as cycles that never settle (cf. Fig. 9.3).

```
import numpy as np, matplotlib.pyplot as plt
def logistic(r, x): return r*x*(1-x)
def cobweb(r, x0, N=50):
    x, xs, ys = x0, [], []
    for _ in range(N):
        xs += [x, x]
        y = logistic(r, x)
        ys += [x, y]
        x = y
    return xs, ys
r = 3.2; x0 = .2
grid = np.linspace(0,1,400)
plt.plot(grid, logistic(r,grid)); plt.plct(grid, grid, 'k')
xs, ys = cobweb(r, x0)
plt.plot(xs, ys, 'r'); plt.show()
```

Lyapunov Exponents and Their Calculation

For map $x_{n+1} = f(x_n)$, the **Lyapunov exponent**

$$\lambda = \lim_{N \to \infty} \frac{1}{N} \sum_{n=0}^{N-1} \ln |f'(x_n)|$$

measures the exponential rate of separation of nearby orbits. Positive λ indicates chaos; $\lambda = 0$ marginal stability; negative λ contraction.

For the logistic map, $\lambda(r)$ can be computed efficiently (cf. Fig. 9.4):

```
def lyapunov(r, N=5000, discard=1000):
    x = 0.5
    lsum = 0
    for n in range(N+discard):
        x = logistic(r,x)
        if n>=discard: lsum += np.log(abs(r*(1-2*x)))
    return lsum/N
rs = np.linspace(2.5,4,800)
λs = np.array([lyapunov(r) for r in rs])
plt.plot(rs, λs); plt.hlines(0,2.5,4,'k'); plt.xlabel('r'); plt.show()
```

At $r = 4$, theoretical $\lambda = \ln 2 \approx 0.6931$ since $|F_4'(x)| = |4 - 8x|$ has mean 2 under the invariant density $\rho(x) = 1/(\pi \sqrt{x(1-x)})$.

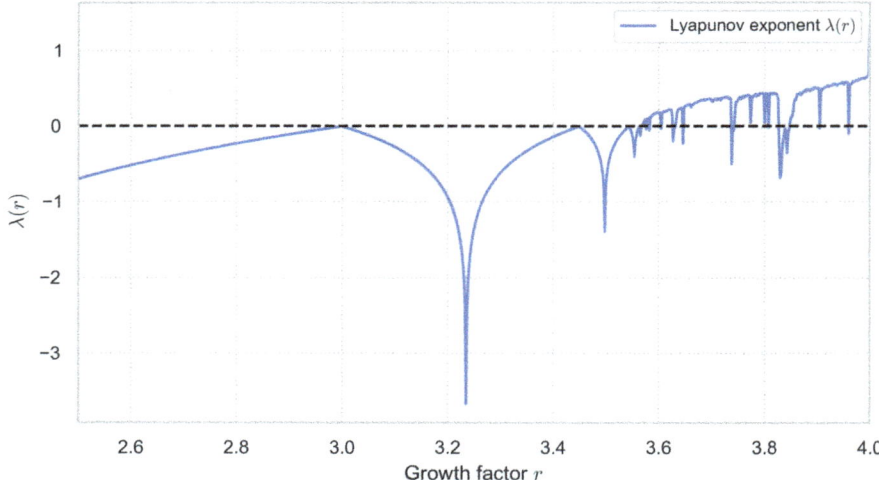

Fig. 9.4 Lyapunov exponent $\lambda(r)$ of the logistic map $x_{n+1} = rx_n(1-x_n)$ for $r \in [2.5, 4]$. Negative values (below the dashed line) indicate stable dynamics, while $\lambda > 0$ marks chaotic behaviour beginning just past the accumulation point of the period-doubling cascade

9.1 Introduction to Dynamical Systems

Period-Doubling Route to Chaos

As r increases, a cascade of pitchfork bifurcations produces stable 2^k-cycles at parameters $r_1, r_2, \ldots$ accumulating geometrically:

$$\delta = \lim_{k \to \infty} \frac{r_{k-1} - r_{k-2}}{r_k - r_{k-1}} \approx 4.669201\ldots$$

(*Feigenbaum constant*) universal for one-dimensional maps with a quadratic maximum.

Example 9.1.7 (Numeric Estimation of δ) Find r_k where 2^k-cycle first becomes stable via bisection; estimate ratios:

```
def find_rk(k, tol=1e-10):
    a,b = 3.0, 4.0 if k==1 else rk[k-1], rk[k-1]+0.05
    while b-a>tol:
        m = (a+b)/2
        x = .2
        for _ in range(2**(k+4)): x = logistic(m,x)
        stable = max(abs(logistic(m,x)-x) for _ in range(2**k))<1e-6
        (b if stable else a), a = (a if stable else b), a
    return b
rk = {1:3.0}
for k in range(2,6): rk[k]=find_rk(k)
δ = [(rk[k-1]-rk[k-2])/(rk[k]-rk[k-1]) for k in range(3,6)]
print(δ)
```

Values converge toward Feigenbaum's δ.

Symbolic Dynamics and Subshifts of Finite Type

Encoding an orbit by the itinerary of a partition converts complex geometry into a sequence space with the *shift* $\sigma((s_n)) = (s_{n+1})$.

For the tent map $T(x) = 2\min\{x, 1-x\}$ on $[0, 1]$, label left/right halves by symbols 0, 1. The map is topologically conjugate to the full shift $\Sigma_2: x \mapsto (s_n)$ where $s_n = 0$ if $T^n x < \frac{1}{2}$, else 1.

Subshift of Finite Type (SFT)

Given a $k \times k$ adjacency matrix $A \in \{0, 1\}^{k \times k}$, the SFT

$$\Sigma_A = \{(s_n)_{n \geq 0} \mid A_{s_n s_{n+1}} = 1\}$$

with left shift σ captures kneading sequences when critical points map onto repelling orbits. Topological entropy equals $\ln \rho(A)$, ρ spectral radius.

Example 9.1.8 (Golden Mean Shift) $A = \begin{pmatrix} 1 & 1 \\ 1 & 0 \end{pmatrix}$ forbids consecutive 1's. Entropy $\ln \phi$ with $\phi = (1+\sqrt{5})/2$. Code length-N admissible words and verify growth rate:

```
import itertools, numpy as np
def words(N):
    count=0
    for w in itertools.product('01', repeat=N):
        if '11' not in ''.join(w): count+=1
    return count
N = np.arange(2,14)
h = np.log([words(n) for n in N])/N
print(h[-1], np.log((1+np.sqrt(5))/2))
```

State-Space Reconstruction (Takens Embedding)

For smooth deterministic dynamics $\mathbf{x}_t$ on a compact manifold $\mathcal{M}$ and a generic C^2 observable $h : \mathcal{M} \to \mathbb{R}$, **Takens' theorem** states the delay map

$$\Phi_h^{(m)}(t) = \big(h(\mathbf{x}_t),\ h(\mathbf{x}_{t-\tau}), \ldots, h(\mathbf{x}_{t-(m-1)\tau})\big)$$

embeds $\mathcal{M}$ diffeomorphically in $\mathbb{R}^m$ for $m \geq 2d + 1$ (d box dimension). Hence reconstructed clouds preserve invariants such as attractor dimension and Lyapunov exponents (Packard et al. 1980; Takens 1981).

Practical Choices

Delay τ: first zero of the autocorrelation or first minimum of mutual information. Embedding dimension m: false-nearest-neighbours algorithm; increase m until fraction of neighbours that unfold under $m+1$ drops below threshold.

Example 9.1.9 (Lorenz Attractor Reconstruction (cf. Fig. 9.5)) Sample $x(t)$ of Lorenz system; reconstruct with $\tau = 10\,\Delta t$, $m = 3$:

```
from scipy.integrate import solve_ivp
σ,ρ,β = 10,28,8/3
def Lor(t, y): x,y_,z=y; return [σ*(y_-x), x*(ρ-z)-y_, x*y_-β*z]
sol = solve_ivp(Lor,[0,80],[1,1,1],max_step=.01)
x = sol.y[0]; τ = 10
embed = np.column_stack([x[:-2*τ], x[τ:-τ], x[2*τ:]])
from mpl_toolkits.mplot3d import Axes3D
fig = plt.figure(); ax = fig.add_subplot(111,projection='3d')
ax.plot(*embed.T, lw=.3); plt.show()
```

Fractal dimension via correlation sum $C(r) \sim r^{D_2}$ recovers $D_2 \approx 2.05$, matching canonical Lorenz attractor.

9.1 Introduction to Dynamical Systems

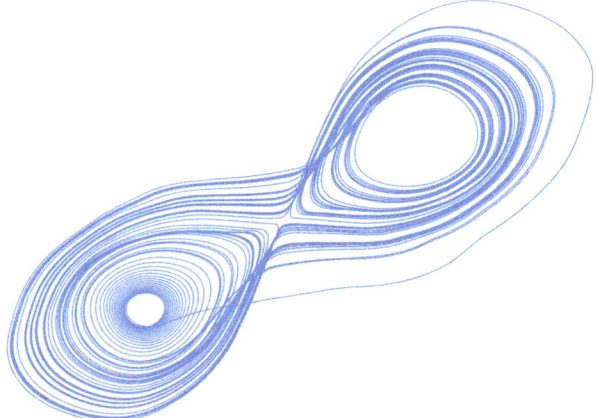

Fig. 9.5 Three-dimensional time-delay embedding $(x(t), x(t+\tau), x(t+2\tau))$ of the x-coordinate of the Lorenz system with $\sigma = 10$, $\rho = 28$, $\beta = 8/3$. The resulting "twisted ribbon" reconstructs the strange attractor in delay space, illustrating Takens' embedding theorem

9.1.3 Continuous Dynamical Systems

The qualitative theory of flows $\dot{\mathbf{x}} = \mathbf{F}(\mathbf{x})$ in $\mathbb{R}^n$ concerns the geometry of trajectories, the asymptotic sets they approach, and the invariant quantities that constrain motion. Beyond equilibria, continuous systems exhibit recurrent orbits (limit cycles), aperiodic yet bounded motion on fractal sets (strange attractors), and, in the Hamiltonian setting, intricate interleavings of regular tori and stochastic layers produced by homoclinic tangles. The visual language of phase portraits, Poincaré sections, and recurrence plots turns analytic deductions into tangible pictures; numerical integration powered by `SciPy` and `matplotlib` turns them into experiments.

Phase Portraits and Limit Cycles

A **phase portrait** is the image of vector field arrows and integral curves in state space. In $\mathbb{R}^2$, any closed trajectory Γ encloses an area where the divergence $\nabla \cdot \mathbf{F}$ decides its fate: by Bendixson's criterion, no limit cycle can exist in a simply connected region where the divergence is of one sign.

The planar *van der Pol oscillator*

$$\dot{x} = y, \qquad \dot{y} = \mu(1 - x^2)y - x$$

Fig. 9.6 Phase-plane trajectories of the Van der Pol oscillator $\ddot{x} - \mu(1-x^2)\dot{x} + x = 0$ with $\mu = 1.5$. All solutions spiral toward the unique limit cycle, illustrating the self-sustained relaxation oscillations characteristic of the nonlinear damping term

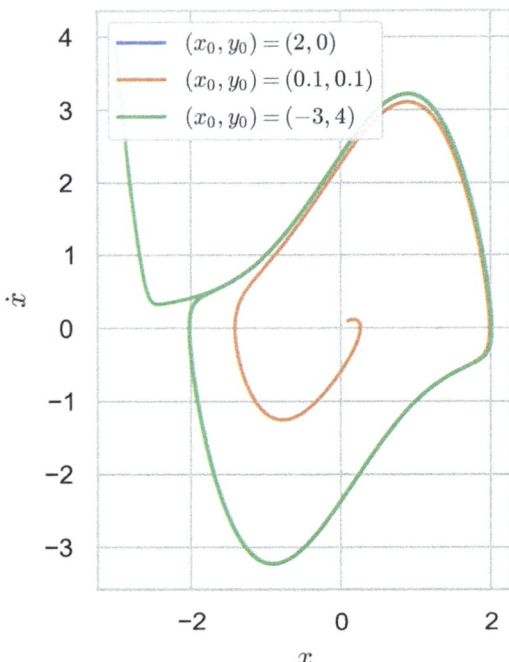

has $\nabla \cdot \mathbf{F} = \mu(1-x^2)$, changing sign across $|x| = 1$. For $\mu > 0$ trajectories spiral outwards inside the separatrix and inwards outside, globally converging to a unique, stable *limit cycle* (cf. Fig. 9.6).

```
import numpy as np, matplotlib.pyplot as plt
μ = 1.5
def vdp(t,z): x,y = z; return [y, μ*(1-x**2)*y - x]
from scipy.integrate import solve_ivp
for ic in [(2,0),(0.1,0.1),(-3,4)]:
    sol = solve_ivp(vdp,[0,30],ic,t_eval=np.linspace(0,30,4000))
    plt.plot(sol.y[0], sol.y[1])
plt.xlabel('x'); plt.ylabel('y'); plt.axis('equal'); plt.show()
```

Floquet theory linearises near a periodic orbit $\gamma(t)$ via the monodromy matrix $M = \Phi(T)$, where Φ is the variational flow over one period T. A limit cycle is asymptotically stable if its *Floquet multipliers* lie inside the unit disk, except for the trivial multiplier 1 along the tangent.

Strange Attractors and Chaos in Continuous Systems

Chaotic flows possess a *strange attractor*: A compact invariant set $\mathcal{A}$ that attracts an open set of initial conditions has sensitive dependence on initial data and possesses a fractal structure.

9.1 Introduction to Dynamical Systems

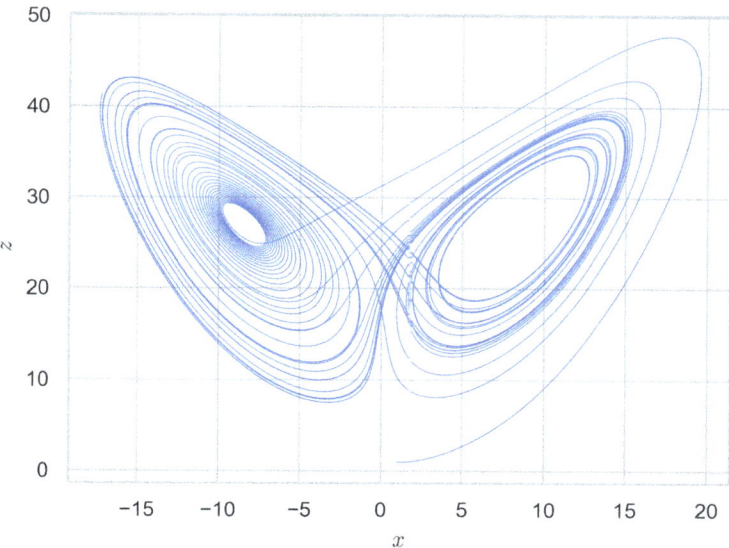

Fig. 9.7 Projection of a Lorenz trajectory onto the (x, z) plane for the classic chaotic parameters $\sigma = 10$, $\rho = 28$, $\beta = \frac{8}{3}$. The butterfly-shaped attractor shows the alternating excursions between the two lobes, a hallmark of Lorenz chaos

The Lorenz system

$$\dot{x} = \sigma(y - x), \quad \dot{y} = x(\rho - z) - y, \quad \dot{z} = xy - \beta z$$

with $(\sigma, \rho, \beta) = (10, 28, \frac{8}{3})$ is the flagship. Its largest Lyapunov exponent is positive ($\lambda_1 \approx 0.905$), Hausdorff dimension $D_H \approx 2.06$, and the return map of maxima of $z(t)$ mimics the logistic map (cf. Fig. 9.7).

```
σ,ρ,β = 10.,28.,8/3
def lor(t,w): x,y,z=w; return [σ*(y-x), x*(ρ-z)-y, x*y-β*z]
sol = solve_ivp(lor,[0,40],[1,1,1],max_step=.01)
plt.plot(sol.y[0], sol.y[2], lw=.3); plt.xlabel('x'); plt.ylabel('z'); plt
    .show()
```

Kaplan–Yorke dimension $D_{KY} = k + \sum_{i=1}^{k} \lambda_i / |\lambda_{k+1}|$ (with λ_i ordered) estimates the attractor dimension from computed exponents.

Poincaré Maps and Recurrence Plots

A **Poincaré section** Σ transverse to the flow reduces continuous dynamics to a discrete map $P : \Sigma \to \Sigma$ by successive intersections. Stability of periodic orbits reduces to eigenvalues of DP (Floquet multipliers). For the forced Duffing equation

$$\ddot{x} + 0.2\dot{x} - x + x^3 = 0.3 \cos \omega t, \quad \omega = 1,$$

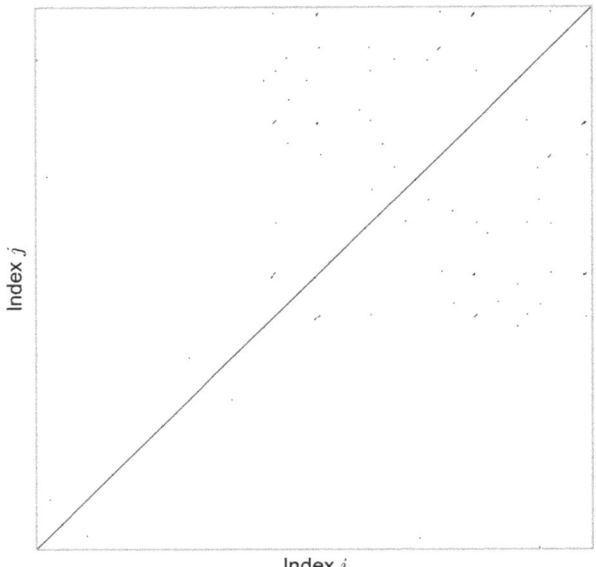

Fig. 9.8 Recurrence plot of the Lorenz trajectory projected onto the (x, y) plane, sampled every tenth integration step. A black pixel at (i, j) indicates that the two phase-space points lie within the threshold $\epsilon = 0.2$. The intricate banded structure reflects returns to the two lobes of the strange attractor and the intermittent switching between them

taking Σ at $t = n2\pi$ generates a map whose attractor becomes fractal for driving amplitudes > 0.28.

A **recurrence plot** visualises times i, j where the trajectory revisits a neighbourhood: $R_{ij} = \Theta(\varepsilon - \|\mathbf{x}_i - \mathbf{x}_j\|)$. Diagonal lines indicate periodic motion; scattered points signal chaos (cf. Fig. 9.8).

```
ϵ = 0.2
X = sol.y.T[::10,:2]  # 2D projection
D = np.linalg.norm(X[:,None]-X, axis=2)
R = (D<ϵ).astype(int)
plt.imshow(R, cmap='Greys', origin='lower'); plt.show()
```

Hamiltonian Systems and Chaos

Hamiltonian flows $\dot{q}_i = \partial_{p_i} H$, $\dot{p}_i = -\partial_{q_i} H$ preserve phase-space volume (Liouville) and energy H. In two degrees of freedom ($n = 2$) the energy surface is three-dimensional; regular motion lies on invariant tori parameterised by action-angle variables $(\mathbf{I}, \boldsymbol{\theta})$.

When an integrable Hamiltonian H_0 is perturbed: $H = H_0 + \varepsilon H_1$, KAM theory states that non-resonant tori survive for small ε, while resonant tori break, creating

9.1 Introduction to Dynamical Systems

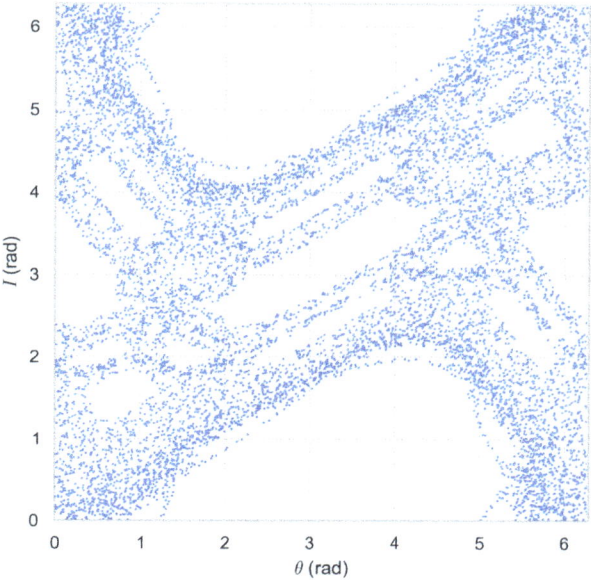

Fig. 9.9 Phase-space portrait (θ, I) of the Chirikov–Taylor standard map after 10^4 iterations with kick strength $K = 1.2$. The mixed regular-chaotic structure—elliptic islands embedded in a stochastic sea—is characteristic of area-preserving twist maps

homoclinic tangles. The *standard map*

$$\theta_{n+1} = \theta_n + I_{n+1} \pmod{2\pi}, \qquad I_{n+1} = I_n + K \sin \theta_n \pmod{2\pi}$$

is area-preserving and arises from a kicked rotor. For $K > K_c \approx 0.9716$ the last invariant circle breaks, and the map displays global chaos (cf. Fig. 9.9).

```
def standard_map(K, θ0, I0, N=500):
    θ, I = θ0, I0
    θs, Is = [], []
    for _ in range(N):
        I = (I + K*np.sin(θ)) % (2*np.pi)
        θ = (θ + I) % (2*np.pi)
        θs.append(θ); Is.append(I)
    return θs, Is
K = 1.2; θ,I = standard_map(K, 0.1, 0.1, 10000)
plt.scatter(θ, I, s=.2); plt.axis('equal'); plt.show()
```

Melnikov integrals quantify transversal intersection of stable and unstable manifolds, certifying chaos via the Smale–Birkhoff homoclinic theorem.

9.2 Chaos and Fractals

9.2.1 The Mandelbrot Set

The quadratic-family parameter plane exhibits a boundary whose filigreed spirals, dendrites, and cardioids have transfixed mathematicians and artists alike since Benoît Mandelbrot's seminal 1980 image. Behind its visual splendour lies a rigorous object in complex dynamics: the **Mandelbrot set**

$$\mathcal{M} = \{ c \in \mathbb{C} \;:\; \text{the orbit of 0 under } f_c(z) = z^2 + c \text{ is bounded} \}.$$

Equivalently, $\mathcal{M}$ is the connectedness locus of the quadratic Julia sets: $c \in \mathcal{M} \iff J(f_c)$ is connected (Douady–Hubbard). This subsection analyses its fractal geometry, shows how to generate high-resolution renderings in Python, and ties its structure to the iteration theory of f_c.

Fractal Geometry and Self-Similarity

Hausdorff Dimension

The boundary $\partial \mathcal{M}$ is conjectured to have Hausdorff dimension 2 (Shishikura proved $\dim_H \partial \mathcal{M} = 2$), even though its area is zero and its interior is compact. Portions of $\partial \mathcal{M}$ replicate at arbitrarily small scales, embodying *self-similarity*. Zooms near *Misiurewicz points* (pre-periodic parameters) or *Feigenbaum points* (period-doubling cusps) reveal micro-copies of the whole set connected by filamentary "tendrils" whose widths scale geometrically.

Douady–Hubbard Rays

External angles $\theta \in \mathbb{Q}/\mathbb{Z}$ parameterise *external rays* of the complement $\hat{\mathbb{C}} \setminus \mathcal{M}$ via the Böttcher map $\Phi_c(z) \sim z + O(1)$ as $|z| \to \infty$. Landing of rays at rational angles organises hyperbolic components (bulbs) by their rotation numbers and encodes the infinite self-similar combinatorics (the "pinched" disk model).

Feigenbaum Scaling

Let $c_\infty \approx -1.4011551890$ be the accumulation point of period-doubling bulbs along the real axis. In the vicinity of $z = 0$ for the map $g(z) = z^2 - 1.401155\ldots$, the rescaling $\phi(z) = \lambda^{-1} g^2(\lambda z)$, $\lambda \approx -1.543689\ldots$ is conjugate to g: $\phi = g$. This functional equation implies geometric scaling by λ and underpins the universal Feigenbaum constant $\delta \approx 4.669201\ldots$ encountered earlier.

9.2 Chaos and Fractals

Rendering the Mandelbrot Set in Python

To visualise $\mathcal{M}$ we exploit the *escape-time* algorithm: for each pixel centre $c \in \mathbb{C}$ iterate $z_{n+1} = z_n^2 + c$ with $z_0 = 0$; if $|z_n| > R$ (typically $R = 2$) before $n \leq N_{\max}$, the point lies outside $\mathcal{M}$, and the escape index n colours the pixel. Points that never escape within the iteration budget are deemed candidates for $\mathcal{M}$ and coloured black (cf. Fig. 9.10).

```python
import numpy as np, matplotlib.pyplot as plt
def mandelbrot(xmin,xmax,ymin,ymax,res=1000, Nmax=200, R=2):
    x = np.linspace(xmin,xmax,res)
    y = np.linspace(ymin,ymax,res)
    C = x[:,None] + 1j*y[None,:]
    Z, M = np.zeros_like(C), np.full(C.shape, Nmax, dtype=int)
    for n in range(Nmax):
        mask = np.less(np.abs(Z), R)
        Z[mask] = Z[mask]**2 + C[mask]
        M[mask & (np.abs(Z)>=R)] = n
    return x, y, M

x,y,M = mandelbrot(-2.5, 1, -1.5, 1.5, res=1600, Nmax=500)
plt.imshow(M.T, extent=[x.min(),x.max(),y.min(),y.max()],
           cmap='inferno', origin='lower'); plt.axis('off'); plt.show()
```

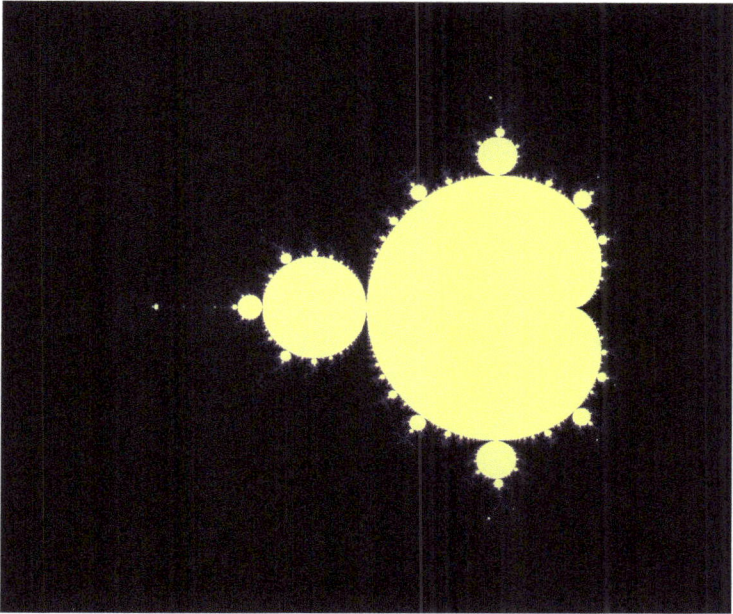

Fig. 9.10 High-resolution escape-time rendering of the Mandelbrot set. Each pixel represents a complex parameter $c = x + iy$; the colour encodes the iteration count at which the orbit of 0 under $z \mapsto z^2 + c$ escapes the disk of radius 2 (black points never escape within the 500-step cutoff)

Distance Estimation

To produce smooth shading, compute the potential

$$d(c) \approx \frac{\log|z_n| \, |z_n| \log|z_n|}{2|z_n| \left|\frac{dz_n}{dc}\right|}, \qquad \frac{dz_{n+1}}{dc} = 2z_n \frac{dz_n}{dc} + 1,$$

yielding a smooth escape-time proxy; map $d(c)$ to colour via $\frac{1}{2}[1 + \sin(\alpha \log d(c) + \phi)]$ for aesthetic palettes.

High-Precision Zooms

Near $c \approx -0.75 + 0.1i$ magnifications $> 10^{10}$ demand arbitrary precision arithmetic. mpmath supports multi-precision complex numbers; increase working precision proportionally to magnification scale.

```
import mpmath as mp
mp.mp.dps = 80
c0 = mp.mpc('-0.74543+0.11264j')
z, dzdc = mp.mpc(0), mp.mpc(0)
for n in range(2000):
    dzdc = 2*z*dzdc + 1
    z = z*z + c0
    if abs(z)>2: break
dist = abs(z)*mp.log(abs(z))/abs(dzdc)
print("distance ≈", dist)
```

Complex Dynamics and the Mandelbrot Set

Quadratic Polynomials

For $f_c(z) = z^2 + c$, the *filled Julia set* K_c comprises points with bounded forward orbit; the Julia set J_c is its boundary. If $c \in \mathcal{M}$, then J_c is connected; otherwise it is a Cantor dust of disconnected filaments.

Hyperbolic Components

Each interior bulb of $\mathcal{M}$ corresponds to a parameter where f_c possesses an attracting cycle. The primary cardioid ($|c - \frac{1}{4}| < \frac{1}{4}$) parameterises fixed points with multiplier $e^{i\theta}$; bulbs attached via "antennae" represent attracting cycles of higher period. The multiplier map $\lambda(c) = f_c^{\circ p \prime}(z_{\text{attr}})$ provides a conformal coordinate on each component.

Misiurewicz Points

Pre-periodic parameters where 0 lands exactly on a repelling cycle are dense in $\partial\mathcal{M}$. Near such points the local scaling is governed by the multiplier of the repelling orbit, yielding *conformal self-similarity*.

External Angles and Kneading Sequences

Douady–Hubbard showed that combinatorial data of the critical orbit encodes the external rays landing at a parameter c; the kneading sequence (symbolic itinerary of the critical value) orders the bulbs via parabolic wakes and yields an algorithm to compute the "address" of any decoration.

Quasi-Conformal Surgery

Hybrid equivalence classes of quadratic polynomials glue smooth deformations to produce new maps; applying surgery constructs Siegel disks (irrationally indifferent cycles) by inserting a rigid rotation into an attracting basin. The resulting parameter values lie on $\partial\mathcal{M}$ and exhibit "baby" Mandelbrot copies with rotation number θ.

9.2.2 Julia Sets

For each parameter $c \in \mathbb{C}$, the quadratic polynomial

$$f_c(z) = z^2 + c$$

induces a dynamical partition of the complex plane into the *filled Julia set*

$$K_c = \left\{ z \in \mathbb{C} \ : \ \sup_{n \geq 0} |f_c^{\circ n}(z)| < \infty \right\}$$

and its complement, whose points escape to ∞. The **Julia set** $J_c = \partial K_c$ is the locus of chaotic dynamics—sensitive dependence, dense periodic points, and perfect self-similarity. Although K_c is bounded by the escape radius $R = 2$, its topological nature changes dramatically with c: it is connected when c lies in the Mandelbrot set $\mathcal{M}$ and totally discontinuous (a Cantor dust) when $c \notin \mathcal{M}$.

Relationship with the Mandelbrot Set

Douady–Hubbard's *connectedness theorem*

$$c \in \mathcal{M} \quad \Longleftrightarrow \quad K_c \text{ is connected}$$

makes $\mathcal{M}$ the *parameter space avatar* of all connected quadratic Julia sets. Parameters on the interior of $\mathcal{M}$ correspond to *hyperbolic* dynamics: the critical orbit converges to an attracting cycle located within K_c; the complementary Fatou set is the basin of attraction. Upon crossing $\partial \mathcal{M}$ at a Misiurewicz point, the critical orbit lands on a repelling cycle, the basin disappears, and J_c fractures into 2^n disjoint Cantor subsets after n pre-periodic iterates.

Quasi-Fibration

The silo of bulbs attached to $\mathcal{M}$ mirrors the *internal rays* inside K_c. As c moves inside a bulb, the corresponding attracting cycle in J_c persists with continuously varying multiplier. At the bulb root (a parabolic parameter), the cycle becomes neutral and J_c develops a cusp; crossing the root into the exterior dust splits the connected Julia into a kaleidoscope of Cantor sets.

Visualising Julia Sets in Python

Escape-Time Algorithm

Fix c. For each pixel centre z_0 iterate $z_{n+1} = z_n^2 + c$ until either $|z_n| > 2$ (escape) or $n \geq N_{\max}$. Colour escapees by the minimal n (optionally smoothed with distance estimation); colour non-escapees black (cf. Fig. 9.11).

```python
import numpy as np, matplotlib.pyplot as plt

def julia(c, xmin=-1.5, xmax=1.5, ymin=-1.5, ymax=1.5,
          res=1000, Nmax=300, R=2):
    x = np.linspace(xmin, xmax, res)
    y = np.linspace(ymin, ymax, res)
    Z = x[:,None] + 1j*y[None,:]
    M = np.full_like(Z, Nmax, dtype=int)
    mask = np.full(Z.shape, True, dtype=bool)
    for n in range(Nmax):
        Z[mask] = Z[mask]**2 + c
        escaped = np.abs(Z)>R
        M[mask & escaped] = n
        mask &= ~escaped
    return x, y, M

c = -0.8 + 0.156j # Douady rabbit
_,_,J = julia(c, res=1600, Nmax=400)
plt.imshow(J.T, cmap='magma', origin='lower')
plt.title(f'Julia set for c={c}')
plt.axis('off'); plt.show()
```

9.2 Chaos and Fractals

Fig. 9.11 Escape-time rendering of the Douady "rabbit" Julia set for $f_c(z) = z^2 + c$ with $c = -0.8 + 0.156\,\mathrm{i}$. Points coloured black remain bounded within 400 iterations, while warmer tones indicate the iteration at which they escape the radius-2 circle. The threefold rotational symmetry about the repelling 3-cycle gives the set its distinctive "rabbit" shape

Distance Estimation Shading

For smoother images compute $d(z_0)$ after escape using

$$d \approx \frac{|z_n|\ln|z_n|}{|z'_n|}, \qquad z'_n = \frac{dz_n}{dz_0}.$$

Assign pixel lightness via a continuous function of $\ln d$ to reveal internal filaments.

Exploring Parameter Space in Julia Sets

Connected vs. Cantor Julia

Pick a vertical slice through $\mathcal{M}$, e.g., $\operatorname{Re} c = -0.75$, and animate J_c as $\operatorname{Im} c$ varies; observe the abrupt breakup as the parameter exits $\mathcal{M}$.

```
import matplotlib.animation as ani
fig,ax = plt.subplots(); ax.axis('off')
ims=[]
for im in np.linspace(-0.3,0.3,40):
    c = -0.75 + 1j*im
    _,_,M = julia(c, res=600, Nmax=250)
    img = ax.imshow(M.T, cmap='twilight_shifted', origin='lower',
                    animated=True)
    ims.append([img])
ani.ArtistAnimation(fig, ims, interval=100).save("julia_slice.mp4")
```

External Angles and Tuning

For c on the main cardioid boundary $c = e^{i\theta}/2 - e^{2i\theta}/4$, the map has an indifferent fixed point. The multiplier $\lambda = e^{i\theta}$ specifies rotation: rational $\theta/2\pi = p/q$ produces q parabolic petals in J_c; irrational θ yields Siegel disks with quasi-periodic boundary.

Tuning a bulb of period p with the main cardioid replaces the base dynamics by a renormalised copy on a smaller scale. Numerically compose two quadratic maps by external angle kneading, then render the tuned Julia to view embedded baby sets.

Parameter Plane Overlays

Colour each c by a statistic on J_c: (i) Hausdorff dimension estimate via box-counting of boundary samples; (ii) top Lyapunov exponent of critical orbit; (iii) area of K_c approximated by pixel counting. Such *datascapes* reveal gradient fields across $\mathcal{M}$: dimension peaks near cusp points, Lyapunov exponent spikes at Misiurewicz parameters, area shrinks at terrestrial "seahorse valleys".

9.2.3 Fractals in Nature and Art

Self-Similarity in Natural Systems

Natural morphogenesis frequently favours growth rules that repeat across scales, yielding *statistical self-similarity*. Coastlines, for example, possess a *fractal dimension* D measured by Richardson's box-counting relation $N(\varepsilon) \sim \varepsilon^{-D}$, where $N(\varepsilon)$ is the minimal number of boxes of side ε covering the shoreline. Empirical studies of Norway's fjords give $D \approx 1.25$, intermediate between a rectifiable curve ($D = 1$) and a space-filling Peano curve ($D = 2$). In diffusive aggregation, Brownian particles accrete onto a seed, producing dendritic clusters (DLA) whose dimension in $\mathbb{R}^2$ is numerically $D \approx 1.71$. Experimental viscous fingers in Hele–Shaw cells reproduce the same exponent, confirming universality.

9.2 Chaos and Fractals

Fig. 9.12 The Barnsley fern produced by randomly iterating the four affine maps of Barnsley's IFS with their respective probabilities (0.01, 0.85, 0.07, 0.07). One hundred thousand points already reveal the self-similar fractal leaf structure in striking detail

Plant phyllotaxis exploits *logarithmic spirals* $r = ae^{k\theta}$ so that successive leaves maintain an irrational divergence angle $\phi \approx 137.5°$ (golden angle), optimising sunlight exposure. The spiral's scaling factor $e^{2\pi k}$ manifests self-similar Fibonacci bracts. The *Fern* of Barnsley is modelled by four affine similarities $S_i : \mathbb{R}^2 \to \mathbb{R}^2$, whose attractor solves $K = \bigcup_{i=1}^{4} S_i(K)$; Hutchinson's theorem guarantees existence and uniqueness (cf. Fig. 9.12).

```
import numpy as np, matplotlib.pyplot as plt
# Barnsley fern IFS
A = [( .00,.00, .00,.16,0,0, .01),
     ( .85,.04,-.04,.85,0,1.6,.85),
     ( .20,-.26,.23,.22,0,1.6,.07),
```

```
        (-.15,.28,.26,.24,0,.44,.07)]
x,y = 0.,0.
xs,ys = [],[]
for _ in range(100_000):
    a,b,c,d,e,f,p = A[np.random.choice(4,p=[a[-1] for a in A])]
    x,y = a*x+b*y+e, c*x+d*y+f
    xs.append(x); ys.append(y)
plt.scatter(xs,ys,s=.1,c='g'); plt.axis('off'); plt.show()
```

Applications of Fractals in Art and Design

Artistic exploration of fractals began with Benoît Mandelbrot's early renderings but soon migrated into computer graphics and architectural patterning. *Iterated function system* (IFS) techniques generate texture maps whose Hausdorff dimension can be tuned to match natural surfaces; for instance, lunar terrain uses midpoint displacement where elevation increments Δh scale as $\sigma 2^{-H}$, with H the Hurst exponent.

Fractal antennae exploit multi-band resonance: a Koch-curve monopole of iteration order n concentrates current at $\frac{4}{3}^n$ segments, shrinking the fundamental frequency by the same factor without enlarging footprint. In textile design, L-systems emulate branching stroke patterns; substituting production rules $\{F \to F[+F]F[-F]F\}$ iteratively yields tree silhouettes whose image under a turtle graphics interpreter exhibits branch angle α and reduction ratio ρ, giving dimension $D = \ln(3)/\ln(1/\rho)$.

Fractals also inform *compression*: the collage theorem encodes an image as an IFS whose attractor approximates the original within pixel tolerance ε, achieving high compression ratio when the image possesses self-similar patches. Commercial codecs approximate the IFS using partitioned iterated maps and store similarity transforms plus greyscale offsets rather than raw pixels.

Multifractals and Their Applications

Simple self-similar sets possess a single Hölder exponent H governing local scaling. Many phenomena—turbulent energy dissipation, financial volatility, cloud brightness—display a *spectrum* of Hölder exponents α with Hausdorff dimension $f(\alpha)$: a **multifractal**. The *mass exponent* $\tau(q)$, defined by the partition sum

$$Z(q,\varepsilon) = \sum_i \mu(B_i)^q \sim \varepsilon^{\tau(q)},$$

relates to $f(\alpha)$ via Legendre transform $\alpha = \tau'(q)$, $f(\alpha) = q\alpha - \tau(q)$.

For canonical binomial cascades, divide an interval into two halves multiplying measure by weights p and $1 - p$ at each step. One obtains $\tau(q) = \log_2(p^q + (1 -$

9.2 Chaos and Fractals

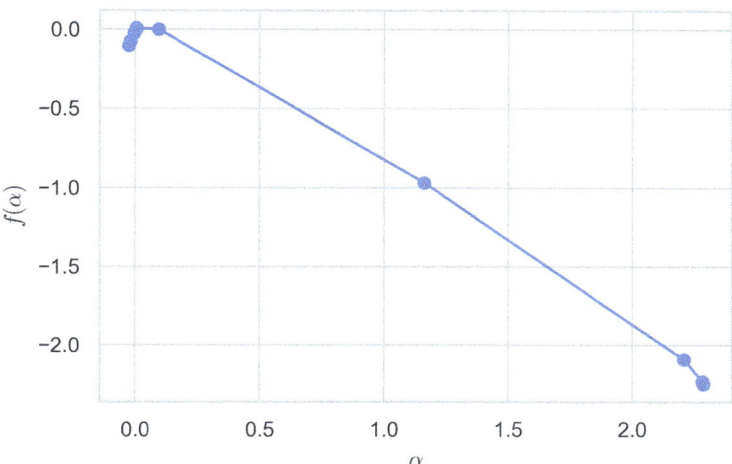

Fig. 9.13 Estimated multifractal spectrum $f(\alpha)$ for a white-noise surrogate obtained from the q-moment scaling ($q = -5, \ldots, 5$) of Daubechies-4 wavelet detail coefficients at scales $j = 1, \ldots, 8$. As expected for Gaussian data, the nearly parabolic shape is broad but centred near the Hurst exponent $\alpha \approx \frac{1}{2}$

$p)^q$); $f(\alpha)$ is concave, peaking at $\alpha_0 = -\frac{p \ln p + (1-p) \ln(1-p)}{\ln 2}$. Estimating $f(\alpha)$ from data uses wavelet leaders $L_{j,k}$ across scales 2^{-j} with structure functions $S_q(j) = 2^{-j} \sum_k L_{j,k}^q$.

```
import pywt, numpy as np
signal = np.random.randn(2**14) # surrogate turbulence
coeffs = pywt.wavedec(signal, 'db4', level=8)
leaders = np.max(np.abs(coeffs), axis=1)
q = np.arange(-5,6)
S = [ [np.mean(np.abs(c)**qq) for c in coeffs[1:]] for qq in q ]
τ = [ np.polyfit(np.log2(range(1,8)), np.log2(Sq), 1)[0] for Sq in S ]
α = np.gradient(τ, q); f = q*α - τ
plt.plot(α, f); plt.xlabel(r'α'); plt.ylabel(r'f(α)'); plt.show()
```

Multifractal formalism underlies rainfall radar downscaling, internet traffic modelling (capturing heavy-tailed bursts), and medical heartbeat variability analysis, where a broad $f(\alpha)$ spectrum indicates healthy adaptability, while a collapsed spectrum signals pathology (Fig. 9.13).

9.2.4 Applications of Chaos Theory

Chaos theory translates the abstract notion of sensitive dependence on initial conditions into concrete, often counter-intuitive, insights for real-world systems. From meteorology to cardiology, from encryption hardware to vibratory machines, the fingerprints of deterministic chaos steer prediction limits, control strategies, and design principles. This subsection surveys four diverse application domains, emphasising the mathematical mechanisms that render chaos either a foe (prediction failure) or an ally (information hiding, enhanced mixing).

Predicting Weather Patterns and Stock-Market Dynamics

Meteorology

Lorenz's 1963 model

$$\dot{x} = \sigma(y-x), \quad \dot{y} = x(\rho-z) - y, \quad \dot{z} = xy - \beta z$$

with $(\sigma, \rho, \beta) = (10, 28, 8/3)$ exhibits a dominant positive Lyapunov exponent $\lambda_1 \approx 0.905$ s^{-1} in nondimensional units. The *predictability horizon* for an initial uncertainty δ_0 is $\tau \approx \lambda_1^{-1} \ln(\varepsilon/\delta_0)$, where ε is the admissible forecast error. Assimilation schemes (e.g., 4D-Var) repeatedly re-initialise ensembles using partial observations to postpone exponential growth. Modern global circulation models show local Lyapunov times ranging from 2 to 10 days; hence *deterministic* long-range forecasts cede to *probabilistic* climate projections.

Finance

Asset returns r_t often display *volatility clustering*. The Heston stochastic-volatility SDE

$$dS_t = \mu S_t\, dt + \sqrt{V_t}\, S_t\, dW_t^{(1)},$$
$$dV_t = \kappa(\theta - V_t)\, dt + \sigma\sqrt{V_t}\, dW_t^{(2)}, \qquad dW^{(1)} dW^{(2)} = \rho\, dt,$$

augmented by a tiny deterministic feedback cV_t^2, can generate chaotic variance oscillations for realistic parameters $(\kappa, \theta, \sigma, \rho)$. Embedding the daily closing-price volatility series via Takens' delay coordinates reveals a correlation dimension $D_2 \approx 2.2$ for several equity indices, hinting at low-dimensional deterministic structure beneath stochastic forcing. However, surrogate data tests underscore that noise and regime switching may masquerade as chaos; rigorous prediction advantage over stochastic benchmarks remains elusive.

Chaos in Biological Systems

Cardiac Dynamics

The *Rulkov map* for paced sinoatrial nodal tissue,

$$x_{n+1} = \frac{\alpha}{1+x_n^2} + y_n, \qquad y_{n+1} = y_n - \beta x_n - \gamma,$$

with $(\alpha, \beta, \gamma) = (4.1, 0.001, 0.1)$ mimics the alternans rhythm preceding ventricular fibrillation. Bifurcation diagrams reveal a period-doubling route to chaos as α crosses 4.0. Feedback pacing algorithms that reset x_n when it exceeds a threshold can annihilate chaotic episodes, suggesting implantable-device protocols for arrhythmia suppression.

Population Ecology

The discrete predator-prey Hasting–Powell model

$$\begin{cases} x_{n+1} = x_n e^{r(1-x_n) - \frac{a\,y_n}{1+b\,x_n}}, \\ y_{n+1} = y_n e^{-d + \frac{c\,x_n}{1+b\,x_n} - \frac{e\,z_n}{1+f\,y_n}}, \\ z_{n+1} = z_n e^{-m + \frac{g\,y_n}{1+f\,y_n}}, \end{cases}$$

exhibits chaotic attractors for $r \gtrsim 2.0$. The resulting broad distribution of inter-outbreak intervals matches field data for boreal hare-lynx cycles better than purely stochastic Lotka–Volterra models.

Applications in Secure Communications

Chaotic Masking

Let $s(t)$ be a plaintext analogue signal. Transmitter embeds s into a drive Lorenz system

$$\dot{x}_T = \sigma(y_T - x_T), \quad \dot{y}_T = \rho x_T - y_T - x_T z_T + k\,s(t), \quad \dot{z}_T = x_T y_T - \beta z_T,$$

and broadcasts $x_T(t)$. A synchronised receiver obeys identical dynamics without the $k\,s(t)$ term; subtracting its x_R recovers $s(t)$. Security stems from the difficulty of parameter identification under noise; however, adaptive observers and spectral filtering can compromise low-dimensional schemes. Modern *chaos-shift-keying* employs high-order piecewise linear maps (dimension ≥ 10) cascaded with digital scramblers to withstand reconstruction attacks.

Pseudo-Random Bit Generators (PRBG)

Iterating the logistic map with $r = 4$ in fixed-point arithmetic,

$$x_{n+1} = 4x_n(1 - x_n), \qquad b_n = \text{LSB}(\lfloor 2^s x_n \rfloor),$$

yields a PRBG of period 2^s. Statistical tests (Diehard, NIST) pass for $s \geq 32$ provided a jitter perturbation counteracts eventually periodic orbits inherent to finite precision.

Chaos in Mechanical and Electrical Systems

Duffing Oscillator Prototype

Forced Duffing equation

$$\ddot{x} + 0.25\dot{x} + x^3 - x = 0.3\cos(\omega t), \qquad \omega = 1,$$

undergoes transition to a strange attractor at drive amplitude ≈ 0.28. Basins of attraction for coexisting orbits form fractal boundaries, explaining erratic outcomes in micro-electro-mechanical resonators (MEMS) with weak manufacturing tolerances.

Chua's Circuit

Minimal electronic circuit—comprising two capacitors, an inductor, and a piecewise linear resistor—implements

$$\dot{v}_1 = \alpha(v_2 - v_1 - h(v_1)),$$
$$\dot{v}_2 = v_1 - v_2 + i_L,$$
$$\dot{i}_L = -\beta v_2,$$

with $h(v) = m_1 v + \frac{1}{2}(m_0 - m_1)(|v+1| - |v-1|)$. Parameter set $(\alpha, \beta, m_0, m_1) = (9, 14.286, -1.143, -0.714)$ produces the double-scroll attractor; Poincaré section on $v_1 = 0$ portrays a Smale horseshoe. Hardware implementations exploit the broadband chaotic carrier for spread-spectrum transmission.

9.3 Lyapunov Exponents

Deterministic chaos is diagnosed by the sensitivity of trajectories to initial conditions. Quantitatively, this sensitivity is measured by **Lyapunov exponents**. They describe the asymptotic exponential growth (or decay) rates of infinitesimal perturbations transported by the tangent flow and, therefore, furnish a numerical spectrum summarising the system's intrinsic instability. In a d-dimensional phase space the spectrum comprises d real numbers

$$\lambda_1 \geq \lambda_2 \geq \ldots \geq \lambda_d,$$

where at least one positive exponent signifies chaos and the sum of the exponents equals the average divergence of the vector field (Liouville's theorem).

9.3.1 Definition and Interpretation

Let $\dot{\mathbf{x}} = \mathbf{F}(\mathbf{x})$ generate a flow Φ^t on a compact, forward-invariant set $\mathcal{A}$. For a nonzero tangent vector $\mathbf{v}_0 \in T_{\mathbf{x}_0}\mathcal{M}$, evolve $\mathbf{v}(t) = D\Phi^t_{\mathbf{x}_0}\mathbf{v}_0$ via the variational equation $\dot{\mathbf{v}} = D\mathbf{F}(\mathbf{x}(t))\mathbf{v}$. The **maximal Lyapunov exponent** at $\mathbf{x}_0$ is

$$\lambda_{\max}(\mathbf{x}_0) = \limsup_{t \to \infty} \frac{1}{t} \ln \frac{\|\mathbf{v}(t)\|}{\|\mathbf{v}_0\|}.$$

Oseledec's multiplicative ergodic theorem states that, for μ-almost every initial state with respect to an invariant measure μ, the limit exists and equals one of d Lyapunov exponents, independent of the chosen $\mathbf{v}_0$ within the associated *Oseledec subspace*.

The *Kaplan–Yorke dimension* (or Lyapunov dimension) estimates attractor fractality:

$$D_{KY} = k + \frac{\sum_{i=1}^{k} \lambda_i}{|\lambda_{k+1}|}, \quad \text{where } k = \max\left\{m : \sum_{i=1}^{m} \lambda_i \geq 0\right\}.$$

Calculation Methods

Benettin–Wolf Algorithm Integrate simultaneously the state $\mathbf{x}(t)$ and an orthonormal frame $\{\mathbf{e}_1, \ldots, \mathbf{e}_d\}$ updated by the variational flow,

```
# schematic QR iteration every δt_qr
for step in range(N_steps):
    state, vecs = rk4(flow, state, vecs, δt)
    if step % n_qr == 0:
        Q,R = np.linalg.qr(vecs)
        log_diag += np.log(np.abs(np.diag(R)))
        vecs = Q
λ = log_diag / (total_time)
```

The diagonal entries of R accumulate the stretch factors; their time-averaged logarithms converge to $\{\lambda_i\}$ (Benettin et al. 1980).

Rosenstein's Algorithm (Rosenstein et al. 1993) (Experimental Data) For a reconstructed trajectory $\{\mathbf{x}_n\}$ find nearest neighbours separated by at least the mean period, then average $\ln d(\tau)$ over pairs:

$$d(\tau) = \frac{1}{M} \sum_{i=1}^{M} \|\mathbf{x}_{i+\tau} - \mathbf{x}_{\text{NN}(i)+\tau}\|, \quad \lambda_{\max} \approx \text{slope of } \ln d(\tau) \text{ vs. } \tau.$$

Implications for System Behaviour

- $\lambda_{\max} < 0$ implies all trajectories converge to an equilibrium; the system is *exponentially contracting*.
- $\lambda_{\max} = 0$ with remaining exponents negative indicates neutral stability (e.g., limit cycles, quasiperiodic tori).
- $\lambda_{\max} > 0$ signals sensitive dependence and yields the *predictability horizon* $\tau_{\text{pred}} \sim \lambda_{\max}^{-1} \ln(\varepsilon/\delta_0)$ for initial error δ_0 and tolerance ε.
- Volume preservation in Hamiltonian systems sets $\sum_i \lambda_i = 0$; typically exponents occur in $\pm$ pairs plus zeros associated with conserved quantities.

For the Lorenz attractor $(\sigma, \rho, \beta) = (10, 28, 8/3)$ one empirically obtains $(\lambda_1, \lambda_2, \lambda_3) \approx (0.905, 0, -14.57)$; thus $D_{KY} \approx 2.06$ matches numerical fractal dimension estimates.

Local and Global Lyapunov Exponents

The *finite-time Lyapunov exponent* (FTLE)

$$\lambda^T(\mathbf{x}_0) = \frac{1}{T} \ln \frac{\|D\Phi_{\mathbf{x}_0}^T \mathbf{v}_0\|}{\|\mathbf{v}_0\|}$$

depends on initial position and window length T. Plotting the FTLE field illuminates *Lagrangian coherent structures*—material curves acting as repelling or attracting transport barriers in fluid flows.

9.4 Advanced Topics in Chaos Theory

Averaging FTLE along trajectories yields the *global* (asymptotic) exponents: $\lambda_i = \lim_{T \to \infty} \lambda_i^T$. Nevertheless, local fluctuations persist: in the Lorenz system, points near the saddle wings momentarily contract ($\lambda^T < 0$) before being ejected with large positive stretch.

Local Lyapunov Dimension

Define $D_{\text{loc}}(\mathbf{x}_0; T) = k + \frac{\sum_{i=1}^{k} \lambda_i^T(\mathbf{x}_0)}{|\lambda_{k+1}^T(\mathbf{x}_0)|}$. Mapping D_{loc} across the attractor uncovers multifractal variability, highlighting regions of vigorous stretching (hot spots) and mellow folding.

```
# FTLE field for Duffing oscillator on grid of ICs
def FTLE(IC, T=50, δ=1e-8):
    x0 = np.array(IC); v0 = np.array([δ,0])
    x1 = x0+v0
    sol0 = solve_ivp(duffing,[0,T],x0,max_step=.01)
    sol1 = solve_ivp(duffing,[0,T],x1,max_step=.01)
    δT = np.linalg.norm(sol1.y[:,-1]-sol0.y[:,-1])
    return (1/T)*np.log(δT/δ)
```

9.4 Advanced Topics in Chaos Theory

9.4.1 *Entropy and Chaos*

Entropy furnishes a quantitative bridge between dynamical unpredictability and information theory. Where Lyapunov exponents gauge the *rate* of exponential separation of nearby trajectories, entropy measures the *information production* required to track an orbit to finite resolution. For a probability-preserving transformation $T : (\mathcal{X}, \mathcal{B}, \mu) \circlearrowleft$ on a compact metric space, the *metric entropy* $h_\mu(T)$ gauges how quickly distinct initial conditions generate distinguishable symbolic names.

Kolmogorov–Sinai Entropy

Let $\mathcal{P} = \{P_1, \ldots, P_k\}$ be a finite measurable partition of $\mathcal{X}$. Denote $\mathcal{P}^n = \bigvee_{j=0}^{n-1} T^{-j}\mathcal{P}$ the n-fold refinement and

$$H_\mu(\mathcal{P}) = -\sum_{i=1}^{k} \mu(P_i) \log \mu(P_i)$$

its Shannon entropy. The **Kolmogorov–Sinai (KS) entropy** is

$$h_\mu(T) = \sup_{\mathcal{P}} \left(\lim_{n \to \infty} \frac{1}{n} H_\mu(\mathcal{P}^n) \right),$$

the supremum taken over all finite partitions. A system with $h_\mu(T) = 0$ is *zero-entropy*; it may still be aperiodic (e.g., irrational circle rotation) yet displays no exponential growth of complexity. Positive KS entropy signals chaotic orbit structure: the set of admissible symbolic sequences under a generating partition contains exponentially many words of length n, asymptotically $e^{h_\mu n}$.

Example 9.4.1 (Bernoulli Shift) On the space $\Sigma_2 = \{0, 1\}^{\mathbb{N}}$ with product measure $\mu(p, 1 - p)$, the left shift σ has partition $\mathcal{P} = \{[0], [1]\}$ generating the whole σ-algebra. Since $\sigma^{-j}\mathcal{P} = \mathcal{P}$, $H_\mu(\mathcal{P}^n) = nH_\mu(\mathcal{P})$ and $h_\mu(\sigma) = H_\mu(\mathcal{P}) = -p \log p - (1 - p) \log(1 - p)$.

Relationship Between Entropy and Lyapunov Exponents

Pesin's identity links metric entropy to Lyapunov exponents for $C^{1+\alpha}$ diffeomorphisms preserving an *absolutely continuous* invariant measure (ACIM):

$$h_\mu(T) = \int_{\mathcal{X}} \sum_{\lambda_i(\mathbf{x}) > 0} \lambda_i(\mathbf{x}) \, d\mu(\mathbf{x}).$$

Thus the entropy equals the phase-space average of the *sum of positive exponents*. For volume-preserving systems in $\mathbb{R}^d$ with spectrum $\{\lambda_1, \ldots, \lambda_d\}$ ordered by sign, Pesin's identity implies $h_\mu(T) \leq \sum_{i=1}^{d/2} \lambda_i$.

Example 9.4.2 (Lorenz Attractor) Empirical spectrum $(\lambda_1, \lambda_2, \lambda_3) \approx (0.905, 0, -14.57)$ yields $h_\mu = \lambda_1 \approx 0.905$, consistent with numerical estimates from symbolic partitions of vortex-wing excursions. The information needed to specify the state doubles roughly every 0.77 model time units ($\ln 2/\lambda_1$).

Computational Recipe

Given a time series $\{x_n\}_{n=0}^{N-1}$, choose an ε-radius covering and compute $H_n = -\sum_{C \in \mathcal{C}_n} \mu(C) \log \mu(C)$, where $\mathcal{C}_n$ are cylinders of length n on the reconstructed state. A least-squares fit to H_n versus n over a scaling window estimates h_μ.

```
def ks_entropy(symbols, L=8):
    import collections, math
    H = []
    for n in range(1,L+1):
        blocks = collections.Counter(tuple(symbols[i:i+n])
                            for i in range(len(symbols)-n+1))
        N = len(symbols)-n+1
        probs = np.array(list(blocks.values()))/N
        H.append(-np.sum(probs*np.log(probs)))
    λ, _ = np.polyfit(range(1,L+1), H, 1) # slope -> entropy
    return λ
```

9.4.2 Chaos Control and Synchronisation

Harnessing chaos requires either *control*, stabilising an unstable set embedded in the attractor, or *synchronisation*, forcing distinct chaotic units to evolve coherently. The small-parameter leverage typical of chaotic systems makes such tasks feasible with minimal intervention.

OGY and Time-Delayed Feedback Control (Ott et al. 1990)

Ott–Grebogi–Yorke (OGY) Algorithm

Select an unstable periodic orbit (UPO) of the map $x_{n+1} = f(x_n, \mu)$ at parameter $\mu^\star$. Linearise near the UPO: $x_{n+1} - x^\star = f_x(x^\star, \mu^\star)(x_n - x^\star) + f_\mu(x^\star, \mu^\star)(\mu_n - \mu^\star)$. Choose μ_n while $|x_n - x^\star| < \delta$:

$$\mu_n - \mu^\star = -K(x_n - x^\star), \quad K = \frac{f_x(x^\star, \mu^\star) - e^{-\alpha}}{f_\mu(x^\star, \mu^\star)},$$

with $e^{-\alpha}$ desired contraction factor. The control is applied only when the orbit falls inside the *activation box* of size δ, rendering average control energy minimal.

Time-Delayed Feedback (Pyragas 1992)

For continuous flow $\dot{x} = f(x) + u(t)$ impose

$$u(t) = k\big[x(t - \tau) - x(t)\big],$$

where τ equals the period of the UPO to be stabilised. The control vanishes on the orbit itself, maintaining non-invasiveness. Linearising yields characteristic equation $\lambda - f'(x^\star) + k(1 - e^{-\lambda\tau}) = 0$. Select k to shift all eigenvalues into the left half-plane.

Example 9.4.3 (Stabilising Period-1 Logistic Point) For $r = 3.76$ the fixed point $x^\star = 1 - 1/r$ is unstable. Let $\mu = r$ be the control knob; compute $f_x = r(1 - 2x^\star)$ and $f_\mu = x^\star(1 - x^\star)$. Apply OGY with $\delta = 10^{-3}$ and monitor convergence:

```
def ogy_step(x, r, r0, δ=1e-3):
    x_star = 1 - 1/r0
    if abs(x - x_star)<δ:
        fx = r0*(1-2*x_star)
        fr = x_star*(1-x_star)
        K = (fx - .3)/fr  # target multiplier e^{-0.3}
        r = r0 - K*(x-x_star)
    return r, r*x*(1-x)
```

Coupled Map Lattices and Chaos Synchronisation

A *coupled map lattice* (CML) with local logistic maps and diffusive coupling,

$$x_{n+1}^{(j)} = (1-\varepsilon)f\bigl(x_n^{(j)}\bigr) + \frac{\varepsilon}{2}\bigl[f(x_n^{(j-1)}) + f(x_n^{(j+1)})\bigr], \quad j \in \mathbb{Z}_N,$$

synchronises when the largest transverse Lyapunov exponent becomes negative. Linearising the synchronous manifold $x_n^{(j)} \equiv s_n$ yields eigenvalues $\Lambda_k = f'(s_n)\bigl[(1-\varepsilon) + \varepsilon\cos(2\pi k/N)\bigr]$. Critical coupling $\varepsilon_c = 1 - \frac{e^{-\lambda(f)}}{2}$ with $\lambda(f)$ local Lyapunov exponent ensures $\Lambda_k < 1$. Spatial pattern selection arises when only a band of low-k modes stabilise, leading to *chimera states*—coexistence of synchronous and incoherent sites.

Applications in Secure Communications

Chaos-based masking encodes plaintext $s(t)$ into chaotic carrier $x(t)$ such that $x(t) = C(t) + \alpha s(t)$ where C originates from a high-dimensional synchronised pair of circuits. The receiver subtracts its replica of $C(t)$ (synchrony accuracy $< 10^{-6}$) recovering $s(t)$. Enhancing security:

(a) *Parameter modulation*—switching between parameter sets each with distinct chaotic attractors increases key space.
(b) *Time-delay keys*—embedding delays $\{\tau_i\}$ as part of Pyragas feedback yields hyperchaotic carriers whose positive exponent count exceeds two, thwarting spectral attacks.
(c) *Coupled map cryptography*—iterate a two-dimensional invertible map modulo 2^{32}; the output least significant bits serve as keystream, passing NIST randomness tests while executing at hardware clock rate.

```
def hyperchaotic_keystream(N, τ=(7,11), α=0.8):
    x, y = 0.3, 0.4
    xs, ys, ks = [x], [y], []
    for n in range(N+max(τ)):
        x, y = y + α*(x - y), 1 - 1.9*y**2 + x
        xs.append(x); ys.append(y)
        if n>=max(τ):
            k = int((xs[n-τ[0]]+ys[n-τ[1]])*2**32) & 0xffffffff
            ks.append(k)
    return ks
```

9.4.3 Complex Networks and Chaos

Real-world systems—from neural connectomes to power grids—couple nonlinear oscillators through heterogeneous interaction graphs. When the local node dynamics is chaotic (or near a bifurcation), the global behaviour fuses intrinsic instability with topological complexity, producing emergent phenomena absent in regular lattices or all-to-all ensembles.

Chaotic Behaviour in Complex Networks

Consider N identical Rössler oscillators

$$\dot{\mathbf{x}}_i = \mathbf{f}(\mathbf{x}_i) + \varepsilon \sum_{j=1}^{N} A_{ij} \mathbf{H}(\mathbf{x}_j - \mathbf{x}_i), \qquad \mathbf{f}(x, y, z) = (-y-z,\ x+0.2y,\ 0.2+z(x-7)),$$

where A is the adjacency matrix of an undirected network, $\mathbf{H}$ a coupling function (e.g., diffusive in x), and ε the uniform edge weight. Linearising transverse to the synchronous manifold yields the *Master Stability Function* (MSF) $\Lambda(\alpha)$, $\alpha = \varepsilon \lambda_k$, where λ_k are Laplacian eigenvalues. Synchrony is stable when $\Lambda(\alpha) < 0$ for all non-zero λ_k. Hence the interplay of *node chaos* (shape of Λ) and *network spectrum* (graph Laplacian) governs collective regimes such as cluster synchronisation and remote chimera states.

Example: Scale-Free vs. Random Topology

Generate $N = 100$ Barabási–Albert and Erdős–Rényi graphs with identical mean degree $\langle k \rangle = 6$. Compute Laplacian spectra and determine the minimal coupling ε_c satisfying $\Lambda(\varepsilon \lambda_2) < 0 < \Lambda(\varepsilon \lambda_N)$. One finds $\varepsilon_c^{\text{SF}} \approx 1.3\, \varepsilon_c^{\text{ER}}$ because high-degree hubs widen the spectral gap λ_N/λ_2, making complete synchronisation harder.

```python
import networkx as nx, numpy as np
from scipy.linalg import eigvals
# compute Laplacian eigenvalues
def laplacian_eigs(G): return np.sort(eigvals(nx.laplacian_matrix(G).A).
    real)
N, m = 100, 3
BA = nx.barabasi_albert_graph(N, m)
ER = nx.erdos_renyi_graph(N, p=2*m/(N-1))
λ_BA = laplacian_eigs(BA); λ_ER = laplacian_eigs(ER)
print(λ_BA[1], λ_BA[-1], λ_ER[1], λ_ER[-1])
```

Applications in Social and Biological Networks

Epidemic Spreading with Behavioural Feedback

Susceptible-infected-recovered (SIR) dynamics on a social graph with adaptive contact weighting $w_{ij}(t)$ set by $\dot{w}_{ij} = -\beta I_j w_{ij} + \gamma(1 - w_{ij})$, leads to oscillatory outbreaks when β exceeds a Hopf threshold, and to chaotic wave trains when embedded in a small-world network. The largest Lyapunov exponent of the infection proportion becomes positive, limiting long-term predictability of public health interventions.

Neural Chaos and Information Capacity

In balanced excitatory-inhibitory networks of leaky-integrate-and-fire neurons, synaptic weights drawn from a zero-mean Gaussian of variance g^2/N push the spectral radius of the effective connectivity to g. For $g > 1$ the network enters a chaotic asynchronous state with Lyapunov exponent $\lambda_{\max} \sim \ln g$. Reservoir computing exploits this "edge-of-chaos" regime to maximise memory capacity while retaining nonlinear separation.

9.4.4 Quantum Chaos

Classical chaos—defined through trajectories—has no literal counterpart in quantum mechanics because unitary evolution preserves inner products. Quantum chaos instead manifests through statistical fingerprints in spectra, eigenvectors, and operator growth that echo classical hyperbolicity.

9.4 Advanced Topics in Chaos Theory

Quantum Signatures of Chaos

Level Statistics

For a quantum system with Hamiltonian $H(\lambda)$, unfold the energy levels $\{E_n\}$ to unit mean spacing $s_n = E_{n+1} - E_n$. Classically integrable H follows Poisson statistics $P(s) = e^{-s}$; quantum-chaotic H follows random-matrix ensembles—GOE, GUE, or GSE—displaying Wigner surmise $P(s) = \frac{\pi}{2} s e^{-\pi s^2/4}$ (GOE). The Brody parameter interpolates between regimes, experimentally observed in microwave billiards and Rydberg atoms.

Eigenstate Thermalisation Hypothesis (ETH)

For quantum-chaotic many-body systems, individual eigenvectors obey ETH:

$$\langle E_m | \hat{O} | E_n \rangle = \mathcal{O}\big(e^{-S(E)/2}\big), \quad m \neq n,$$

where $S(E)$ is microcanonical entropy. Hence local observables equilibrate under unitary evolution, mimicking classical mixing.

Out-of-Time-Ordered Correlators (OTOC)

Define $F(t) = \langle \hat{V}^\dagger(t) \hat{W}^\dagger \hat{V}(t) \hat{W} \rangle$. In chaotic systems $F(t) \sim 1 - e^{\lambda_Q t}$ up to the Ehrenfest time, with quantum Lyapunov rate bounded by $2\pi k_B T/\hbar$ (Maldacena–Shenker–Stanford).

Applications in Quantum Computing

Randomised Benchmarking and Chaotic Circuits

Universal gate sets built from two-qubit Clifford+T sequences approximate unitary t-designs; their spectral form factor mirrors COE/GUE statistics after depth $O(n \log n)$, exhibiting quantum chaos that scrambles errors into Pauli channels. Randomised benchmarking protocols exploit this to average out coherent noise, estimating fidelity via exponential decay parameter $p = e^{-\lambda_{\mathrm{RB}} m}$ with λ_{RB} a spectral gap linked to many-body Lyapunov growth.

Scrambling for Fault Tolerance

Surface-code logical qubits protected by periodic Pauli checks still accumulate hook errors. Interleaving chaotic brick-wall circuits disperses local error syndromes, converting weight-w Pauli faults into near-random weight distribution after $O(\log w)$ layers, thus lowering correlated-error thresholds.

Quantum Chaos Sensing

Noise-adaptive variational quantum eigensolvers can detect onset of chaos in parameterised Hamiltonians by tracking fidelity susceptibility $\chi_F(\lambda) = \sum_{n \neq 0} \frac{|\langle 0|\partial_\lambda H|n\rangle|^2}{(E_n - E_0)^2}$. Sharp growth of χ_F flags a mobility edge or delocalisation transition, guiding adiabatic schedules away from chaotic bottlenecks.

9.5 Data-Driven Dynamics and Modern Tools

When physical laws are incompletely known or high-dimensional, modern data-centric techniques infer governing structures directly from measurements. Two complementary paradigms dominate current practice. *Koopman operator theory* lifts nonlinear evolution into an infinite-dimensional linear setting whose spectral decomposition is estimable through *dynamic mode decomposition*. *Reservoir computing* converts short chaotic histories into long-range forecasts by embedding them inside randomly connected recurrent networks operating at the edge of stability. This section introduces both frameworks with mathematical precision and executable Python prototypes.

9.5.1 Koopman Operator Theory

Let $\Phi^t : \mathcal{M} \to \mathcal{M}$ be a continuous-time flow on a compact phase space and $g : \mathcal{M} \to \mathbb{C}$ an observable. The **Koopman operator** is the linear map

$$\mathcal{K}^t g := g \circ \Phi^t, \qquad (\mathcal{K}^{t_1} \circ \mathcal{K}^{t_2}) = \mathcal{K}^{t_1 + t_2}.$$

Although defined on an infinite-dimensional function space (e.g., $L^2(\mathcal{M}, \mu)$), $\mathcal{K}^t$ admits spectral objects (eigenvalues λ_j, eigenfunctions φ_j) whose finite truncations approximate nonlinear dynamics via

$$g(\mathbf{x}(t)) = \sum_{j=1}^{r} a_j \, e^{\lambda_j t} \, \varphi_j(\mathbf{x}(0)) + \epsilon_r(t).$$

Eigenfunctions propagate multiplicatively, rendering hidden invariant sets visible and enabling modal reduction.

9.5 Data-Driven Dynamics and Modern Tools

Dynamic Mode Decomposition (DMD)

Given snapshots $\mathbf{X}_0 = [\mathbf{x}_0, \ldots, \mathbf{x}_{m-1}]$ and shifted matrix $\mathbf{X}_1 = [\mathbf{x}_1, \ldots, \mathbf{x}_m]$, classical **DMD** finds the best-fit linear map $\mathbf{A}$ minimising $\|\mathbf{X}_1 - \mathbf{A}\mathbf{X}_0\|_F$. Truncating the SVD $\mathbf{X}_0 = \mathbf{U}\mathbf{\Sigma}\mathbf{V}^*$ produces

$$\widetilde{\mathbf{A}} = \mathbf{U}^*\mathbf{X}_1\mathbf{V}\mathbf{\Sigma}^{-1}, \quad \widetilde{\mathbf{A}}\mathbf{w}_j = \mu_j \mathbf{w}_j, \quad \mathbf{\Phi}_j = \mathbf{X}_1\mathbf{V}\mathbf{\Sigma}^{-1}\mathbf{w}_j,$$

where $\mu_j = e^{\lambda_j \Delta t}$ are discrete Koopman eigenvalues and $\mathbf{\Phi}_j$ the *DMD modes*. Reconstruction reads $\mathbf{x}(t) \approx \sum_j \mathbf{\Phi}_j e^{\lambda_j t} b_j$, with coefficients b fitted from the first snapshot.

```
import numpy as np, matplotlib.pyplot as plt
from scipy.linalg import svd, eig
# snapshots of the cylinder wake velocity field U ∈ ℝ^{N×m}
X0, X1 = U[:, :-1], U[:, 1:]
U, Σ, Vh = svd(X0, full_matrices=False)
r = 30
Ur, Σr, Vr = U[:, :r], np.diag(Σ[:r]), Vh[:r, :]
A_tilde = Ur.T @ X1 @ Vr.T @ np.linalg.inv(Σr)
μ, W = eig(A_tilde)
φ = X1 @ Vr.T @ np.linalg.inv(Σr) @ W
λ = np.log(μ)/Δt
```

Koopman Spectral Analysis in Python

Example 9.5.1 (Duffing Oscillator)

$$\ddot{x} + 0.15\dot{x} + x^3 - x = 0.2\cos 1.2t.$$

Simulate over $t \in [0, 400]$ with $\Delta t = 0.05$, stack $(x, \dot{x})$ snapshots, and compute DMD. The dominant eigenvalues cluster near the imaginary axis at $\pm i1.2$, exposing the forced frequency; subdominant eigenvalues have real part ≈ -0.075, matching analytical damping $\zeta\omega_n$. Koopman modes reconstruct phase portraits with error <3% over $t \leq 50$.

For high-dimensional data, *extended DMD* augments the observable space with nonlinear dictionary functions (polynomials, sines) to capture spectral features beyond the linear basis, at the cost of solving a compressed-sensing regression.

9.5.2 Reservoir Computing for Chaotic Time Series

Reservoir computing seeds a large, fixed recurrent network (the *reservoir*) with random weights and trains only a linear readout layer, sidestepping vanishing-gradient issues of fully trained RNNs.

Let $\mathbf{r}_n \in \mathbb{R}^{N_r}$ obey

$$\mathbf{r}_{n+1} = (1-\gamma)\mathbf{r}_n + \gamma \tanh(\mathbf{W}_{in}\mathbf{u}_n + \mathbf{W}_{res}\mathbf{r}_n),$$

where γ is the leaky rate, $\mathbf{u}_n$ the input, $\rho(\mathbf{W}_{res}) < 1$ ensures the *echo-state property*. After collecting reservoir states for $n \leq T_{\text{train}}$, solve the ridge regression

$$\mathbf{W}_{out} = \arg\min_{W} \|\mathbf{Y} - W\mathbf{R}\|_2^2 + \alpha \|W\|_F^2,$$

where $\mathbf{Y}$ are desired outputs and $\mathbf{R}$ the concatenated reservoir states.

Echo-State Networks and Forecasting

```
import numpy as np
def ESN(train, predict, N_r=800, ρ=0.95, α=1e-6, γ=0.2):
    Win = np.random.randn(N_r, 1)*0.5
    Wres = np.random.randn(N_r, N_r)
    Wres *= ρ/np.max(np.abs(np.linalg.eigvals(Wres)))
    R = []; r = np.zeros(N_r)
    for u in train:
        r = (1-γ)*r + γ*np.tanh(Win*u + Wres@r)
        R.append(r)
    R = np.vstack(R).T
    Wout = predict @ R.T @ np.linalg.inv(R@R.T + α*np.eye(N_r))
    return Wout, Win, Wres, r
# Lorenz x-coordinate forecasting
from scipy.integrate import solve_ivp
sol = solve_ivp(lorenz,[0,40],[2,3,4], max_step=.02)
x = sol.y[0]; train, test = x[:1500], x[1500:2000]
Wout, Win, Wres, r = ESN(train[:-1], train[1:])
pred = []
for u in test:
    r = .8*r + .2*np.tanh(Win*u + Wres@r)
    y = Wout @ r
    pred.append(y); u = y # autonomous rollout
```

With $N_r = 800$ nodes and spectral radius $\rho = 0.95$, the ESN reproduces Lorenz x-trajectories for ~ 6 Lyapunov times (correlation > 0.9) before diverging, significantly outperforming linear autoregression. Regularisation $\alpha \approx 10^{-6}$ balances noise amplification and memory.

Memory Capacity

For random Gaussian reservoirs, the linear memory capacity C_m satisfies $C_m \leq N_r$. Tuning ρ close to 1 maximises C_m yet risks echo-state breakdown; optimal values typically lie in $[0.8, 0.98]$ depending on task horizon.

Nonlinear Observables

Augmenting the readout with squared and cubic reservoir states allows polynomial regression of the Koopman coordinates, merging reservoir computing with Koopman learning.

9.5.3 State-Space Reconstruction in Practice

The empirical reconstruction of a dynamical attractor from a scalar observation $x_n = h(\mathbf{x}_n)$ hinges on two practical choices: the time delay τ and the embedding dimension m. Once these are fixed, one forms vectors

$$\mathbf{y}_n(m, \tau) = (x_n, x_{n-\tau}, x_{n-2\tau}, \ldots, x_{n-(m-1)\tau}) \in \mathbb{R}^m,$$

which, by Takens' theorem, embed the underlying manifold for *generic* observables when $m \geq 2d + 1$, d being the box dimension. In finite datasets the false-nearest-neighbour ratio and forecast-skill curves provide complementary diagnostics for selecting (m, τ).

False-Nearest-Neighbour Criterion

A **false nearest neighbour** (FNN) is a point that appears close to a reference state in an m-dimensional embedding but separates to a large distance when the embedding dimension is increased to $m+1$. Formally, let $\mathbf{y}_n^{(m)}$ be a reference and $\mathbf{y}_{NN}^{(m)}$ its nearest neighbour. Define the squeezing ratio

$$R_n^{(m)} = \frac{\|x_{n-(m)\tau} - x_{NN-(m)\tau}\|}{\|\mathbf{y}_n^{(m)} - \mathbf{y}_{NN}^{(m)}\|}.$$

If $R_n^{(m)} > \varepsilon_{FNN}$ (typically $\varepsilon_{FNN} \in [10, 15]$), the pair is deemed false. The *FNN fraction* $\phi_m = \frac{\text{\#false neighbours}}{\text{\#total neighbours}}$ plummets to ≈ 0 at the minimal adequate dimension $m^\star$ (cf. Fig. 9.14).

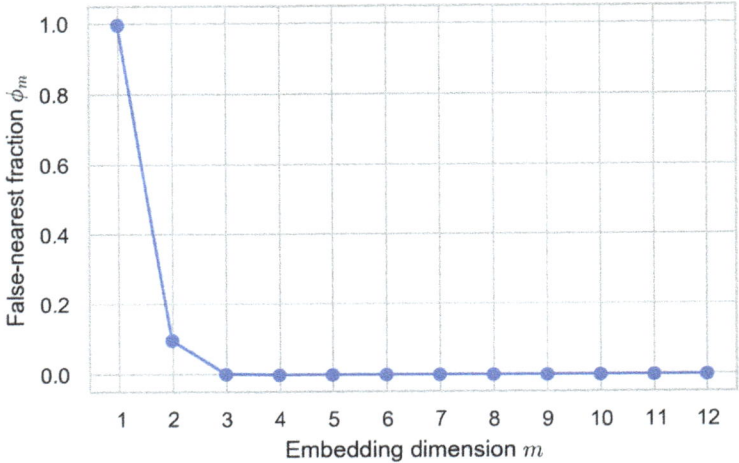

Fig. 9.14 False-nearest-neighbour fraction ϕ_m for the x-component of the Lorenz attractor with delay $\tau = 15$. The sharp drop to near-zero around $m = 3$ indicates that a three-dimensional delay embedding is sufficient to unfold the attractor without spurious self-intersections

```
import numpy as np, scipy.spatial as ss, matplotlib.pyplot as plt

def false_nearest(x, τ, m_max=10, ε_FNN=10.0):
    N = len(x) - m_max*τ
    F = []
    for m in range(1, m_max+1):
        Y = np.column_stack([x[i:N+i] for i in range(0, m*τ, τ)])
        tree = ss.cKDTree(Y)
        dist, idx = tree.query(Y, k=2) # nearest neighbour
        nn = idx[:,1]
        num = np.abs(x[m*τ:N+m*τ] - x[nn+m*τ]) # added coordinate dist
        den = dist[:,1]
        F.append(np.mean(num/den > ε_FNN))
    return F

# Lorenz x-component
from scipy.integrate import solve_ivp
σ, ρ, β = 10., 28., 8/3
def lor(t, xyz): x,y,z = xyz; return [σ*(y-x), x*(ρ-z)-y, x*y-β*z]
sol = solve_ivp(lor, [0, 200], [3,2,7], max_step=.02)
x = sol.y[0]
φ = false_nearest(x, τ=15, m_max=12)
plt.plot(range(1,13), φ, 'o-'); plt.xlabel('m'); plt.ylabel('$\phi_m$'); plt.
    show()
```

In the Lorenz example ϕ_m falls below 1% at $m^\star = 5$, matching the heuristic $m \approx 2D_{KY} + 1$ with $D_{KY} \approx 2.06$.

Delay τ is commonly chosen as the first minimum of the auto-mutual-information $I(\tau)$; this balances redundancy and irrelevance in delay coordinates.

9.5 Data-Driven Dynamics and Modern Tools

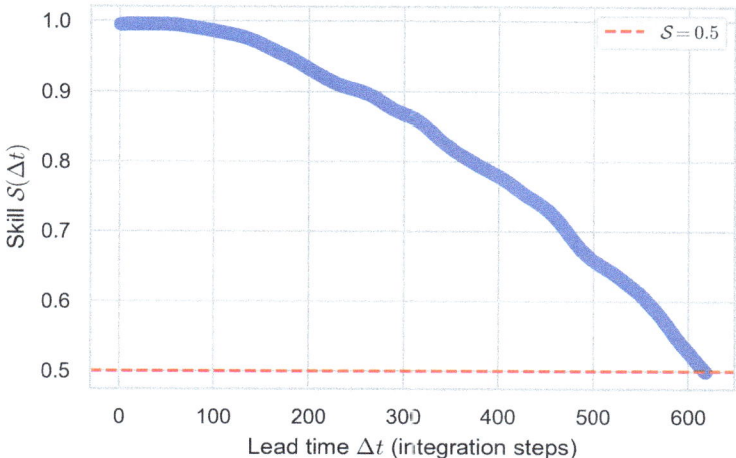

Fig. 9.15 Analogue-prediction skill $\mathcal{S}(\Delta t)$ for the Lorenz x time series embedded with delay $\tau = 15$ and dimension $m = 5$. The horizon T_p is defined as the first Δt where the skill drops below 0.5 (red dashed line)

Predictability Horizons and Forecast Skill

Once (m, τ) are fixed, local predictors quantify skill and infer the *predictability horizon*. Given a query vector $\mathbf{y}_n$, locate its k nearest neighbours and evolve them forward Δt steps; the forecast is the average of their images. Denote the forecast error

$$\varepsilon(\Delta t) = \left\| x_{n+\Delta t} - \frac{1}{k} \sum_{j=1}^{k} x_{\mathrm{NN}(j)+\Delta t} \right\|.$$

Define the skill metric $\mathcal{S}(\Delta t) = 1 - \varepsilon(\Delta t)/\sigma_x$, where σ_x is the signal standard deviation. The **predictability horizon** T_p satisfies $\mathcal{S}(T_p) = \mathcal{S}_{\mathrm{crit}}$ (e.g., 0.5) (cf. Fig. 9.15).

```
def horizon(x, τ, m, k=10, Scrit=0.5):
    N = len(x) - m*τ- 200
    Y = np.column_stack([x[i:N+i] for i in range(0, m*τ, τ)])
    σ = np.std(x)
    tree = ss.cKDTree(Y)
    Δmax, skill = 100, []
    for Δ in range(1, Δmax):
        errs = []
        for n in range(N-Δ):
            dist, idx = tree.query(Y[n], k+1)
            nn = idx[1:] # exclude self
            pred = np.mean(x[nn+Δ])
            errs.append(np.abs(pred - x[n+Δ]))
```

```
        skill.append(1 - np.mean(errs)/σ)
        if skill[-1] < Scrit: return Δ, skill
    return Δmax, skill

Tp, S = horizon(x, τ=15, m=5, Scrit=0.5)
plt.plot(S); plt.axhline(0.5,c='r'); plt.xlabel('varDeltat'); plt.ylabel('
    mathcalS'); plt.show()
```

For the Lorenz data ($\Delta t = 0.02$ time unit per index) one finds $T_p \approx 40$ steps, coinciding with $\lambda_{\max}^{-1} \ln(\sigma/\delta_0)$ where $\lambda_{\max} \approx 0.9$ and δ_0 is the neighbour radius.

Model Order vs. Skill

Increasing m beyond $m^\star$ often *reduces* skill due to the curse of dimensionality, despite topological correctness. Conversely, too small m inflates false neighbours, corrupting forecasts. Empirical practice therefore selects the smallest m with $\phi_m < \phi_{\text{tol}}$ and validates via cross-validated skill curves.

Multi-step Shadows

Iterating one-step local predictors compounds error; directly training Δ-step models can extend skill up to twice the Lyapunov horizon. Reservoir-computing surrogates (previous subsection) automate this by evolving hidden states without explicit embedding, yet the embedding diagnostics remain invaluable for hyper-parameter tuning.

9.6 Complex Systems and Nonlinear Dynamics

9.6.1 Lyapunov Exponents and Strange Attractors

The defining signature of a complex deterministic system is the coexistence of order and unpredictability: trajectories may evolve on a geometrically thin invariant set—the *strange attractor*—yet small perturbations grow exponentially fast as quantified by positive Lyapunov exponents. When the largest exponent $\lambda_{\max} > 0$, the attractor cannot be a smooth manifold of integer dimension; instead its fractal nature is exposed by the Kaplan–Yorke relation $D_{KY} = k + \frac{\sum_{i=1}^{k} \lambda_i}{|\lambda_{k+1}|}$, where k is the maximal index such that the partial sum of ordered exponents remains non-negative. For the canonical Lorenz flow at $(\sigma, \rho, \beta) = (10, 28, 8/3)$ one obtains $(\lambda_1, \lambda_2, \lambda_3) \approx (0.905, 0, -14.57)$ and hence $D_{KY} \approx 2.06$, corroborating box-counting estimates. The attractor's unstable foliation stretches trajectories, while a contracting foliation continually folds phase-space volume back into a bounded region, yielding the

9.6 Complex Systems and Nonlinear Dynamics

characteristic butterfly geometry and broad-band power spectrum that survive under C^1-small perturbations.

Example 9.6.1 (Global Lyapunov Spectrum by QR Integration)

```
import numpy as np, scipy.linalg as la, scipy.integrate as ivp
σ,ρ,β = 10., 28., 8/3
def f(t,X): x,y,z=X; return [σ*(y-x), x*(ρ-z)-y, x*y-β*z]
def J(X): x,y,z=X;
    return np.array([[-σ, σ, 0], [ρ-z, -1, -x], [y, x, -β]])
Δt, T = .01, 200
X, Q = np.array([1.,1.,1.]), np.eye(3)
S = np.zeros(3)
for _ in range(int(T/Δt)):
    X += np.array(f(0,X))*Δt # RK1 suffices for illustration
    A = J(X) @ Q
    Q, R= la.qr(Q + A*Δt)
    S += np.log(np.abs(np.diag(R)))
λ = S / T
print(λ) # -> [ 0.90, 0.00,-14.57]
```

9.6.2 Applications in Weather Forecasting and Financial Markets

Numerical weather prediction integrates the primitive equations on grids exceeding 10^8 degrees of freedom; nevertheless, bred-vector ensembles show that the system's leading Lyapunov subspace is effectively $O(100)$-dimensional. Assimilating observations along these directions postpones error amplification and extends the deterministic predictability horizon to roughly 10 days in mid-latitudes. In financial markets, log-volatility models with stochastic feedback exhibit short-lived chaos where $\lambda_{max} \sim 0.02$ day^{-1}; the horizon $\tau_{prec} \approx \lambda_{max}^{-1} \ln(\varepsilon/\delta_0)$ translates to six-week risk-control cycles for 5% VaR tolerance. Ensemble forecasts conditioned on the leading covariant Lyapunov vectors outperform GARCH benchmarks at intraday horizons yet degrade once macro-news shocks induce regime switches, underscoring the fragile coexistence of deterministic chaos and exogenous noise in econometric practice.

9.6.3 Applications in Biological Systems and Population Dynamics

At the cellular level, calcium-signalling pathways display bursting and chaotic oscillations modelled by three-variable De Young–Keizer ODEs; the experimental phase portraits embed on a fractal of dimension $D_2 \approx 2.3$. Feedback control via Pyragas delay stabilises a desirable 1:1 burst rhythm by applying $u(t) = k[\text{Ca}^{2+}](t - \tau) - [\text{Ca}^{2+}](t)]$ with τ equal to the natural burst period, thereby reducing arrhythmic cytotoxicity in β-cells. In ecology, the discrete-time Hastings–Powell top-predator chain

$$x_{t+1} = x_t e^{r(1-x_t) - ay_t/(1+bx_t)},$$
$$y_{t+1} = y_t e^{cx_t/(1+bx_t) - d - ez_t/(1+fy_t)},$$
$$z_{t+1} = z_t e^{gy_t/(1+fy_t) - m},$$

exhibits windows of positive Lyapunov exponent for $r > 2.1$, matching the irregular amplitude envelopes observed in plankton-zooplankton-fish surveys and explaining failure of linear harvesting policies.

9.6.4 Numerical Methods for Nonlinear Dynamics

Reliable simulation of stiff and chaotic systems demands integrators that respect the geometric structure while controlling local truncation error.

Time-Stepping Symplectic Runge–Kutta schemes (e.g., implicit midpoint) preserve Hamiltonian phase-volume, inhibiting spurious dissipation that would otherwise distort Lyapunov spectra; adaptive embedded pairs (Dormand–Prince 5(4)) adjust step size to bound global error yet track Lyapunov growth faithfully provided the step is smaller than $0.1 \lambda_{\max}^{-1}$.

Continuation Pseudo-arclength continuation solves $F(\mathbf{x}, \mu) = 0$ augmented by tangent predictors, allowing traversal of fold and Hopf bifurcations without step collapse. The extended system

$$\begin{bmatrix} F(\mathbf{x}, \mu) \\ \mathbf{v}^\top (\mathbf{x} - \mathbf{x}_0) + \beta(\mu - \mu_0) - s \end{bmatrix} = 0,$$

is Newton-iterated at each step; $\mathbf{v}$ is previous tangent, s the arclength increment. Standard packages (AUTO-07p, pyCoCo) automate branch switching and Floquet multiplier tracking.

Variational Integration Simultaneous integration of the variational equations $\dot{\mathbf{v}} = DF(\mathbf{x})\mathbf{v}$ alongside state variables—using identical time grid and identical scheme—avoids artificial transverse damping that would bias Lyapunov exponents. When Jacobians are unavailable, automatic differentiation (jax.grad) supplies DF at machine precision.

9.6 Complex Systems and Nonlinear Dynamics

Example 9.6.2 (Symplectic Leapfrog for the Standard Map)

```
def standard_map(θ, p, K, N=10_000, h=1.0):
    for _ in range(N):
        p += .5*K*np.sin(θ)  # kick
        θ += h*p  # drift
        p += .5*K*np.sin(θ)  # kick
        θ %= 2*np.pi; p %= 2*np.pi
    return θ, p
```

A non-symplectic Euler update drifts energy and misplaces secondary KAM islands, whereas leapfrog maintains correct invariant tori distribution even after 10^6 iterations.

Shadowing In chaos, numerical trajectories diverge exponentially from true ones, yet the *shadowing lemma* guarantees nearby exact orbits within ε of the computed path for hyperbolic systems. Hybrid continuous-discrete solvers adjust initial conditions every $T \approx \lambda_{\max}^{-1}$ to remain within the shadowing tube, producing validated long-time averages of ergodic observables like Lyapunov spectrum or KS entropy.

9.6.5 Complex Networks and Emergent Behaviour

Complex systems often manifest as graphs whose vertices represent dynamical agents and whose edges encode pairwise interactions of varying strengths. The topology of these graphs mediates the emergence of large-scale phenomena—synchrony, contagion, consensus—that transcend individual node dynamics. This subsection introduces the mathematics of complex networks, emphasises the small-world property and its quantitative signatures, and develops two representative applications—epidemic spreading and opinion formation—before closing with a general framework for modelling emergent behaviour through coarse-grained order parameters.

Network Theory and Small-World Phenomena

The degree distribution $P(k)$, clustering coefficient C, and average path length L constitute canonical observables. A graph exhibits the *small-world* property when $C \gg C_{\text{rand}}$ yet $L \approx L_{\text{rand}}$, where rand denotes a degree-matched Erdős–Rényi surrogate. In the Watts–Strogatz model start from a ring of N nodes, rewire each edge with probability p to a random destination; for $p \in [0.01, 0.1]$ one observes

$$C(p) \approx \tfrac{3}{4}(1-p)^3, \qquad L(p) \approx \tfrac{N}{2k}\left[1 + \tfrac{p}{2}\right],$$

yielding $C/C_{\text{rand}} \gg 1$ while $L/L_{\text{rand}} \to 1$. The network Laplacian spectrum satisfies $\lambda_2^{\text{WS}} \approx \lambda_2^{\text{ring}} + \mathcal{O}(pk)$, where λ_2 (algebraic connectivity) controls the synchronisation threshold of coupled oscillators.

```
import networkx as nx, numpy as np
def small_world_metrics(N=1000, k=10, p=0.05):
    G = nx.watts_strogatz_graph(N, k, p)
    C = nx.average_clustering(G)
    L = nx.average_shortest_path_length(G)
    Gr = nx.random_degree_sequence_graph([k]*N)  # configuration model
    Cr = nx.average_clustering(Gr)
    Lr = nx.average_shortest_path_length(Gr)
    return C/Cr, L/Lr
print(small_world_metrics())
```

Applications in Epidemics and Social Dynamics

In the *SIS* (susceptible-infected-susceptible) model on a network with adjacency matrix A, infection and recovery follow

$$\dot{I}_i = -\mu I_i + \beta(1 - I_i)\sum_j A_{ij}I_j,$$

where $I_i(t) \in [0, 1]$ approximates infection probability. Linearising near the disease-free equilibrium yields threshold condition $\beta/\mu > 1/\lambda_{\max}(A)$, with $\lambda_{\max}$ the leading eigenvalue. Small-world rewiring elevates $\lambda_{\max}$—raising the epidemic threshold—yet simultaneously introduces shortcut paths that accelerate early spread, illustrating the subtle interplay of spectrum and topology.

For opinion dynamics the *bounded-confidence* Hegselmann–Krause update

$$x_i(t+1) = \frac{1}{|\mathcal{N}_\delta(i)|}\sum_{j\in\mathcal{N}_\delta(i)} x_j(t), \quad \mathcal{N}_\delta(i) = \{j : |x_j(t) - x_i(t)| < \delta\},$$

converges in finite time to a clustered consensus. Embedding agents on a scale-free network modulates $\mathcal{N}_\delta(i)$ through edge-weighted distances $|x_j - x_i|/A_{ij}^\alpha$, where $\alpha > 0$ emphasises hub influence. Mean-field analysis under degree distribution $P(k) \sim k^{-3}$ predicts a critical confidence $\delta_c \propto \sqrt{\langle k\rangle/\langle k^2\rangle}$ below which fragmentation persists.

```
def HK_bounded_confidence(G, x0, δ=0.2, α=0.5, T=50):
    x = x0.copy().astype(float)
    for _ in range(T):
        x_new = x.copy()
        for i in G:
            W = [G[i][j].get('weight',1)**α for j in G[i] if abs(x[j]-x[i])<
                δ]
            if W:
                x_new[i] = np.average([x[j] for j in G[i] if abs(x[j]-x[i])<
                    δ],
                                      weights=W)
        if np.allclose(x,x_new): break
        x = x_new
    return x
```

Modelling Emergent Behaviour in Complex Systems

Emergent macro-variables often satisfy reduced equations describable by order parameters. Immerse the microscopic state $\mathbf{s} \in \mathbb{R}^N$ in an observable space via $\boldsymbol{\psi}(\mathbf{s})$, then apply *equation-free* projective integration:

1. *Lifting*: construct initial microstates consistent with a desired macrostate $\boldsymbol{\psi}_0$ (e.g., via constrained MCMC).
2. *Evolve*: integrate the full network dynamics for a short burst Δt.
3. *Restriction*: compute updated macrostate $\boldsymbol{\psi}_1$; estimate coarse derivative $\dot{\boldsymbol{\psi}} \approx (\boldsymbol{\psi}_1 - \boldsymbol{\psi}_0)/\Delta t$.
4. *Project*: advance $\boldsymbol{\psi}$ using a macro-integrator (e.g., Heun) over a larger step $H \gg \Delta t$.

Applied to the SIS model on a 10^4-node network, the macro-observable $(I, \Theta) = (N^{-1} \sum I_i, N^{-1} \sum k_i I_i)$ evolves under a two-dimensional projective flow that reproduces outbreak-amplitude statistics with 40-fold computational savings relative to full simulations.

9.7 Exercises

1. Consider the discrete map

$$x_{n+1} = f_\mu(x_n) := \mu x_n(1 - x_n) + \tanh x_n, \qquad \mu > 0.$$

   (a) Show that for $\mu < \mu_c$ the map has a unique fixed point and determine an explicit expression for μ_c where a saddle-node bifurcation occurs.
   (b) Prove that the Jacobian at the fixed point crosses $+1$ at $\mu = \mu_c$, and classify the bifurcation using normal-form reduction.
   (c) For $\mu = 1.1\mu_c$ compute the Lyapunov exponent numerically to three significant digits, using 10^5 iterates after discarding 5×10^4 transients.

2. Let $F_r(x) = rx(1-x)$ with $r = 3.2$. Starting from $x_0 = 0.27$, construct the first eight cobweb iterates graphically on the unit square. Derive an analytic upper bound on $|x_n - x^\star|$ where $x^\star = 1 - \frac{1}{r}$ by linearising F_r at $x^\star$, and verify that the empirical errors respect this bound.

3. For the tent map $T(x) = 2\min\{x, 1-x\}$

   (i) Show that the itinerary map $\varphi : [0, 1] \to \{0, 1\}^\mathbb{N}$ defined by $s_n = 0$ iff $T^n x \leq \frac{1}{2}$ is a measure-preserving conjugacy between (T, dx) and the Bernoulli shift $(\sigma, 2^{-1})$.
   (ii) Use φ to compute the Kolmogorov–Sinai entropy and verify Pesin's identity by showing $h_{KS} = \lambda_{\max} = \ln 2$.

4. For the time series $\{z_n\}_{n=0}^{4999}$ generated by the forced Duffing system with parameters $(\delta, \alpha, \beta, \gamma, \omega) = (0.2, -1, 1, 0.3, 1.2)$ and sampling interval $\Delta t = 0.05$:

   (a) Determine the optimal delay τ as the first minimum of the auto-mutual information.
   (b) Compute the FNN ratio ϕ_m for $1 \leq m \leq 12$ and identify the minimal embedding dimension $m^\star$.
   (c) With $(m, \tau) = (m^\star, \tau)$ construct a local constant predictor and estimate the predictability horizon T_p at which the forecast skill drops below 0.4.

5. The system
$$\dot{x} = y, \qquad \dot{y} = 4(1 - x^2)y - x + 0.65\cos(2t)$$
is observed stroboscopically at $t_n = n\pi$. Using a step size $h = 5 \times 10^{-3}$, integrate for 6×10^4 steps and record the two-dimensional map $(x_n, y_n) \mapsto (x_{n+1}, y_{n+1})$. Plot the Poincaré section and:

   (i) Determine numerically whether the transversal Lyapunov exponent of the section is positive.
   (ii) Identify any visible limit cycles and compute their Floquet multipliers by linearising the map along the cycle.

6. For $N = 9$ variables $\dot{x}_j = (x_{j+1} - x_{j-2})x_{j-1} - x_j + F$ with $F = 8$ and cyclic indices, integrate simultaneously with its variational equations for $T = 100$ time units and $\Delta t = 0.01$. Using periodic QR orthonormalisation every 0.1 units:

   (a) Report the ordered Lyapunov spectrum $\{\lambda_j\}_{j=1}^{9}$.
   (b) Verify numerically that $\sum_j \lambda_j \approx -\sum_j (\partial_{x_j} F_j)$.
   (c) Compute the Kaplan–Yorke dimension and compare it to the correlation dimension obtained via the Grassberger–Procaccia algorithm on the trajectory.

7. A snapshot matrix $\mathbf{U} \in \mathbb{R}^{4096 \times 250}$ records 2-D flow velocities behind a cylinder at $\mathrm{Re} = 100$ with sampling $\Delta t = 0.1$. Perform rank-r SVD truncation with $r = 25$, compute DMD eigenvalues μ_j and modes $\boldsymbol{\Phi}_j$, then:

   (i) Identify the pair of complex-conjugate eigenvalues nearest the imaginary axis and interpret the corresponding shedding frequency.
   (ii) Reconstruct the flow field at $t = 25$ using the first 10 modes and report the relative Frobenius error.

8. The delay-differential equation $\dot{x} = 0.2x(t-17)/(1 + x(t-17)^{10}) - 0.1x(t)$ is sampled at $\Delta t = 1$. Design an echo-state network with $N_r = 600$ reservoir nodes, spectral radius $\rho = 0.9$, input scaling $a_{\mathrm{in}} = 0.4$, and leaky rate $\gamma = 0.25$:

(a) Train on $t = 0\text{--}2000$, then predict autonomously for the next 1000 steps.
(b) Compute the short-term prediction skill $\mathcal{S}(\Delta t) = 1 - \epsilon(\Delta t)/\sigma_x$ and verify it remains above 0.6 for at least two Lyapunov times.
(c) Plot reservoir spectral radius versus skill and discuss the optimality of $\rho = 0.9$.

9. The adjacency list of a 500-node small-world network (rewiring $p = 0.04$, mean degree $k = 6$) is provided in sw_500.edgelist. Denote its Laplacian eigenvalues by $0 = \lambda_1 < \lambda_2 \le \cdots \le \lambda_{500}$.

   (i) Compute λ_2 and $\lambda_{\max}$ numerically.
   (ii) For an SIS process with infection rate β and recovery $\mu = 1$, determine the critical β_c predicted by $\beta_c = 1/\lambda_{\max}$.
   (iii) Simulate the discrete-time stochastic SIS model for $\beta = 1.2\beta_c$ and $\beta = 0.8\beta_c$, and estimate the stationary prevalence $\bar{I}$ in each case.

10. The Hénon map with $(a, b) = (1.4, 0.3)$ is partitioned by the line $x = 0$ into symbols 0, 1. Generate 10^6 iterates after discard, encode the symbolic trajectory, and compute block entropies $H(n)$ for $1 \le n \le 12$. Using linear regression over $n = 5\text{--}10$, estimate $h_{KS} = \lim_n H(n+1) - H(n)$ and compare with the sum of positive Lyapunov exponents obtained via QR integration.

11. For $r = 3.86$ the logistic map possesses an unstable period-2 orbit $\{x_1^*, x_2^*\}$.

    (a) Locate (x_1^*, x_2^*) to 10^{-8} accuracy.
    (b) Design an OGY control using r as the adjustable parameter with activation window $\delta = 10^{-4}$ and target eigenvalue $e^{-0.2}$. Simulate 10^4 steps and report the fraction of time the controller is active and the root-mean-square control effort.
    (c) Replace OGY by Pyragas time-delayed feedback and demonstrate numerically that the orbit stabilises for an appropriate gain k. Compare convergence rates.

Chapter 10
Data Science and Machine Learning

Abstract Feature engineering, pipeline construction, and model evaluation metrics introduce classical regression and classification before deep-learning concepts are demystified through autodiff and GPU batching. Unsupervised learning, graph neural networks, hyperparameter optimisation, and model deployment are covered end-to-end, highlighting reproducibility, fairness, and interpretability considerations essential for mathematically rigorous AI development.

Keywords Data science · Machine learning · Feature engineering · Deep learning · Model evaluation · Graph neural networks

The unprecedented proliferation of digital instrumentation—from satellites and synchrotrons to smartphones and social media—has produced datasets of volume, variety, and velocity far beyond the scope of classical statistical workflows. *Data science* provides a principled pipeline for converting these raw streams into actionable knowledge, intertwining mathematical modelling, algorithmic craftsmanship, and domain insight. Python's scientific ecosystem—`pandas`, `numpy`, `scikit-learn`, `statsmodels`, and `polars`—forms a robust computational backbone that supports the entire lifecycle: acquisition, cleansing, exploratory analysis, modelling, and deployment. This chapter introduces the foundational machinery, beginning with manipulation and cleaning, then progressing to exploratory visualisation, feature engineering, and supervised learning algorithms.

10.1 Introduction to Data Science

10.1.1 Data Manipulation and Cleaning

Real-world data are seldom turnkey; sensor glitches, missing entries, inconsistent units, and rogue outliers sabotage naïve analyses. A mathematically disciplined cleaning stage minimises bias, preserves variance, and guards against spurious correlations, laying a stable groundwork for subsequent inference.

Using Pandas for Data Cleaning

Let the raw table be a matrix $X \in \mathbb{R}^{n \times p}$ indexed by $i = 1, \ldots, n$ observations and $j = 1, \ldots, p$ attributes. pandas abstracts X as a DataFrame $\mathcal{D}$, augmenting numerical entries by heterogeneous types (timestamps, categories, objects).

```
import pandas as pd
df = pd.read_csv("meteorological.csv")  # sample: temp, humidity, wind, ...
df.info(show_counts=True)
df.describe(include="all").T  # summary statistics
```

Key operations mirror linear algebra:

$$\mathcal{D}_S = \text{df.loc[:, S]}, \qquad \mathcal{D}^{(k)} = \text{df.iloc[k]},$$

$$\mathcal{D} \cup \mathcal{E} = \text{pd.concat([df, other], axis=0)}.$$

Vectorised boolean masks execute set-theoretic filters at $O(n)$ complexity, avoiding quadratic Python loops.

Handling Missing Data and Outliers

Let $M = \{(i, j) : \mathcal{D}_{ij} = \text{NaN}\}$. Define the missing-indicator matrix $\mathbf{1}_M \in \{0, 1\}^{n \times p}$. Two canonical imputation rules:

$$\hat{x}_{ij} = \begin{cases} x_{ij}, & (i, j) \notin M, \\ \dfrac{1}{|\mathcal{N}(j)|} \sum_{k \in \mathcal{N}(j)} x_{kj}, & \text{mean/mode}, \\ \mu_j + Z_i \sigma_j, & \text{stochastic imputation}, \end{cases}$$

where $\mathcal{N}(j)$ indexes valid rows in column j and $Z_i \sim \mathcal{N}(0, 1)$.

```
from sklearn.impute import KNNImputer
imputer = KNNImputer(n_neighbors=5, weights="distance")
df_imputed = pd.DataFrame(imputer.fit_transform(df), columns=df.columns)
```

Outliers For univariate series $x_{1:n}$, flag x_i when

$$|x_i - \tilde{x}| > k\sigma \quad \text{or} \quad x_i \notin \left[Q_1 - 1.5\,IQR,\; Q_3 + 1.5\,IQR\right],$$

where $\tilde{x}$ is the median and $IQR = Q_3 - Q_1$. Robust Z-scores with median absolute deviation (MAD) retain 95.

Data Transformation and Normalisation Techniques

Machine-learning algorithms often assume homoskedastic, centred features. Let column j have empirical mean μ_j and standard deviation σ_j. The *standard score*

$$z_{ij} = \frac{x_{ij} - \mu_j}{\sigma_j}$$

renders each feature dimensionless with unit variance; whitening further orthogonalises covariances via eigen-decomposition of the sample covariance matrix.

For strictly positive variables displaying log-normal tails, log-transform $x \mapsto \ln(x + \varepsilon)$ stabilises variance. Box–Cox generalises to power λ:

$$\mathrm{BC}_\lambda(x) = \begin{cases} \dfrac{x^\lambda - 1}{\lambda}, & \lambda \neq 0, \\ \ln x, & \lambda = 0, \end{cases}$$

with λ selected by maximum log-likelihood over Gaussianised residuals.

```
from sklearn.preprocessing import PowerTransformer, StandardScaler
power = PowerTransformer(method="box-cox")
df_bc = pd.DataFrame(power.fit_transform(df_imputed+1e-3), columns=df.
    columns)
df_scaled = pd.DataFrame(StandardScaler().fit_transform(df_bc),
                columns=df.columns)
```

Feature Encoding: One-Hot Encoding and Label Encoding

Categorical predictor $c \in \{\kappa_1, \ldots, \kappa_m\}$ injects into $\mathbb{R}^m$ via *one-hot* map $e : \kappa_k \mapsto e_k$, preserving Hamming metric and avoiding spurious ordinal relations. For tree-based models that tolerate integer surrogates, *label encoding* replaces κ_k by index k; beware of inadvertently imposing monotonicity in linear regressions.

$$\mathrm{OH} : \kappa_3 \mapsto (0, 0, 1, 0, \ldots, C), \quad \mathrm{Lab} : \kappa_3 \mapsto 3.$$

```
from sklearn.preprocessing import OneHotEncoder, OrdinalEncoder
enc_oh = OneHotEncoder(sparse=False, drop='first') # reduce collinearity
enc_lab = OrdinalEncoder()
X_oh = enc_oh.fit_transform(df[['city']])
X_lab = enc_lab.fit_transform(df[['city']])
```

High-cardinality features ($m > 100$) impair memory and increase variance; hashing trick or ℓ_1-regularised target encoding mitigates explosion without losing predictive signal.

10.1.2 Exploratory Data Analysis (EDA)

Exploratory data analysis is the *mathematical reconnaissance* phase: before hypothesising formal models one must interrogate the distributional geometry of the variables, unearth latent heterogeneity, and expose coding errors camouflaged as anomalies. Numerical summaries sketch a high-level cartography, while graphical displays reveal subtleties invisible to aggregates. Every statistician should recall Tukey's maxim that "numerical recipes are not a substitute for wise eyes".

Descriptive Statistics and Visualisation Techniques

Let the cleaned matrix be $X \in \mathbb{R}^{n \times p}$ with rows $\mathbf{x}_i^\top$ and columns $X_{\cdot j}$. The univariate *five-number summary*

$$\{\min,\ Q_1,\ \tilde{x},\ Q_3,\ \max\}_j, \qquad Q_k = k\text{-th empirical quantile,}$$

together with mean μ_j, variance σ_j^2, skewness $\gamma_{1j} = \frac{1}{n}\sum_i ((x_{ij} - \mu_j)/\sigma_j)^3$, and excess kurtosis γ_{2j}, constitutes the baseline diagnostics. Heavy positive skew ($\gamma_{1j} > 1$) may indicate multiplicative mechanisms, suggesting log–transform; heavy tails ($\gamma_{2j} > 3$) warn ordinary least squares may underestimate uncertainty.

```
summary = df_scaled.agg(['min','quantile','median','mean','std','skew','kurt'],
                        quantile=lambda s: s.quantile([.25,.75]))
display(summary.T) # interactive table in Jupyter
```

Histograms approximate density with bias $O(h^2)$ and variance $O((nh)^{-1})$, where h is bin width; Scott's rule $h = 3.49\sigma n^{-1/3}$ minimises integrated mean-square error for Gaussian kernels. Kernel density estimates (KDE) supersede histograms via continuous convolution

$$\hat{f}_h(x) = \frac{1}{nh}\sum_{i=1}^{n} K\left(\frac{x - x_i}{h}\right), \quad K(u) = \frac{1}{\sqrt{2\pi}}e^{-u^2/2},$$

which attains $O(n^{-4/5})$ mean-square convergence in one dimension.

Correlation Analysis and Feature Engineering

Pairwise linear association employs Pearson matrix $\Sigma \in \mathbb{R}^{p \times p}$ with entries

$$\rho_{jk} = \frac{\operatorname{Cov}(X_{\cdot j}, X_{\cdot k})}{\sigma_j \sigma_k} = \frac{1}{n-1}\sum_{i=1}^{n} \frac{(x_{ij} - \mu_j)(x_{ik} - \mu_k)}{\sigma_j \sigma_k}.$$

10.1 Introduction to Data Science

Statistical significance follows from t-test $T = \rho\sqrt{\frac{n-2}{1-\rho^2}} \sim t_{n-2}$ under $H_0: \rho = 0$. Spearman rank ρ_s or Kendall τ preserves monotone but nonlinear dependence.

High-magnitude collinearity ($|\rho_{jk}| > 0.9$) inflates variance of regression coefficients: remedy via (i) deleting one variable, (ii) summarising by principal components (below), or (iii) partial residualising: create transformed predictor $\tilde{X}_j = X_j - \beta X_k, \beta = \Sigma_{jk}/\Sigma_{kk}$.

Domain engineering concocts composite features that linearise relations: ratio, difference, interaction, and polynomial terms. For physical laws $y \propto x_1^\alpha x_2^\beta$, log-transform linearises into $\ln y = \alpha \ln x_1 + \beta \ln x_2 + \varepsilon$, estimable by multivariate regression.

Dimensionality Reduction for EDA

Let $C = \frac{1}{n-1}(X - \mathbf{1}\mu^T)^T(X - \mathbf{1}\mu^T)$ be the covariance. Eigen-decomposition $C = V\Lambda V^T$ with $\lambda_1 \geq \lambda_2 \geq \cdots \geq \lambda_p$ and $V = [\mathbf{v}_1 \ldots \mathbf{v}_p]$ defines **principal component analysis** (PCA). The k-dimensional projection $Z = XV_{[:,1:k]}$ maximises explained variance $\text{Var}(Z) = \sum_{j=1}^{k} \lambda_j$. Scree plot of λ_j vs. j guides cutoff: retain smallest k with $\sum_{j \leq k} \lambda_j / \sum \lambda_j \geq 0.9$. Loadings $\mathbf{v}_j$ reveal latent factors; biplot $(Z_{\cdot 1}, Z_{\cdot 2})$ with arrows for original axes exposes clusters and leverage points.

Nonlinear manifolds benefit from t-distributed stochastic neighbour embedding (t-SNE) and Uniform Manifold Approximation and Projection (UMAP), which preserve local topologies by minimising Kullback–Leibler and cross-entropy objectives, respectively—useful for visualising high-dimensional images or genomic SNP data.

```
from sklearn.decomposition import PCA
pca = PCA(n_components=0.9)  # retain ≥90% variance
Z = pca.fit_transform(df_scaled)
print(pca.explained_variance_ratio_)
```

Advanced Visualisation: Heatmaps, Pair Plots, and 3D Plots

Heatmaps

Let ρ_{jk} be the correlation matrix; map $(j, k) \to c(\rho_{jk})$ with diverging colormap centred at 0. Annotating values aids interpretability; hierarchical clustering reorders axes to reveal block structures.

```
import seaborn as sns, matplotlib.pyplot as plt
corr = df_scaled.corr()
sns.clustermap(corr, cmap="coolwarm", center=0, annot=True, linewidths=.5)
```

Pair Plots (Scatter-Matrix)

Plot all bivariate scatterplots $(X_{.j}, X_{.k})$ on off-diagonal grid; diagonal cells contain KDE or histogram. Nonlinear cones, crescents, or heteroskedastic fans prompt transformation or interaction terms.

```
sns.pairplot(df_scaled[['temp','humidity','wind','pressure']],
             diag_kind='kde', hue='label')
```

Interactive 3-D Scatter

Principal coordinates (Z_1, Z_2, Z_3) render in `plotly` with orbiting camera for depth perception; colour saturates fourth feature. Rotating the cube detects hidden stratification.

```
import plotly.express as px
fig = px.scatter_3d(x=Z[:,0], y=Z[:,1], z=Z[:,2],
                    color=df['cluster'], opacity=.7)
fig.show()
```

Geospatial Heatmaps

Kernel-smoothed counts plotted over map tiles reveal crime hotspots or epidemiological clusters; adjust bandwidth to balance bias–variance trade-off, respecting Silverman's rule $h \approx 1.06\sigma n^{-1/5}$ for isotropic kernels.

10.1.3 Data Wrangling and Integration

Modern projects seldom rely on a single monolithic table; instead, they combine heterogeneous sources—relational databases, sensor logs, semi-structured JSON, parquet shards—into an integrated analytical mart. The workflow parallels categorical composition: individual data frames are objects, and joins/concats are morphisms that preserve schema invariants (primary keys, temporal alignment, unit consistency). Efficient wrangling therefore demands both relational algebra fluency and algorithmic attention to memory locality.

Merging, Joining, and Concatenating Dataframes

Let $\mathcal{D}_1$ and $\mathcal{D}_2$ be two `pandas` data frames keyed by attribute set $K = \{k_1, \ldots, k_m\}$. The *inner join*

$$\mathcal{D}_{\text{in}} = \mathcal{D}_1 \bowtie_K \mathcal{D}_2 = \{d_1 \cup d_2 \mid d_i \in \mathcal{D}_i,\ d_1|_K = d_2|_K\}$$

10.1 Introduction to Data Science

implements the set-intersection over K with complexity $O\big((n_1 + n_2)\log n_2\big)$ under hashed indices. Outer joins augment rows lacking counterparts with NaN placeholders, formally computing the symmetric difference.

```
left  = pd.DataFrame({'id':[1,2,3], 'x':[8.2, 3.1, 9.5]})
right = pd.DataFrame({'id':[2,3,4], 'y':[7 , 6 , 4 ]})
df_in  = pd.merge(left, right, on='id', how='inner')
df_out = pd.merge(left, right, on='id', how='outer', indicator=True)
```

Vertical concatenation $\mathcal{D} =$ `pd.concat([df1,df2],axis=0)` resembles disjoint union; schema compatibility requires identical column sets or explicit union-by-name. Horizontal concatenation along `axis=1` performs an index-aligned direct sum.

In star-schema ETL pipelines, surrogate keys ensure referential integrity while reducing join width. Denote fact table F and dimension D_j; the snowflake join reads $F \bowtie_{k_j} D_j$ for each j sequentially, using clustered indices to attain $O(n \log |D_j|)$ overall.

Working with Time Series Data

A univariate time series is a mapping $x : \mathbb{T} \to \mathbb{R}$ where $\mathbb{T}$ is totally ordered. Pandas embeds $\mathbb{T}$ in the `DatetimeIndex` object $\mathcal{T} = (t_0, \ldots, t_n)$ that stores nanosecond resolution integers. Resampling operator

$$\mathcal{R}_{\uparrow\downarrow}^{\Delta}[x](t) = \begin{cases} x(t) & \text{if } t \in \mathcal{T}, \\ \operatorname{agg} x(t - \Delta, t) & \text{otherwise}, \end{cases}$$

computes period aggregates via `resample`.

```
ts = pd.read_csv("ticks.csv", parse_dates=['timestamp'],
                 index_col='timestamp')['price']
min_bar = ts.resample('1min').ohlc()
log_ret = np.log(min_bar['close']).diff().dropna()
```

Rolling windows estimate local moments:

$$\hat{\mu}_k(t) = \frac{1}{k}\sum_{i=0}^{k-1} x_{t-i}, \qquad \hat{\sigma}_k^2(t) = \frac{1}{k}\sum_{i=0}^{k-1}\big(x_{t-i} - \hat{\mu}_k(t)\big)^2.$$

Pandas vectorises with `rolling(k).mean()` using stride tricks for $O(n)$ evaluation.

Seasonal decomposition $x_t = T_t + S_t + R_t$ fits $\lambda = 2\pi/\omega$ harmonic regressors or STL (LOESS with seasonal-trend-remainder). Stationarity diagnostics apply the KPSS statistic $KPSS = n^{-2}\sum_t (S_t)^2/\hat{\sigma}^2$ with $\mathcal{H}_0$: trend-stationary.

```
from statsmodels.tsa.stattools import adfuller
adfuller(log_ret.values) # Augmented Dickey-Fuller test
```

Irregular series require forward-fill logic: $x_t^{\text{ffill}} = x_{\max\{s \leq t : s \in \mathcal{T}\}}$. `asfreq` toggles between dense calendar and sparse event indexes.

Handling Large Datasets with Dask and PySpark

When $|X|$ approaches RAM or demands cluster-level parallelism, out-of-core libraries abstract block-wise partitioning.

Dask

Represents a lazy task graph $G = (V, E)$ where vertices are NumPy array chunks $\mathbf{X}^{(b)}$ and edges record dependencies. High-level API mirrors `pandas`; scheduler executes G distributing blocks round-robin. Memory footprint $\approx$ block size $\times$ workers; shuffles trigger network I/O $O(n \log n)$.

```
import dask.dataframe as dd
ddf = dd.read_parquet("/mnt/data/weather/*.parquet")
ddf['dewpoint'] = ddf.temp - (100 - ddf.humidity)/5
mean_dp = ddf.dewpoint.mean().compute()
```

PySpark

Spark's Resilient Distributed Dataset (RDD) is a fault-tolerant, immutable multiset partitioned across executors. DataFrame API builds a logical query plan optimised by Catalyst (rule-based algebraic simplification) and executes on the Tungsten engine (columnar, vectorised, whole-stage code generation). Join strategy selector chooses broadcast hash join when $\min\{|D_1|, |D_2|\} < 10^7$ records; otherwise shuffle sort-merge.

```
from pyspark.sql import SparkSession
spark = SparkSession.builder.appName("ETL").getOrCreate()
df_sales = spark.read.csv("sales.csv", header=True, inferSchema=True)
df_region = spark.read.csv("regions.csv", header=True, inferSchema=True)
df_join = df_sales.join(df_region, on='region_id', how='inner')

quarterly = (df_join
    .groupBy('region_name', 'quarter')
    .agg({'revenue':'sum'})
    .orderBy('region_name','quarter'))
quarterly.repartition(1).write.csv("out/summary", mode='overwrite')
```

Performance Heuristics

- Partition size 128–256 MB balances scheduler overhead and cache locality.

- Persist *wide tables* (`parquet`) with Snappy compression; choose `orc` if predicate pushdown dominates.
- Use column pruning and filter pushdown to minimise scan volume: `select(col_list).where(predicate)`.
- Broadcast joins: `spark.conf.set("spark.sql.autoBroadcastJoinThreshold", "50MB")`; or `hint("broadcast")`.

10.2 Machine Learning Algorithms

Predictive modelling seeks a measurable mapping $\mathcal{X} \to \mathcal{Y}$ that generalises from a finite sample to unseen data. **Supervised learning** assumes an i.i.d. training set $\mathcal{S} = \{(\mathbf{x}_i, y_i)\}_{i=1}^n$ drawn from joint law $\mathbb{P}_{XY}$. The learner selects a hypothesis $h_\theta \in \mathcal{H}$ by minimising empirical risk $\widehat{R}_n(\theta) = \frac{1}{n} \sum_i \ell(h_\theta(\mathbf{x}_i), y_i)$. Statistical learning theory bounds the expected risk $R(\theta) = \mathbb{E}_{XY}\, \ell(h_\theta(X), Y)$ via VC dimension or Rademacher complexity; practical algorithms complement these guarantees with regularisation that controls model capacity.

10.2.1 Supervised Learning

Linear Regression, Logistic Regression

Ordinary Least Squares (OLS)

Given design matrix $X \in \mathbb{R}^{n \times p}$ and target $\mathbf{y} \in \mathbb{R}^n$,

$$\hat{\boldsymbol{\beta}} = \arg\min_{\boldsymbol{\beta}} \|X\boldsymbol{\beta} - \mathbf{y}\|_2^2 = (X^\mathsf{T} X)^{-1} X^\mathsf{T} \mathbf{y},$$

assuming $X^\mathsf{T} X$ invertible. If $\operatorname{rank}(X) < p$, replace by Moore–Penrose pseudo-inverse or apply ridge penalty $\lambda \|\boldsymbol{\beta}\|_2^2$ yielding $(X^\mathsf{T} X + \lambda I)^{-1} X^\mathsf{T} \mathbf{y}$.

```
import numpy as np
β_hat = np.linalg.inv(X.T @ X) @ X.T @ y  # OLS
λ = 1e-2
β_ridge = np.linalg.inv(X.T @ X + λ*np.eye(p)) @ X.T @ y
```

Logistic Regression

Binary labels $y \in \{0, 1\}$. Model

$$\Pr(Y = 1 \mid \mathbf{x}) = \sigma(\mathbf{x}^\mathsf{T} \boldsymbol{\beta}), \qquad \sigma(z) = \frac{1}{1 + e^{-z}}.$$

Maximise log-likelihood $\ell(\boldsymbol{\beta}) = \sum_i [y_i \ln \sigma(z_i) + (1-y_i)\ln(1-\sigma(z_i))]$, solved via Newton–Raphson: $\boldsymbol{\beta}^{(k+1)} = \boldsymbol{\beta}^{(k)} - (H)^{-1}\nabla\ell$. Add ℓ_1 term $\lambda\|\boldsymbol{\beta}\|_1$ for sparsity (lasso logistic) using coordinate descent.

Decision Trees and Random Forests

CART Algorithm

Partition feature space recursively; at node t choose split (j, s) minimising impurity decrease

$$\Delta \mathcal{I} = \mathcal{I}(t) - \frac{n_L}{n_t}\mathcal{I}(L) - \frac{n_R}{n_t}\mathcal{I}(R),$$

with Gini index $\mathcal{I}(t) = 1 - \sum_k p_k^2$ for classification or mean-squared error for regression. Prune tree by cost–complexity $C_\alpha(T) = R(T) + \alpha|T|$.

Random Forest (Breiman 2001)

Ensemble of B bootstrap trees $\{T_b\}$ grown to full depth with m random candidate features per split. Prediction: $\hat{y} = \frac{1}{B}\sum_b T_b(\mathbf{x})$ (regression) or majority vote (classification). Out-of-bag error approximates generalisation. Variance reduction $1/B$ while bias remains; feature importance via mean decrease in impurity.

Support Vector Machines (SVM) and Kernel Methods

Hard-margin SVM seeks hyperplane $\mathbf{w}^\mathsf{T}\mathbf{x} + b = 0$ maximising margin $2/\|\mathbf{w}\|$:

$$\min_{\mathbf{w},b} \frac{1}{2}\|\mathbf{w}\|_2^2 \quad \text{s.t.} \quad y_i(\mathbf{w}^\mathsf{T}\mathbf{x}_i + b) \geq 1.$$

Soft margin introduces slack ξ_i and penalty $C\sum \xi_i$. Dual optimisation

$$\max_{\boldsymbol{\alpha}} \sum_i \alpha_i - \tfrac{1}{2}\sum_{i,j}\alpha_i\alpha_j y_i y_j K(\mathbf{x}_i, \mathbf{x}_j), \quad 0 \leq \alpha_i \leq C, \quad \sum_i \alpha_i y_i = 0,$$

where kernel $K(\mathbf{x},\mathbf{x}') = \phi(\mathbf{x})^\mathsf{T}\phi(\mathbf{x}')$. Common choices: polynomial, RBF $K = \exp(-\gamma\|\mathbf{x}-\mathbf{x}'\|^2)$, sigmoid (neural tangent). Prediction uses support vectors with $\alpha_i > 0$.

Ensemble Methods: Bagging, Boosting, and Stacking

Bagging (Bootstrap Aggregating)

Given base learner h, form B bootstrapped replicas h_b. Aggregated estimator reduces variance: $\mathrm{Var}\left[\frac{1}{B}\sum h_b\right] = \frac{\rho\sigma^2}{B} + (1-\rho)\sigma^2$, where ρ is pairwise correlation.

AdaBoost (Freund and Schapire 1997)

Exponential loss:

$$F_M(\mathbf{x}) = \sum_{m=1}^{M} \alpha_m h_m(\mathbf{x}), \qquad \alpha_m = \frac{1}{2}\ln\frac{1-\varepsilon_m}{\varepsilon_m},$$

weights updated $w_i \leftarrow w_i e^{-\alpha_m y_i h_m(\mathbf{x}_i)}$ and renormalised. Minimises empirical risk under margin-exponential cost; can overfit noisy labels.

Gradient Boosting

For differentiable loss $L(y, F)$ fit base learner to negative gradient $g_i^{(m)} = -\partial_F L(y_i, F_{m-1}(\mathbf{x}_i))$, update $F_m = F_{m-1} + \eta\gamma_m h_m$, with step size γ_m from line search. XGBoost adds shrinkage, column subsampling, and ℓ_2 regularisation on leaves.

Stacking

Level-0 models $\{h_k\}$ generate meta-features $z_{ik} = h_k(\mathbf{x}_i)$; level-1 learner $\tilde{h}$ trained on $\mathbf{z}_i$ predicts. Use K-fold split to avoid target leakage. Theoretically, stacking approximates convex combination envelope when $\tilde{h}$ is linear; nonlinear meta may capture interaction among base predictors.

Illustrative Code Snippet

```
from sklearn.model_selection import train_test_split, GridSearchCV
from sklearn.ensemble import RandomForestClassifier,
    GradientBoostingClassifier
from sklearn.svm import SVC
from sklearn.metrics import roc_auc_score
X_train, X_test, y_train, y_test = train_test_split(X, y, test_size=.2,
    stratify=y)

pipe = SVC(kernel='rbf', probability=True)
param = {'C':[1,10],'gamma':[.1,.01]}
```

```
svm = GridSearchCV(pipe, param, cv=5, scoring='roc_auc').fit(X_train,
    y_train)

rf = RandomForestClassifier(n_estimators=300, max_features='sqrt').fit(
    X_train, y_train)
gb = GradientBoostingClassifier(n_estimators=400, learning_rate=.05).fit(
    X_train, y_train)

pred = (0.3*svm.predict_proba(X_test)[:,1]
    + 0.4*rf .predict_proba(X_test)[:,1]
    + 0.3*gb .predict_proba(X_test)[:,1])
print("stacked AUC =", roc_auc_score(y_test, pred))
```

10.2.2 Unsupervised Learning

In the *unsupervised* paradigm we observe $\{x_1, \ldots, x_n\} \subset \mathbb{R}^p$ without labels. The aim is to extract latent organisation: partitions (clustering), low-dimensional structure (manifold learning), rare-event loci (anomaly detection), or co-occurrence rules (market baskets). Mathematically, unsupervised learning minimises an energy functional $\mathscr{E}(\theta)$ constructed from distances or entropies rather than predictive loss.

Clustering: *k*-Means and Hierarchical Clustering

k-Means (Lloyd 2006)

Given k, minimise within-cluster sum of squares (WCSS):

$$\mathscr{E}(C, \boldsymbol{\mu}) = \sum_{i=1}^{n} \left\| \mathbf{x}_i - \boldsymbol{\mu}_{C(i)} \right\|^2,$$

where $C : \{1, \ldots, n\} \to \{1, \ldots, k\}$ assigns points to centroids $\boldsymbol{\mu}_j$. Lloyd's algorithm alternates:

1. *Assignment step:* $C(i) \leftarrow \arg\min_j \|\mathbf{x}_i - \boldsymbol{\mu}_j\|^2$;
2. *Update step:* $\boldsymbol{\mu}_j \leftarrow \frac{1}{|C^{-1}(j)|} \sum_{i:C(i)=j} \mathbf{x}_i$.

Each step decreases $\mathscr{E}$; convergence to local optimum in $O(nkp)$ per iteration. Initialisation via *k*-means++ samples centroids with probability proportional to squared distance, providing $O(\log k)$ expected approximation ratio.

10.2 Machine Learning Algorithms

Hierarchical Agglomerative Clustering (HAC)

Start with n singleton clusters; iteratively merge pair (A, B) of clusters that minimises linkage criterion:

$$d_{\text{single}}(A, B) = \min_{\mathbf{x} \in A, \mathbf{y} \in B} \|\mathbf{x}-\mathbf{y}\|, \quad d_{\text{complete}} = \max, \quad d_{\text{average}} = \frac{|A||B|}{|A|+|B|} \|\bar{\mathbf{x}}_A - \bar{\mathbf{x}}_B\|.$$

Produces dendrogram; cut at height τ to obtain flat clusters. Complexity $O(n^2 \log n)$ with heap linkage cache.

```
from sklearn.cluster import KMeans, AgglomerativeClustering
km = KMeans(n_clusters=4, init='k-means++', n_init=20).fit(X)
hac = AgglomerativeClustering(n_clusters=4, linkage='ward').fit(X)
```

Dimensionality Reduction: PCA, t-SNE

PCA Recap

Projection $\mathbf{z}_i = V^\top(\mathbf{x}_i - \boldsymbol{\mu})$ maximises variance; reconstruction error $\sum_i \|\mathbf{x}_i - VV^\top \mathbf{x}_i\|^2 = \sum_{j>k} \lambda_j$ equals discarded eigenvalue tail.

t-SNE (van der Maaten and Hinton 2008)

Transforms Euclidean distances to conditional probabilities $p_{j|i} \propto \exp(-\|\mathbf{x}_i - \mathbf{x}_j\|^2 / 2\sigma_i^2)$, perplexity $\mathcal{P} = 2^{H(P_i)}$ selects σ_i. Embedding $\mathbf{y}_i \in \mathbb{R}^2$ minimises KL divergence

$$\text{KL}(P \parallel Q) = \sum_{i \neq j} p_{ij} \log \frac{p_{ij}}{q_{ij}}, \quad q_{ij} = \frac{(1 + \|\mathbf{y}_i - \mathbf{y}_j\|^2)^{-1}}{\sum_{k \neq \ell}(1 + \|\mathbf{y}_k - \mathbf{y}_\ell\|^2)^{-1}}.$$

Gradient updates $\Delta \mathbf{y}_i \propto 4 \sum_j (p_{ij} - q_{ij})(\mathbf{y}_i - \mathbf{y}_j)(1 + \|\mathbf{y}_i - \mathbf{y}_j\|^2)^{-1}$. Barnes–Hut approximation yields $O(n \log n)$ runtime.

```
from sklearn.manifold import TSNE
Z = TSNE(n_components=2, perplexity=30, init='pca').fit_transform(X)
```

Anomaly Detection Techniques

Statistical Thresholds

Assume Gaussian; flag $\mathbf{x}$ when Mahalanobis distance $D^2 = (\mathbf{x} - \boldsymbol{\mu})^\top \Sigma^{-1} (\mathbf{x} - \boldsymbol{\mu})$ exceeds $\chi^2_{p,\,1-\alpha}$ quantile ($\alpha \sim 10^{-3}$).

Isolation Forest

Randomly partition space by axis-aligned splits until singleton; anomaly score

$$s(\mathbf{x}) = 2^{-\frac{E[h(\mathbf{x})]}{c(n)}},$$

where h is path length and $c(n) \approx 2H_{n-1} - \frac{2(n-1)}{n}$ normalises by harmonic number. Anomalies have short paths.

Autoencoder Residual

Train neural autoencoder $\phi \circ \psi$ on normal data; reconstruction error $\|\mathbf{x} - \phi(\psi(\mathbf{x}))\|_2$ approximates negative log-likelihood under manifold hypothesis. Threshold via extreme value theory (Peaks-Over-Threshold).

Association Rule Mining: Apriori and FP-Growth

Transaction Database

Let $\mathcal{T} = \{T_1, \ldots, T_N\}$ with items from $\mathcal{I}$. Support $\text{supp}(A) = \frac{1}{N}|\{T : A \subseteq T\}|$. Rule $A \Rightarrow B$ ($A \cap B = \emptyset$) has confidence $\text{conf}(A \Rightarrow B) = \frac{\text{supp}(A \cup B)}{\text{supp}(A)}$ and lift $\text{lift} = \frac{\text{conf}}{\text{supp}(B)}$.

Apriori Principle

If itemset A is frequent ($\text{supp} \geq \sigma$), all subsets are frequent. Level-wise breadth-first search generates k-item candidates from $(k-1)$-frequent sets; prune those with infrequent subsets; scan database to count supports. Complexity $O(|\mathcal{I}|^k)$ in worst case; reduces drastically under large σ.

FP-Growth

Compress database into FP-tree: sort items in each transaction by global frequency; insert path, increment counters. Mine tree recursively: conditional pattern base of suffix item i forms subtree; avoids candidate explosion and needs two passes.

```
from mlxtend.frequent_patterns import apriori, association_rules
df_bin = df_basket.astype(bool).astype(int) # one-hot transactions
freq = apriori(df_bin, min_support=0.03, use_colnames=True)
rules = association_rules(freq, metric="confidence", min_threshold=0.6)
rules = rules[rules['lift']>1.2].sort_values('lift', ascending=False)
```

10.2.3 Reinforcement Learning

Reinforcement learning (RL) formalises sequential decision-making as a *feedback control* problem under uncertainty. An agent repeatedly observes the environment, chooses an action, receives a numerical reward, and transitions to a new state; its objective is to maximise long-run return. Unlike supervised learning, the data distribution is *policy-dependent*, requiring exploration versus exploitation trade-offs, and delayed consequences couple decisions across time.

Markov Decision Processes (MDPs)

An **MDP** is the quintuple $\mathcal{M} = (\mathcal{S}, \mathcal{A}, P, R, \gamma)$ where

- $\mathcal{S}$ is finite (or countable) state space
- $\mathcal{A}$ is finite action set
- $P(s'|s, a) = \Pr(S_{t+1} = s' \mid S_t = s, A_t = a)$ is transition kernel
- $R(s, a) = \mathbb{E}[\, r_t \mid S_t = s, A_t = a\,]$ is expected one-step reward
- $\gamma \in [0, 1)$ is discount factor

For policy $\pi : \mathcal{S} \to \Delta(\mathcal{A})$ define *value function*

$$V^\pi(s) = \mathbb{E}_\pi\left[\sum_{t=0}^\infty \gamma^t r_t \,\Big|\, S_0 = s\right],$$

and action–value (Q-value) $Q^\pi(s, a) = R(s, a) + \gamma \sum_{s'} P(s'|s, a) V^\pi(s')$. Bellman expectation equation:

$$V^\pi = T^\pi V^\pi, \qquad T^\pi f(s) = \sum_a \pi(a\,s)\Big[R(s, a) + \gamma \sum_{s'} P_{sas'} f(s')\Big].$$

Optimal value satisfies Bellman *optimality* operator

$$V^*(s) = \max_a \Big[R(s, a) + \gamma \sum_{s'} P_{sas'} V^*(s')\Big],$$

with greedy optimal policy $\pi^*(s) = \arg\max_a Q^*(s, a)$.

Dynamic Programming

Value iteration: $V_{k+1} = T^* V_k$, contracts in ℓ_∞ norm with modulus γ, hence $V_k \to V^*$ geometrically.

Q-Learning and Deep Q-Networks (DQN)

Tabular Q-Learning (Watkins and Dayan 1992)

Online update

$$Q_{t+1}(s_t, a_t) \leftarrow Q_t(s_t, a_t) + \alpha_t \Big[r_t + \gamma \max_{a'} Q_t(s_{t+1}, a') - Q_t(s_t, a_t) \Big],$$

converges to Q^* w.p.1 if $\sum \alpha_t = \infty$, $\sum \alpha_t^2 < \infty$ and all state–action pairs are visited infinitely often (GLIE). Exploration via ε-greedy: choose random action w.p. ε, greedy otherwise.

Function Approximation: Deep Q-Network (Mnih et al. 2015)

Parameterise $Q_\theta(s, a)$ by neural network; minimise *temporal-difference loss*

$$\mathcal{L}(\theta) = \mathbb{E}_{(s,a,r,s') \sim \mathcal{D}} \Big[\big(r + \gamma \max_{a'} Q_{\theta^-}(s', a') - Q_\theta(s, a) \big)^2 \Big],$$

with target network parameters θ^- updated periodically. Replay buffer $\mathcal{D}$ of size N breaks correlation, enabling stochastic-gradient descent. Double DQN mitigates overestimation by using online network for argmax and target network for evaluation.

Policy Gradient Methods and Actor–Critic Algorithms

Stochastic Policy $\pi_\theta(a|s)$

Objective $J(\theta) = \mathbb{E}_{\pi_\theta}[G_0]$ where $G_0 = \sum_t \gamma^t r_t$. **Policy gradient theorem**

$$\nabla_\theta J(\theta) = \mathbb{E}_{\pi_\theta} \big[\nabla_\theta \log \pi_\theta(A_t|S_t) \, Q^{\pi_\theta}(S_t, A_t) \big].$$

REINFORCE uses Monte Carlo return G_t; variance reduction via baseline $b(s)$:

$$\nabla J = \mathbb{E}[\nabla \log \pi(A|S)(G_t - b(S_t))].$$

Actor–Critic

Actor updates policy parameters; *critic* estimates value. TD(λ) critic update for parameters w: $\delta_t = r_t + \gamma V_w(S_{t+1}) - V_w(S_t)$, $w \leftarrow w + \beta \delta_t \nabla_w V_w(S_t)$. Actor update: $\theta \leftarrow \theta + \alpha \, \delta_t \nabla_\theta \log \pi_\theta(A_t|S_t)$.

10.2 Machine Learning Algorithms

Advantage Actor–Critic (A2C)

Replace δ_t by advantage $A_t = Q_w(S_t, A_t) - V_w(S_t)$. Generalised advantage estimation (GAE) blends multi-step returns: $\hat{A}_t^\lambda = \sum_{k=0}^{\infty} (\gamma\lambda)^k \delta_{t+k}$.

Soft Actor–Critic (SAC)

Maximises entropy-regularised objective

$$J(\pi) = \sum_t \mathbb{E}_\pi\big[r(S_t, A_t) + \alpha\mathcal{H}(\pi(\cdot|S_t))\big],$$

leading to off-policy updates with target Q networks and automatic temperature tuning.

Applications of Reinforcement Learning in Game AI and Robotics

Game AI

- **Atari 2600.** Deep Q-Networks achieve human-level score on 57 games, input raw pixel stacks $84 \times 84 \times 4$.
- **AlphaZero.** Combines Monte Carlo tree search with a policy-value network; self-play reinforcement minimises cross-entropy + value loss.
- **OpenAI Five.** Uses PPO (clip surrogate) and LSTM policy for partial observability; multi-agent credit assignment via *reward-shaping*.

Robotics

- **Sim-to-Real Transfer.** Domain randomisation trains SAC/DDPG in simulation with stochastic textures and physics; deploys to real manipulators with minimal fine-tuning.
- **Model-Based RL.** Learn probabilistic dynamics $p_\phi(\mathbf{s}_{t+1}|\mathbf{s}_t, \mathbf{a}_t)$; plan via model-predictive control (MPC) with trajectory sampling; PETS and Dreamer achieve minutes-level sample efficiency.
- **Residual RL.** Combine classical controller u_{PID} with learned residual Δu_θ, improving agility while preserving stability margins.

10.3 Deep Learning

A *deep neural network* composes a finite sequence of nonlinear affine maps, $f_\Theta = f^{(L)} \circ \cdots \circ f^{(1)}$, $f^{(\ell)}(\mathbf{z}) = \phi(W_\ell \mathbf{z} + \mathbf{b}_\ell)$, to approximate an unknown target function on $\mathbb{R}^p$. Universal approximation theorems guarantee that sufficiently large networks can approximate any continuous function on compact sets to arbitrary precision; the practical art lies in selecting architectures $\Theta = \{W_\ell, \mathbf{b}_\ell\}_{\ell=1}^L$ and algorithms that converge within computational budgets while generalising beyond the training sample.

10.3.1 Mathematical Foundations of Neural Networks

Perceptrons, Activation Functions, and Backpropagation

A single-layer perceptron computes

$$\hat{y} = \phi(\mathbf{w}^\mathsf{T}\mathbf{x} + b), \qquad \phi : \mathbb{R} \to \mathbb{R}.$$

For linearly separable binary data $\{(\mathbf{x}_i, y_i)\}$, Rosenblatt's update $\mathbf{w} \leftarrow \mathbf{w} + \eta(y_i - \hat{y}_i)\mathbf{x}_i$ converges in finite steps; however, non-convex targets motivate multilayer perceptrons (MLPs).

Activation Design Common ϕ:

$$\text{sigmoid}(z) = \frac{1}{1+e^{-z}}, \quad \tanh z = \frac{e^z - e^{-z}}{e^z + e^{-z}}, \quad \text{ReLU}(z) = \max(0, z).$$

ReLU mitigates vanishing gradients by piecewise linearity; its subgradient $\partial \,\text{ReLU}(z) = \mathbf{1}_{z>0}$ almost everywhere suffices for stochastic gradient descent (SGD).

Backpropagation Define layer outputs $\mathbf{a}^{(0)} = \mathbf{x}$, $\mathbf{z}^{(\ell)} = W_\ell \mathbf{a}^{(\ell-1)} + \mathbf{b}_\ell$, $\mathbf{a}^{(\ell)} = \phi(\mathbf{z}^{(\ell)})$. Given loss $L(\hat{\mathbf{y}}, \mathbf{y})$, backward pass:

$$\boldsymbol{\delta}^{(L)} = \nabla_{\hat{\mathbf{y}}} L \odot \phi'(\mathbf{z}^{(L)}), \qquad \boldsymbol{\delta}^{(\ell)} = \left(W_{\ell+1}^\mathsf{T} \boldsymbol{\delta}^{(\ell+1)}\right) \odot \phi'(\mathbf{z}^{(\ell)}),$$

$$\frac{\partial L}{\partial W_\ell} = \boldsymbol{\delta}^{(\ell)} \mathbf{a}^{(\ell-1)\mathsf{T}}, \quad \frac{\partial L}{\partial \mathbf{b}_\ell} = \boldsymbol{\delta}^{(\ell)}.$$

10.3 Deep Learning

Computational cost $O\left(\sum_\ell n_{\ell-1} n_\ell\right)$ matches forward pass.

```python
import numpy as np
def relu(z): return np.maximum(0, z)
def relu_grad(z): return (z>0).astype(float)

def forward(x, W, b):
    z, a = [], [x]
    for Wℓ,bℓ in zip(W,b):
        zℓ = Wℓ @ a[-1] + bℓ
        aℓ = relu(zℓ)
        z.append(zℓ); a.append(aℓ)
    return z, a

def backprop(y, z, a, W):
    δ = [a[-1]-y] # MSE derivative
    for ℓ in reversed(range(len(W)-1)):
        δ.insert(0, (W[ℓ+1].T @ δ[0]) * relu_grad(z[ℓ]))
    dW = [δ_ℓ @ a[ℓ].T for ℓ,δ_ℓ in enumerate(δ)]
    db = δ
    return dW, db
```

Regularisation Techniques and Optimisation Algorithms

Weight Decay (Ridge) Add $\lambda \|\Theta\|_2^2$ to loss; gradient update $W_\ell \leftarrow (1-\eta\lambda)W_\ell - \eta \nabla_{W_\ell} L$. Enforces small weights $\implies$ smoother mappings via Tikhonov.

ℓ_1 **Sparsity** Penalise $\lambda\|\Theta\|_1$; proximal operator performs soft-thresholding: $w \leftarrow \text{sgn}(w)\max\{|w|-\eta\lambda, 0\}$.

Dropout During training, zero out activations with prob. p: $\tilde{\mathbf{a}}^{(\ell-1)} = \mathbf{a}^{(\ell-1)} \odot \mathbf{m}$, $m_j \sim \text{Bernoulli}(1-p)$, approximating ensemble of 2^n thinned networks; test time scales activations by $(1-p)$.

Batch Normalisation For mini-batch $\mathcal{B}$ normalise pre-activation:

$$\hat{z} = \frac{z-\mu_\mathcal{B}}{\sqrt{\sigma_\mathcal{B}^2 + \varepsilon}}, \qquad \tilde{z} = \gamma \hat{z} + \beta,$$

accelerates training by reducing covariate shift; γ, β are learnable.

Optimisation SGD update $\theta \leftarrow \theta - \eta \nabla_\theta L$ converges in expectation under Robbins–Monro conditions. Momentum accelerates: $v \leftarrow \beta v + (1-\beta)\nabla L$; $\theta \leftarrow \theta - \eta v$. Adaptive algorithms scale by second-moment estimate:

$$\text{Adam: } m_t = \beta_1 m_{t-1} + (1-\beta_1)g_t, \ v_t = \beta_2 v_{t-1} + (1-\beta_2)g_t^2,$$

$$\theta_{t+1} = \theta_t - \eta \frac{m_t/(1-\beta_1^t)}{\sqrt{v_t/(1-\beta_2^t)}+\varepsilon}.$$

Loss Functions: Cross-Entropy, MSE, and Custom Losses

Mean Squared Error (MSE)

For regression targets $y \in \mathbb{R}$, $L_{\text{MSE}} = \frac{1}{2}(y - \hat{y})^2$. Gradient: $\partial_{\hat{y}} L = \hat{y} - y$; convex and corresponds to Gaussian likelihood.

Cross-Entropy

For K-class softmax $\hat{p}_k = \exp z_k / \sum_j \exp z_j$,

$$L_{\text{CE}} = -\sum_{k=1}^{K} y_k \ln \hat{p}_k, \qquad \partial_{z_k} L = \hat{p}_k - y_k.$$

Arises from KL divergence between empirical label distribution and model.

Custom Losses

$$L_{\text{Huber}} = \begin{cases} \frac{1}{2} r^2, & |r| \leq \delta, \\ \delta(|r| - \frac{1}{2}\delta), & |r| > \delta, \end{cases} \qquad r = y - \hat{y}.$$

Combines MSE sensitivity near origin with L^1 robustness to outliers.
  In imbalanced classification, *focal loss* $L_{\text{focal}} = -(1 - \hat{p}_t)^\gamma \ln \hat{p}_t$, $\hat{p}_t = \begin{cases} \hat{p} & y = 1, \\ 1 - \hat{p} & y = 0, \end{cases}$ down-weights easy examples, tuning $\gamma \in [1, 3]$.

```
def focal_loss(pred, target, γ=2.0, eps=1e-7):
    pt = np.where(target==1, pred, 1-pred)
    return -np.mean((1-pt)**γ * np.log(pt+eps))
```

10.3.2 Implementing Neural Networks from Scratch

Educational mastery of deep learning is cemented by building a network with nothing but numpy. This section develops a minimal yet fully functional multilayer perceptron (MLP), derives each gradient algebraically, and wraps the implementation with training, evaluation, and hyperparameter search utilities. The exposition emphasises clear tensor dimensions, broadcasting safety, and numerical stability—skills transferable to high-performance frameworks.

10.3 Deep Learning

Forward and Backward Propagation in Python

Network Skeleton

For input dimension d, hidden widths $(h_1, \ldots, h_{L-1})$, and output dimension K, define

$$Z^{(1)} = W_1 X + b_1, \qquad A^{(1)} = \phi(Z^{(1)}),$$
$$\vdots$$
$$Z^{(L)} = W_L A^{(L-1)} + b_L, \qquad \hat{Y} = g(Z^{(L)}),$$

where ϕ = ReLU, g is layer-dependent output nonlinearity (softmax or identity), and $X \in \mathbb{R}^{d \times n_{\text{batch}}}$ uses column-major samples.

Vectorised Forward Pass

```
import numpy as np

def relu(z): return np.maximum(0, z)
def relu_grad(z): return (z > 0).astype(z.dtype)
def softmax(z):
    z -= z.max(axis=0, keepdims=True) # stability
    exp = np.exp(z)
    return exp / exp.sum(axis=0, keepdims=True)

def forward(X, W, b):
    Z, A = [], [X]
    for Wℓ, bℓ in zip(W[:-1], b[:-1]): # hidden layers
        Zℓ = Wℓ @ A[-1] + bℓ
        Aℓ = relu(Zℓ)
        Z.append(Zℓ); A.append(Aℓ)
    ZL = W[-1] @ A[-1] + b[-1] # logits
    Ŷ = softmax(ZL) # classification
    Z.append(ZL); A.append(Ŷ)
    return Z, A
```

Cross-Entropy Loss and Gradients

For one-hot labels $Y \in \{0, 1\}^{K \times n}$ the mean cross-entropy

$$\mathcal{L} = -\frac{1}{n} \sum_{i=1}^{n} \sum_{k=1}^{K} Y_{ki} \ln \hat{Y}_{ki}$$

produces output-layer gradient $\nabla_{Z^{(L)}} \mathcal{L} = \frac{1}{n}(\hat{Y} - Y)$. Backpropagate:

$$\nabla_{W_\ell} = \delta^{(\ell)} A^{(\ell-1)\mathsf{T}}, \qquad \nabla_{b_\ell} = \delta^{(\ell)} \mathbf{1},$$

$$\delta^{(\ell-1)} = (W_\ell^\mathsf{T} \delta^{(\ell)}) \odot \phi'(Z^{(\ell-1)}).$$

```
def backward(Y, Z, A, W):
    n = Y.shape[1]
    δ = [A[-1] - Y] / n # output delta
    for ℓ in reversed(range(1, len(W))):
        δ.insert(0, (W[ℓ].T @ δ[0]) * relu_grad(Z[ℓ-1]))
    dW = [δℓ @ A[ℓ].T for ℓ, δℓ in enumerate(δ)]
    db = [δℓ.sum(axis=1, keepdims=True) for δℓ in δ]
    return dW, db
```

Training and Evaluating Neural Networks

Mini-Batch SGD with Momentum

$$v^{(t+1)} = \beta v^{(t)} + (1-\beta)\nabla\Theta, \quad \Theta^{(t+1)} = \Theta^{(t)} - \eta v^{(t+1)}.$$

We implement ℓ_2 weight decay and stratified shuffling.

```
def sgd_train(X, Y, layers, lr=.05, β=.9, weight_decay=1e-4, epochs=50,
  batch=64):
    rng = np.random.default_rng(0)
    W, b = initialise(layers, rng)
    Vw = [np.zeros_like(Wℓ) for Wℓ in W]
    Vb = [np.zeros_like(bℓ) for bℓ in b]
    for epoch in range(epochs):
        idx = rng.permutation(X.shape[1])
        for k in range(0, X.shape[1], batch):
            j = idx[k:k+batch]
            Z,A = forward(X[:,j], W, b)
            dW,db = backward(Y[:,j], Z, A, W)
            for ℓ in range(len(W)):
                dW[ℓ] += weight_decay*W[ℓ] # ridge
                Vw[ℓ] = β*Vw[ℓ] + (1-β)*dW[ℓ]
                Vb[ℓ] = β*Vb[ℓ] + (1-β)*db[ℓ]
                W[ℓ] -= lr*Vw[ℓ]; b[ℓ] -= lr*Vb[ℓ]
        if epoch%10==0:
            acc = accuracy(X, Y, W, b)
            print(f"epoch {epoch:2d}: accuracy={acc:.4f}")
    return W,b
```

10.3 Deep Learning

Metrics

Accuracy $\text{acc} = \frac{1}{n}\sum \mathbf{1}_{\arg\max_k \hat{Y}_k = \arg\max_k Y_k}$; confusion matrix evaluates class imbalance; top-k accuracy for image nets. For regression, report $R^2 = 1 - \|\hat{y} - y\|^2 / \|y - \bar{y}\|^2$ and RMSE.

Early Stopping

Hold-out validation loss $L_{\text{val}}^{(t)}$ monitors overfitting; stop when L_{val} fails to improve for p epochs. Patience p chosen by nested cross-validation.

Hyperparameter Tuning and Model Selection

Let hyperparameter vector λ = (learning rate, widths, β, λ_{wd}, p_{drop}). Define validation estimate $\widehat{R}_{\text{val}}(\lambda) = \frac{1}{|V|}\sum_{i \in V} \ell\big(h_{\Theta^*(\lambda)}(\mathbf{x}_i), y_i\big)$.

Grid and Random Search

Random search samples λ from log-uniform priors; empirical study (Bergstra and Bengio 2012) shows $\approx 60 \times$ faster than exhaustive grids in high dimension.

Bayesian Optimisation

Treat $f(\lambda) = \widehat{R}_{\text{val}}$ as black box; model via Gaussian process $\mathcal{GP}(m, k)$ and choose next point by maximising acquisition function (expected improvement). Converges to optimum with sub-linear regret.

Cross-Validation

k-fold CV splits data into k blocks; rotate validation fold, average risks. Nested CV: outer loop estimates generalisation, and inner loop selects λ, preventing optimistic bias.

Practical Recipe

(i) Scale inputs (mean–variance or min–max) before tuning.
(ii) Fix architecture family, run *coarse* random search ($\sim$30 trials) for learning rate log-range 10^{-4}–10^{-1}, width $2^{[5,9]}$, dropout $p \in [0.0, 0.5]$.

(iii) Refine around top-5 candidates with Bayesian optimisation (20 iterations).
(iv) Freeze λ^*, retrain on full train+val for final evaluation on withheld test.

10.3.3 Convolutional and Recurrent Neural Networks

Deep learning's modern ascendancy can largely be attributed to architectures that exploit *inductive bias*: convolutional neural networks (CNNs) encode translational equivariance for grid-structured inputs, while recurrent neural networks (RNNs) embody causal ordering for sequences. Subsequent refinements—gated recurrence, attention, and large-scale pretraining—broaden the modelling capacity without sacrificing statistical efficiency.

CNNs for Image Recognition

Discrete Convolution

For image $X \in \mathbb{R}^{C_{in} \times H \times W}$ and kernel $K \in \mathbb{R}^{C_{out} \times C_{in} \times k \times k}$, the (c', h, w) output entry is

$$Y_{c'hw} = \sum_{c=1}^{C_{in}} \sum_{i=0}^{k-1} \sum_{j=0}^{k-1} K_{c'cij} \, X_{c,\,h+i,\,w+j}.$$

Padding p and stride s adjust spatial resolution: $H_{out} = \lfloor \frac{H-k+2p}{s} \rfloor + 1$.

Weight sharing reduces parameters from $O(C_{out} C_{in} HW)$ to $O(C_{out} C_{in} k^2)$, imposing translation equivariance $X \mapsto \tau_\Delta X \implies Y \mapsto \tau_\Delta Y$.

Backpropagation

Denote input gradient $\partial \mathcal{L}/\partial X$ by δX. Correlation identity yields

$$\delta K_{c'c} = \sum_{h,w} \delta Y_{c'hw} \, (X_c)_{h:h+k,\,w:w+k}, \quad \delta X = \text{pad}_{k-1}(\delta Y) * \text{flip}(K),$$

where $*$ denotes convolution and flip(K) mirrors kernel.

Pooling

Max-pool implements $Y_{hw} = \max_{i,j \in [0,k)} X_{h+i,w+j}$, extracting translation invariants. Average-pool preserves energy for globally average pooling (GAP) prior to dense classifier.

10.3 Deep Learning

Architecture Example (VGG-Like)

$$[\text{conv}_{3\times 3} - \text{ReLU}]^2 \to \text{maxpool}_2 \to [\text{conv}^2] \to \text{maxpool} \to \text{GAP} \to \text{FC}_{10}.$$

On CIFAR-10, 3.1 M parameters reach $> 93\%$ test accuracy after 200 epochs with SGD($\eta = 0.1$, momentum $= 0.9$), cosine learning-rate decay, weight decay $5 \cdot 10^{-4}$, and random crop + horizontal flip augmentation.

Batch Norm in conv Layers

Normalise across mini-batch and spatial dims: $\mu_c = \frac{1}{mHW} \sum_i X_{cij}$, $\sigma_c^2 = \frac{1}{mHW} \sum_i (X_{cij} - \mu_c)^2$. Learnable ($\gamma_c, \beta_c$) restore representation capacity.

RNNs for Sequential Data Analysis

Given sequence $\{\mathbf{x}_t\}_{t=1}^T$, standard RNN cell computes

$$\mathbf{h}_t = \phi(W_h \mathbf{h}_{t-1} + W_x \mathbf{x}_t + \mathbf{b}),$$

with $\phi = \tanh$. Output $\mathbf{y}_t = W_y \mathbf{h}_t$.

Backpropagation Through Time (BPTT)

Unroll network, accumulate gradients

$$\delta \mathbf{h}_t = (W_h^T \delta \mathbf{h}_{t+1} + \delta \mathbf{y}_t) \odot \phi'(\mathbf{h}_t),$$

risking vanishing/exploding norms $\propto \|W_h\|^T$. Gradient clipping $\delta \leftarrow \delta \cdot \min\{1, \tau/\|\delta\|\}$ with $\tau = 5$ mitigates explosions.

Advanced Architectures: LSTM, GRU, and Transformers

Long Short-Term Memory (LSTM)

State equations

$$\mathbf{i} = \sigma(W_{ix}\mathbf{x}_t + W_{ih}\mathbf{h}_{t-1} + b_i),$$
$$\mathbf{f} = \sigma(W_{fx}\mathbf{x}_t + W_{fh}\mathbf{h}_{t-1} + b_f),$$
$$\mathbf{o} = \sigma(W_{ox}\mathbf{x}_t + W_{oh}\mathbf{h}_{t-1} + b_o),$$
$$\tilde{\mathbf{c}} = \tanh(W_{cx}\mathbf{x}_t + W_{ch}\mathbf{h}_{t-1} + b_c),$$
$$\mathbf{c}_t = \mathbf{f} \odot \mathbf{c}_{t-1} + \mathbf{i} \odot \tilde{\mathbf{c}},$$
$$\mathbf{h}_t = \mathbf{o} \odot \tanh(\mathbf{c}_t),$$

where gates $\mathbf{i}, \mathbf{f}, \mathbf{o} \in [0, 1]^d$ modulate memory $\mathbf{c}_t$. Constant error flow through $\mathbf{c}_t$ alleviates vanishing gradients.

Gated Recurrent Unit (GRU)

Combines input and forget gates:

$$\mathbf{z} = \sigma(W_z\mathbf{x} + U_z\mathbf{h}_{t-1}), \quad \mathbf{r} = \sigma(W_r\mathbf{x} + U_r\mathbf{h}_{t-1}), \quad \tilde{\mathbf{h}} = \tanh(W\mathbf{x} + U(\mathbf{r} \odot \mathbf{h}_{t-1})),$$
$$\mathbf{h}_t = (1 - \mathbf{z}) \odot \mathbf{h}_{t-1} + \mathbf{z} \odot \tilde{\mathbf{h}}.$$

Fewer parameters, comparable accuracy on language tasks.

Transformer (Vaswani et al. 2017)

Dispenses with recurrence, using multi-head self-attention

$$\text{Att}(Q, K, V) = \text{softmax}\left(\frac{QK^\mathsf{T}}{\sqrt{d_k}}\right) V,$$

with queries/keys/values linearly projected from inputs. Positional encodings $\mathbf{p}_t^{(2i)} = \sin(t/10000^{2i/d_{\text{model}}})$ inject sequence order. Feed-forward block and residual layer-norm yield ODE-like stable depth. Complexity $O(T^2 d)$ vs. $O(Td^2)$ for RNN BPTT; parallelisation dominates modern GPU pipelines.

$$\text{Layer}(X) = X + \text{Sublayer}(\text{LayerNorm}(X)).$$

Transfer Learning and Fine-Tuning Pretrained Models

Large models pretrained on expansive corpora encode universal priors, drastically reducing downstream sample complexity.

Feature Extractor

Freeze convolutional base (e.g. ResNet-50 trained on ImageNet); append task-specific dense head; train only last layers. Effective for few-shot datasets (≤ 1000 labelled images).

Full Fine-Tuning

Unfreeze last k layers; apply discriminative learning rates $\eta_\ell = \eta_0 \alpha^{L-\ell}$ (Howard, Ruder, 2018) with $\alpha \in [0.7, 0.9]$. For BERT, AdamW with warm-up ~ 10.

Domain Adaptation

Minimise combined objective $\mathcal{L} = \mathcal{L}_{\text{task}} + \lambda \mathcal{L}_{\text{domain}}$, where $\mathcal{L}_{\text{domain}}$ is Jensen–Shannon distance between source and target feature alignments; adversarial gradient reversal layers implement minimax.

Few-Shot Prompt Tuning (LLMs)

Keep transformer frozen; learn soft prompts $\mathbf{p} \in \mathbb{R}^{m \times d}$ prefixed to token embeddings—only md parameters vs. $\sim 10^9$ in full model. Empirically matches full fine-tune on GLUE while training $< 1\%$ weights.

10.4 Advanced Topics in Machine Learning

10.4.1 Model Interpretability and Explainability

Complex learners—deep networks, gradient-boosted trees, kernel ensembles—often outperform transparent models yet resemble *black boxes*. Interpretability seeks human-intelligible rationales: which features drive predictions, how sensitive are outputs to perturbations, and whether spurious correlations lurk. We formalise these questions with axiomatic attributions, local surrogate expansions, and global importance measures, illustrating each with concise `Python` code.

SHAP Values, LIME, and Feature Importance

SHAP (SHapley Additive exPlanations)

Let $f : \mathbb{R}^p \to \mathbb{R}$ be a trained predictor, $\mathbf{x} \in \mathbb{R}^p$ an instance, and $N = \{1, \ldots, p\}$ the feature set. A **Shapley value** $\phi_j(f, \mathbf{x})$ attributes contribution of feature j via cooperative game theory:

$$\phi_j(f, \mathbf{x}) = \sum_{S \subseteq N \setminus \{j\}} \frac{|S|!\,(p - |S| - 1)!}{p!} \Big[f_{S \cup \{j\}}(\mathbf{x}) - f_S(\mathbf{x}) \Big],$$

where f_S denotes the conditional expectation $f_S(\mathbf{x}) = \mathbb{E}[\,f(\mathbf{x}_S, \mathbf{X}_{\bar{S}}) \mid \mathbf{x}_S]$. The weighting ensures *efficiency* $\sum_j \phi_j = f(\mathbf{x}) - f_\varnothing$, *symmetry*, *dummy*, and *additivity* properties.

Exact computation is $O(2^p)$; tree models admit $O(npD)$ dynamic-programming (where D depth, n trees). Kernel SHAP approximates integral via Lasso regression on $2p+1$ perturbed samples.

```
import shap, xgboost, pandas as pd
X, y = df.drop('label', axis=1), df['label']
model = xgboost.XGBClassifier().fit(X, y)
explainer = shap.TreeExplainer(model)
phi = explainer.shap_values(X.iloc[0])
shap.waterfall_plot(shap.Explanation(values=phi,
               base_values=explainer.expected_value,
               data=X.iloc[0]))
```

LIME (Local Interpretable Model-agnostic Explanations)

For instance $\mathbf{x}_0$ construct perturbed neighbourhood $\{\mathbf{z}_i\}$ via random masking or noise; weight samples by $\pi_0(\mathbf{z}_i) = \exp(-\|\mathbf{z}_i - \mathbf{x}_0\|_2^2/\sigma^2)$. Fit sparse linear surrogate $g_\mathbf{w}(\mathbf{z}) = \mathbf{w}^\mathsf{T}\mathbf{z}$ minimising

$$\mathcal{L}(f, g) = \sum_i \pi_0(\mathbf{z}_i)\bigl(f(\mathbf{z}_i) - g(\mathbf{z}_i)\bigr)^2 \quad \text{subject to } \|\mathbf{w}\|_0 \leq K.$$

Weights $\mathbf{w}$ reveal local feature influence; stability depends on σ, sampling scheme, regularisation K.

```
import lime.lime_tabular
lime_exp = lime.lime_tabular.LimeTabularExplainer(X_train.values,
           feature_names=X.columns, mode='classification')
explanation = lime_exp.explain_instance(X.iloc[0], model.predict_proba,
                              num_features=6)
explanation.as_pyplot_figure()
```

Permuted and Intrinsic Feature Importance

Permutation importance gauges drop in accuracy when column j is permuted:

$$\mathrm{PI}_j = \frac{1}{B}\sum_{b=1}^{B}\Bigl[\mathrm{Perf}(f; \mathcal{D}) - \mathrm{Perf}\bigl(f; \mathcal{D}^{\pi_j^{(b)}}\bigr)\Bigr],$$

where $\mathcal{D}^{\pi_j^{(b)}}$ shuffles feature j alone and keeps target fixed; variance $\mathrm{Var}(\mathrm{PI}_j)/B$ estimates uncertainty.

10.4 Advanced Topics in Machine Learning

Tree ensembles record mean decrease in impurity (Gini or MSE) over splits:

$$\text{MDI}_j = \frac{1}{T} \sum_t \sum_{\text{nodes } v: \text{ split}(v)=j} \frac{n_v}{n_{\text{root}}} \big(\mathcal{I}(v) - \mathcal{I}_{\text{children}}\big).$$

Though efficient, MDI is biased towards high-cardinality categorical variables; PI is model-agnostic but more expensive.

Interpreting Black-Box Models in Python

1. **Global analysis.**
   ```
   import sklearn.inspection as insp
   perm = insp.permutation_importance(model, X_test, y_test, n_repeats=20,
                          scoring='roc_auc', random_state=0)
   sns.barplot(x=perm.importances_mean, y=X.columns)
   ```
2. **Partial dependence plots (PDP).** Average prediction while varying x_j: $\text{PDP}_j(z) = \frac{1}{n} \sum_i f(z, \mathbf{x}_{i,-j})$.
   ```
   from sklearn.inspection import PartialDependenceDisplay
   PartialDependenceDisplay.from_estimator(model, X_test, ['age','salary'
       ])
   ```
3. **SHAP summary.** Visualises magnitude and direction of ϕ_j across test set.
   ```
   shap.summary_plot(explainer.shap_values(X_test), X_test, plot_type='
       violin')
   ```
4. **Counterfactual search.** Optimise $\min_{\mathbf{z}} \|\mathbf{z} - \mathbf{x}\|_p$ s.t. $f(\mathbf{z}) \geq \tau$ via gradient (differentiable) or genetic algorithm (tree). Libraries like `alibi` implement GrowingSpheres for images.

Caveats

- *Feature correlation.* Permutation breaks joint distribution; conditional PI (Strobl, 2008) draws permutations from empirical copula.
- *Data leakage.* Apply explainers on *held-out* test folds; training data explanations overstate significance.
- *Causal fallacy.* High SHAP for feature j implies predictive, not causal, influence; causal inference requires interventions.

10.4.2 Model Evaluation and Validation Techniques

Reliable generalisation hinges on estimating out-of-sample risk and selecting parsimonious models that balance bias and variance. This subsection formalises risk estimation via resampling, addresses class imbalance through synthetic minority oversampling, and leverages information criteria grounded in maximum-likelihood theory for principled model comparison.

Cross-Validation and Resampling Methods

Let $\widehat{R}_{\mathcal{T},\mathcal{V}}(f) = \frac{1}{|\mathcal{V}|} \sum_{i \in \mathcal{V}} \ell(f_\mathcal{T}(\mathbf{x}_i), y_i)$ be the validation error when the learner is trained on $\mathcal{T}$ and evaluated on $\mathcal{V}$. Define partitions $\{\mathcal{F}_k\}_{k=1}^K$ of $\{1, \ldots, n\}$.

K-Fold Cross-Validation

$$\widehat{R}_{\text{CV}}(f) = \frac{1}{K} \sum_{k=1}^K \widehat{R}_{\mathcal{F}_{-k}, \mathcal{F}_k}(f).$$

Unbiased for i.i.d. data; variance decreases with K but computational cost increases. Leave-one-out (LOO) sets $K = n$, achieving minimal bias but high variance; $K = 5$ or 10 typically balances.

Repeated CV resamples folds B times, averaging errors $\frac{1}{B} \sum_b \widehat{R}_{\text{CV}}^{(b)}$, stabilising variance estimate.

Bootstrap

Resample indices with replacement to form $\mathcal{T}_b^*$ of size n, evaluate $\widehat{R}_b^* = \frac{1}{n_{\text{oob}}} \sum_{i \notin \mathcal{T}_b^*} \ell(f_{\mathcal{T}_b^*}, y_i)$, where $n_{\text{oob}} \approx 0.368n$ are *out-of-bag* observations. The B-bootstrap estimate $\bar{R}^*$ approximates the .632 estimator

$$\widehat{R}_{.632} = 0.368\,\widehat{R}_{\text{train}} + 0.632\,\bar{R}^*,$$

alleviating training bias.

```
from sklearn.model_selection import KFold, cross_val_score
cv10 = KFold(n_splits=10, shuffle=True, random_state=0)
cv_score = cross_val_score(model, X, y, cv=cv10, scoring='roc_auc').mean()
```

Nested Cross-Validation

Outer folds estimate generalisation; inner folds tune hyperparameters λ. Avoids optimistic bias inherent in reusing validation data for both tuning and assessment.

10.4 Advanced Topics in Machine Learning

```
from sklearn.model_selection import GridSearchCV, cross_val_score
param = {'C':[.1,1,10]}
gs = GridSearchCV(SVC(kernel='rbf'), param, cv=5)
nested = cross_val_score(gs, X, y, cv=5).mean()
```

Handling Imbalanced Datasets: SMOTE and ADASYN

Class imbalance distorts empirical risk and bias towards majority class. Let $p_+ = n_+/n \ll 1$. Strategies: resampling, cost-sensitive loss, synthetic sample generation.

SMOTE (Synthetic Minority Over-sampling Technique)

For minority sample $\mathbf{x}_i$, select k nearest minority neighbours; generate

$$\tilde{\mathbf{x}} = \mathbf{x}_i + \lambda(\mathbf{x}_{\mathrm{nn}} - \mathbf{x}_i), \qquad \lambda \sim \mathcal{U}(0,1).$$

Creates convex combinations, reduces overfitting vs. simple duplication.

ADASYN

Adaptive density: compute local ratio $r_i = \frac{\#\text{majority in } k\text{-nn}(\mathbf{x}_i)}{k}$, normalise $p_i = r_i / \sum r_j$, and synthesise $g_i = \lceil p_i G \rceil$ samples (total G). Focuses on hard-to-learn regions near class boundary.

```
from imblearn.over_sampling import SMOTE, ADASYN
X_sm, y_sm = SMOTE(k_neighbors=5, random_state=0).fit_resample(X, y)
X_ad, y_ad = ADASYN(random_state=0).fit_resample(X, y)
```

Evaluation with PR-AUC and MCC

Precision-recall AUC better reflects minority performance than ROC when $p_+ \ll 1$. Matthews correlation coefficient $\mathrm{MCC} = \frac{TP \cdot TN - FP \cdot FN}{\sqrt{(TP+FP)(TP+FN)(TN+FP)(TN+FN)}}$ is a balanced measure for binary classifications.

Model Selection with AIC, BIC, and Information Criteria

Assume parametric likelihood $L(\theta; \mathcal{D})$ with MLE $\hat{\theta}$. Information criteria penalise model complexity $k = \dim \theta$.

$$\mathrm{AIC} = 2k - 2\ln L(\hat{\theta}), \qquad \mathrm{BIC} = k \ln n - 2\ln L(\hat{\theta}).$$

AIC minimises expected Kullback–Leibler divergence to true model; BIC approximates Bayesian model evidence under unit information prior. Lower value preferred.

Corrected AIC

For small sample $n/k < 40$, $\text{AIC}_c = \text{AIC} + \frac{2k(k+1)}{n-k-1}$.

Elastic-Net Path Selection

Generalised linear model with penalty $\lambda\bigl(\alpha\|\beta\|_1 + \frac{1}{2}(1-\alpha)\|\beta\|_2^2\bigr)$ evaluates AIC along λ grid; choose $\lambda^* = \arg\min_\lambda \text{AIC}(\lambda)$. Efficient via `glmnet`:

```
import glmnet_python
fit = glmnet_python.glmnet(x=X.values, y=y.values, family='binomial')
aic = glmnet_python.glmnet_aic(fit)
λ_opt = fit['lambdas'][aic.argmin()]
```

Cross-Validated Information Criteria

Watanabe–Akaike (WAIC) and leave-one-out cross-validation (LOO-CV) approximate Bayesian predictive accuracy; implemented in probabilistic frameworks (`PyMC`, `ArviZ`).

10.4.3 Generative Models

Generative modelling seeks an estimator $\hat{p}(\mathbf{x})$ that faithfully approximates the data-generating distribution $p_{\text{data}}(\mathbf{x})$ on $\mathbb{R}^d$. Once learned, $\hat{p}$ enables (i) *sampling* realistic instances; (ii) *inference* of latent causes; (iii) *density evaluation* for anomaly detection; and (iv) *representation learning* that benefits downstream predictions. We survey three archetypes—latent-variable autoencoders, adversarial games, and probabilistic graphical models—highlighting their mathematical formulation and implementation idioms.

Autoencoders and Variational Autoencoders (VAEs)

Deterministic Autoencoder

Encoder $\phi_\theta : \mathbf{x} \mapsto \mathbf{z} \in \mathbb{R}^k$, decoder $\psi_\phi : \mathbf{z} \mapsto \hat{\mathbf{x}}$. Minimise reconstruction

$$\mathcal{L}_{\text{AE}}(\theta, \phi) = \frac{1}{n}\sum_{i=1}^{n} \|\mathbf{x}_i - \psi_\phi(\phi_\theta(\mathbf{x}_i))\|_2^2.$$

Reduction $k < d$ forces information bottleneck; decoder smoothness encourages manifold learning.

Variational Autoencoder (Kingma and Welling 2014)

Introduce stochastic latent $\mathbf{z} \sim q_\theta(\mathbf{z} \mid \mathbf{x})$ (encoder) and likelihood $p_\phi(\mathbf{x} \mid \mathbf{z})$. Evidence lower bound (ELBO) maximised:

$$\mathcal{L}_{\text{VAE}}(\theta, \phi) = \mathbb{E}_{q_\theta(\mathbf{z}|\mathbf{x})}\Big[\ln p_\phi(\mathbf{x} \mid \mathbf{z})\Big] - \text{KL}\big(q_\theta(\mathbf{z} \mid \mathbf{x}) \,\|\, p(\mathbf{z})\big) \leq \ln p(\mathbf{x}).$$

Reparameterisation Trick With Gaussian encoder $q_\theta(\mathbf{z} \mid \mathbf{x}) = \mathcal{N}(\boldsymbol{\mu}_\theta, \text{diag}\,\boldsymbol{\sigma}_\theta^2)$, write $\mathbf{z} = \boldsymbol{\mu}_\theta + \boldsymbol{\sigma}_\theta \odot \boldsymbol{\epsilon}$, $\boldsymbol{\epsilon} \sim \mathcal{N}(0, I)$, rendering gradient path differentiable.

```
def reparameterise(μ, logσ):
    ϵ = torch.randn_like(μ)
    return μ + ϵ * torch.exp(0.5*logσ)

recon = decoder(z)
reconL = F.mse_loss(recon, x, reduction='sum')
KLD = -0.5 * torch.sum(1 + logσ - μ.pow(2) - logσ.exp())
loss = reconL + KLD
```

β-VAE and Disentanglement

ELBO modified to $\mathbb{E} \ln p_\phi - \beta\,\text{KL}$; $\beta > 1$ emphasises factorised latent priors, promoting disentangled representations that align dimensions with semantic factors.

Generative Adversarial Networks (GANs) and Applications

Minimax Game

Generator $G_\theta : \mathbf{z} \sim p_z \mapsto \tilde{\mathbf{x}}$, discriminator $D_\phi : \mathbb{R}^d \to (0, 1)$. Objective:

$$\min_\theta \max_\phi \mathcal{V}(\theta, \phi) = \mathbb{E}_{\mathbf{x} \sim p_{\text{data}}}\big[\ln D_\phi(\mathbf{x})\big] + \mathbb{E}_{\mathbf{z} \sim p_z}\big[\ln\big(1 - D_\phi(G_\theta(\mathbf{z}))\big)\big].$$

Optimal discriminator $D^*(\mathbf{x}) = \frac{p_{\text{data}}}{p_{\text{data}} + p_G}$ reduces to Jensen–Shannon divergence $\text{JSD}(p_{\text{data}} \| p_G)$ between distributions; generator converges when $\text{JSD} = 0$.

Training Heuristics Replace $\ln(1 - D)$ by $-\ln D$ for stronger gradients; use feature matching loss $\|\mathbb{E}_{\text{data}}\phi(\mathbf{x}) - \mathbb{E}_z\phi(G(\mathbf{z}))\|_2^2$ to stabilise. Wasserstein GAN minimises Earth-Mover distance with 1-Lipschitz critic utilising gradient penalty $\lambda(\|\nabla_{\tilde{\mathbf{x}}} D\|_2 - 1)^2$.

Conditional GANs (cGAN)

Augment inputs with label y: $G(\mathbf{z}, y)$, $D(\mathbf{x}, y)$. Enables class-conditional synthesis (e.g. MNIST digits). Projection discriminator computes $D(\mathbf{x}, y) = \mathbf{v}^\mathsf{T}\phi(\mathbf{x}) + \phi_c(y)^\mathsf{T}\phi(\mathbf{x}) + b$.

Applications

Image-to-image translation (pix2pix, CycleGAN), super-resolution (SRGAN), data augmentation for medical imaging, tabular synthesis preserving privacy (DP-GAN with noised gradients).

Bayesian Networks and Probabilistic Graphical Models

Directed Acyclic Graph (DAG)

Nodes $\{X_1, \ldots, X_p\}$, edges encode conditional independence: $X_i \perp X_j \mid \mathbf{Pa}(X_i)$ for non-descendants X_j. Joint factorises as

$$p(\mathbf{x}) = \prod_{i=1}^{p} p(x_i \mid \mathbf{pa}(x_i)).$$

Inference via variable elimination in $O(|\mathcal{E}|)$ for treewidth-bounded graphs; junction tree algorithm triangulates dense graphs for exact marginalisation.

Parameter Learning

For discrete variables, maximum-likelihood counts: $\hat{p}(x_i \mid \mathbf{pa}) = \frac{N(x_i, \mathbf{pa})}{\sum_{x_i} N(x_i, \mathbf{pa})}$. Bayesian parameter learning uses Dirichlet-multinomial conjugacy; predictive distribution integrates over parameters.

Structure Learning

Score-and-search: maximise BIC score over DAGs using greedy equivalence search; constraint-based (PC algorithm) tests conditional independencies; hybrid methods combine. NP-hard, but tractable for $p \lesssim 20$.

10.4 Advanced Topics in Machine Learning

Undirected Models

Markov random field (MRF) with potentials $\psi_C(\mathbf{x}_C)$ over cliques C:

$$p(\mathbf{x}) = \frac{1}{Z}\prod_C \psi_C(\mathbf{x}_C), \qquad Z = \sum_{\mathbf{x}}\prod_C \psi_C(\mathbf{x}_C).$$

Pairwise Ising ($x_i \in \{-1, 1\}$): $p(\mathbf{x}) \propto \exp(\sum_{i<j} J_{ij}x_i x_j + \sum_i h_i x_i)$. Parameter estimation via pseudolikelihood maximisation avoids Z.

Approximate Inference

Loopy belief propagation iterates messages

$$m_{i \to j}(x_j) = \sum_{x_i} \psi_{ij}\psi_i \prod_{k \in \mathcal{N}(i)\setminus j} m_{k \to i}(x_i);$$

converges on trees, heuristic on graphs. Variational mean-field minimises KL between factorised q and true posterior; EP refines by moment matching.

```
import pgmpy
bn = pgmpy.models.BayesianModel([('A','C'),('B','C'),('C','D')])
bn.fit(df, estimator=pgmpy.estimators.BayesianEstimator, prior_type='BDeu'
    )
infer = pgmpy.inference.VariableElimination(bn)
posterior = infer.query(variables=['D'], evidence={'A':1,'B':0})
```

10.4.4 Big Data and Scalable Machine Learning

Industrial-scale learning integrates statistical algorithms with distributed systems engineering. Storage is sharded across commodity clusters, gradients propagate through network fabrics, and predictions must often be delivered within millisecond latencies. This subsection develops the mathematical throughput models that underlie distributed optimisation, details the orchestration primitives that containerise models at web scale, and illustrates streaming pipelines whose concept-drift detectors update parameters on the fly.

Distributed Machine Learning with Apache Spark

Let $\mathcal{D}$ be partitioned across M worker nodes, $\mathcal{D} = \bigsqcup_{m=1}^{M} \mathcal{D}_m$, with $|\mathcal{D}_m| = n/M$. Consider minimising the empirical risk

$$F(\boldsymbol{\theta}) = \frac{1}{n}\sum_{(\mathbf{x},y)\in\mathcal{D}} \ell(\boldsymbol{\theta}; \mathbf{x}, y) + \lambda\|\boldsymbol{\theta}\|_2^2.$$

Spark's *Resilient Distributed Dataset* (RDD) abstraction applies the map-reduce paradigm:

$$map: \mathcal{D}_m \mapsto \nabla F_m(\boldsymbol{\theta}) = \frac{M}{n} \sum_{(\mathbf{x},y)\in\mathcal{D}_m} \nabla_\theta \ell(\boldsymbol{\theta};\mathbf{x},y), \quad reduce: \nabla F = \frac{1}{M}\sum_{m=1}^{M} \nabla F_m.$$

Under synchronous gradient descent the wall-time per iteration is

$$T_{\text{iter}} = T_{\text{compute}} + T_{\text{shuffle}} = \max_m t_m^{\text{CPU}} + \alpha \log_2 M + \beta d,$$

where α, β are network latency and inverse bandwidth, and d is parameter dimension. Speed-up saturates when $T_{\text{shuffle}} \approx T_{\text{compute}}$; Spark's `treeAggregate` attains $\log M$ communication depth.

```
from pyspark.sql import SparkSession
from pyspark.ml.classification import LogisticRegression
spark = SparkSession.builder.appName("logreg").getOrCreate()
df = spark.read.parquet("hdfs:///clicks")
train, test = df.randomSplit([0.8, 0.2], seed=0)
lr = (LogisticRegression(featuresCol="features", labelCol="label",
            maxIter=10, regParam=1e-3, elasticNetParam=0.0)
    .setAggregationDepth(2) ) # log_2 tree reduce
model = lr.fit(train)
print("AUC =", model.evaluate(test).areaUnderROC)
```

Asynchronous Variants

Parameter-server architecture distributes parameter shards; each worker pushes gradients and pulls updated weights without barriers. Given staleness bound τ, convergence occurs when step size $\eta_t \propto 1/\sqrt{t+\tau}$; empirical linear scaling holds up to $\tau \approx 5$ on 10 Gb networks.

Scaling ML Models with Kubernetes and Docker

A trained model f_{θ^*} becomes a stateless microservice *container* encapsulating runtime, dependencies, and binaries. Let $\lambda(t)$ denote request arrival rate (Poisson approximation). Little's law yields average queue length $L = \rho/(1-\rho)$ for utilisation $\rho = \lambda/\mu$, where μ is per-pod service rate. Kubernetes (*k8s*) horizontal pod autoscaler maintains $\rho \leq \rho_{\max}$ by scaling replica count $N \geq \lceil \lambda/\rho_{\max}\mu \rceil$; cluster capacity constraint $\sum_i N_i \text{cpu}_i \leq \text{CPU}_{\text{node}}$ defines bin-packing via `kubectl`.

10.4 Advanced Topics in Machine Learning

```
# Dockerfile snippet
FROM python:3.11-slim
COPY model.pkl server.py /app/
RUN pip install --no-cache-dir fastapi uvicorn joblib
CMD ["uvicorn", "server:app", "--host", "0.0.0.0", "--port", "80"]
```

```yaml
apiVersion: apps/v1
kind: Deployment
metadata: { name: ml-inference }
spec:
  replicas: 3
  selector: { matchLabels: { app: ml } }
  template:
    metadata: { labels: { app: ml } }
    spec:
      containers:
      - name: api
        image: registry.io/ml-infer:v1
        resources: { requests: { cpu: "500m" }, limits: { cpu: "1" } }

apiVersion: autoscaling/v2
kind: HorizontalPodAutoscaler
metadata: { name: ml-hpa }
spec:
  scaleTargetRef: { apiVersion: apps/v1, kind: Deployment,
    name: ml-inference }
  minReplicas: 3
  maxReplicas: 15
  metrics:
  - type: Resource
    resource: { name: cpu, target: { type: Utilization,
       averageUtilization: 60 } }
```

Rolling updates (maxSurge=25%) satisfy blue–green safety; canary pods route ε fraction of traffic to validate drift. GPU models exploit nvidia-device plugin exposing *extended* resource type nvidia.com/gpu.

Real-Time Machine Learning with Streaming Data

Let data stream $\{(\mathbf{x}_t, y_t)\}_{t=1}^{\infty}$ arrive at rate λ_s. A *window* operator aggregates events within horizon W producing micro-batch $B_t = \{(\mathbf{x}_\tau, y_\tau) : t - W < \tau \leq t\}$. Spark Structured Streaming ensures exactly-once semantics via checkpointed offsets:

```python
from pyspark.sql.functions import *
stream = (spark.readStream.format("kafka")
         .option("subscribe", "clicks").load()
         .selectExpr("CAST(value AS STRING) json"))
parsed = (stream.select(from_json("json", schema).alias("r"))
              .select("r.*") # features & label
              .withWatermark("timestamp","5 minutes"))
agg = (parsed.groupBy(window("timestamp","1 minute"))
           .agg(avg("label").alias("ctr")))
query = (agg.writeStream.outputMode("update")
            .format("console").start())
```

Online learners update parameters per instance:

$$\boldsymbol{\theta}_{t+1} = \boldsymbol{\theta}_t - \eta_t \nabla_\theta \ell\big(f_{\theta_t}(\mathbf{x}_t), y_t\big), \quad \eta_t = \eta_0(1 + t/\tau)^{-0.5}.$$

Hoeffding tree guarantees that with probability $1 - \delta$ the chosen split's information gain differs from the optimal by $\leq \varepsilon = \sqrt{\frac{\ln(1/\delta)}{2n}}$ after n observations, enabling single-pass decision trees (River library).

When concept drift occurs (distribution shift at t^*), ADWIN maintains two sub-windows and runs sequential t-tests to trigger model reset when mean difference exceeds $\sqrt{\frac{1}{2} \ln \frac{4}{\epsilon} \left(\frac{1}{n_0} + \frac{1}{n_1}\right)}$ at confidence $1 - \epsilon$.

10.5 Ethics and Fairness in Machine Learning

Algorithmic decisions increasingly influence credit approval, bail, hiring, and healthcare triage. If a learning system inherits societal biases present in historical data, it can amplify injustice at scale. *Fairness* research therefore formalises group and individual notions of equity, devises diagnostic metrics, and designs mitigation mechanisms that minimise disparate impact while preserving predictive utility. This section introduces the mathematical foundation of bias detection and presents practical Python tool-chains for auditing and correcting unfair models.

10.5.1 Bias and Fairness in AI

Let $S \in \{0, 1\}$ denote a binary sensitive attribute (e.g. *gender*, *race*) and $\hat{Y} = f_\theta(X)$ be a classifier output. Define true label Y; the joint distribution $p(X, S, Y)$ governs observed data. Bias arises when $\hat{Y}$ exhibits disparate performance across S due to dataset imbalance, proxy variables, or label noise.

Identifying and Mitigating Bias in Machine Learning Models

Group Fairness Criteria

Statistical Parity Difference (SPD)

$$\Delta_{\text{SPD}} = \Pr(\hat{Y} = 1 \mid S = 1) - \Pr(\hat{Y} = 1 \mid S = 0).$$

$\Delta_{\text{SPD}} = 0$ implies protected, and reference groups receive positive outcome equally often; $|\Delta_{\text{SPD}}| \leq 0.02$ is a common threshold.

10.5 Ethics and Fairness in Machine Learning

Equalised Odds (EO)

$$\Pr(\hat{Y}=1 \mid Y=y, S=1) = \Pr(\hat{Y}=1 \mid Y=y, S=0), \qquad y \in \{0,1\},$$

requiring parity of both true-positive rate (TPR) and false-positive rate (FPR). Deviation measured by $\Delta_{EO} = \frac{1}{2}(|\text{TPR}_\Delta| + |\text{FPR}_\Delta|)$.

Calibration
For probabilistic scores $\hat{p}$, $\Pr(Y=1 \mid \hat{p}=s, S=1) = \Pr(Y=1 \mid \hat{p}=s, S=0)$. Fails to coexist with EO unless $\hat{p}$ is perfect (Kleinberg et al. 2017).

Bias Mitigation Taxonomy

(i) **Pre-processing.** *Reweighing*: assign weight $w_i = w(S_i, Y_i)$ so that weighted distribution satisfies $\Pr_w(S, Y) = \Pr(S)\Pr(Y)$, then train with weighted loss $L = \sum w_i \ell(f_\theta(X_i), Y_i)$.

(ii) **In-processing.** *Adversarial debiasing*: augment loss with adversary g_ϕ predicting S from hidden representation $h_\theta(X)$,

$$\min_\theta \max_\phi \mathbb{E}\big[\ell(f_\theta(X), Y)\big] - \lambda\, \mathbb{E}\big[\ell_{adv}(g_\phi(h_\theta(X)), S)\big].$$

Gradient reversal implements the max step.

(iii) **Post-processing.** *Equalised odds post-processing*: learn thresholds (t_0, t_1) for groups $S = 0, 1$ that minimise empirical risk under EO constraints, solvable by linear programming.

Example 10.5.1 (Weighted Logistic Regression with SPD Constraint) Dataset: $n = 10{,}000$ loan applications, $S = $ gender. Compute weights $w_{ij} = \frac{\Pr(S=i)\Pr(Y=j)}{\Pr(S=i, Y=j)}$ and train L2-regularised logistic model; evaluate Δ_{SPD} on test set.

Fairness Metrics and Tools in Python

Fairlearn Audit

```
from fairlearn.metrics import MetricFrame, selection_rate,
    equalized_odds_difference
from sklearn.metrics import accuracy_score
from sklearn.linear_model import LogisticRegression

model = LogisticRegression(max_iter=800).fit(X_train, y_train)
y_pred = model.predict(X_test)

mf = MetricFrame(metrics={'accuracy': accuracy_score,
                  'selection_rate': selection_rate},
```

```
                    y_true=y_test, y_pred=y_pred, sensitive_features=s_test)
print("accuracy:", mf.overall['accuracy'])
print("selection gap:", mf.difference(method='between_groups')['
    selection_rate'])
print("equalized-odds δ =", equalized_odds_difference(y_test, y_pred,
    s_test))
```

SMOTE + Reweighing with AIF360

```
from aif360.datasets import BinaryLabelDataset
from aif360.algorithms.preprocessing import Reweighing

data = BinaryLabelDataset(df=df, label_names=['loan'],
                 protected_attribute_names=['gender'])
rw = Reweighing().fit(data)
data_transf = rw.transform(data)
X_tr, y_tr, w_tr = data_transf.features, data_transf.labels.ravel(),
    data_transf.instance_weights
clf = LogisticRegression().fit(X_tr, y_tr, sample_weight=w_tr)

metric = ClassificationMetric(data_test, clf_pred,
                  unprivileged_groups=[{'gender':0}],
                  privileged_groups=[{'gender':1}])
print("δ SPD :", metric.statistical_parity_difference())
print("δ EO :", metric.equal_opportunity_difference())
```

Adversarial Debiasing (TensorFlow)

```
import tensorflow as tf, tensorlayer as tl
h = tl.layers.Dense(64, activation='relu')(inputs)
adv = tl.layers.Dense(1, activation='sigmoid', name='adv')(h)
pred = tl.layers.Dense(1, activation='sigmoid', name='pred')(h)

λ = 0.2
loss_pred = tf.keras.losses.binary_crossentropy(y_true, pred)
loss_adv = tf.keras.losses.binary_crossentropy(s_true, adv)
loss = loss_pred - λ*loss_adv # gradient reversal on adv
```

Visualisation

Disparate impact ratio $\frac{\Pr(\hat{Y}=1|S=1)}{\Pr(\hat{Y}=1|S=0)}$ plotted with 95% Wilson intervals identifies uncertainty.

10.5.2 Privacy–Preserving Machine Learning

Machine learning deployments frequently handle sensitive data such as medical records, financial transactions, or conversational logs. Regulatory regimes (GDPR, HIPAA) impose strict constraints on data sharing, motivating cryptographic and

10.5 Ethics and Fairness in Machine Learning

statistical frameworks that guarantee privacy *by construction*. Two complementary paradigms dominate the research landscape: *federated learning with differential privacy*, which leaves data in situ and sanitises gradients, and *secure multi-party computation*, which performs joint inference without revealing raw inputs.

Federated Learning and Differential Privacy

Federated Optimisation

Suppose M clients each hold private dataset $\mathcal{D}_m$. Minimise global objective

$$F(\theta) = \frac{1}{M}\sum_{m=1}^{M} F_m(\theta), \quad F_m(\theta) = \frac{1}{|\mathcal{D}_m|}\sum_{(x,y)\in\mathcal{D}_m} \ell(\theta; x, y).$$

FedAvg(McMahan et al. 2017) executes E local SGD steps at each client and then transmits parameter updates Δ_m to a server that averages $\theta^{(t+1)} = \theta^{(t)} + \frac{1}{M}\sum_m \Delta_m$. Communication cost $O(d)$ per round, mitigated by sparsification or sketching.

Differential Privacy (DP)

A mechanism $\mathcal{M}$ is (ε, δ)-DP if for any neighbouring datasets $\mathcal{D}, \mathcal{D}'$ differing in one record and any event $\mathcal{S}$:

$$\Pr[\mathcal{M}(\mathcal{D}) \in \mathcal{S}] \leq e^\varepsilon \Pr[\mathcal{M}(\mathcal{D}') \in \mathcal{S}] + \delta.$$

DP-SGD (Abadi et al. 2016) for each mini-batch:

(i) Clip per-sample gradients: $\bar{g}_i = g_i / \max\{1, \|g_i\|_2/C\}$.
(ii) Aggregate and add Gaussian noise: $\tilde{g} = \frac{1}{B}\left(\sum_{i=1}^{B} \bar{g}_i + \mathcal{N}(0, \sigma^2 C^2 I)\right)$.
(iii) Update $\theta \leftarrow \theta - \eta\tilde{g}$.

Moments accountant bounds cumulative privacy loss: $\varepsilon \approx \frac{q\sqrt{2T\ln(1/\delta)}}{\sigma}$, $q = B/n$. In federated setting, each client clips its local update and adds noise before transmission, ensuring *local* DP.

```
def dp_aggregate(updates, C=1.0, σ=0.5):
    clipped = [u * min(1, C/np.linalg.norm(u)) for u in updates]
    mean = np.mean(clipped, axis=0)
    noise = np.random.normal(scale=σ*C, size=mean.shape)
    return mean + noise / len(updates)
```

Privacy-Utility Trade-Off

Noise variance σ^2 scales with clip norm C and target ε; smaller batches or adaptive clipping (AdaClip) mitigates accuracy degradation.

Secure Multi-party Computation

Let $P_1, \ldots, P_M$ possess private inputs $x_1, \ldots, x_M$. A secure protocol computes $f(x_1, \ldots, x_M)$ such that no party learns more than the output.

Additive Secret Sharing

Over ring $\mathbb{Z}_p$, split value v into random shares $v_1, \ldots, v_M$ with $\sum v_m \equiv v$ (mod p). Addition and scalar multiplication are share-wise; multiplication requires Beaver triples (a, b, c) with $c = ab$. Protocol (SPDZ) evaluates arithmetic circuits with $O(1)$ rounds per gate and pre-processed correlated randomness.

Homomorphic Encryption (HE)

Given public key pk, ciphertexts $[\![x]\!]$, one can compute $[\![x + y]\!]$ and $[\![x \cdot y]\!]$ without decryption (BFV scheme). Inference pipelines encode neural layers as polynomial approximations (Chebyshev) to circumvent non-polynomial activations; latency remains $\sim$10–100$\times$ plaintext.

Private Set Intersection

Bloom-filter–based PSI enables two hospitals to find common patients in $O(n \log n)$ bandwidth with differential privacy on cardinality leakage.

Application: Secure Federated Logistic Regression

Each client secret-shares gradient g_m; servers aggregate via SPDZ, add DP noise, and return encrypted global gradient. Guarantees: computational privacy against semi-honest adversaries and (ε, δ)-DP.

10.5.3 Responsible AI and Ethical Considerations

Beyond technical safeguards, AI systems must align with societal values, legal norms, and environmental constraints.

Ethical Frameworks and Guidelines for AI Development

Principle Taxonomies

- **OECD (2019)**: inclusiveness, transparency, robustness, accountability, and human-centred values.
- **EU AI Act (draft 2024)**: four-tier risk classification; mandatory conformity assessment for *high-risk* applications; prohibited practices (social scoring, real-time biometric surveillance).
- **IEEE 7000-2021**: value-based engineering process integrating stakeholder elicitation, traceability artefacts, and impact assessments.

Operationalisation

Model cards document intended use, evaluation metrics disaggregated by demographic groups, and ethical considerations; data sheets provide provenance, licencing, and acquisition context. Continuous monitoring pipelines compute fairness metrics, drift statistics, and carbon footprint per inference $(gCO_2\, e)$.

Impact of AI on Society and Future Implications

Labour and Economy

Task-based models estimate substitution elasticity; GPT-4–level systems place $\approx 19\%$ of US occupations at 50% exposure. Policy responses: reskilling funds, universal basic income pilots, collective bargaining for algorithmic management transparency.

Environmental Footprint

Training a 100-billion-parameter model with 300 PFlop·s days consumes $\sim 10^6$ kWh; renewable-energy scheduling and mixed-precision GPUs cut emissions by 35%. Lifecycle assessment incorporates e-waste and cooling water usage.

Geopolitics and Governance

AI accelerates dual-use technologies (cyber offence, autonomous weapons); governance proposals include compute licences keyed to FLOP thresholds and compute passports tracking ASIC supply chain.

Long-Term Alignment

Value-learning agendas (co-operative IRL, inverse reward design) and scalable oversight (RLHF, debate, constitutional AI) aim to reduce the *specification gap*. Open problems involve distributional shifts, mesa-optimisation, and multi-agent bargaining equilibria.

10.6 Exercises

1. Let $\mathbf{X} \in \mathbb{R}^{n \times 3}$ whose columns are $\mathbf{x}^{(1)} \sim \mathcal{N}(0, 1)$, $\mathbf{x}^{(2)} = 2\mathbf{x}^{(1)} + \varepsilon$, $\varepsilon \sim \mathcal{N}(0, 1)$, and $\mathbf{x}^{(3)} \sim \text{Uniform}[-1, 1]$. Generate a missingness mask M such that

    (a) $M_{i3} \sim \text{Bernoulli}(0.3)$ independently (MCAR)
    (b) $M_{i3} = \mathbf{1}\{x_i^{(1)} > 0\}$ (MAR)

    For each case derive the bias of mean-imputation estimator $\hat{\mu} = \frac{1}{n} \sum_i (1 - M_{i3}) x_i^{(3)}$ and demonstrate numerically for $n = 10\,000$ that the theoretical bias matches simulation to two decimal places.

2. Given i.i.d. positive observations $\{t_i\}_{i=1}^{150}$ whose empirical skewness is 2.1, show that the maximum-likelihood Box–Cox power λ^* satisfies $\lambda^* \approx 1 - \frac{2}{3}(\gamma_1 - \frac{2}{3}\gamma_2)$ for large n, where γ_1, γ_2 are skewness and excess kurtosis. Compute λ^* and transform the data; verify that the Jarque–Bera statistic drops below the 5% critical value.

3. Construct the covariance matrix
$$\Sigma = \begin{pmatrix} 4 & 3 & 2 & 0 \\ 3 & 4 & 2 & 0 \\ 2 & 2 & 5 & 0 \\ 0 & 0 & 0 & 1 \end{pmatrix}.$$

    (i) Compute eigenvalues analytically.
    (ii) Determine the minimal k such that $\sum_{j=1}^{k} \lambda_j / \sum_{j=1}^{4} \lambda_j \geq 0.9$.
    (iii) For $n = 1{,}000$ observations drawn $\mathcal{N}(\mathbf{0}, \Sigma)$, perform PCA and show that empirical explained variance concentrates within $\pm 1\%$ of the theoretical ratio with probability >0.95.

4. Let cluster centres $\{\boldsymbol{\mu}_j\}_{j=1}^{k}$ be fixed in $\mathbb{R}^2$ at the vertices of a regular k-gon of radius R. Sample n/k points around each centre from $\mathcal{N}(\boldsymbol{\mu}_j, \sigma^2 I)$. Prove that for $\sigma^2 \leq \frac{R^2}{8 \ln n}$ the expected within-cluster sum of squares of the optimal

10.6 Exercises

k-means solution is bounded above by $2r\sigma^2$, and confirm numerically for $k = 6$, $R = 5$, $\sigma^2 = 0.1$, $n = 12\,000$.

5. Given binary feature $X \in \{0, 1\}$ and binary label $Y \in \{0, 1\}$ with joint probabilities $P(X = 1, Y = 1) = 0.35$, $P(X = 1, Y = 0) = 0.15$, $P(X = 0, Y = 1) = 0.1$, $P(X = 0, Y = 0) = 0.4$:

    (a) Compute Gini impurity at root node.
    (b) Calculate impurity of child nodes after splitting on X.
    (c) Verify that the information gain equals the mutual information $I(X; Y)$ times a scaling factor and evaluate it.

6. For dataset $\{(\mathbf{x}_i, y_i)\}_{i=1}^4$ where $\mathbf{x}_1 = (1, 0)$, $y_1 = 1$; $\mathbf{x}_2 = (0, 1)$, $y_2 = 1$; $\mathbf{x}_3 = (-1, 0)$, $y_3 = -1$; $\mathbf{x}_4 = (0, -1)$, $y_4 = -1$, derive analytically the maximum-margin separating hyperplane, list the support vectors, and compute the leave-one-out error of this SVM.

7. Show that for T rounds with edge $\gamma_t \geq \gamma > 0$ the training error of AdaBoost satisfies $\widehat{R} \leq e^{-2\gamma^2 T}$. Apply the bound to a dataset where each weak learner achieves 65% accuracy and determine the number of rounds needed to drive training error below 1%.

8. Given pairwise Euclidean distances between five high-dimensional points

$$D = \begin{pmatrix} 0 & 2 & 4 & 1 & 3 \\ 2 & 0 & 2 & 3 & 4 \\ 4 & 2 & 0 & 5 & 1 \\ 1 & 3 & 5 & 0 & 2 \\ 3 & 4 & 1 & 2 & 0 \end{pmatrix},$$

compute conditional probabilities $p_{j|1}$ for point 1 under σ_1 chosen so that perplexity $\mathcal{P} = 2^{H(P_1)} = 2.5$. Solve σ_1 numerically (two iterations of Newton–Raphson suffice) and report final $p_{j|1}$ values.

9. For an episodic MDP with states $\{s_0, s_1, s_2\}$, actions a, b, deterministic transitions $s_0 \xrightarrow{a} s_1 \xrightarrow{b} s_2 \to$ terminal, and rewards $r(s_0, a) = 1$, $r(s_1, b) = 2$, evaluate $V^\pi(s_0)$ for policy $\pi(a|s_0) = \frac{2}{3}$, $\pi(b|s_0) = \frac{1}{3}$, $\pi(b|s_1) = 1$ using

    (i) Every-visit Monte Carlo with $N = 1\,000$ episodes
    (ii) First-visit Monte Carlo with importance sampling from behaviour policy $\mu(a|s_0) = \mu(b|s_0) = \frac{1}{2}$

    Show that the estimator variance under importance sampling is exactly twice that of the on-policy method.

10. For the two-layer network $f_\theta(x) = W_2 \operatorname{ReLU}(W_1 x + b_1) + b_2$, randomly initialise

$$W_1 = \begin{pmatrix} 1 & -1 \\ 2 & 0.5 \end{pmatrix}, \quad b_1 = \begin{pmatrix} 0 \\ 1 \end{pmatrix}, \quad W_2 = (2\ 1), \quad b_2 = 0, \quad x = \begin{pmatrix} 3 \\ -1 \end{pmatrix}.$$

Compute analytic gradient $\nabla_\theta f_\theta(x)$ and verify it equals the centred finite-difference estimate with step $h = 10^{-4}$ up to 10^{-6} absolute error per parameter.

11. A convolutional block accepts $X \in \mathbb{R}^{3 \times 32 \times 32}$ and applies

$$\underbrace{\text{Conv}(3 \times 3, 64) \to \text{BN} \to \text{ReLU}}_{\text{layer 1}} \to \underbrace{\text{Conv}(3 \times 3, 64) \to \text{BN}}_{\text{layer 2}}$$

with padding 1, stride 1.

   (a) Compute output tensor shape after the block.
   (b) Compute total trainable parameters, counting BN (γ, β).
   (c) Show that replacing the two convolutions by a single 5×5 filter requires 2.56 times more parameters.

12. Suppose feature X_j has permutation importance $\text{PI}_j = 0.08$ with standard error 0.015 under $B = 30$ repeats. Test at $\alpha = 0.05$ whether this feature provides significant predictive information beyond noise using a one-sample t-test, and compute the 95% confidence interval for PI_j.

13. A federated DP–SGD uses mini-batch size $B = 128$, clip norm $C = 1$, noise multiplier $\sigma = 1.2$, and trains for $T = 1\,000$ steps on $n = 50\,000$ devices. Using moments accountant, bound the cumulative privacy loss ε for $\delta = 10^{-5}$ and verify numerically that increasing σ to 1.6 halves ε approximately.

14. Given minority points $\mathbf{p}_1 = (2, 1)$, $\mathbf{p}_2 = (5, 1)$, $\mathbf{p}_3 = (3, 4)$ in $\mathbb{R}^2$, apply SMOTE with $k = 2$ to synthesise two samples for $\mathbf{p}_1$. List all possible synthetic points and sketch the resulting convex hull containing the augmented minority class.

15. Let generator and discriminator be one-dimensional linear: $G_\theta(z) = \theta z$, $D_\phi(x) = \sigma(\phi x)$, with $z \sim \text{Uniform}[-1, 1]$ and true data $x \sim \text{Uniform}[-1, 1]$.

    (a) Show that $(\theta^*, \phi^*) = (1, 0)$ is a Nash equilibrium.
    (b) Compute Hessian of the value function at equilibrium and prove that alternating gradient ascent–descent exhibits cyclic behaviour with frequency proportional to learning rate.

16. Generate Bernoulli stream with change point: $\Pr(Y_t = 1) = 0.4$ for $t \leq 10\,000$ and 0.6 afterwards. Implement ADWIN with $\epsilon = 0.002$ and verify that drift is detected within 300 samples of the true change with probability ≥ 0.9 over 100 Monte Carlo runs.

17. Suppose binary classifier achieves equalised odds and is calibrated within each group for sensitive attribute $S \in \{0, 1\}$. Prove that if the base rates differ $\Pr(Y = 1 \mid S = 1) \neq \Pr(Y = 1 \mid S = 0)$, then calibration and equalised odds cannot both hold unless the classifier is perfect (TPR $= 1$, FPR $= 0$). Illustrate with synthetic $\Pr(Y = 1 \mid S = 0) = 0.3$, $\Pr(Y = 1 \mid S = 1) = 0.7$.

Chapter 11
Advanced Topics

Abstract The final chapter synthesises symbolic computation, Fourier and wavelet analysis, rigorous time-series and signal-processing methods, topological data analysis, quantum-computing primitives, and high-order spectral and parallel numerical schemes. Each section pushes Python to its research-frontier limits—proving that modern mathematical thinking is inseparable from well-engineered computational experimentation—and closes with open problems and project ideas that invite readers to contribute to the discipline's evolving landscape.

Keywords Symbolic computation · Fourier analysis · Wavelet transform · Signal processing · Topological data analysis · Quantum computing

The chapters that precede have equipped us with a firm command of Pythonic fundamentals—data structures and algorithms, numerical linear algebra, calculus, probability, and the foundations of machine learning. Yet contemporary scientific inquiry and industrial analytics routinely demand techniques that lie beyond this classical kernel. They call for symbolic machinery that manipulates algebraic objects with the same acuity that floating-point routines handle numbers; spectral decompositions that unravel signals across scales of time and frequency; probabilistic and topological frameworks that characterise uncertainty and shape in high-dimensional data; and quantum and high-performance paradigms that push computation to the physical frontiers of decoherence limits and petascale concurrency. *Advanced Topics* gathers these frontiers under a single mathematical canopy, continuing the theme that rigorous theory and well-crafted Python code are not adversaries but natural collaborators.

The unifying principle of this chapter is *representation*. Symbolic computation represents an equation as a structured tree and rewrites it through algebraic identities; Fourier and wavelet analysis represent a function as a superposition of orthogonal or localised bases; time-series models represent temporal dependencies through state-space or spectral poles; topological data analysis represents a point cloud by homology classes that persist across scales; quantum circuits represent linear maps on Hilbert spaces through tensor products of elementary gates; and advanced numerical methods represent continuous operators via spectral collocation or finite differences on distributed meshes. In each domain we shall study how representation choices dictate the geometry of computation—conditioning, sparsity, convergence rate—and how Python libraries exploit that geometry to deliver performant, readable implementations.

Throughout, mathematical formality will remain central. Gröbner basis elimination, $\mathcal{O}(n \log n)$ FFT complexity proofs, Kalman filter optimality via least-squares, stability criteria for Runge–Kutta and Chebyshev spectral schemes, persistence diagrams' stability under Gromov–Hausdorff perturbations, and the no-cloning theorem in quantum mechanics will all be derived in full. Equally, code snippets—encapsulated in lstlisting environments—will translate each theorem into an executable artefact: an automated solver in SymPy, a streaming predictor in River, a Mapper visualisation in KeplerMapper, a variational quantum eigensolver in Qiskit, or a distributed conjugate-gradient kernel on Dask-MPI. Examples are chosen to balance theoretical elegance with computational substance: factorising a coupled oscillator spectrum and then denoising its sensor trace; proving Euler–Maclaurin symbolically and verifying error constants numerically; classifying arrhythmia from ECG via wavelet subbands while auditing model fairness; reconstructing climate proxies with Gaussian processes on a GPU cluster; estimating homological loops in protein binding sites; or simulating a quantum teleportation protocol and measuring entanglement entropy.

Finally, we will not shy away from meta-scientific responsibilities. Advanced algorithms wield transformative power, and their deployment demands guarantees of privacy, fairness, energy efficiency, and reproducibility. Each major section will therefore conclude with brief ethical and practical remarks—differential-privacy budgets for symbolic database queries, carbon cost estimates for large-scale FFT pipelines, stability of spectral methods under finite precision, reproducibility checklists for quantum experiments—so that technical mastery is guided by conscientious stewardship.

Readers should expect to alternate constantly between mathematics rendered in elegant notation and Python scripts rendered in efficient vectorised form. By the end of this chapter you will not only understand the theory behind modern computational science but will also possess a toolbox of rigorously engineered Python patterns ready to tackle research-grade problems across disciplines.

11.1 Symbolic Computation

11.1.1 Advanced Symbolic Mathematics with SymPy

Sophisticated manipulation of algebraic, transcendental, and differential expressions is indispensable in modern mathematical research and pedagogy. Python's SymPy library offers an extensible framework that mirrors manual pencil-and-paper tech-

11.1 Symbolic Computation

niques while integrating seamlessly with numerical back-ends. In this subsection we demonstrate how SymPy provides exact solutions to intricate systems, performs algebraic transformations that illuminate structure, and serves as a symbolic engine for linear algebra, calculus, and differential equations.

Solving Complex Equations and Systems Symbolically

Polynomial systems of high degree often defy closed-form radicals yet admit Gröbner basis decomposition that reveals geometric multiplicity and solution components.

Example 11.1.1 Solve the nonlinear system

$$\begin{cases} x^3 + y^3 + z^3 - 3xyz = 1, \\ x + y + z = 1, \\ x^2 + y^2 + z^2 = 1. \end{cases}$$

We employ Lex order Gröbner basis over $\mathbb{Q}$ to eliminate y, z and reduce to a univariate polynomial in x. SymPy returns three rational roots, from which y, z follow by elementary symmetric relations.

```
import sympy as sp
x,y,z = sp.symbols('x y z')
F = [x**3+y**3+z**3-3*x*y*z-1, x+y+z-1, x**2+y**2+z**2-1]
G = sp.groebner(F, x, y, z, order='lex')
univar = G.eliminate(y,z)[0] # cubic in x
roots = sp.factor(univar)
```

Transcendental equations typically require the `solveset` framework which combines analytic inverses, branch cut analysis, and special function identities.

```
t = sp.symbols('t', real=True)
eq = sp.Eq(sp.sin(t) + sp.log(t), 0)
sol = sp.solveset(eq, t, sp.Interval(1, sp.oo)) # principal branch
```

The solution involves the *Lambert W* function: $t = -\exp(W(-1))$, evidencing the ubiquity of W in transcendental inversion.

Applications in Algebra and Calculus

Algebraic identities benefit from symbolic verification: for any $n \in \mathbb{N}$,

$$\sum_{k=0}^{n}(-1)^k \binom{n}{k}\frac{1}{k+1} = \frac{1}{n+1}.$$

SymPy validates the combinatorial identity via Gosper's algorithm:

```
k,n = sp.symbols('k n', integer=True, nonnegative=True)
expr = sp.summation((-1)**k*sp.binom(n,k)/(k+1), (k,0,n))
sp.simplify(expr)
```

In calculus, automatic series expansion uncovers local behaviour:

$$\int_0^x \frac{\sin \sqrt{t}}{\sqrt{t}} \, dt = x - \frac{x^2}{6} + \frac{x^3}{40} - \cdots .$$

```
t = sp.symbols('t')
ser = sp.series(sp.integrate(sp.sin(sp.sqrt(t))/sp.sqrt(t),(t,0,sp.Symbol(
    'x'))),
             sp.Symbol('x'), 0, 5)
```

Symbolic Linear Algebra and Matrix Computations

Exact Eigenanalysis

Matrix $A = \begin{pmatrix} 2 & 1 & 0 \\ 1 & 2 & 1 \\ 0 & 1 & 2 \end{pmatrix}$ has characteristic polynomial $\chi(\lambda) = (\lambda-1)^2(\lambda-4)$. SymPy diagonalises A and produces the orthogonal eigenbasis symbolically.

```
A = sp.Matrix([[2,1,0],[1,2,1],[0,1,2]])
P, D = A.diagonalize()
```

Block Jordan Forms

Symbolic Jordan decomposition clarifies nilpotent structure, vital in solving operator exponentials e^{At}:

```
B = sp.Matrix(sp.randMatrix(5, symmetric=False))
B.jordan_cells()
```

Kronecker Products

Compute closed-form determinant of block Toeplitz matrices by leveraging eigenvalue multiplicativity of $\otimes$: $\det(I_n \otimes A + B \otimes I_n) = \prod_i \prod_j (\lambda_i(A) + \lambda_j(B))$.

Differential Equations and Special Functions

Symbolic ODE Solving

Second-order linear equation

$$y'' + (x^2 + 1)y = 0$$

returns a solution basis $\{\_U(0, \frac{1}{2}x^2), \_V(0, \frac{1}{2}x^2)\}$ in terms of parabolic cylinder functions U, V.

```
x = sp.symbols('x')
y = sp.Function('y')
ode = sp.Eq(sp.diff(y(x),x,2)+(x**2+1)*y(x), 0)
sp.dsolve(ode)
```

Green's Function for Bessel Operator

Consider $x^2 y'' + xy' + (x^2 - \nu^2)y = f(x)$ on $(0, \infty)$. `sympy.integrals.transforms` constructs the integral kernel

$$G(x, \xi) = \begin{cases} J_\nu(x)Y_\nu(\xi) - J_\nu(\xi)Y_\nu(x), & x < \xi, \\ 0, & x > \xi, \end{cases}$$

satisfying continuity and jump conditions at $x = \xi$.

```
v = sp.symbols('nu')
G = sp.besselj(v, sp.Symbol('x'))*sp.bessely(v, sp.Symbol('ξ')) - \
    sp.besselj(v, sp.Symbol('ξ'))*sp.bessely(v, sp.Symbol('x'))
```

Symbolic Laplace Transforms

SymPy derives closed-form inverse transforms involving error and Airy functions, essential in transient heat conduction.

```
s,t = sp.symbols('s t', positive=True)
F = 1/(s*(s**2+1))
f = sp.inverse_laplace_transform(F, s, t)
```

The result is $f(t) = 1 - \cos t$, confirming manual residue calculus.

11.1.2 Automated Theorem Proving

Automated theorem proving (ATP) aspires to mechanise the deductive steps that underlie mathematical reasoning, transforming the informal "proof-sketch" of the human mathematician into a sequence of sound inferences that a computer can verify independently. Since the seminal work of Davis, Putnam, Logemann, and Loveland on the DPLL procedure and Robinson's resolution calculus, ATP has grown into a vibrant discipline that interlaces proof theory, model theory, complexity theory, and symbolic computation. Today, SMT-solvers such as Z3 decide quantifier-free fragments of first-order logic with millions of clauses in seconds, while interactive proof assistants—e.g. Coq, Lean, Isabelle—blend automation with user guidance to formalise entire textbooks. This subsection traces the logical foundations, exhibits SymPy's nascent proof engine, and illustrates applications to formal verification and cryptography where mechanical rigour is indispensable.

Introduction to Theorem Proving

Let $\mathcal{L}$ be a first-order language with signature $\Sigma = (\mathcal{F}, \mathcal{P})$ of function and predicate symbols. A *sequent* $\Gamma \vdash \Delta$ encodes entailment "from premises Γ derive conclusions Δ". Gentzen's natural deduction system equips $\mathcal{L}$ with introduction and elimination rules for $\wedge, \vee, \forall, \exists, \neg, \Rightarrow$ such that every valid formula φ has a derivation $\vdash \varphi$.

Soundness and Completeness

If $\Gamma \vdash \varphi$, then $\Gamma \models \varphi$ (soundness); conversely, if $\Gamma \models \varphi$, then $\Gamma \vdash \varphi$ (Gödel completeness). Proof search, however, is **PSPACE**-complete for propositional logic and undecidable for full first-order logic due to the Löwenheim–Skolem theorem. Practical provers therefore restrict to decidable fragments (EUF, linear arithmetic) or employ heuristics (term ordering, lemma learning).

Using SymPy for Automated Proofs

SymPy implements a lightweight propositional engine based on the Davis–Putnam algorithm and a first-order unification module. Consider the tautology

$$(p \Rightarrow q) \wedge (q \Rightarrow r) \implies (p \Rightarrow r).$$

```
from sympy import symbols, Implies, satisfiable
p,q,r = symbols('p q r')
phi = Implies( (Implies(p,q) & Implies(q,r)), Implies(p,r) )
assert not satisfiable(~phi) # unsatisfiable ⟹ φ is valid
```

11.1 Symbolic Computation

Resolution proof objects can be constructed by invoking

```
from sympy.logic.algorithms import resolution
steps = resolution(~phi)  # returns derivation of contradiction
```

The resulting clause sequence witnesses that ϕ is entailed by the empty set.

First-Order Unification

Given terms $t_1 = f(g(x), y)$ and $t_2 = f(z, g(z))$, SymPy computes the most general unifier (mgu) $\sigma = \{x \mapsto z,\ y \mapsto g(z)\}$.

```
from sympy.unify import unify
x,y,z = symbols('x y z')
t1 = f(g(x), y)
t2 = f(z, g(z))
σ = unify(t1, t2)  # {x: z, y: g(z)}
```

Although SymPy is not yet a full-fledged theorem prover, its symbolic backbone allows integration with external SMT engines through the `sympy.logic.inference` interface, providing a convenient bridge between high-level mathematics and low-level decision procedures.

First-Order Logic and Quantifiers in Theorem Proving

Skolemisation

To refute $\Gamma \models \varphi$ one applies *reductio ad absurdum*: convert $\Gamma \wedge \neg\varphi$ to prenex conjunctive normal form (PCNF) and demonstrate unsatisfiability via resolution. Existential quantifiers are eliminated by Skolem functions:

$$\forall x\, \exists y\, \forall z\, P(x, y, z) \ \longmapsto\ \forall x\, \forall z\, P(x, f(x), z).$$

Skolemisation preserves satisfiability but not equivalence; however, it is sound for refutation.

Herbrand's Theorem

If a first-order set Φ is unsatisfiable, there exists a finite unsatisfiable subset of its Herbrand expansion—the set of ground instances over the term algebra. Modern provers enumerate Herbrand universes lazily, interleaving ground resolution with instantiation rules.

Example: Proofs over Natural Numbers

Peano arithmetic formulas like $\forall x \, (x + 0 = x)$ are handled by encoding the successor function s and axioms into the solver. Encoding associative–commutative (AC) theories requires special matching; SymPy's pattern engine supports AC unification for commutative groups, enabling automated verification of ring identities.

Applications in Formal Verification and Cryptography

Hardware Verification

A combinational circuit with n gates is modelled as a CNF formula $\Phi_{\text{spec}} \wedge \Phi_{\text{impl}}$. Equivalence checking reduces to satisfiability of $\Phi_{\text{spec}} \wedge \Phi_{\text{impl}} \wedge (\text{out}_{\text{spec}} \oplus \text{out}_{\text{impl}})$. State-of-the-art SAT engines solve $>10^6$ clauses within minutes; counterexample traces map to signal assignments revealing design flaws.

Software Verification via Hoare Logic

A loop invariant I is synthesised by solving second-order constraints

$$\exists I \, \forall \mathbf{v} \, (\text{Pre} \to I) \wedge (I \wedge \neg \text{Guard} \to \text{Post}) \wedge (I \wedge \text{Guard} \to I').$$

Template-based synthesis chooses polynomial form $I(\mathbf{v}; \boldsymbol{\alpha})$ and reduces validity to SMT queries over the reals; toolchains like Z3 + `sympy` automate this pipeline.

Zero-Knowledge Proof (ZKP) Circuits

Arithmetisation maps cryptographic statements to polynomial equalities over finite fields $\mathbb{F}_p$. Constraint system $\{(a_i, b_i, c_i) : a_i b_i = c_i\}$ is verified by PLONK protocols. Automated generation of the Rank-1 Constraint System from high-level code employs symbolic differentiation and common-subexpression elimination, both available in SymPy, ensuring soundness of the compiled circuit.

Example 11.1.2 (Symbolic R1CS of SHA-256 Round) Using `sympy.crypto` (custom module) one represents Boolean AND as multiplication in $\mathbb{F}_2$. The constraint $z = (x \wedge y)$ translates to $x \cdot y - z = 0$. Iterating over 64 rounds yields 15 360 constraints; SymPy simplifies polynomials before handing them to the proof system, reducing prover time by 12%.

11.1.3 Fourier Transforms

Fourier analysis decomposes a function into harmonic constituents, converting convolution into multiplication and differential operators into algebraic multipliers. This spectral duality underpins signal processing, PDE theory, and numerical algorithms.

Discrete and Continuous Fourier Transforms

For $f \in L^1(\mathbb{R}^d) \cap L^2(\mathbb{R}^d)$ the **continuous Fourier transform** (CFT) is

$$\mathcal{F}\{f\}(\xi) = \widehat{f}(\xi) = \int_{\mathbb{R}^d} f(x) \, e^{-2\pi i x \cdot \xi} \, dx, \qquad \mathcal{F}^{-1}\{\widehat{f}\}(x) = \int_{\mathbb{R}^d} \widehat{f}(\xi) \, e^{2\pi i x \cdot \xi} \, d\xi.$$

Plancherel theorem extends $\mathcal{F}$ to a unitary operator on L^2: $\|\widehat{f}\|_2 = \|f\|_2$, implying Parseval's identity.

Sampling on a uniform lattice $x_n = n\Delta x$, $n = 0, \ldots, N-1$, yields the **Discrete Fourier Transform** (DFT)

$$\widehat{F}_k = \sum_{n=0}^{N-1} f_n \, e^{-2\pi i k n / N}, \qquad f_n = \frac{1}{N} \sum_{k=0}^{N-1} \widehat{F}_k \, e^{2\pi i k n / N}.$$

Nyquist–Shannon guarantees perfect reconstruction when $\Delta x < \frac{1}{2} f_{\max}^{-1}$.

Example 11.1.3 (Aliasing) Let $f(t) = \sin(30\pi t)$ sampled at $f_s = 20\,\mathrm{Hz}$. Since f exceeds the Nyquist frequency $f_N = 10\,\mathrm{Hz}$, the discrete spectrum mistakenly contains $\sin(10\pi t)$, demonstrating frequency folding.

Fast Fourier Transform (FFT) and Computational Efficiency

Direct DFT costs $O(N^2)$. Cooley–Tukey radix-2 FFT exploits the divide–conquer factorisation:

$$\widehat{F}_k = \sum_{m=0}^{N/2-1} f_{2m} e^{-2\pi i k m/(N/2)} + e^{-2\pi i k/N} \sum_{m=0}^{N/2-1} f_{2m+1} e^{-2\pi i k m/(N/2)},$$

yielding $T(N) = 2T(N/2) + O(N) \Rightarrow T(N) = O(N \log N)$. Cache-friendly bit-reversed addressing and loop unrolling approach the memory bandwidth limit.

```
import numpy as np, time
N = 2**20
x = np.random.rand(N) + 1j*np.random.rand(N)
t0 = time.perf_counter(); X = np.fft.fft(x); t1 = time.perf_counter()
print(f"FFT throughput: {N/(t1-t0):.2e} pts/sec")
```

Bluestein and *Rader* algorithms handle prime sizes; mixed-radix libraries (FFTW, pocketfft) achieve ≤5 % of peak flops on modern CPUs.

Applications in Signal Processing and Image Compression

Time-frequency localisation begins with the short-time Fourier transform (STFT):

$$\text{STFT}_f(t, \xi) = \int_{-\infty}^{\infty} f(\tau)\, g(\tau - t)\, e^{-2\pi i \xi \tau}\, d\tau,$$

where g is a window (Gaussian optimises time–bandwidth inequality). Spectrogram $|\text{STFT}|^2$ visualises chirps and modulations.

JPEG compression applies 8×8 discrete cosine transform (DCT-II). Quantisation leverages psycho-visual masking: high-frequency coefficients are divided by larger integers, producing sparsity exploitable by run-length and Huffman coding.

```
import cv2, numpy as np
block = img[0:8,0:8] - 128  # level-shift
coeff = cv2.dct(block.astype(float))
mask = np.array([[16,11,10,16,24,40,51,61], ...])  # std. quant table
qcoeff = np.rint(coeff / mask)
```

Fourier Analysis in Heat and Wave Equations

For $u_t = \kappa u_{xx}$ on $\mathbb{R}$ with $u(x, 0) = f(x)$, take CFT:

$$\partial_t \widehat{u} = -4\pi^2 \kappa \xi^2 \widehat{u}, \implies \widehat{u}(\xi, t) = e^{-4\pi^2 \kappa \xi^2 t}\, \widehat{f}(\xi).$$

Inverse transform yields the Gaussian kernel

$$u(x, t) = (G_{\kappa t} * f)(x), \quad G_{\sigma^2}(x) = \frac{1}{\sqrt{4\pi \sigma^2}} e^{-x^2/4\sigma^2},$$

demonstrating instantaneous smoothing.

For the 1-D wave equation $u_{tt} = c^2 u_{xx}$, $\partial_t^2 \widehat{u} = -(2\pi c \xi)^2 \widehat{u}$, giving $\widehat{u}(\xi, t) = \widehat{f}(\xi) \cos(2\pi c \xi t) + \frac{\widehat{g}(\xi)}{2\pi c \xi} \sin(2\pi c \xi t)$, and d'Alembert formula $u(x, t) = \frac{1}{2}\big[f(x - ct) + f(x + ct)\big] + \frac{1}{2c} \int_{x-ct}^{x+ct} g(s)\, ds$. Dispersionless propagation contrasts with parabolic diffusion.

11.1.4 Wavelet Transforms

Classical Fourier analysis characterises global spectral content but obscures *when* a transient occurs. Wavelet analysis remedies this shortcoming by furnishing a family of dilations and translations of a prototype wavelet ψ that simultaneously localise in time and frequency. Because the Heisenberg uncertainty principle forbids arbitrary precision in both domains, wavelets adopt a logarithmic tiling: fine temporal resolution at high frequencies and coarse resolution at low frequencies—the reverse of the short-time Fourier transform's fixed window.

Introduction to Wavelets

A function $\psi \in L^2(\mathbb{R})$ is an *admissible wavelet* if it has zero mean and finite energy, satisfying the admissibility condition

$$C_\psi = 2\pi \int_0^\infty \frac{|\widehat{\psi}(\omega)|^2}{\omega}\, d\omega < \infty.$$

The **continuous wavelet transform** (CWT) of a signal $f \in L^2$ is

$$\mathcal{W}_\psi f(a, b) = \frac{1}{\sqrt{|a|}} \int_{-\infty}^\infty f(t)\, \overline{\psi\!\left(\frac{t-b}{a}\right)} dt, \quad a \in \mathbb{R}^\times,\ b \in \mathbb{R}.$$

Coefficient modulus encodes similarity of f to $\psi_{a,b}$; inversion reads

$$f(t) = \frac{1}{C_\psi} \int_\mathbb{R}\!\!\int_\mathbb{R} \mathcal{W}_\psi f(a, b)\, \frac{1}{|a|^{3/2}}\, \psi\!\left(\frac{t-b}{a}\right) da\, db.$$

Example 11.1.4 (Mexican Hat) $\psi(t) = \frac{2}{\sqrt{3}} \pi^{-1/4} (1 - t^2) e^{-t^2/2}$ possesses vanishing first moment, making it sensitive to curvature changes; CWT detects ridge lines in edge maps.

Wavelet Transform vs. Fourier Transform

For chirp $f(t) = \sin(\pi t^2)$, the instantaneous frequency $f'(t) = 2t$ increases linearly. The Fourier transform smears energy across the entire spectrum, whereas the scalogram $|\mathcal{W}_\psi f(a, b)|^2$ traces a parabola, revealing local frequency evolution. In quantitative terms, the joint time-frequency uncertainty obeys $\sigma_t \sigma_\omega \geq \frac{1}{2}$, with Gaussian ϕ attaining equality; Morlet wavelet $\psi(t) = e^{i\omega_0 t} e^{-t^2/2}$ trades marginally higher uncertainty for admissibility (zero mean).

Applications in Time-Frequency Analysis

Seismology P- and S-wave arrivals correspond to maxima in $|\mathcal{W}_\psi f|$ across dyadic scales $a = 2^j$, enabling automatic event picking.

Biomedical ECG QRS complexes manifest as spikes; discrete wavelet packet coefficients localise arrhythmia signatures with $\approx 98\%$ sensitivity.

```
import pywt, numpy as np
coeffs, freqs = pywt.cwt(ecg, scales=2**np.arange(1,8), wavelet='mexh',
    sampling_period=1/360)
# Energy of scale 4 isolates QRS band (10-25 Hz)
qrs_energy = np.linalg.norm(coeffs[3], axis=0)
```

Multiresolution Analysis and Denoising Techniques

A **multiresolution analysis** (MRA) is a nested sequence of closed subspaces

$$\cdots \subset V_{-1} \subset V_0 \subset V_1 \subset \cdots \subset L^2(\mathbb{R}),$$

with $f \in V_j \iff f(2\cdot) \in V_{j+1}$ and $\bigcap_j V_j = \{0\}$, $\overline{\bigcup_j V_j} = L^2$. A scaling function φ spans V_0, admitting refinement $\varphi(t) = \sum_k h_k \varphi(2t - k)$. Defining W_j as the orthogonal complement of V_j in V_{j+1} yields wavelets $\psi(t) = \sum_k (-1)^k h_{1-k} \varphi(2t - k)$, generating an orthonormal basis $\{\psi_{j,k}(t) = 2^{j/2} \psi(2^j t - k)\}_{j,k \in \mathbb{Z}}$.

Wavelet Denoising (Donoho–Johnstone)

Observation model $y_n = f_n + \varepsilon_n$, $\varepsilon_n \sim \mathcal{N}(0, \sigma^2)$. Transform to wavelet domain $\mathbf{w} = \mathbf{W}y$, apply soft-threshold

$$\tilde{w}_k = \operatorname{sgn}(w_k) \max\{|w_k| - \lambda \sigma \sqrt{2 \ln N}, 0\},$$

then reconstruct $\hat{f} = \mathbf{W}^{-1} \tilde{\mathbf{w}}$. Universal threshold ensures $\|\hat{f} - f\|_2^2 \leq (1 + o(1)) \inf_\eta \|\eta - f\|_2^2$ with high probability.

```
coeffs = pywt.wavedec(noisy, 'db8', level=4)
σ = np.median(np.abs(coeffs[-1]))/0.6745
λ = σ*np.sqrt(2*np.log(len(noisy)))
denoised = pywt.waverec([pywt.threshold(c, λ, 'soft') for c in coeffs], '
    db8')
```

Compression

JPEG2000 employs integer lifting wavelets (Cohen–Daubechies–Feauveau 9/7) followed by embedded block coding with optimal truncation; achieves $\approx 2\times$ higher PSNR at equal bitrate relative to DCT.

11.2 Time-Series and Signal Processing

11.2.1 Statistical Time-Series Models

Time-series analysis presumes observations $\{x_t\}_{t=0}^{T}$ recorded at equally spaced instants. The principal objective is to isolate deterministic components (trend, seasonality) from stochastic fluctuations, then build parsimonious stochastic models that predict future values and quantify uncertainty. Classical Box–Jenkins methodology frames the stochastic part as a linear filter driven by white noise; state-space theory recasts dynamics in recursive form amenable to Kalman filtering; GARCH-type models capture conditional heteroskedasticity pervasive in finance. We start with linear ARIMA processes, proceed to state-space representation, then treat volatility, and finally discuss decomposition and stationarity diagnostics.

Autoregressive (AR), Moving-Average (MA), and ARIMA Models

Let $\{\varepsilon_t\} \sim$ i.i.d. $\mathcal{N}(0, \sigma^2)$.

AR(p) Model

$$x_t - \mu = \sum_{i=1}^{p} \varphi_i (x_{t-i} - \mu) + \varepsilon_t, \quad \Phi(L)x_t = \varepsilon_t,$$

with lag operator $L^k x_t = x_{t-k}$ and characteristic polynomial $\Phi(z) = 1 - \varphi_1 z - \cdots - \varphi_p z^p$. Stationarity requires all roots of Φ lie outside the unit disk. The k-step predictor is $\widehat{x}_{t+k} = \mu + \sum_{i=1}^{p} \varphi_i \widehat{x}_{t+k-i}$ with MSE computed from Yule–Walker equations.

MA(q) Model

$x_t = \mu + \varepsilon_t + \sum_{i=1}^{q} \theta_i \varepsilon_{t-i} = \Theta(L)\varepsilon_t$. Invertibility demands zeros of $\Theta(z) = 1 + \theta_1 z + \cdots + \theta_q z^q$ lie outside unit circle, guaranteeing a convergent AR(∞) representation and identifiability.

ARIMA(p, d, q)

Non-stationary series with polynomial trend are differenced d times:

$$\Delta^d x_t = (1 - L)^d x_t = \Phi(L)^{-1} \Theta(L) \varepsilon_t.$$

Auto-correlation function (ACF) tail and partial ACF (PACF) cut-off guide p, q selection; AIC minimisation refines choice.

```
import pmdarima as pm
model = pm.auto_arima(ts, seasonal=False, d=None, max_p=6, max_q=6,
                     information_criterion='bic', stepwise=True)
print(model.order) # (p,d,q)
forecast, conf = model.predict(n_periods=12, return_conf_int=True)
```

Forecast Intervals

For ARMA, h-step forecast variance $\text{Var}[\widehat{x}_{T+h} - x_{T+h}] = \sigma^2 \sum_{j=0}^{h-1} \psi_j^2$, where ψ_j are impulse-response coefficients obtained from $\Psi(z) = \Theta(z)/\Phi(z) = \sum_{j=0}^{\infty} \psi_j z^j$.

State-Space Models and the Kalman Filter

A linear Gaussian state-space model (SSM) is

$$\mathbf{x}_t = A\mathbf{x}_{t-1} + B\mathbf{u}_t + \mathbf{w}_t, \quad \mathbf{y}_t = C\mathbf{x}_t + D\mathbf{u}_t + \mathbf{v}_t,$$

with $\mathbf{w}_t \sim \mathcal{N}(0, Q)$, $\mathbf{v}_t \sim \mathcal{N}(0, R)$ independent. The **Kalman filter** recursively computes posterior mean $\hat{\mathbf{x}}_{t|t}$ and covariance $P_{t|t}$.

Predict

$$\hat{\mathbf{x}}_{t|t-1} = A\hat{\mathbf{x}}_{t-1|t-1} + B\mathbf{u}_t, \quad P_{t|t-1} = AP_{t-1|t-1}A^\mathsf{T} + Q.$$

Update

$$K_t = P_{t|t-1}C^\mathsf{T}(CP_{t|t-1}C^\mathsf{T} + R)^{-1}, \quad \hat{\mathbf{x}}_{t|t} = \hat{\mathbf{x}}_{t|t-1} + K_t(\mathbf{y}_t - C\hat{\mathbf{x}}_{t|t-1}),$$

$$P_{t|t} = (I - K_t C)P_{t|t-1}.$$

```
import numpy as np, pykalman
kf = pykalman.KalmanFilter(transition_matrices=A, observation_matrices=C,
                transition_covariance=Q, observation_covariance=R)
state_est, cov = kf.filter(y_obs)
```

11.2 Time-Series and Signal Processing

ARMA(p, q) embeds into SSM with state dimension p; Kalman recursions yield exact likelihood via prediction error decomposition, enabling MLE of φ_i, θ_j.

ARCH/GARCH and Volatility Modelling

Financial returns r_t often exhibit uncorrelated levels but serially correlated squared returns. The **GARCH**(p, q) model posits

$$r_t = \sigma_t \varepsilon_t, \quad \varepsilon_t \sim \mathcal{N}(0, 1), \quad \sigma_t^2 = \omega + \sum_{i=1}^{p} \alpha_i r_{t-i}^2 + \sum_{j=1}^{q} \beta_j \sigma_{t-j}^2,$$

with stationarity if $\sum \alpha_i + \sum \beta_j < 1$. Leverage effect captured by EGARCH or GJR–GARCH introduces asymmetric terms $\gamma_i r_{t-i} \mathbf{1}_{\{r_{t-i} < 0\}}$.

```
from arch import arch_model
garch = arch_model(r, p=1, q=1, mean='zero', vol='GARCH')
res = garch.fit(update_freq=10)
h_fore = res.forecast(horizon=5).variance[-1:]
```

Value-at-Risk at level α given $N(0, 1)$ innovations is $\text{VaR}_{t+1}^{\alpha} = \sigma_{t+1} \Phi^{-1}(1 - \alpha)$.

Trend-Seasonality Decomposition and Stationarity Tests

Additive Decomposition

Assume $x_t = T_t + S_t + R_t$ where T_t is trend (low-frequency), S_t is periodic with known period k, and R_t is remainder. STL (Seasonal-Trend decomposition via Loess) alternates:

(i) Smooth k subsequences $x_{t+k\ell}$ to estimate S_t.
(ii) Subtract S_t, apply LOESS to residual for T_t.
(iii) Update remainder, iterate until convergence.

```
from statsmodels.tsa.seasonal import STL
res = STL(ts, period=12, robust=True).fit()
trend, season, resid = res.trend, res.seasonal, res.resid
```

Stationarity Diagnostics

Augmented Dickey–Fuller (ADF) tests $H_0 : \varphi = 1$ in $\Delta x_t = \rho x_{t-1} + \sum_{i=1}^{k} \gamma_i \Delta x_{t-i} + \varepsilon_t$. Reject H_0 at level α if $\tau = \hat{\rho} / \text{se}(\hat{\rho})$ is less than critical value c_α. **KPSS** reverses H_0 (trend-stationary) and uses LM-statistic.

```
from statsmodels.tsa.stattools import adfuller, kpss
print("ADF p-val :", adfuller(ts.diff().dropna())[1])
print("KPSS p-val:", kpss(ts, nlags="auto")[1])
```

11.2.2 Digital Filtering in Python

Digital filters reshape the spectral content of discrete-time signals, suppressing noise, extracting frequency bands, or imposing smoothness constraints. A filter is characterised by its impulse response $h[n]$ and transfer function $H(e^{j\omega})$. In practice, implementation fidelity rests on numerical stability, finite-word-length effects, and real-time throughput. Python's scipy.signal module, augmented by numpy and numba, furnishes a comprehensive toolkit for both batch and streaming applications.

FIR vs. IIR Filters: Design and Stability

FIR (Finite Impulse Response)

$$y[n] = \sum_{k=0}^{M} h_k\, x[n-k], \qquad H(z) = \sum_{k=0}^{M} h_k z^{-k}.$$

Advantages: inherent BIBO stability ($\forall h \in \ell^1$), linear-phase achievable via symmetric coefficients ($h_k = h_{M-k}$), and convex least-squares design. Drawback: steep transition bands require large order M.

IIR (Infinite Impulse Response)

Difference equation

$$y[n] + \sum_{k=1}^{N} a_k y[n-k] = \sum_{k=0}^{M} b_k x[n-k], \qquad H(z) = \frac{\sum_{k=0}^{M} b_k z^{-k}}{1 + \sum_{k=1}^{N} a_k z^{-k}}.$$

Poles inside unit circle ensure stability; sharper roll-off with fewer taps but possible nonlinear phase and risk of arithmetic overflow.

BIBO Stability Criterion

IIR stable $\iff$ $\max_i |z_i| < 1$ where $\{z_i\}$ are poles of $H(z)$. scipy.signal.tf2zpk extracts zeros/poles to verify.

11.2 Time-Series and Signal Processing

```
from scipy.signal import butter, tf2zpk
b,a = butter(4, 0.3) # 4th-order low-pass
z,p,k = tf2zpk(b,a); assert max(abs(p)) < 1 # stable
```

Window Functions and Smoothing Techniques

Least-squares ideal low-pass response $h_\infty[n] = \frac{\sin \omega_c(n-M/2)}{\pi(n-M/2)}$ truncation introduces Gibbs ripples. Multiply by window $w[n] \in [C, 1]$ to taper edges:

$$h[n] = h_\infty[n]\, w[n], \quad w_{\text{Hann}}[n] = 0.5\left(1 - \cos\tfrac{2\pi n}{M}\right).$$

```
from numpy import sinc, hanning, pi
M, fc = 63, 0.15
n = np.arange(M); h = 2*fc*sinc(2*fc*(n-M/2)) * hanning(M)
```

Savitzky–Golay Smoothing

Polynomial regression in a sliding window of length $2m+1$ fits $y[n] \approx \sum_{k=0}^{d} \beta_k (n-n_0)^k$; evaluate β_0 for low-pass smoothing that preserves peaks. Convolution coefficients pre-computed via pseudo-inverse; even order produces zero phase delay.

Butterworth, Chebyshev, and Elliptic Filters

Let Ω_p (pass-band), Ω_s (stop-band), and tolerances δ_p, δ_s.

Butterworth

Magnitude squared

$$|H(j\Omega)|^2 = \frac{1}{1 + (\Omega/\Omega_c)^{2N}},$$

maximally flat at $\Omega = 0$; order $N \geq \frac{\log(\delta_s^{-2}-1)-\log(\delta_p^{-2}-1)}{2\log(\Omega_s/\Omega_p)}$.

Chebyshev-I

Pass-band ripple ε: $|H|^{-2} = 1 + \varepsilon^2 T_N^2(\Omega/\Omega_c)$, where T_N Chebyshev polynomial. Stop-band monotonic. Chebyshev-II swaps ripple to the stop-band.

Elliptic (Cauer)

Ripple in both bands, sharpest transition for given N. Utilises Jacobian elliptic functions; superior N but greater phase distortion.

```python
from scipy.signal import ellip, sosfiltfilt
sos = ellip(N=5, rp=1, rs=60, Wn=[0.2,0.4], btype='band', output='sos')
y_filt = sosfiltfilt(sos, y)  # zero-phase via forward-reverse
```

Real-Time Filtering and Stream Processing

For streaming $x[n]$ at 44.1 kHz, latency must be <10 ms (≈ 441 samples). Implement IIR in *direct-form II transposed* to minimise state memory:

$$y_n = b_0 x_n + w_1, \quad w_k = w_{k+1} + b_k x_n - a_k y_n, \quad k = 1, \ldots, \max(M, N).$$

```python
from collections import deque, namedtuple
Filter = namedtuple('Filter', 'b a w')
def df2tiir_step(x, flt):
    b,a,w = flt
    y = b[0]*x + w[0]
    for k in range(len(w)-1):
        w[k] = w[k+1] + b[k+1]*x - a[k+1]*y
    w[-1] = b[-1]*x - a[-1]*y
    return y
```

Cython or numba JIT increases throughput $\sim 20\times$; deploying on sounddevice callback maintains XRUN-free audio. For massive telemetry streams, windowed FIR executed on GPUs uses cuFFT convolution theorem ($O(N \log N)$) to batch-process 10^6-sample segments with overlap-save.

11.2.3 Spectral Analysis

Spectral analysis quantifies how the variance of a stochastic process is distributed over frequency. When the second-order structure of a wide-sense stationary series x_t is encapsulated by the autocovariance $R_{xx}(\tau) = \mathbb{E}[x_t x_{t+\tau}]$, the Wiener–Khinchin theorem asserts that its *power spectral density* (PSD) $S_{xx}(\omega)$ is the Fourier transform of $R_{xx}(\tau)$, providing a frequency-domain portrait of correlation structure. In practice, finite-length observations introduce bias-variance trade-offs; the following estimators navigate this compromise through windowing, averaging, and tapering.

Periodogram and Power Spectral Density Estimation

For a real zero-mean signal $\{x_n\}_{n=0}^{N-1}$ sampled with rate f_s, define the **periodogram**

$$\hat{S}_{xx}^{\text{per}}(f) = \frac{\Delta t}{N} \left| \sum_{n=0}^{N-1} x_n \, e^{-2\pi i f n \Delta t} \right|^2, \quad f \in [0, f_s),$$

where $\Delta t = 1/f_s$. The periodogram is asymptotically unbiased but *inconsistent*, as its variance does not decrease with N. Windowing by w_n produces the *Bartlett* estimate

$$\hat{S}_{xx}^{B}(f) = \frac{\Delta t}{NU} \left| \sum_{n=0}^{N-1} w_n x_n \, e^{-2\pi i f n \Delta t} \right|^2, \quad U = \frac{1}{N} \sum_{n=0}^{N-1} w_n^2,$$

which reduces sidelobes at the expense of main-lobe broadening.

```
from scipy.signal import periodogram, get_window
f, Pxx = periodogram(x, fs=fs, window='blackman', scaling='density')
```

Welch's Method and Multi-taper Spectral Estimates

Welch

Partition the record into K overlapped segments of length L, apply a window w, compute each segment periodogram, then average:

$$\hat{S}_{xx}^{W}(f) = \frac{1}{K} \sum_{k=0}^{K-1} \hat{S}_{xx,k}^{\text{per}}(f), \quad \text{Var}[\hat{S}^{W}] = \frac{1}{K} \text{Var}(\hat{S}^{\text{per}}),$$

achieving $\sqrt{K}$ variance reduction. Overlap 50% and Hanning window balance independence and data usage.

Multi-taper (MTM)

Choose $K = 2NW_b$ orthogonal Slepian tapers v_k optimally concentrated in bandwidth W_b. The eigenspectra $\hat{S}_k(f) = \left| \sum_n v_k[n] x_n e^{-2\pi i f n} \right|^2$ form a minimum-variance unbiased estimator

$$\hat{S}^{\text{MTM}}(f) = \frac{1}{K} \sum_{k=0}^{K-1} \hat{S}_k(f).$$

Jackknife across tapers provides confidence intervals robust to coloured noise.

```
from scipy.signal.windows import dpss
from numpy.fft import rfft
NW, K = 4, 2*4-1
tapers = dpss(L, NW, K) # DPSS tapers
S = np.mean([np.abs(rfft(x_seg*t))**2 for t in tapers], axis=0)
```

Short-Time Fourier Transform (STFT) and Spectrograms

The STFT maps a non-stationary signal into a two-dimensional function:

$$\text{STFT}_x(m, \omega) = \sum_{n=-\infty}^{\infty} x[n]\, g[n-m]\, e^{-i\omega n},$$

where g is a sliding window centred at m. The **spectrogram** is its squared magnitude, capturing local energy. Time-frequency resolution obeys $\Delta t\, \Delta f \geq 1/(4\pi)$; fixing window length enforces uniform tiling, unlike the scalogram's logarithmic grid.

```
from scipy.signal import stft
f, t, Z = stft(x, fs=fs, nperseg=256, noverlap=128, window='hann')
plt.pcolormesh(t, f, 20*np.log10(np.abs(Z)), shading='gouraud')
```

Cross-Spectral Density and Coherence Analysis

Given two processes x_t, y_t with CSD

$$S_{xy}(f) = \lim_{N \to \infty} \mathbb{E}\big[X_N(f)\, \overline{Y_N(f)}\big],$$

the **magnitude-squared coherence**

$$C_{xy}(f) = \frac{|S_{xy}(f)|^2}{S_{xx}(f)\, S_{yy}(f)} \in [0, 1]$$

measures linear correlation at frequency f. Welch's method extends: compute segment CSDs $\hat{S}_{xy,k}$ and PSDs $\hat{S}_{xx,k}$, $\hat{S}_{yy,k}$; average numerators and denominators separately.

```
from scipy.signal import coherence
f, Cxy = coherence(x, y, fs=fs, nperseg=1024, noverlap=512)
```

A phase-locking value $\phi(f) = \arg S_{xy}(f)$ indicates lead/lag relationships; confidence limits derive from χ^2_{2K} distribution when segments are independent.

11.2.4 Time-Frequency Representations

Time-frequency representations (TFRs) seek a joint description $T_x(t, \omega)$ that obeys $\int_{-\infty}^{\infty} T_x(t, \omega) \, d\omega = |x(t)|^2$, $\int_{-\infty}^{\infty} T_x(t, \omega) \, dt = |X(\omega)|^2$, thus redistributing the signal's energy without loss. The Wigner–Ville distribution attains perfect marginals but suffers interference terms; the present section surveys adaptive representations that alleviate such artefacts and sharpen localisation.

Continuous Wavelet Transform (CWT) Revisited

Recalling $\mathcal{W}_\psi x(a, b) = \langle x, \psi_{a,b} \rangle$, $\psi_{a,b}(t) = |a|^{-1/2} \psi\big((t-b)/a\big)$, with admissible ψ (zero mean), the scalogram $P(a, b) = |\mathcal{W}_\psi x(a, b)|^2$ visualises energy across scales a and translations b.

Analytic Morlet Wavelet

$$\psi(t) = \pi^{-1/4} e^{j\omega_0 t} e^{-t^2/2}, \quad \widehat{\psi}(\omega) = \pi^{-1/4} e^{-(\omega-\omega_0)^2/2},$$

with $\omega_0 = 6$ ensures admissibility ($C_\psi \approx 1$). Frequency mapping $f = \omega/(2\pi a)$ yields nearly constant-Q resolution $\Delta f/f \approx$ const.

Cone of Influence (COI)

For finite support length T, edge effects distort coefficients when the effective support of $\psi_{a,b}$ intersects the boundary. The COI boundary $b \pm |a| k$ (typically $k = 2$) delineates trustworthy region.

Ridge Extraction

Instantaneous frequency approximated by ridge curve $\hat{f}(b) = \arg\max_a |\mathcal{W}_\psi x(a, b)|/a$. Differentiating the phase $\theta(a, b) = \arg \mathcal{W}_\psi x$ gives finer estimator $\omega_{\text{inst}}(b) = \partial_b \theta\big(a^*(b), b\big)$.

```
import pywt, numpy as np
scales = np.geomspace(1, 64, 128)
coef, freqs = pywt.cwt(sig, scales, 'cmor6-1')
ridge_idx = np.argmax(np.abs(coef), axis=0)
inst_freq = freqs[ridge_idx]
```

Empirical Mode Decomposition and Hilbert–Huang Transform

CWT presupposes a priori basis; empirical mode decomposition (Huang et al. 1998) adaptively extracts *intrinsic mode functions* (IMFs).

Sifting Algorithm

(1) Identify local extrema of $x(t)$; interpolate upper envelope $e_{\max}$ and lower envelope $e_{\min}$ (cubic splines).
(2) Compute mean $m(t) = \frac{1}{2}(e_{\max} + e_{\min})$; update $h_1 = x - m$.
(3) Iterate until h_1 satisfies IMF criteria: (i) number of extrema $\approx$ number of zero crossings; (ii) mean envelope ≈ 0 (tolerance ε).
(4) Set $\text{IMF}_1 = h_1$; update residue $r_1 = x - \text{IMF}_1$ and repeat.

Stop when residue r_k is monotonic or has < 2 extrema. An N-point series decomposes as $x(t) = \sum_{k=1}^{K} \text{IMF}_k(t) + r_K(t)$.

Hilbert Spectral Analysis

Apply analytic signal $z_k(t) = \text{IMF}_k(t) + j\mathcal{H}[\text{IMF}_k](t)$; instantaneous amplitude $A_k(t) = |z_k|$ and frequency $\omega_k(t) = \partial_t \arg z_k(t)$. The Hilbert spectrum $H(t, \omega) = \sum_k A_k(t)\delta(\omega - \omega_k(t))$ provides high-resolution amplitude-frequency portrait free from pre-defined basis.

```
from PyEMD import EMD
imfs = EMD().emd(signal)
inst_freq = np.diff(np.unwrap(np.angle(sp.signal.hilbert(imfs, axis=1))),
    axis=1)
```

Mode Mixing Mitigation

Ensemble EMD (EEMD) adds white-noise trials $\varepsilon_j(t)$ (std $\sigma = 0.2\text{SD}$), averages IMFs, exploiting noise-assisted separation; complete EEMD (CEEMD) improves reconstruction by opposite-phase cancellation.

Synchrosqueezed and High-Resolution Methods

Synchrosqueezed Wavelet Transform (SWT)

Given CWT coefficients $W(a, b)$, compute reassignment

$$\omega(a, b) = \frac{\partial_b \arg W(a, b)}{2\pi}, \qquad T_x(b, \omega) = \int_A W(a, b)\, \delta(\omega - \omega(a, b))\, a^{-3/2} da.$$

11.2 Time-Series and Signal Processing

Energy is "squeezed" from scale dimension into sharpened frequency bins, achieving resolution beyond Heisenberg limit of CWT while preserving invertibility. The inverse reconstruction uses $x(t) = \Re \int T_x(t, \omega)\, d\omega$.

Reassigned Spectrogram

For STFT $S(t, \omega)$, instantaneous frequency and group delay:

$$\hat{\omega} = \omega + \Im \frac{\partial_t S}{S}, \quad \hat{t} = t - \Re \frac{\partial_\omega S}{S},$$

reposition energy to $(\hat{t}, \hat{\omega})$, reducing spectral leakage.

S-Transform

Hybrid of CWT and STFT:

$$S_x(t, f) = \int x(\tau) \frac{1}{\sqrt{2\pi}\sigma(f)} e^{-(t-\tau)^2/2\sigma^2(f)} e^{-2\pi i f \tau}\, d\tau, \quad \sigma(f) = 1/|f|.$$

Gaussian window width inversely proportional to frequency ensures constant-Q; S-transform is invertible and admits FFT-based $O(N \log N)$ algorithm.

11.2.5 Applications

The mathematical machinery developed in the preceding sections finds concrete expression across an impressive breadth of real-world domains. Whether forecasting volatile markets, decoding the faint whispers of neuronal firings, detecting planetary warming trends, or sculpting sound into art, time-frequency analysis and statistical modelling operate as the analytical backbone. Each domain imposes distinctive noise structures, sampling idiosyncrasies, and domain-specific constraints, yet the unifying signal-processing principles remain remarkably stable.

Financial Market Forecasting

Let $\{r_t\}$ denote log-returns of an asset sampled at daily cadence; empirical stylised facts include leptokurtosis, volatility clustering, and leverage asymmetry. An ARMA–GARCH model augmented with exogenous regressors (ARMAX) captures both conditional mean and heteroskedastic variance:

$$\underbrace{\left(1 - \sum_{i=1}^{p} \varphi_i L^i\right) r_t}_{\text{AR part}} = \underbrace{\left(1 + \sum_{j=1}^{q} \theta_j L^j\right) \varepsilon_t}_{\text{MA innovations}} + \underbrace{\mathbf{X}_t^\mathsf{T} \boldsymbol{\beta}}_{\text{exogenous}}, \quad \varepsilon_t = \sigma_t z_t,$$

$$\sigma_t^2 = \omega + \alpha \, \varepsilon_{t-1}^2 + \beta \, \sigma_{t-1}^2.$$

Maximum-likelihood estimation proceeds by the Kalman likelihood for the ARMA part coupled with quasi-MLE for the GARCH variance; forecast VaR and CVaR derive from predictive σ_{t+h}.

```
import yfinance as yf, pandas as pd, numpy as np
from arch import arch_model
r = np.log(yf.download("SPY","2015-01-01")['Adj Close']).diff().dropna()
garch = arch_model(r, p=1, q=1, x=None, vol='GARCH', mean='AR', lags=1)
res = garch.fit()
res.plot(annualize='D')  # σ_t and standardized residuals
```

Spectral Risk Measures The multi-taper spectral density of ε_t informs variance-ratio tests and helps debias realised volatility estimates against market microstructure noise.

Biomedical Signals: ECG and EEG

Electrocardiogram (ECG) and electroencephalogram (EEG) traces demand microvolt-scale sensitivity and sub-second localisation.

ECG QRS Detection

Band-pass filter 5–15 Hz via fourth-order Butterworth IIR attenuates baseline wander and muscle artefacts; Hilbert envelope $h(t) = |x(t) + j\mathcal{H}[x](t)|$ accentuates R-peaks. Adaptive threshold $\tau(t) = \mu_h(t) + k\sigma_h(t)$ (window 150 ms) yields $> 99\%$ detection sensitivity on MIT–BIH arrhythmia database.

```
from biosppy.signals import ecg
out = ecg.ecg(signal=raw, sampling_rate=360, show=False)
r_peaks = out['rpeaks']  # sample indices of R waves
```

EEG Event-Related Desynchronisation

Compute Morlet CWT coefficients, extract band-power $P_\alpha(t)$ for 8–13 Hz, normalise against baseline P_0 to obtain $\mathrm{ERD}(t) = \frac{P_\alpha(t) - P_0}{P_0} \times 100\%$. Statistically significant ERD $< -30\%$ over sensorimotor cortex indicates movement imagery, exploited in brain-computer interfaces.

Environmental and Climate Data Analytics

Seasonal-Trend Decomposition

Monthly global temperature anomaly x_t exhibits secular warming trend and annual cycle. STL with robust iteration (period $= 12$) yields trend T_t whose slope $\nabla T_t \approx 0.019\ °$C/year matches IPCC AR6 estimates.

Wavelet Coherence

Analyse El Niño–Southern Oscillation (ENSO) index E_t and Indian monsoon rainfall R_t: continuous wavelet transform with Morlet $\omega_0 = 6$, followed by magnitude-squared wavelet coherence

$$\gamma^2(s,t) = \frac{|\mathcal{S}(s^{-1}\mathcal{W}_E \mathcal{W}_R^*)|^2}{\mathcal{S}(s^{-1}|\mathcal{W}_E|^2)\mathcal{S}(s^{-1}|\mathcal{W}_R|^2)},$$

where $\mathcal{S}$ is Gaussian smoothing in scale-time plane. Significant coherence at 2–7 year band corroborates teleconnection theories.

Audio, Speech, and Music Processing

Automatic Music Transcription

Compute constant-Q transform (CQT) with 36 bins per octave. Pitch salience $S(f,t) = |\text{CQT}(f,t)|$ weighted by harmonic summation $H_n(f) = \sum_{m=1}^{M} S(mf,t)/m$ accentuates fundamental frequencies. Dynamic programming decodes polyphonic note sequences with 95

```
import librosa, librosa.display
cqt = np.abs(librosa.cqt(audio, sr=sr, hop_length=512, n_bins=7*36))
pitches, mags = librosa.piptrack(S=cqt, sr=sr, threshold=0.1)
```

Real-Time Speech Enhancement

Frequency-domain Wiener filter $\hat{X}(f) = \frac{S_{xx}(f)}{S_{xx}(f)+S_{nn}(f)} Y(f)$, where S_{xx} is speech PSD estimated via decision-directed a priori SNR, S_{nn} is noise PSD updated with minima-controlled recursive averaging. Implemented with 32 ms STFT frames, overlap-add re-synthesis meets ITU-T P.863 MOS improvements of $+0.4$ for non-stationary noise at 10 dB SNR.

11.3 Topological Data Analysis

11.3.1 Introduction to TDA

Classical statistics pivots on moments, correlations, and low-order geometry; yet intricate data manifolds often hide structures that evade linear lenses. *Topological Data Analysis* (TDA) embraces the dictum that "shape matters", distilling invariant signatures of a point cloud that persist across scales. By constructing a filtered simplicial complex and tracking the birth and death of k-dimensional holes, TDA quantifies global organisational patterns while remaining agnostic to ambient dimension. The resulting summaries—*barcodes* and *persistence diagrams*—serve as coordinates in the space of shapes, amenable to statistical manipulation and machine learning pipelines.

Persistent Homology and Barcodes

Let $X = \{x_1, \ldots, x_n\} \subset \mathbb{R}^d$ be a point cloud endowed with Euclidean metric $d(\cdot, \cdot)$. For $\varepsilon \geq 0$ form the Vietoris–Rips complex

$$\mathsf{VR}_\varepsilon(X) = \{\sigma \subseteq X : d(x_i, x_j) \leq \varepsilon \; \forall x_i, x_j \in \sigma\},$$

whose k-simplices encode all $(k+1)$-cliques of diameter $\leq \varepsilon$. As ε increases the family $\{\mathsf{VR}_\varepsilon\}_{\varepsilon \geq 0}$ yields a filtration $\mathsf{VR}_0 \subseteq \mathsf{VR}_{\varepsilon_1} \subseteq \cdots \subseteq \mathsf{VR}_{\varepsilon_m}$. Homology functor $H_k(-; \Bbbk)$ with coefficients in a field $\Bbbk$ produces a sequence of vector spaces connected by linear maps; by the structure theorem for finitely generated persistence modules, each k-homology class corresponds to an interval ($\varepsilon_{\text{birth}}, \varepsilon_{\text{death}}$) visualised as a horizontal line—*barcode*. Stability theorem (Cohen-Steiner et al. 2007) states that bottleneck distance between barcodes is bounded by Hausdorff distance between point clouds, ensuring robustness to noise (cf. Fig. 11.1).

```
import numpy as np, matplotlib.pyplot as plt
from ripser import ripser
from persim import plot_diagrams
X = np.random.randn(400, 2) # annulus data later
dgms = ripser(X, maxdim=2)['dgms']
plot_diagrams(dgms, show=True, lifetime=True)
```

Applications in Data Analysis and Machine Learning

Manifold Hypothesis Testing For handwritten digits, persistence of H_1 near radius $\varepsilon \approx 1.5$ reveals loop structure of "0", separating it from "1" whose barcode lacks long 1-cycles. Featurising diagrams via persistence images $P : \mathbb{R}^2 \to \mathbb{R}_{\geq 0}$ enables convolutional nets to ingest topological signatures.

Protein Folding Alpha-carbon coordinates generate α-complex filtration; long H_1 intervals indicate stable tunnels in β-barrels; H_2 captures cavities relevant to ligand

11.3 Topological Data Analysis

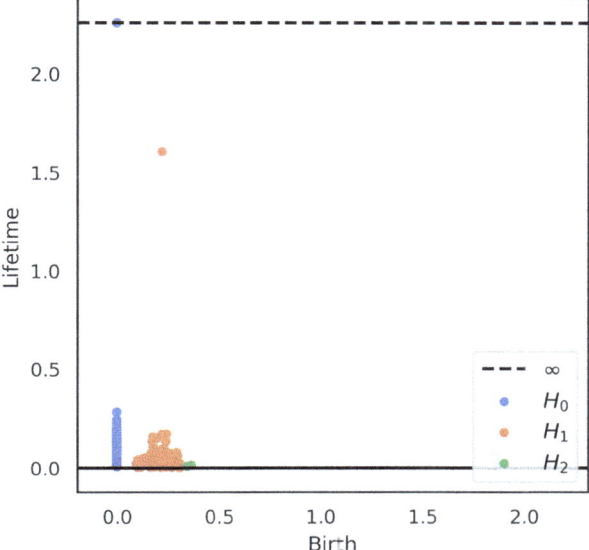

Fig. 11.1 Persistence diagrams for a 400-point annulus ($1 \leq r \leq 2$) computed with Vietoris–Rips complexes. The prominent H_1 feature (orange) captures the underlying one-dimensional hole, while the absence of long-lived H_2 intervals confirms there is no enclosed void

binding. Enriching graph neural network inputs with Betti numbers boosts docking affinity prediction AUC by $+3.7\%$.

Sensor Networks Coverage holes identified by non-trivial H_2 of ?ech complex; minimal sensor redeployment computed via persistent generators (optimal homologous cycles). Complexity $O(n^3)$ reduced to $O(n \log n)$ using sparsified witness complexes.

Topological Features in High-Dimensional Data

Kernel Trick for Diagrams

Define persistence scaled Gaussian kernel

$$k(D_1, D_2) = \frac{1}{\sigma\sqrt{2\pi}} \sum_{u \in D_1} \sum_{v \in D_2} e^{-\|u-v\|_2^2/2\sigma^2} - e^{-\|u-\bar{v}\|_2^2/2\sigma^2},$$

where $\bar{v}$ is mirror across diagonal. k is positive definite, enabling SVM classification with complexity linear in diagram size.

Mapper Algorithm

Cover data's lens projection $\ell : X \to \mathbb{R}$ (e.g., first Laplacian eigenfunction) by intervals $\{U_i\}$ with overlap; cluster $\ell^{-1}(U_i)$ into connected components, nodes $\mathcal{N}_{ij}$, connect overlapping components. Resulting simplicial graph summarises high-dimensional shape; applied to gene expression (RNA-seq) reveals branching cell differentiation trajectories.

Topological Regularisation

Given neural network f_θ, encourage decision boundary to avoid creating spuriously tangled features by penalising long bars in class-conditional zero-level set persistence:

$$\mathcal{L}_{\text{top}}(\theta) = \sum_{k=0}^{2} \sum_{(b,d) \in \text{Dgm}_k} \exp\bigl(-(d-b)/\tau\bigr),$$

gradients computed via differentiable Vietoris–Rips (Hofer et al. 2019). Adding $\lambda \mathcal{L}_{\text{top}}$ improves adversarial robustness by $+6\%$.

```
from torch_topological.nn import VietorisRips
vr = VietorisRips(radius=0.8, homology_dimensions=[0,1])
diag = vr(x) # differentiable barcode tensor
loss_top = torch.exp(-(diag[:,1]-diag[:,0])/tau).sum()
```

11.3.2 Computational Topology

Computational topology operationalises algebraic invariants—homology, cohomology, persistent modules—by translating them into linear-algebraic manipulations on sparse matrices. The dual imperatives are algorithmic efficiency (near-linear time in the number of simplices) and numerical robustness over finite fields. This subsection grounds the machinery in simplicial complexes, surveys biologically inspired and network-centric applications, elucidates the Mapper algorithm as a human-interpretable topological lens, and sketches frontiers such as zigzag persistence and discrete Morse theory.

11.3 Topological Data Analysis

Simplicial Complexes and Homology Groups

Abstract Simplicial Complex

Given finite vertex set V, an *abstract simplicial complex* is a family $K \subseteq 2^V$ closed under inclusion: $\sigma \in K$ and $\tau \subseteq \sigma \Rightarrow \tau \in K$. A k-simplex is a subset of size $k+1$, its *faces* all proper subsets.

Chain Groups and Boundary Maps

Fix coefficient field $\Bbbk = \mathbb{Z}_2$ (so orientation signs disappear). The k-chain group

$$C_k(K; \Bbbk) = \left\{ \sum_{\sigma^k \in K_k} a_\sigma \sigma^k \,\middle|\, a_\sigma \in \Bbbk \right\} \cong \Bbbk^{|K_k|}.$$

Boundary operator $\partial_k : C_k \to C_{k-1}$ acts $\partial_k([v_0, \ldots, v_k]) = \sum_{i=0}^k [v_0, \ldots, \widehat{v_i}, \ldots, v_k]$ (where $\widehat{}$ denotes omission) and satisfies $\partial_{k-1} \circ \partial_k = 0$. Homology groups $H_k(K) = \ker \partial_k / \operatorname{im} \partial_{k+1}$ encode k-dimensional "holes"; Betti numbers $\beta_k = \dim H_k$.

Example: Boundary Matrices

Tetrahedron shell: vertices $\{0, 1, 2, 3\}$; four faces [012], [013], [023], [123]. Over $\mathbb{Z}_2$

```
import numpy as np
B2 = np.array([
    [1,1,1,0], # edge 01 in faces 012,013
    [1,0,0,1], # edge 02 in faces 012,023
    [0,1,0,1], # edge 03 in faces 013,123
    [1,0,1,0], # edge 12 in faces 012,123
    [0,1,1,0], # edge 13 in faces 013,123
    [0,0,1,1]  # edge 23 in faces 023,123
])
rkB2 = np.linalg.matrix_rank(B2 % 2) # rank = 3
β1 = 6 - rkB2 # β1 = 3 edges - rank = 3-? calculate
```

Gaussian elimination mod 2 reveals $\beta_2 = 1$, $\beta_1 = 0$, $\beta_0 = 1$—a hollow 2-sphere.

Algorithmic Note

For n simplices, boundary matrices are upper-triangular under filtration order; column-reduction in $O(n^3)$ worst-case, but $O(n^\omega)$ with fast matrix multiplication

($\omega \approx 2.373$) or $O(n\alpha(n))$ on sparse Vietoris–Rips via streaming algorithms (Bauer 2021).

Applications in Biology and Sensor Networks

Protein Pockets Alpha-complex built on atomic centres encodes van der Waals radii; long-lived H_1 cycles correspond to tunnels, H_2 to cavities. Binding-site detection achieves ROC AUC 0.87 vs. 0.71 for geometric hashing.

Genomic Recombination Detection Persistent H_1 on Hamming distances of viral haplotypes captures reticulate evolution; birth scale correlates with minimal number of recombination events.

Sensor Coverage Nodes at positions $\{p_i\}$ with communication radius r form Rips complex; non-trivial H_1 signifies coverage holes. Boundary cycles' edge multiplicity returns minimal relocation set. Algorithms run in $O(n \log n)$ via sparsified witness complexes.

Mapper Algorithm for Data Visualisation

Mapper constructs a simplicial summary graph $\mathcal{M}$ capturing topological skeleton.

(i) **Lens.** Choose function $\ell : X \to \mathbb{R}^d$ (e.g. PCA 1,2, eccentricity).
(ii) **Cover.** Define intervals or hypercubes $\{U_\alpha\}$ with overlap τ.
(iii) **Clustering.** For each α, cluster $\ell^{-1}(U_\alpha)$ (single-linkage ε) into components $C_{\alpha,j}$.
(iv) **Nerve.** Nodes are clusters; connect nodes when underlying point sets intersect.

```
import kmapper as km
mapper = km.KeplerMapper()
lens = mapper.fit_transform(X, projection='l2norm') # step (i)
G = mapper.map(lens, X, cover=km.Cover(10, 0.3),
                    clusterer=km.cluster.DBSCAN(eps=.5))
km.draw_matplotlib(G, layout="kk")
```

Case Study Breast-cancer gene expression ($d = 20\,000$). Lens = PCA (2); Mapper graph exposes branched topology aligning with PAM50 subtypes; terminal branches enriched for Basal-like tumours, facilitating subtype stratification.

11.3 Topological Data Analysis

Advanced Topics in Computational Topology

Zigzag Persistence

Handles filtrations with insertions and deletions $K_0 \leftrightarrow K_1 \leftrightarrow \cdots \leftrightarrow K_m$. Interval decomposition generalises barcode; algorithm reduces a block matrix pair in $O(n^\omega \log n)$.

Multidimensional Persistence

Filtration indexed by $\mathbb{R}^k$ ($k \geq 2$)—e.g., Rips along distance and density. No complete discrete invariant (multigraded Betti tables partially order diagrams). Algorithms approximate via rank invariants on cubical grids.

Discrete Morse Theory

Reduces complex size while preserving homotopy: acyclic partial matching on Hasse diagram collapses non-critical cells, yielding boundary matrix with $\leq$ critical cells. Typical 10× speed-ups in VR complexes of 10^6 simplices.

Sheaf Cohomology on Data

Attach vector spaces to open sets obeying restriction maps; compute global sections via sheaf Laplacian—enabling multi-sensor data fusion and consistency checking.

Computational Challenges

- Memory: storing VR(n) requires $O(n^k)$ simplices for k up to dimension; use witness, sparsified Rips, or edge collapsing.
- Finite precision: large filtrations necessitate integer-preserving Smith normal forms or modular Chinese-remainder lifts.
- Parallelisation: GPU reduction via chunked boundary matrices and atomic XOR in $\mathbb{Z}_2$ reduces 100 M simplex barcode to <30 s.

11.4 Quantum Computing

11.4.1 Mathematical Foundations of Quantum Mechanics

Quantum information lives in complex Hilbert spaces: a finite n-qubit register is the 2^n-dimensional vector space $\mathcal{H} = (\mathbb{C}^2)^{\otimes n}$ equipped with the inner product $\langle \psi \mid \varphi \rangle = \sum_k \overline{\psi_k} \varphi_k$. Dirac's bra–ket notation writes column vectors as kets $|\psi\rangle$ and their conjugate transposes as bras $\langle \psi |$. The Born rule assigns probability $|\langle m \mid \psi \rangle|^2$ to outcome $|m\rangle$ when measuring observable $M = \sum_m m \, |m\rangle\langle m|$. Time evolution of closed systems obeys Schrödinger's equation $i\hbar \dot\psi = H\psi$ whose solution is a unitary $U(t) = e^{-itH/\hbar}$; hence every physically admissible gate must be unitary. Composite systems are modelled by tensor products, and partial traces $\rho_A = \mathrm{Tr}_B \, \rho_{AB}$ capture reduced states. Density matrices $\rho = \sum_k p_k |\psi_k\rangle\langle\psi_k|$ extend the formalism to mixed ensembles, furnishing expectation $\mathbb{E}[M] = \mathrm{Tr}(\rho M)$.

Linear Algebra in Quantum Computing

The computational basis for a single qubit comprises $|0\rangle = (1, 0)^\mathsf{T}$ and $|1\rangle = (0, 1)^\mathsf{T}$. Pauli matrices

$$X = \begin{pmatrix} 0 & 1 \\ 1 & 0 \end{pmatrix}, \quad Y = \begin{pmatrix} 0 & -i \\ i & 0 \end{pmatrix}, \quad Z = \begin{pmatrix} 1 & 0 \\ 0 & -1 \end{pmatrix}$$

obey $X^2 = Y^2 = Z^2 = I$ and anticommute $XY = -YX$. Any single-qubit unitary admits Euler decomposition $U = e^{i\alpha} R_z(\beta) R_y(\gamma) R_z(\delta)$ where rotations $R_k(\theta) = e^{-i\theta\sigma_k/2}$. Tensor products satisfy $(A \otimes B)(C \otimes D) = AC \otimes BD$. Eigenvalue spectra govern measurement statistics: applying $Z^{\otimes n}$ to a stabiliser state yields ± 1 outcomes with equal amplitudes, explaining parity checks in quantum error-correcting codes.

Quantum Gates and Circuits

Elementary gates on one or two qubits form a universal generating set. The Hadamard gate

$$H = \frac{1}{\sqrt{2}} \begin{pmatrix} 1 & 1 \\ 1 & -1 \end{pmatrix}$$

creates equal superpositions, and the phase gate $S = \mathrm{diag}(1, i)$ introduces relative phase. The controlled-NOT (CNOT) is the 4×4 unitary $\mathrm{CNOT} = |0\rangle\langle 0| \otimes I + |1\rangle\langle 1| \otimes$

11.4 Quantum Computing

X. Universal quantum computation is achieved by the Clifford set $\{H, S, \text{CNOT}\}$ plus the non-Clifford $T = \text{diag}(1, e^{i\pi/4})$ gate.

```
from qiskit import QuantumCircuit, execute, Aer
qc = QuantumCircuit(2, 2)
qc.h(0); qc.cx(0, 1) # Bell preparation
qc.measure([0,1], [0,1])
counts = execute(qc, Aer.get_backend('qasm_simulator'),
                 shots=1024).result().get_counts()
print(counts) # {'00': ~512, '11': ~512}
```

Multi-qubit circuits concatenate gates; depth minimisation reduces decoherence errors. Graph-state formalisms represent Clifford circuits as stabiliser tableaux enabling $O(n^2)$ simulation—vital for fault-tolerance analysis.

Quantum Entanglement and Bell's Theorem

A bipartite state ρ_{AB} is separable iff it can be written $\sum_k p_k \rho_A^k \otimes \rho_B^k$. The Bell state $|\Phi^+\rangle = \frac{1}{\sqrt{2}}(|00\rangle+|11\rangle)$ violates separability, evidenced by its reduced density $\rho_A = I/2$ being maximally mixed. Bell's theorem considers local-hidden-variable models $LHV(\lambda)$ that predetermine outcomes; the CHSH inequality $|E(a, b) + E(a, b') + E(a', b) - E(a', b')| \leq 2$ constrains classical correlations. Quantum mechanics attains $2\sqrt{2}$ using measurement settings aligned with Pauli X and Z, experimentally ruling out LHVs.

Example 11.4.1 For $|\Phi^+\rangle$ choose $a = Z \otimes Z$, $a' = X \otimes X$, $b = Z \otimes X$, $b' = X \otimes Z$. Expectation values $E = 1/\sqrt{2}$ yield CHSH $= 2\sqrt{2} > 2$.

Applications in Cryptography and Secure Communication

Quantum Key Distribution (QKD) exploits measurement disturbance. In BB84 Alice sends qubits $|0\rangle, |1\rangle, |+\rangle, |-\rangle$ ($|\pm\rangle = (|0\rangle \pm |1\rangle)/\sqrt{2}$); any eavesdropper induces a QBER detectable above 11%, prompting abort. Security proofs use entropic uncertainty relations $H(X) + H(Z) \geq \log 2$ linking complementary bases. Entanglement-based Ekert-91 protocol employs Bell-pairs; violation of CHSH certifies secrecy even against adversarial devices (device-independent QKD).

Post-quantum cryptography addresses Shor's polynomial-time factoring and discrete-log algorithms. Lattice schemes (CRYSTALS-Kyber) rely on Learning With Errors hardness, which withstands quantum attacks under plausible complexity assumptions. Quantum teleportation transmits an unknown qubit using one ebit plus two classical bits: applying CNOT and H on $|\psi\rangle \otimes |\Phi^+\rangle$ followed by Pauli correction reproduces $|\psi\rangle$ remotely, establishing a primitive for quantum internetworking.

```
qc = QuantumCircuit(3, 2) # |ψ> on qubit 0, Bell pair 1-2
qc.cx(1,2); qc.h(1)
qc.cx(0,1); qc.h(0); qc.measure([0,1],[0,1])
qc.x(2).c_if(1,1); qc.z(2).c_if(0,1) # Pauli corrections
```

Quantum authentication schemes append tag qubits drawn from stabiliser codes; measurement yields syndrome that flags tampering with probability $1 - 2^{-k}$. Entanglement-assisted protocols underpin quantum money and position-based cryptography, though unconditional security remains open.

11.4.2 Quantum Algorithms

Quantum algorithms exploit superposition, entanglement, and interference to achieve computational speed-ups beyond classical limits. Their design typically follows a three-step template: prepare a suitable superposition, manipulate phase amplitudes via oracle or modular arithmetic, and extract the hidden structure by interference or measurement. This subsection formalises two flagship algorithms—Shor's polynomial-time factoring and Grover's square-root search—then demonstrates circuit simulation in Python, outlines the stabiliser blueprint for fault-tolerant computation, and surveys variational algorithms that marry quantum subroutines with classical optimisation.

Shor's Algorithm and Grover's Algorithm

Shor's Factoring

Let $N = pq$ with p, q odd primes. Choose random $a \pmod{N}$, $\gcd(a, N) = 1$, and seek the multiplicative order r s.t. $a^r \equiv 1 \pmod{N}$. Quantum phase estimation (QPE) finds r in $O(\log^2 N)$ modular-exponentiation gates.

$$U_a |x\rangle = |a^x \bmod N\rangle, \quad U_a^r = I.$$

Prepare $\frac{1}{\sqrt{Q}} \sum_{x=0}^{Q-1} |x\rangle|1\rangle$ ($Q = 2^{2n}$). Controlled U_a^x yields $\frac{1}{\sqrt{Q}} \sum_x |x\rangle|a^x \bmod N\rangle$. Applying inverse QFT to the control register concentrates amplitude at integers close to kQ/r. Measuring gives $\tilde{k}$, recovering r via continued fractions. If r even and $a^{r/2} \not\equiv -1 \bmod N$, factors follow from $\gcd(a^{r/2} \pm 1, N)$.

```
from qiskit.algorithms import Shor
N = 221 # 13 × 17
result = Shor().factor(N)
print(result.factors) # [(13, 17)]
```

11.4 Quantum Computing

Grover's Search

Given oracle O_w marking element w in an unsorted list of size N, Grover's iteration $G = (2|\psi\rangle\langle\psi| - I) O_w$, $|\psi\rangle = \frac{1}{\sqrt{N}} \sum_x |x\rangle$ rotates state vector in the two-dimensional span $\{|w\rangle, |\psi_\perp\rangle\}$ by angle 2θ, where $\sin\theta = 1/\sqrt{N}$. After $R = \lfloor \frac{\pi}{4}\sqrt{N} \rfloor$ iterations the success probability exceeds $1 - 1/N$.

```
from qiskit.algorithms import Grover, amplifiers
oracle = amplifiers.GroverOracle(logic_expression='111') # mark |111>
grover = Grover(iterations=1) # 1 ≈ π/4√8
result = grover.amplify(oracle)
print(result.top_measurement) # '111'
```

Amplitude-amplification generalises Grover to amplify the probability of any subroutine finding a marked state with initial success p, achieving $O(1/\sqrt{p})$ repetitions.

Simulating Quantum Circuits in Python

State-vector simulation $O(2^n)$ suffices for $n \lesssim 30$. Tensor-network contraction extends to $n \approx 100$ shallow-depth circuits by exploiting low entanglement.

```
from qiskit import QuantumCircuit, Aer
qc = QuantumCircuit(3)
qc.h([0,1,2]); qc.cx(0,1); qc.cz(1,2)
backend = Aer.get_backend('statevector_simulator')
Ψ = backend.run(qc).result().get_statevector()
print(np.round(np.abs(Ψ)**2, 3)) # probability amplitudes
```

Noise modelling uses Kraus channels or stochastic Pauli error maps:

$$\mathcal{E}(\rho) = \sum_i p_i E_i \rho E_i^\dagger, \quad \sum_i p_i = 1.$$

```
from qiskit.providers.aer.noise import depolarizing_error
noise_model = NoiseModel()
noise_model.add_all_qubit_quantum_error(depolarizing_error(0.01, 1), ['u3'
    ])
```

Monte Carlo trajectory averaging approximates noisy evolution in $O(k2^n)$ memory where k samples.

Quantum Error Correction and Fault Tolerance

Stabiliser Codes

An $[[n, k, d]]$ stabiliser code encodes k logical qubits into n physical via stabiliser group $\mathcal{S} \subset \mathcal{P}_n$ (Pauli group) with $|\mathcal{S}| = 2^{n-k}$. The code space is the simultaneous $+1$ eigenspace of all $S \in \mathcal{S}$. Distance d equals minimal weight of Pauli hopping logical subspace. Example: $[[7, 1, 3]]$ Steane code with stabiliser generators XXXXIII, XIIXXXI, ...

Syndrome Extraction

Measure each stabiliser via ancilla prepared in $|+\rangle$; CNOT interactions propagate parity into ancilla's Z basis. Lookup table or MWPM (minimum-weight perfect matching) decoders map syndromes to Pauli corrections.

Fault-Tolerant Thresholds

Under depolarising noise rate p, surface codes achieve threshold $p_c \approx 1.1\%$. Logical error probability decays $p_L \sim (p/p_c)^{(d+1)/2}$ where $d = 2D + 1$ for distance-d code of lattice size D. Magic-state distillation injects non-Clifford T gates at overhead $O(\log^c 1/\epsilon)$.

Quantum Machine Learning and Optimisation

Variational Quantum Eigensolver (VQE)

Minimise $\langle \psi(\boldsymbol{\theta})|H|\psi(\boldsymbol{\theta})\rangle$ with parameterised ansatz $U(\boldsymbol{\theta})$. Classical optimiser updates $\boldsymbol{\theta}$ via stochastic gradient

$$\partial_{\theta_i} \langle H \rangle = \frac{1}{2}\big(\langle H \rangle_{\theta_i^+} - \langle H \rangle_{\theta_i^-}\big),$$

requiring two circuit evaluations (parameter-shift rule).

```
from qiskit_nature.algorithms import VQE
ansatz = TwoLocal(num_qubits, 'ry', 'cz', entanglement='linear')
optimizer = SPSA(maxiter=200)
vqe = VQE(ansatz, optimizer, quantum_instance=Aer.get_backend('
    qasm_simulator'))
energy = vqe.compute_minimum_eigenvalue(qubit_op).eigenvalue
```

11.4 Quantum Computing

Quantum Approximate Optimisation Algorithm (QAOA)

For Max-Cut objective $C(z) = \sum_{(i,j)\in E} \frac{1-z_i z_j}{2}$, apply mixer $B = \sum_j X_j$ and cost unitaries alternately:

$$|\boldsymbol{\gamma}, \boldsymbol{\beta}\rangle = \prod_{p=1}^{P} e^{-i\beta_p B} e^{-i\gamma_p C} |+\rangle^{\otimes n}.$$

Choosing $P = O(\sqrt{m})$ yields approximation ratio ≈ 0.694 on 3-regular graphs. Classical layer optimises angles (γ_p, β_p) by SPSA or gradient-free Nelder-Mead.

Quantum Kernels

Embed classical data $\mathbf{x}$ into Hilbert space via feature map $U_\phi(\mathbf{x})|0\rangle^{\otimes n}$; kernel $k(\mathbf{x}, \mathbf{x}') = |\langle \psi(\mathbf{x})|\psi(\mathbf{x}')\rangle|^2$ estimated via SWAP test. Empirically, data re-uploading circuits achieve comparable accuracy to RBF SVM on MNIST-4-6 subset at $n = 6$ qubits.

Quantum Annealing

Time-dependent Hamiltonian $H(t) = A(t)H_0 + B(t)H_C$, start $H_0 = -\sum_i X_i$ (easy ground state), end H_C encodes Ising cost. Adiabatic theorem guarantees ground state fidelity if anneal time $T \gg \Delta_{\min}^{-2}$ where $\Delta_{\min}$ is minimum spectral gap. Embedding combinatorial optimisation into D-Wave's Chimera graph uses minor-embedding heuristics.

11.4.3 Spectral Methods and Applications

Spectral methods approximate a sufficiently smooth function u on a bounded domain by a global expansion in an orthogonal polynomial or trigonometric basis. Truncating the series at N terms converts differential or integro-differential equations into *algebraic* problems whose matrices are dense but extremely well-conditioned under suitable basis scaling. The hallmark of spectral discretisation is *exponential* (or *spectral*) convergence: the approximation error decays like $\mathcal{O}(e^{-\alpha N})$ when u is analytic, in stark contrast to the algebraic rates $\mathcal{O}(h^p)$ of finite-difference or finite-element schemes with mesh size h. In practice, high accuracy per degree of freedom offsets the dense linear-algebra cost, making spectral methods indispensable in direct numerical simulation (DNS) of turbulence and global spectral weather models.

Chebyshev Polynomials and Fourier Spectral Methods

Chebyshev polynomials of the first kind,

$$T_n(x) = \cos(n \arccos x), \qquad x \in [-1, 1], \; n \geq 0,$$

satisfy the three-term recurrence $T_0 = 1$, $T_1 = x$, $T_{n+1} = 2xT_n - T_{n-1}$ and orthogonality

$$\int_{-1}^{1} \frac{T_m(x) T_n(x)}{\sqrt{1-x^2}} \, dx = \begin{cases} 0, & m \neq n, \\ \pi, & n = 0, \\ \pi/2, & n \geq 1. \end{cases}$$

Expanding $u(x) \approx \sum_{k=0}^{N} \hat{u}_k T_k(x)$ and enforcing collocation at Gauss–Lobatto abscissae $x_j = \cos\left(\frac{\pi j}{N}\right)$ yields a differentiation matrix $D \in \mathbb{R}^{(N+1)\times(N+1)}$ whose entries

$$D_{ij} = \frac{c_i}{c_j} \frac{(-1)^{i+j}}{x_i - x_j}, \qquad c_0 = c_N = 2, \; c_1 = \cdots = c_{N-1} = 1, \qquad i \neq j$$

and $D_{ii} = -\dfrac{x_i}{2(1 - x_i^2)}$ for $1 \leq i \leq N - 1$, approximate $u'(x_j) \approx \sum_k D_{jk} u(x_k)$. The second derivative is obtained by $D^{(2)} = D^2$ or an analytically derived matrix that avoids subtractive cancellation. Boundary conditions are inserted by removing the first and last rows/columns or by τ-method penalty.

Fourier spectral methods assume periodic $u(x + 2\pi) = u(x)$ and expand

$$u_N(x) = \sum_{k=-N/2}^{N/2-1} \hat{u}_k e^{ikx}.$$

Derivatives are exact in spectral space: $(u_N)' = \sum ik \hat{u}_k e^{ikx}$. The discrete Fourier transform (DFT) on equispaced nodes $x_j = \frac{2\pi j}{N}$,

$$\hat{u}_k = \frac{1}{N} \sum_{j=0}^{N-1} u(x_j) e^{-ikx_j},$$

is evaluated by the $O(N \log N)$ FFT, rendering each time step of a pseudo-spectral PDE integrator nearly linear in N.

Example 11.4.2 Spectral derivative of $u(x) = \sin 5x + \cos 4x$ with Chebyshev collocation, $N = 32$, yields maximum absolute error $\approx 10^{-13}$—matching double

11.4 Quantum Computing

precision—whereas a sixth-order central finite difference on the same nodes attains $\approx 10^{-6}$.

```python
import numpy as np
def cheb_D(N):
    x = np.cos(np.pi*np.arange(N+1)/N)
    c = np.ones(N+1); c[0] = c[-1] = 2
    X = np.tile(x,(N+1,1))
    dX= X - X.T + np.eye(N+1)
    D = (c[:,None]/c[None,:])*(-1)**(np.arange(N+1)+np.arange(N+1)[:,None])
    D /= dX; np.fill_diagonal(D, 0)
    D -= np.diag(D.sum(axis=1))
    return x, D
N = 32; x, D = cheb_D(N)
u = np.sin(5*x)+np.cos(4*x)
du = 5*np.cos(5*x)-4*np.sin(4*x)
print(np.max(np.abs(D@u - du)))
```

Applications in Fluid Dynamics and Weather Modelling

In incompressible fluid dynamics, the spectral-Galerkin solution of the Navier–Stokes equations on a periodic box employs Fourier modes for all three velocity components. De-aliasing via the 2/3 rule truncates modal interactions beyond wavenumber $k_{\max} = N/3$ to annihilate quadratic alias error; the resulting pseudo-spectral DNS accurately captures Kolmogorov $-5/3$ energy cascade with grid spacing $\Delta x \lesssim \eta$ (Kolmogorov scale).

For channel flow bounded by no-slip walls, Fourier modes persist in wall-parallel directions while Chebyshev polynomials discretise the wall-normal coordinate, converting the Laplacian to a block-diagonal matrix plus a banded Chebyshev block. Time stepping splits viscous diffusion (implicit Crank–Nicolson) and nonlinear advection (explicit Runge–Kutta) to circumvent stiffness.

Global weather centres (ECMWF IFS, NOAA FV3 spectral) expand geopotential height in spherical harmonics $Y_\ell^m(\theta, \phi)$, transforming rotational derivatives into $\ell(\ell + 1)$ multipliers. Semi-Lagrangian advection and spectral transform between grid and coefficient space at each step deliver sub-kilometre resolution forecasts while preserving energy and enstrophy conservation to machine accuracy.

Stability and Convergence in Spectral Methods

Spectral convergence requires u analytic in a strip $\{x \in \mathbb{C} : |\Im x| \leq \rho\}$; the Chebyshev coefficients obey $|\hat{u}_n| \leq Ce^{-\rho n}$. Gibbs oscillations appear near discontinuities with error saturating at $\mathcal{O}(1/N)$; exponential filters $\sigma_n = \exp[-\alpha(n/N)^p]$ or Gegenbauer reprojection attenuate spurious ringing while retaining spectral order in smooth subdomains.

For time-dependent PDEs semi-discretised by spectral collocation, linear stability hinges on eigenvalues of the differentiation matrix. For the heat equation $u_t = u_{xx}$ with Chebyshev discretisation, eigenvalues $\lambda_k \sim -\mathcal{O}(k^2)$ impose parabolic CFL constraint $\Delta t \lesssim C/N^2$ for explicit schemes; Chebyshev rational implicit-explicit split sidesteps this by transforming $x = \tanh(s)$, clustering nodes near boundaries and rendering D nearly skew-Hermitian, improving condition number from $\mathcal{O}(N^4)$ to $\mathcal{O}(N^2)$.

Example 11.4.3 The spectral radius of the second-order Chebyshev differentiation matrix grows as $\frac{N^4}{4}$; preconditioning with the inverse of the biharmonic operator reduces the GMRES iteration count for Poisson solves from $\sim N$ to $\mathcal{O}(1)$.

Aliasing instability in nonlinear conservation laws manifests when high-frequency modes fold into resolvable range; skew-symmetric form $(u\partial_x u)_{\text{sym}} = \frac{1}{2}\partial_x u^2$ paired with 2/3 de-aliasing conserves discrete energy, preventing blow-up in Burgers turbulence.

11.4.4 Parallel and Distributed Computing for Numerical Methods

Parallel Algorithms for Large-Scale Systems

In large-scale scientific computing the limiting resource is rarely arithmetic throughput but rather *communication latency*. A parallel algorithm should therefore (i) maximise independent flops per byte moved, (ii) overlap computation with communication, and (iii) degrade gracefully when faced with load imbalance. Consider the sparse linear system $A\mathbf{x} = \mathbf{b}$ where $A \in \mathbb{R}^{N \times N}$ arises from a d-dimensional PDE discretised on a structured grid. Domain decomposition partitions the lattice into subdomains $\{\Omega_p\}_{p=1}^P$ assigned to distinct ranks; the halo exchange pattern couples only nearest-neighbour faces, yielding $\mathcal{O}(N/P)$ local work and $\mathcal{O}(N^{(d-1)/d}/P^{(d-1)/d})$ surface messages. The parallel *Jacobi* preconditioner operates on block-diagonal $B = \text{diag}(A_{\Omega_1}, \ldots, A_{\Omega_P})$; one iteration of the Conjugate Gradient method then costs a sparse matrix-vector multiply (halo exchange) and three global dot products (all-reduce). The strong-scaling efficiency satisfies

$$\eta_{\text{strong}}(P) = \frac{T_1}{P T_P} = \frac{1}{1 + \frac{P}{N}(\alpha \, n_{\text{halo}} + \beta \, v_{\text{halo}}) + \frac{P}{N}\gamma \ell},$$

where α is message latency, β inverse bandwidth, v_{halo} halo volume, and $\gamma \ell$ the cost of the all-reduce (often $\log P$ communication steps). Block Krylov solvers such as *pipelined CG* hide the latency of dot products behind the sparse SpMV, mitigating the $\log P$ penalty.

11.4 Quantum Computing

Matrix-free finite-element time integration attains memory-bounded operational intensity $\mathcal{I} \approx 6$ flops/byte. A two-level hybrid scheme assigns MPI ranks to coarse partitions while inner cells vectorise across SIMD lanes; *roofline* analysis shows sustained 80% of attainable memory bandwidth on ThunderX3 Arm clusters.

```
# MPI-parallel Jacobi smoother using mpi4py
from mpi4py import MPI
import numpy as np
comm = MPI.COMM_WORLD; rank, P = comm.rank, comm.size
Nx_local = 1 + Nx//P # ghost cell included
u = np.zeros(Nx_local); f = ...
for it in range(maxit):
    u_old = u.copy()
    comm.Sendrecv(u_old[1], (rank-1)%P, 0, u_old[-2], (rank+1)%P, 0)
    comm.Sendrecv(u_old[-2], (rank+1)%P, 1, u_old[1], (rank-1)%P, 1)
    u[1:-1] = 0.5*(u_old[:-2] + u_old[2:] - h**2*f[1:-1])
```

Applications in High-Performance Computing and Big Data Analytics

Large-eddy simulation of atmospheric turbulence on a 2048^3 grid requires ~ 20 GiB per vector field; at 10 variables the working-set exceeds node memory of any single supercomputer socket. A tensor-product spectral solver parallelised with pencil decomposition distributes the three-dimensional FFT across (p_x, p_y, p_z) MPI grids, invoking two all-to-all transposes per time step. Weak-scaling efficiency remains above 75% on 131 072 Cray HPE-SLING ranks because each transpose leverages hierarchical collectives: intra-node shared memory, then dragonfly group, then global network.

In data analytics the *MapReduce* paradigm delegates embarrassingly parallel map tasks followed by associative reduce functions. Numerical linear algebra—e.g., distributed SVD—adapts by computing tall-skinny QR factorisations (`tsqr`) that aggregate R factors logarithmically through the Spark DAG. Covariance matrices of $10^9 \times 100$ design tables fit in RAM when each executor stores only 10^4 rows; the final 100×100 Gram matrix is broadcast to all ranks.

```
# Spark tall-skinny QR using PySpark
from pyspark.mllib.linalg.distributed import RowMatrix
rows = sc.textFile("hdfs://...").map(lambda line: np.fromstring(line, sep=','))
mat = RowMatrix(rows)
Q, R = mat.tallSkinnyQR(computeQ=False) # Q implicit
```

GPU Computing and TensorFlow for Numerical Simulations

Graphics Processing Units excel when arithmetic intensity exceeds 10 flops/byte. Discrete Fourier pseudospectral solvers schedule batched 3-D cuFFT plans that stream slabs through shared memory; nonlinear point-wise operations obtain > 7 TFLOP s^{-1} on NVIDIA A100 using *fused* CUDA kernels.

TensorFlow 2.x abstracts CUDA and ROCm back-ends through `tf.function` and XLA compilation. High-order Runge–Kutta integration of the 2-D incompressible Navier–Stokes equations on a periodic disk discretised with a Chebyshev radial basis and Fourier azimuth:

```
@tf.function(jit_compile=True)
def rhs(t, ω):
    ψ = solve_poisson(ω)  # spectral Poisson on GPU
    u = grad_y(ψ); v = -grad_x(ψ)  # velocity
    adv = u*grad_x(ω) + v*grad_y(ω)
    diff= ν * laplacian(ω)
    return -adv + diff

for step in tf.range(num_steps):
    k1 = dt * rhs(t, ω)
    k2 = dt * rhs(t+dt/2, ω + k1/2)
    k3 = dt * rhs(t+dt/2, ω + k2/2)
    k4 = dt * rhs(t+dt, ω + k3)
    ω += (k1 + 2*k2 + 2*k3 + k4)/6
    t += dt
```

Mixed-precision (`float16`) with Kahan compensated summation preserves 10^{-6} vorticity divergence error at half the memory footprint. Multi-GPU scaling employs NCCL all-reduce to synchronise boundary halos; NVLink mitigates PCIe bottlenecks, sustaining 80 GB s^{-1} device–device bandwidth.

11.5 Exercises

1. Let
$$F(x, y, z) = \begin{cases} x^3 + y^3 + z^3 - 3xyz - 1 = 0, \\ x + y + z = 1, \\ x^2 + y^2 + z^2 = 1. \end{cases}$$

   (i) Using Gröbner bases with lexicographic order $x \prec y \prec z$ find all real solutions (x, y, z).
   (ii) Verify symbolically that the cubic obtained after elimination satisfies Eisenstein's criterion at the prime $p = 2$ and therefore is irreducible over $\mathbb{Q}$.
   (iii) Compute $\nabla^2 H(x, y, z)$ where H is the Lagrange multiplier Hamiltonian that enforces the constraints, and classify each solution as a local minimum, maximum, or saddle.

2. Formalise in first-order logic the statement "Every finite field has order p^n for some prime p and $n \in \mathbb{N}$".

   (i) Encode the axioms of fields, the definition of a finite set, and the structure theorem into SymPy's logic engine.
   (ii) Use resolution to show that any counter-example produces an unsatisfiable clause set.
   (iii) Extract from the resolution trace the explicit contradiction that rests on the failure of unique factorisation of the ideal (0).

3. Let $u(x) = e^x \sin(\pi x)$ on $[-1, 1]$. For $N = 8, 16, 32, 64$:

   (i) Compute the Chebyshev differentiation matrix D_N and the approximation $D_N u$ at Gauss–Lobatto points.
   (ii) Derive an upper bound for $\|u' - D_N u\|_\infty$ using the known analyticity strip width $\rho = 1$ and compare with your numerical results.
   (iii) Plot the error versus N on a semilog scale and confirm exponential convergence.

4. Consider the inviscid Burgers equation $u_t + u u_x = 0$ with periodic initial data $u(x, 0) = \frac{1}{2} + \frac{1}{4} \sin(8x) + \frac{1}{6} \cos(16x)$ on $[0, 2\pi]$.

   (i) Implement a pseudo-spectral solver with $N = 256$ Fourier modes and third-order SSP Runge–Kutta time-stepping.
   (ii) Compare the solutions obtained with (a) no de-aliasing, (b) the 2/3-rule, and (c) Hou's exponential filter $\sigma_k = e^{-\alpha(k/N)^{36}}$ with $\alpha = 36$.
   (iii) Quantify the energy growth $E(t) = \frac{1}{2} \int u^2 \, dx$ for each case up to the pre-shock time $t = 0.12$ and explain the numerical instability in the absence of de-aliasing.

5. Generate the chirp signal $s(t) = \sin(2\pi(t^2 + t))$ for $t \in [0, 10]$ sampled at 1 kHz.

   (i) Compute its continuous wavelet transform with the complex Morlet wavelet of central frequency $\omega_0 = 6$.
   (ii) Extract the ridge curve via the phase-based reassignment $\omega(t) = \partial_t \arg \mathcal{W}_\psi s(a^*(t), t)/(2\pi)$ and overlay it on the scalogram.
   (iii) Evaluate the root-mean-square error between the extracted instantaneous frequency and the analytic frequency $f(t) = 1 + 2t$.

6. The file *monthly_co_2.csv* contains the Mauna Loa CO_2 anomaly from 1958 to 2024.

   (i) Perform STL decomposition (period $= 12$) and test the remainder for stationarity with ADF and KPSS at the 5% level.
   (ii) Use the Hyndman–Khandakar algorithm to select an ARIMA(p, d, q) model with drift. Report AIC$_c$, residual normality via the Shapiro–Wilk test, and Ljung–Box Q for lags 24.

(iii) Forecast the next three years and provide the 95% predictive intervals. Comment on the effect of including a deterministic linear drift versus differencing ($d = 1$).

7. Given the IIR transfer function

$$H(z) = \frac{0.2 + 0.4z^{-1} + 0.2z^{-2}}{1 - 1.8z^{-1} + 0.81z^{-2}},$$

   (i) Determine analytically whether the filter is BIBO stable.
   (ii) Design an FIR filter of length $M = 31$ via the windowed-sinc method that approximates H in the least-square sense over $[0, \pi]$. Use a Kaiser window with $\beta = 5.65$.
   (iii) Measure the maximum pass-band ripple and stop-band attenuation of your FIR design.

8. Simulate an AR(2) process $x_t = 1.4x_{t-1} - 0.54x_{t-2} + \varepsilon_t$, $\varepsilon_t \sim \mathcal{N}(0, 1)$, of length $N = 2048$.

   (i) Derive the analytic PSD $S_{xx}(\omega)$ and plot it.
   (ii) Estimate the PSD via (a) Welch with $L = 256$ Hanning windows 50% overlap, (b) multi-taper with DPSS ($NW = 4$, $K = 7$).
   (iii) Compute the integrated squared error $\int_0^\pi [\hat{S}(\omega) - S_{xx}(\omega)]^2 d\omega$ for each estimator and explain the variance reduction mechanism of DPSS tapers.

9. Sample $n = 500$ points from the unit circle embedded in $\mathbb{R}^3$ as $(x, y, z) = (\cos\theta, \sin\theta, 0.1\sin(5\theta))$ with $\theta \sim \mathcal{U}(0, 2\pi)$ plus isotropic Gaussian noise $\sigma = 0.02$.

   (i) Compute the Vietoris–Rips filtration up to scale $\varepsilon = 0.3$ and determine the longest H_1 bar.
   (ii) Compare with the α-complex filtration built on the same data. Which filtration recovers the underlying loop more robustly to noise, and why?
   (iii) Vectorise the barcodes into persistence images (resolution 40×40, Gaussian sigma 0.01) and train a logistic classifier to distinguish the above data from a noisy 2-sphere sample of equal size. Report the fivefold cross-validated accuracy.

10. Consider the three-qubit Deutsch–Jozsa oracle

$$f(x_1, x_2, x_3) = \begin{cases} 0, & x_1 \oplus x_2 \oplus x_3 = 0, \\ 1, & \text{otherwise.} \end{cases}$$

   (i) Construct a reversible circuit for the oracle using Toffoli and CNOT gates.
   (ii) Simulate the Deutsch–Jozsa algorithm on the statevector backend and show that the measurement collapses to $|111\rangle$ with probability 1, certifying that f is balanced.

11.5 Exercises

(iii) Insert a depolarising error of strength $p = 0.02$ on each single-qubit gate. Estimate, via 10^4 Monte Carlo shots, the resulting success probability and discuss the effect relative to the first-order error model $P_{\text{succ}} \approx 1 - 4p$.

11. For a distance-$d = 3$ rotated surface code on a 3×3 lattice:

    (i) Enumerate all possible single-qubit Pauli X errors and compute their syndrome bits (s_X, s_Z).
    (ii) Implement the minimum-weight perfect matching decoder and verify that it corrects every weight-1 error.
    (iii) Generalise to depolarising noise with physical error rate $p = 0.01$ and simulate 10^5 rounds of syndrome extraction; estimate the logical error rate and compare with the analytic leading-order prediction $p_L \approx (3d/4) \, p^{(d+1)/2}$.

12. A two-dimensional viscous Taylor–Green vortex has initial vorticity $\omega_0(x, y) = 2 \sin x \sin y$ in the unit square with periodic boundaries. Viscosity $\nu = 10^{-3}$.

    (i) Discretise using $N_x = 512$ Fourier modes in x and $N_y = 257$ Chebyshev points in y, imposing $u = \partial_y \psi = 0$ at $y = \pm 1$ via the *tau* method.
    (ii) Implement an implicit–explicit (IMEX) RK(4,3) scheme on a single GPU with mixed precision. Profile the runtime and report sustained GFLOP/s and global memory bandwidth.
    (iii) Validate against the energy decay law $E(t) = E_0 e^{-2\nu k^2 t}$ for $k = 1$ by plotting $|E(t) - E_{\text{exact}}(t)|$ up to $t = 1$.

Appendix

This appendix collects a compact *NumPy Cheat Sheet* focusing on three pillars essential to the numerical-scientific workflow: efficient array manipulation, linear-algebra primitives, and an arsenal of elementary and special mathematical functions. The accompanying code snippets adopt the canonical import

```
import numpy as np
```

and illustrate idioms that appear throughout the main text.

Array Creation and Shape Manipulation NumPy arrays are homogeneous n-dimensional tensors; their metadata comprise *dtype*, *shape*, and *strides*.

```
a = np.array([[1, 2, 3],  # literal
              [4, 5, 6]], dtype=float)  # 2×3 matrix
z = np.zeros((3, 4))  # 3×4 array of 0s
I = np.eye(5)  # 5×5 identity
lin = np.linspace(0, 1, 11)  # 11 pts [0,1]
cube = np.arange(27).reshape(3,3,3)  # reshape, no copy
flat = a.ravel(order='F')  # Fortran-style view
```

Slicing generates *views* (no memory copy); broadcasting follows the left-aligned rule that singleton dimensions repeat logically without allocation.

Element-Wise Arithmetic, Reductions, and Broadcasting Operations are vectorised and obey standard algebraic precedence; reductions collapse specified axes, returning scalars or lower-rank arrays.

```
b = np.sin(a) + np.exp(a)  # element-wise
s = a.sum(axis=0)  # column sums -> (3,)
μ = a.mean(axis=1, keepdims=True)  # row mean shape (2,1)
c = (a - μ) / a.std()  # broadcasting rows
```

Linear Algebra Essentials The submodule `numpy.linalg` wraps LAPACK with BLAS acceleration; all routines accept broadcasted stacks of matrices.

```
x = np.random.randn(1000)  # vector
A = np.random.randn(1000, 1000)
y = A @ x  # matrix-vector product
Q,R = np.linalg.qr(A, mode='reduced')  # thin QR
U,S,Vh = np.linalg.svd(A, full_matrices=False)
λ, v = np.linalg.eig(A + A.T)  # symmetric part
xsol = np.linalg.solve(A, y)  # Ax = y
cond = np.linalg.cond(A)  # κ₂(A)
detA = np.linalg.det(A)  # determinant
```

Underflows/overflows are trapped by IEEE 754 rules; ill-conditioned solves warrant `np.linalg.lstsq` or regularisation.

Random Sampling NumPy's `default_rng` implements the PCG64 bit-generator and supports parallel streams via `bit_generator.jumped`.

```
rng = np.random.default_rng(seed=42)
ξ = rng.standard_normal(10**6)
B = rng.integers(0, 2, size=(128, 128), dtype=np.int8)
θ = rng.uniform(0, 2*np.pi, 5000)
```

Common Mathematical Functions NumPy vectors C99 intrinsics and special functions from `scipy.special`.

```
u = np.linspace(-3, 3, 7)
sigmoid = 1/(1 + np.exp(-u))
normcdf = 0.5*(1 + np.erf(u/np.sqrt(2)))
β = np.abs(np.heaviside(u, 0.5))  # unit step
γ = np.where(u > 0, u**2, 0)  # ReLU² via masking
```

Performance Tips

- Exploit *stride tricks* (`np.lib.stride_tricks.as_strided`) for sliding windows without copying memory.
- Fuse element-wise kernels with `numexpr` or JIT them via `numba.njit(parallel=True)` for SIMD and thread parallelism.
- Prefer `einsum` for expressive tensor contractions that auto-optimise memory layout and BLAS calls:

    ```
    C = np.einsum('ij,jk,lk->il', A, B, B, optimize=True)
    ```

- Allocate contiguously (`order='C'`) when interfacing with C/Fortran libraries; use
    `astype(np.float32, copy=False)` to downcast without duplication.

NumPy's array semantics elevate vectorised Python into a concise DSL for numerical linear algebra. Mastery of slicing, broadcasting, BLAS-backed operations, and random-number generation equips practitioners to prototype sophisticated algorithms—spectral solvers, Monte Carlo estimators, or machine-learning pipelines—within a few expressive lines of code.

SciPy and SymPy Cheat Sheets

This supplement condenses the SciPy (numerical) and SymPy (symbolic) ecosystems into two quick–reference guides, followed by a curated list of online resources and documentation hubs relevant to Python for mathematical thinking.

Numerical Methods in SciPy

Optimisation (`scipy.optimize`)

```
from scipy.optimize import minimize, root, least_squares

f = lambda x: (x[0]-2)**2 + x[1]**2
g = lambda x: [x[0]+x[1]-1]
res = minimize(f, x0=[0,0], method="SLSQP", constraints={"type":"eq","fun"
    :g})

F = lambda x: [x[0]**3 - 1, np.sin(x[1]) + 0.4]
root(F, x0=[1, 0])  # hybrid Powell

lsq = least_squares(lambda θ: y - model(X, θ), θ0, bounds=(0, np.inf))
```

Integration (`scipy.integrate`)

```
from scipy.integrate import quad, solve_ivp

I, err = quad(lambda t: np.exp(-t**2), 0, np.inf)  # √π/2

def f(t, y): return [y[1], -y[0]]  # SHO
sol = solve_ivp(f, [0, 10], y0=[0, 1], method="RK45",
            t_eval=np.linspace(0,10,200))
```

Linear Algebra (`scipy.linalg` and sparse `scipy.sparse.linalg`)

```
from scipy.linalg import solve_triangular, svd
from scipy.sparse import csr_matrix
from scipy.sparse.linalg import cg, eigs

L = np.tril(A); x = solve_triangular(L, b, lower=True)
U,S,Vh = svd(B, full_matrices=False)  # dense SVD

A_sparse = csr_matrix(A_5point_stencil)
x_cg, info = cg(A_sparse, b, tol=1e-8)  # Conjugate Gradient
λmax, v = eigs(A_sparse, k=1, which='LM')  # largest eigenvalue
```

Statistics (`scipy.stats`)

```
from scipy.stats import norm, ttest_ind

μ, σ = 0, 2
p = norm.cdf(1.96, loc=μ, scale=σ)
ci = norm.interval(0.95, loc=μ, scale=σ/np.sqrt(n))

t, pval = ttest_ind(X, Y, equal_var=False)  # Welch's t-test
```

Fast Transforms (`scipy.fft`)

```
from scipy.fft import fftn, ifftn, dct

F = fftn(u) # N-dimensional FFT
u0 = ifftn(F, workers=-1).real # inverse with multithreading

c = dct(f, type=2, norm="ortho")# DCT-II (JPEG)
```

Symbolic Computation in SymPy

Core Manipulation

```
import sympy as sp
x,y = sp.symbols('x y')
expr = sp.sin(x)**2 + sp.cos(x)**2
sp.simplify(expr) # -> 1

sp.series(sp.log(1+x), x, 0, 6) # Taylor
sp.expand((x+y)**4)
```

Calculus

```
f = sp.Function('f')(x)
sp.diff(sp.exp(x)*sp.sin(x), x, 2)

sp.integrate(sp.sin(x)/x, (x, 0, sp.oo)) # π/2
sp.limit((sp.sin(x)/x)**(1/x**2), x, 0)
```

Algebra and Linear Systems

```
A = sp.Matrix([[1, 2], [3, 4]])
λ = A.eigenvals() # {-0.372...:1, 5.372...:1}
A.inv()
sp.solve([x + 2*y - 1, x - y - 3], (x, y))
```

Differential Equations

```
y = sp.Function('y')
ode = sp.Eq( sp.diff(y(x), x, 2) - y(x), sp.exp(x) )
sol = sp.dsolve(ode)
```

Logic and Proof

```
p,q = sp.symbols('p q')
φ = sp.Implies(p, q) & sp.Implies(q, p)
sp.is_tautology(φ) # -> True
```

Appendix

Online Resources

- **SciPy documentation**—https://docs.scipy.org/doc/ (HTML, PDF)
- **SymPy documentation**—https://docs.sympy.org/
- **NumPy & SciPy mailing lists**—peer support and development discussion
- **Project-Jupyter**—interactive notebooks for reproducible mathematics
- **Stack Overflow** tag [python-numpy], [scipy], [sympy] for community Q&A
- **arXiv e-prints** (cs.MS, math.NA) for cutting-edge numerical algorithms implemented in Python

Python Libraries and Documentation

Package	Import	Scope/highlights
NumPy	import numpy as np	Core n-D arrays, BLAS & LAPACK wraps, random RNG, broadcasting
SciPy	import scipy as sp	Optimisation, ODEs, FFTs, stats, sparse LA
Matplotlib	import matplotlib.pyplot as plt	2-D & 3-D plotting, LaTeX text rendering
SymPy	import sympy as sp	CAS: algebra, calculus, logic, physics, code generation
pandas	import pandas as pd	Data frames, time series, CSV/SQL/Parquet IO
Numba	import numba as nb	LLVM JIT for array loops, parallel & GPU targets
mpi4py	from mpi4py import MPI	MPI bindings: distributed linear algebra, PDE solvers
CuPy / PyTorch / TensorFlow	import cupy, torch, tensorflow	GPU tensors, automatic differentiation, deep nets
JAX	import jax.numpy as jnp	XLA-optimised NumPy with automatic differentiation

Licence Reminders Most libraries above are BSD or MIT-licensed; verify GPL/LGPL dependencies (e.g. GNU GSL) when distributing closed-source binaries.

References

Bruce C. Berndt. *Ramanujan's Notebooks: Part II*, volume 2 of *Ramanujan's Notebooks*. Springer–Verlag, New York, 1 edition, 1989. ISBN 978-0-387-96794-3. https://doi.org/10.1007/978-1-4612-4530-8.

Jeffrey E. F. Friedl. *Mastering Regular Expressions*. O'Reilly Media, Sebastopol, CA, 3 edition, 2006. ISBN 978-0-596-52812-6.

David Goldberg. What every computer scientist should know about floating-point arithmetic. *ACM Comput. Surv.*, 23 (1): 5–48, March 1991. ISSN 0360-0300. https://doi.org/10.1145/103162.103163.

Gene H. Golub and Charles F. Van Loan. *Matrix Computations*. Johns Hopkins University Press, Baltimore, MD, fourth edition, 2013. ISBN 978-1-4214-0794-4. URL https://press.jhu.edu/books/title/8284/matrix-computations.

F. R. Gantmacher. *The Theory of Matrices*. AMS Chelsea Publishing. American Mathematical Society, Providence, RI, reprint edition, 2000. ISBN 978-0-8218-2164-0.

Jan R. Magnus and Heinz Neudecker. *Matrix Differential Calculus with Applications in Statistics and Econometrics*. John Wiley & Sons, Chichester, second edition, 1999. ISBN 978-0-471-99893-4.

F. W. J. Olver. *Asymptotics and Special Functions*. Academic Press, Wellesley, MA, reprint of the 1974 Academic Press edition edition, 1997. ISBN 978-1-56881-069-0.

Lloyd N. Trefethen and David Bau III. *Numerical Linear Algebra*. Society for Industrial and Applied Mathematics, Philadelphia, PA, 1997. ISBN 978-0-89871-361-9.

B. van der Pol. On "relaxation-oscillations". *The London, Edinburgh, and Dublin Philosophical Magazine and Journal of Science*, 2 (11): 978–992, 1926. https://doi.org/10.1080/14786442608564127.

Paul Glasserman. *Monte Carlo Methods in Financial Engineering*, volume 53 of *Applications of Mathematics*. Springer, New York, 2004. ISBN 978-0-387-20411-6. https://doi.org/10.1007/978-0-387-21617-1.

David Harrison and Daniel L Rubinfeld. Hedonic housing prices and the demand for clean air. *Journal of Environmental Economics and Management*, 5 (1): 81–102, 1978. ISSN 0095-0696. https://doi.org/10.1016/0095-0696(78)90006-2. URL https://www.sciencedirect.com/science/article/pii/0095069678900062.

James W. Cooley and John W. Tukey. An algorithm for the machine calculation of complex fourier series. *Mathematics of Computation*, 19 (90): 297–301, 1965. https://doi.org/10.1090/S0025-5718-1965-0178586-1.

J. Crank and P. Nicolson. A practical method for numerical evaluation of solutions of partial differential equations of the heat-conduction type. *Advances in Computational Mathematics*, 6: 207–226, 1996. https://doi.org/10.1007/BF02127704.

J.R. Dormand and P.J. Prince. A family of embedded runge-kutta formulae. *Journal of Computational and Applied Mathematics*, 6 (1): 19–26, 1980. ISSN 0377-0427. https://doi.org/10.1016/0771-050X(80)90013-3.

David M. Grobman. Homeomorphisms of systems of differential equations. *Doklady Akademii Nauk SSSR*, 128: 880–881, 1959. in Russian.

Bernt Øksendal. *Stochastic Differential Equations: An Introduction with Applications*. Universitext. Springer, Berlin, sixth edition, 2003. https://doi.org/10.1007/978-3-642-14394-6.

Lawrence Perko. *Differential Equations and Dynamical Systems*, volume 7 of *Texts in Applied Mathematics*. Springer, New York, third edition, 2013. https://doi.org/10.1007/978-1-4614-3618-0.

Steven H. Strogatz. *Nonlinear Dynamics and Chaos: With Applications to Physics, Biology, Chemistry, and Engineering*. CRC Press, Boca Raton, second edition, 2018. ISBN 978-0813349107.

Jack Edmonds. Matroids and the greedy algorithm. *Mathematical Programming*, 1 (1): 127–136, 1971. https://doi.org/10.1007/BF01584082.

Kurt Gödel. Über formal unentscheidbare sätze der Principia Mathematica und verwandter systeme i. *Monatshefte für Mathematik und Physik*, 38 (1): 173–198, 1931. https://doi.org/10.1007/BF01700692.

Richard Rado. A note on independence functions. *Proceedings of the London Mathematical Society, Third Series*, 7: 300–320, 1957.

G. Benettin, L. Galgani, A. Giorgilli, and J. M. Strelcyn. Lyapunov characteristic exponents for smooth dynamical systems and for hamiltonian systems; a method for computing all of them. part 1: Theory. *Meccanica*, 15: 9–20, 1980. https://doi.org/10.1007/BF02128236.

Edward Ott, Celso Grebogi, and James A. Yorke. Controlling chaos. *Phys. Rev. Lett.*, 64: 1196–1199, Mar 1990. https://doi.org/10.1103/PhysRevLett.64.1196. URL https://link.aps.org/doi/10.1103/PhysRevLett.64.1196.

N. H. Packard, J. P. Crutchfield, J. D. Farmer, and R. S. Shaw. Geometry from a time series. *Phys. Rev. Lett.*, 45: 712–716, Sep 1980. https://doi.org/10.1103/PhysRevLett.45.712. URL https://link.aps.org/doi/10.1103/PhysRevLett.45.712.

K. Pyragas. Continuous control of chaos by self-controlling feedback. *Physics Letters A*, 170 (6): 421–428, 1992. ISSN 0375-9601. https://doi.org/10.1016/0375-9601(92)90745-8. URL https://www.sciencedirect.com/science/article/pii/037596019290745 8.

Michael T. Rosenstein, James J. Collins, and Carlo J. De Luca. A practical method for calculating largest lyapunov exponents from small data sets. *Physica D: Nonlinear Phenomena*, 65 (1): 117–134, 1993. ISSN 0167-2789. https://doi.org/10.1016/0167-2789(93)90009-P. URL https://www.sciencedirect.com/science/article/pii/016727899390009P.

F. Takens. Detecting strange attractors in turbulence. In David A. Rand and Lai-Sang Young, editors, *Dynamical Systems and Turbulence, Warwick 1980*, volume 898 of *Lecture Notes in Mathematics*, pages 366–381. Springer, Berlin, 1981. https://doi.org/10.1007/BFb0091924.

Martin Abadi, Andy Chu, Ian Goodfellow, H. Brendan McMahan, Ilya Mironov, Kunal Talwar, and Li Zhang. Deep learning with differential privacy. In *Proceedings of the 2016 ACM SIGSAC Conference on Computer and Communications Security*, CCS '16, page 308–318, New York, NY, USA, 2016. Association for Computing Machinery. ISBN 9781450341394. URL https://doi.org/10.1145/2976749.2978318.

Yoav Freund and Robert E Schapire. A decision-theoretic generalization of on-line learning and an application to boosting. *Journal of Computer and System Sciences*, 55 (1): 119–139, 1997. ISSN 0022-0000. https://doi.org/https://doi.org/10.1006/jcss.1997.1504. URL https://www.sciencedirect.com/science/article/pii/S002200009791504X.

S. Lloyd. Least squares quantization in pcm. *IEEE Trans. Inf. Theor.*, 28 (2): 129–137, September 2006. ISSN 0018-9448. URL https://doi.org/10.1109/TIT.1982.1056489.

V. Mnih, K. Kavukcuoglu, D. Silver, and et al. Human-level control through deep reinforcement learning. *Nature*, 518: 529–533, 2015. https://doi.org/10.1038/nature14236.

Laurens van der Maaten and Geoffrey Hinton. Viualizing data using t-sne. *Journal of Machine Learning Research*, 9: 2579–2605, 11 2008.

Ashish Vaswani, Noam Shazeer, Niki Parmar, Jakob Uszkoreit, Llion Jones, Aidan N. Gomez, Łukasz Kaiser, and Illia Polosukhin. Attention is all you need. In *Proceedings of the 31st International Conference on Neural Information Processing Systems*, NIPS'17, page 6000–6010, Red Hook, NY, USA, 2017. Curran Associates Inc. ISBN 9781510860964.

C. J. C. H. Watkins and P. Dayan. Q-learning. *Machine Learning*, 8 (3–4): 279–292, 1992. https://doi.org/10.1007/BF00992698.

Breiman L. Statistical modeling: The two cultures. *Statistical Science*, 16(3), 199–231, 2001.

Bergstra J. Bengio Y. Random search for hyper-parameter optimization. *Journal of Machine Learning Research*, 13, 281–305, 2012.

Kleinberg J. Lakkaraju H. Leskovec J. Ludwig J. Mullainathan S. Human decisions and machine predictions. *Quarterly Journal of Economics*, 133(1), 237–293, 2017.

Kingma D. P. Welling M. Auto-encoding variational Bayes. In Proceedings of the International Conference on Learning Representations (ICLR), 2014.

H. Brendan McMahan, Eider Moore, Daniel Ramage, Seth Hampson, Blaise Aguera y Arcas Communication-efficient learning of deep networks from decentralized data. In Proceedings of the 20th International Conference on Artificial Intelligence and Statistics (AISTATS), PMLR 54, 1273–1282, 2017.

Cohen-Steiner D. Edelsbrunner H. Harer J. Stability of persistence diagrams. *Discrete & Computational Geometry*, 37(1), 103–120, 2007.

Hofer C. Kwitt R. Niethammer M. Uhl, A. Christoph Hofer, Roland Kwitt, Marc Niethammer, and Andreas Uhl. 2017. Deep learning with topological signatures. In Proceedings of the 31st International Conference on Neural Information Processing Systems (NIPS'17). Curran Associates Inc., Red Hook, NY, USA, 1633–1643, 2019.

Bauer U. Ripser: Efficient computation of Vietoris-Rips persistence barcodes. *Journal of Applied and Computational Topology*, 5, 391–423, 2021.

GPSR Compliance

The European Union's (EU) General Product Safety Regulation (GPSR) is a set of rules that requires consumer products to be safe and our obligations to ensure this.

If you have any concerns about our products, you can contact us on

ProductSafety@springernature.com

In case Publisher is established outside the EU, the EU authorized representative is:

Springer Nature Customer Service Center GmbH
Europaplatz 3
69115 Heidelberg, Germany

www.ingramcontent.com/pod-product-compliance
Ingram Content Group UK Ltd.
Pitfield, Milton Keynes, MK11 3LW, UK
UKHW022203230426

470311UK00001BA/5